MEASUREMENT AND DATA ANALYSIS

McGraw-Hill Series in Mechanical Engineering

MEASUREMENT AND DATA ANALYSIS
for Engineering and Science

PATRICK F. DUNN
University of Notre Dame

Boston   Burr Ridge, IL   Dubuque, IA   Madison, WI   New York   San Francisco   St. Louis
Bangkok   Bogotá   Caracas   Kuala Lumpur   Lisbon   London   Madrid   Mexico City
Milan   Montreal   New Delhi   Santiago   Seoul   Singapore   Sydney   Taipei   Toronto

Higher Education

MEASUREMENT AND DATA ANALYSIS FOR ENGINEERING AND SCIENCE

Published by McGraw-Hill, a business unit of The McGraw-Hill Companies, Inc., 1221 Avenue of the Americas, New York, NY 10020.

1 2 3 4 5 6 7 8 9 0 QPF/QPF 0 9 8 7 6 5 4

ISBN 0–07–282538–3

Publisher: *Elizabeth A. Jones*
Senior sponsoring editor: *Suzanne Jeans*
Developmental editor: *Amanda J. Green*
Senior project manager: *Gloria G. Schiesl*
Lead production supervisor: *Sandy Ludovissy*
Lead media project manager: *Audrey A. Reiter*
Media technology producer: *Eric A. Weber*
Senior coordinator of freelance design: *Michelle D. Whitaker*
Cover designer: *Kelly Fassbinder/Imagine Design Studio*
Lead photo research coordinator: *Carrie K. Burger*
Compositor: *Lachina Publishing Services*
Typeface: *10/12 Times Roman*
Printer: *Quebecor World Fairfield, PA*

(USE) Cover images: First row: Joule's Water-Friction Apparatus: *Science Museum/Science & Society Picture Library, London*
Kelvin's Tide Predictor: *Science Museum/Science & Society Picture Library, London*
Celsius and Fahrenheit Scale Thermometer: © *Photodisc*
Metric Carpenter's Ruler: © *Photodisc*

Second row: Kilogram Weight: © *Photodisc*
Data Acquisition Circuit Board: © *Digital Vision*
Computer Keyboard: © *Photodisc*
Cooke and Wheatstone Morse Tape Perforator: *Science Museum/Science & Society Picture Library, London*

Third row: Leeuwenhoek's Microscope: *Author's Photograph, Universiteitsmuseum, Utrecht*
Theodolite: *Science Museum/Science & Society Picture Library, London*
Galton's Quincunx: *Author's Photograph, The Galton Collection, University College London*
Microprocessor: © *Photodisc*

Fourth row: Voltameter: *Science Museum/Science & Society Picture Library, London*
Compound Microscope: © *Photodisc*
Protractor: © *Photodisc*

Library of Congress Cataloging-in-Publication Data

Dunn, Patrick F.
    Measurement and data analysis for engineering and science / Patrick F. Dunn.—1st ed.
        p.   cm.—(McGraw-Hill series in mechanical engineering)
    Includes index.
    ISBN 0–07–282538–3 (acid-free paper)
    1. Physical measurements. 2. Statistics. I. Title. II. Series.

QC39.D86 2005
530.8—dc22

                                    2004044974

www.mhhe.com

To my wife and life partner, Carol,
for always supporting me and keeping me centered.

To my children—Andrea, Brian, Christopher and his wife Jenny—
our greatest joys in life and hope for the future.

To my mother, Irene,
who shared her creative side with me.

To my father, John,
who always wondered when I would get a real job.

ABOUT THE AUTHOR

PATRICK F. DUNN is a Professor of Aerospace and Mechanical Engineering at the University of Notre Dame, where he has been since 1985. Prior to then, he was a Mechanical Engineer at Argonne National Laboratory from 1976 to 1985 and a Postdoctoral Fellow at Duke University from 1974 to 1976. He received his B.S., M.S., and Ph.D. degrees in Engineering from Purdue University (1970, 1971, and 1974, respectively). He is the author of over 100 scientific publications and a licensed Professional Engineer in Indiana and Illinois. Professor Dunn's scientific expertise is in fluid and microparticle mechanics. He is a teacher and experimentalist with over 35 years of experience involving measurements and data analysis. He is the coauthor of another new textbook, *Uncertainty Analysis for Forensic Science* (ISBN 1930056206), involving uncertainty and other issues related to forensic science.

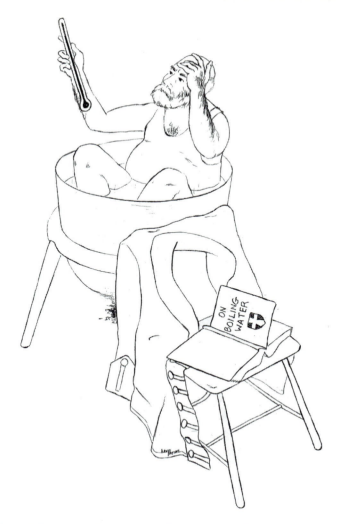

"When I read that water boils at a definite degree of heat, I immediately felt a great desire to make myself a thermometer."

Gabriel Daniel Fahrenheit reportedly said this in 1706 during an inspirational moment when he conceived the thermometer with a calibrated scale. For the scale, he used the reference temperatures of the melting point of ice and the human body. He was the first person to apply the principle of the thermal expansion of a liquid with temperature to the measurement of temperature. This instrument opened new vistas in science by providing the means to perform repeated and accurate measurements of temperature.

This sketch of Fahrenheit was made by David E. Irvine, Ph.D. It appeared originally in the early 1970s in a handout for an undergraduate aerospace engineering laboratory course, AE450, at Purdue University. He and I, fellow graduate students, co-taught AE450, the first laboratory course in our teaching careers.

CONTENTS

Chapter 12
DIGITAL SIGNAL ANALYSIS　401

Appendix A
SYMBOLS　433

Appendix B
GLOSSARY　437

Appendix C
CONVERSIONS　449

Appendix D
LEARNING OBJECTIVE NOMENCLATURE　453

Appendix E
PHYSICAL PRINCIPLES　455

Appendix F
REVIEW PUZZLE SOLUTIONS　461

Appendix G
PROBLEM SOLUTIONS　465

Appendix H
LABORATORY EXERCISES　467

Appendix I
DERIVATIONS　523

INDEX　531

PREFACE

ORGANIZATION

This text presents the fundamentals of measurements and data analysis that are depicted in the accompanying figure, "Road map of the text." The subject matter is divided into twelve chapters organized into three focus areas: basics, hardware, and analyses. Chapter order progresses to the more difficult material. The order of presentation, however, can be changed readily to accommodate the level of the course. The text was written to be flexible enough to be used in an undergraduate introductory course, yet thorough enough to serve as a permanent reference for experimentalists. *Measurement and Data Analysis for Engineering and Science* focuses on essential topics, covers them rigorously, and presents them in a manner leading students to *think* more than to memorize.

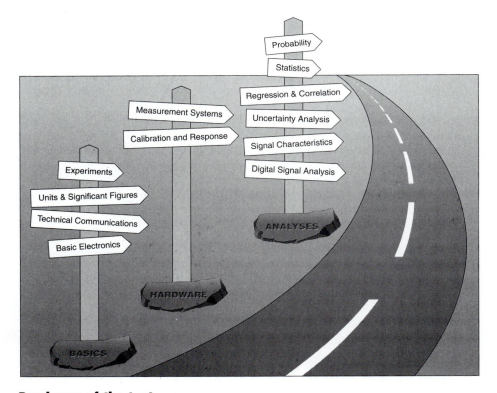

Road map of the text

The subject matter is appropriate for an introductory course in experimental methods, comprising one- or two-undergraduate semesters or one graduate semester. The material is structured to serve as the lecture component of a lecture-laboratory course. Laboratory exercise descriptions and companion data are provided online in the Laboratory Exercises Manual and the Laboratory Exercises Solutions Manual to supplement a stand-alone lecture course.

The text content, more specifically, includes

1. *Basics:* Chapter 1 explores the experiment, including its terminology, role in the scientific process, and classifications. Chapter 2 presents systems of units and significant figures. Chapter 3 covers the technical memo and technical report formats, how to produce professional-looking documents, and how to make oral technical presentations. Basic electronics pertinent to an experiment's hardware are reviewed in chapter 4.

2. *Hardware:* Chapter 5 covers the calibration and response of measurement systems to various inputs. Chapter 6 explores the elements of a measurement system and presents several example measurement systems.

3. *Analyses:* Probability and statistics and their roles in experiments are reviewed in chapters 7 and 8. Chapter 9 extends the use of finite statistics to uncertainty analysis. Regression and correlation are considered in chapter 10. Chapter 11 presents how to characterize signals, how to classify signals, and how to determine the various statistical measures of signals. Chapter 12 discusses digital signal analysis and the techniques that are used to avoid signal processing errors.

4. *Appendices:* A list of symbols for each chapter is presented in Appendix A. Appendix B is a glossary of terms that defines each boldfaced term in the text. Appendix C includes six useful unit conversion tables. Appendix D defines the specific meanings of the action verbs used in the learning objectives presented at the beginning of each chapter. The physical principles behind sensor and transducer designs are given in Appendix E. Appendix F presents the solutions to review puzzles. Appendix G offers brief solutions to all review problems and selected homework problems. Laboratory exercise descriptions are given in Appendix H. Appendix I provides the step-by-step derivations of system response to forcing.

FEATURES

This text has a number of unique features that include the following:

- In-depth coverage of material supported by text references and journal articles
- Historical perspectives on instruments and the people behind experimental methods
- Presentation of laboratory exercises to reinforce text material
- Supplementary sidebars that introduce and reinforce the use of MATLAB as a computational and graphical presentation tool.
- Over fifty commented MATLAB M-files and data files

- Over seventy worked examples
- Chapter review crossword puzzles
- Short-answer review problems and more extensive homework problems

All of the topics covered in most measurements texts are presented in *Measurement and Data Analysis for Engineering and Science*. In addition, some new and unique topics are presented and include the following:

- Role of experimentation in science and engineering
- Design-of-experiments and factorial-design methods
- Rules for oral technical presentations
- Physical-principles approach to sensors and transducer
- Sensor scaling and microelectricalmechanical (MEMS) devices
- Auto and cross-correlation techniques
- Digital filtering

SUPPLEMENTS

The following text supplements can be found on the website at www.mhhe.com/pdunn.

- **Review and Homework Problem Solutions Manual:** This manual contains step-by-step solutions to all review and homework problems, and are presented in a format suitable for lecture, recitation, and handouts. Brief solutions to all review problems and some selected homework problems are presented in Appendix G. This manual is instructor-password-protected.

- **Laboratory Exercises Manual:** This manual presents the descriptions of twelve laboratory exercises that have been conducted as part of Notre Dame's undergraduate measurements course. Actual student handouts are included, in addition to data tables and equipment-specific instructions. More general descriptions of the same exercises are presented in Appendix H. This manual is *not* password-protected.

- **Laboratory Exercises Solutions Manual:** This unique manual contains the data obtained during the exercises and the answers to all of the questions posed in the Laboratory Exercises Manual. It enables the instructor the flexibility of assigning virtual laboratory exercises in which the data is also provided to the students. Parts of the exercises can be used as additional homework or in-class instructional problems. This manual is instructor-password-protected.

- **MATLAB M-files and Data Files:** Over seventy MATLAB sidebars are presented in the text, which refer to standard MATLAB commands and specially written M-files. Each M-file contains comments about the program steps. Five data files are also provided to accompany the homework problems. The MATLAB sidebars supplement the text material, covering topics from simple calculations and data file transfers to advanced digital signal analysis. These files are *not* password-protected.

- **WebCT Course File:** WebCT is an interactive web-based program that can be used to present online quizzes to the students. Most of the text's review problems are included in the WebCT Course File and are organized according to the main subjects of the text. This file is instructor-password-protected.

ACKNOWLEDGMENTS

Several colleagues and many students have contributed to this text. Two colleagues, who have co-taught with me at times, deserve special mention: Professor Flint Thomas for sharing his extensive expertise on experiments and signal processing, and Professor Edmundo Corona for his ability to boil complex topics down to simple examples. I have learned much from them. Professor Raymond Brach, my research cohort, patiently taught me the subtleties of probability and statistics. Professor Mihir Sen was the first to encourage me to form my measurements lecture notes into a text and use LaTeX to do this. The technical staff at Hessert Center also helped: Rod McClain for designing experiments and sharing his laboratory teaching experiences, Joel Preston for doing most of the electronics behind the laboratory exercises, Mike Swadener for fabricating the laboratory exercise setups, and Marilyn Walker for assisting in typing. I have had the pleasure to work with a number of excellent students at Notre Dame. Mike Davis helped me extensively, especially in designing experiments relevant to student interest. Nick Fehring, Alain Pelletier, Denis Lynch, Jason Miller, Tom Szarek, Abdelmaged Ibrahim, Tim Kish, Ward Judson, Brett McMickell, Steve Remis, Susan Olson, Weidong Cheng, and Ashraf Al-Khateeb each contributed to the effort. The photographs presented on the cover and at the beginning of each chapter were acquired with the help of Dr. Klaus B. Staubermann, Ms. Sarah Sykes, and Ms. Bryony Reid. The McGraw-Hill book team was a pleasure to work with, including Jonathan Plant, Amanda Green, Jill Peter, Gloria Schiesl, and Michelle Whitaker. My wife, Carol, deserves special thanks for her understanding, continual encouragement, and fastidious proofreading.

I am indebted to the reviewers of the text manuscript, whose constructive comments and suggestions contributed significantly. They include the following:

David Baldwin *University of Oklahoma*

Jonathan Blotter *Brigham Young University*

Bertram Ezenwa *University of Wisconsin at Milwaukee*

Burford J. Furman *San Jose State University*

Matthew J. Hall *University of Texas at Austin*

R. Glynn Holt *Boston University*

Carl W. Luchies *University of Kansas*

Tim McLain   *Brigham Young University*

Subhash Rakheja   *Concordia University*

Lyndon S. Stephens   *University of Kentucky*

Edward Wheeler   *Rose-Hulman Institute of Technology*

The publisher and I also would like to acknowledge the following people who completed a survey of their Experimentation/Measurements course. The surveys contributed to our market research and development and will be helpful in maximizing the effectiveness of our first edition text. These people include Gary A. Anderson, South Dakota State University; Kendrick Aung, Lamar University; Jay Benziger, Princeton University; Tom A. Bon, North Dakota State University; Donald E. Carlucci, Stevens Institute of Technology; Edgar Conley, New Mexico State University; MaryFran Desrochers, Michigan Technological University; Luc Fréchette, Columbia University; Robert Gao, University of Massachusetts Amherst; John L. Guillory, University of Louisiana at Lafayette; R. Glynn Holt, Boston University; George Johnson, University of California at Berkeley; Ian M. Kennedy, University of California at Davis; Michael R. Kessler, University of Tulsa; Greg Kremer, Ohio University; Jeffrey W. Kysar, Columbia University; Frank Lu, University of Texas at Arlington; Jed S. Lyons, University of South Carolina; Michael E. Momot, University of Wisconsin at Platteville; Marlow D. Moser, University of Alabama at Huntsville; David Pacey, Kansas State University; Eric L. Peterson, University of Central Florida; Stoian Petrescu, Bucknell University; Larry Ricker, University of Washington; Antwone J. Ross, Virginia State University; Roger L. Simpson, Virginia Polytechnic Institute; Barton Smith, Utah State University; Todd Spohnholtz, University of Illinois at Chicago; Jim Tan-atichat, California State University at Chico; Joseph Wehrmeyer, Vanderbilt University; Edward Wheeler, Rose-Hulman Institute of Technology; and Qin Zhang, University of Illinois at Urbana-Champaign.

Patrick F. Dunn
University of Notre Dame
March 2004

Written while at the University of Notre Dame, Notre Dame, Indiana; the University of Notre Dame London Centre, London, England; and Delft University of Technology, Delft, The Netherlands.

MEASUREMENT AND DATA ANALYSIS

Leeuwenhoek's Microscope

It is appropriate that our study of measurement and data analysis fundamentals begins with one of the most famous instruments, which helped to usher in a golden age of experimentation. During the 17th century, it was Galileo Galilei's (1564–1642) telescope and Antony van Leeuwenhoek's (1632–1723) microscope, both made of lenses, that provided the means to begin probing the intricacies of the macroscopic and microscopic worlds.

The microscope pictured is an exact copy, made by J. van Musschenbroek and F. Depouilly, of van Leeuwenhoek's original microscope made before 1673. The original is in possession of the University Museum, Utrecht, The Netherlands. The microscope consists of a double convex lens mounted in a 0.7-mm aperture made in two thin 24-mm × 46-mm brass plates that are riveted together. The object to be examined is held approximately 1 mm in front of the lens on the point of a short rod mounted on a small brass block that is riveted to a long coarse-threaded screw. Its magnification was calibrated and reported by Dr. J. van Zuylen in 1981 and was found to be 266× with a focal length of 0.94 mm and a resolution of 1.35 μm. This magnification was at least one order of magnitude better than any other contemporary device and was not exceeded until over one century later.

Brian J. Ford in his book *Single Lens, The Story of the Simple Microscope* (London: Heinemann, 1985) describes this microscope best as follows: "this postage-stamp-sized object is one of the most important instruments of scientific discovery. It changed our lives, altered our self-image, revolutionized our understanding of the world in which we live. Our modern biology-oriented era of the cell nucleus, bacteria and the whole gamut of microscopic organisms arose through the use of this little object. For it is a microscope—not the first microscope, nor the most powerful; but an instrument with a significance that few people have ever begun to appreciate."

1

EXPERIMENTS

... there is a diminishing return from increased theoretical complexity and ... in many practical situations the problem is not sufficiently well defined to merit an elaborate approach. If basic scientific understanding is to be improved, detailed experiments will be required ...

Graham B. Wallis. 1980.
International Journal of Multiphase Flow 6:97.

The lesson is that no matter how plausible a theory seems to be, experiment gets the final word.

Robert L. Park. 2000.
Voodoo Science. New York: Oxford University Press.

Experiments essentially pose questions and seek answers. A good experiment provides an unambiguous answer to a well-posed question.

Henry N. Pollack. 2003.
Uncertain Science ... Uncertain World. Cambridge: Cambridge University Press.

CHAPTER OUTLINE

1.1   CHAPTER OVERVIEW

Experimentation has been part of the human experience ever since its beginnings. We are born with highly sophisticated data acquisition and computing systems ready to experiment with the world around us. We come loaded with the latest tactile, gustatory, auditory, olfactory, and optical sensor packages. And we even have a central processing unit capable of processing data and performing highly complex operations at incredible rates with a memory far surpassing any that we can purchase.

One of our first rudimentary experiments, although not a conscious one, is to cry and then to observe whether or not a parent will come to aid our discomfort. We *change* the environment and *record* the result. We are *active* participants in the process. Our view of reality is formed by what we sense. But what really are experiments? What roles do they play in the process of understanding the world in which we live? How are they classified? Such questions are addressed in this chapter.

1.2   LEARNING OBJECTIVES

You should be able to do the following after completing this chapter:

- Know the meanings of the scientific method, an experiment, and a variable
- Know the types of variables and be able to identify them for a given experiment
- Know the various purposes of experiments

1.3   ROLE OF EXPERIMENTS

Perhaps the first question to ask is, "Why do we do experiments?" Some of my former students have offered the following answers.

"Experiments are the basis of all theoretical predictions. Without experiments, there would be no results, and without any tangible data, there is no basis for any scientist or engineer to formulate a theory. . . . The advancement of culture, civilization depends on experiments which bring about new technology . . ." (P. Cuadra).

"Making predictions can serve as a guide to what we expect, . . . but to really learn and know what happens in reality, experiments must be done" (M. Clark).

"If theory predicted everything exactly, there would be no need for experiments. NASA planners could spend an afternoon drawing up a mission with their perfect computer models, and then launch a flawlessly executed mission that evening (of course, what would be the point of the mission, since the perfect models could already predict behavior in space anyway?)" (A. Manella).

In the most general sense, people seek to reach a better understanding of the world. In this quest, they rely on the collective knowledge of their predecessors and peers.

If one understood *everything* about nature, there would be no need for experiments. One could predict every outcome (at least for deterministic systems). But, that is not the case. Our understanding is imperfect. We need to experiment in the world.

So how do experiments play a role in our process of understanding? The Greeks were the earliest civilization that attempted to gain a better understanding of their world through observation and reasoning. Previous civilizations functioned within their environment by observing its behavior and then adapting to it. It was the Greeks who first went beyond the stage of simple observation and attempted to arrive at the underlying physical causes of what they observed [1]. Two opposing schools emerged, both of which still exist but in somewhat different forms. Plato (428–347 B.C.) advanced that the highest degree of reality was that which men *think* by reasoning. He believed that better understanding followed from rational thought alone. This is called **rationalism**. On the contrary, Aristotle (384–322 B.C.) believed that the highest degree of reality is that which man *perceives* with his senses. Aristotle argued that better understanding came through careful observation. This is known as **empiricism**. Empiricism maintains that knowledge originates from and is limited to concepts developed from sensory experience. Today it is recognized that both approaches play important roles in advancing scientific understanding.

There are several different roles that experiments play in the process of scientific understanding. Harré [2], who discusses some of the landmark experiments in science, describes three of the most important roles: **inductivism**, **fallibilism**, and **conventionalism**. Inductivism is the process whereby the laws and theories of nature are arrived at based on the facts gained from the experiments. In other words, a greater theoretical understanding of nature is reached through induction. Taking the fallibilistic approach, experiments are performed to test the validity of a conjecture. The conjecture is rejected if the experiments show it to be false. The role of experiments in the conventionalistic approach is illustrative. These experiments do not induce laws or disprove hypotheses but rather show us a more useful or illuminating description of nature. *Tests* fall into the category of conventionalistic experiments.

All three of these approaches are elements of the **scientific method**. Credit for its formalization often is given to Francis Bacon (1561–1626). The seeds of experimental science were sown earlier by Roger Bacon (c. 1220–1292), who was not related to Francis. Roger attempted to incorporate experimental science into the university curriculum, but was prohibited by Pope Clement IV. He wrote of his findings in secrecy. Roger is considered "the most celebrated scientist of the Middle Ages" [3]. Francis argued that our understanding of nature could be increased through a disciplined and orderly approach in answering scientific questions. This approach involved experiments, done in a systematic and rigorous manner, with the goal of arriving at a broader theoretical understanding. Using the approach of Bacon's time, first the results of positive experiments and observations are gathered and considered. A preliminary hypothesis is formed. All rival hypotheses are tested for possible validity. Hopefully, only one correct hypothesis (ours) remains. Today, the scientific method is used mainly to validate a particular hypothesis or to determine the range of validity of a hypothesis. In the end, it is the constant interplay between experiment and theory that leads to advancing our understanding, as illustrated schematically

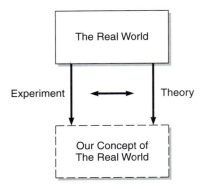

Figure 1.1   The interplay between experiment and theory.

in Figure 1.1. The concept of the real world is developed from the data acquired through experiment and the theories constructed to explain the observations. Often new experimental results improve theory and new theories guide and suggest new experiments. Hopefully, through this process a more refined and realistic concept of the world is developed. Anthony Lewis of the *New York Times* summarizes it well, "The whole ethos of science is that any explanation for the myriad mysteries in our universe is a theory, subject to challenge and experiment. That is the scientific method."

Gale [4], in his treatise on science, advances that there are two goals of science: explanation and understanding, prediction and control. Science is not absolute; it evolves. Its modern basis is the experimental method of proof. Explanation and understanding encompass statements that make causal connections. One example statement is that an increase in the temperature of a perfect gas under constant volume causes an increase in its pressure. Such statements usually lead to an algorithm or law that relates the variables involved in the process under investigation. Prediction and control establish correlations between variables. For the previous example, these would result in the correlation between pressure and temperature. Science is a process in which false hypotheses are disproved and, eventually, the true one remains.

1.4   THE EXPERIMENT

What exactly is an experiment? An **experiment** is an act in which one physically intervenes with the process under investigation and records the results. This is shown schematically in Figure 1.2. Examine this definition more closely. In an experiment one *physically changes in an active manner* the process being studied and then *records* the results of the change. Thus, computational simulations are *not* experiments.

Figure 1.2   The experiment.

Likewise, sole observation of a process is *not* an experiment. An astronomer charting the heavens does not alter the paths of planetary bodies; he does not perform an experiment, he observes. An anatomist who dissects something does not physically change a process (although she physically may move anatomical parts); again, she observes. Yet, it is through the interactive process of observation-experimentation-hypothesizing that understanding advances. All elements of this process are essential. Traditionally, theory explains existing results and predicts new results; experiments validate existing theory and gather results for refining theory.

When conducting an experiment, it is imperative to identify all the variables involved. **Variables** are those physical quantities involved in the process under investigation that can undergo change during the experiment and thereby affect the process. They are classified as **independent**, **dependent**, and **extraneous**. An experimentalist manipulates the independent variable(s) and records the effect on the dependent variable(s). An extraneous variable cannot be controlled and affects the value of what is measured to some extent. A **controlled experiment** is one in which *all* of the variables involved in the process are identified and controlled. In reality, almost all experiments have extraneous variables and, therefore, strictly are not controlled. This inability to precisely control *every* variable is the primary source of experimental uncertainty, which is considered in Chapter 9. The measured variables are called **measurands**.

A variable that is either actively or passively fixed throughout the experiment is called a **parameter**. Sometimes, a parameter can be a specific function of variables. For example, the Reynolds number, which is a nondimensional number used frequently in fluid mechanics, can be a parameter in an experiment involving fluid flow. The Reynolds number is defined as $\text{Re} = \rho U d / \mu$, where U is the fluid velocity, d is a characteristic length, and ρ and μ are the fluid's density and absolute viscosity, respectively. Measurements can be made by conducting a number of experiments for various U, d, ρ, and μ. Then the data can be organized for certain fixed values of Re, each corresponding to a different experiment.

Consider for example a fluid-flow experiment designed to ascertain whether or not there is laminar (Poiseuille) flow through a smooth pipe. If laminar flow is present, then theory (conservation of momentum) predicts that $\Delta p = 8QL\mu/(\pi R^4)$, where Δp is the pressure difference between two locations along the length, L, of a pipe of radius, R, for a liquid with absolute viscosity, μ, flowing at a volumetric flow rate, Q. In this experiment, the volumetric flow rate is varied. Thus, μ, R, and L are parameters; Q is the independent variable; and Δp is the dependent variable. Also R, L, Q, and Δp are measurands. The viscosity is a dependent variable that is determined from the fluid's temperature (another measurand and parameter). If the fluid's temperature is not controlled in this experiment, it could affect the values of density and viscosity, and hence the values of the dependent variables.

Example 1.1

STATEMENT

An experiment is performed to determine the coefficient of restitution, e, of a ball over a range of impact velocities. For impact normal to the surface, $e = v_f/v_i$, where v_i is the normal velocity component immediately before impact with the surface, and v_f is that immediately after impact. The velocity v_i is controlled by dropping the ball from a known, initial height, h_a, and then measuring its return height, h_b. What are the variables in this experiment? List which ones are independent, dependent, parameter, and measurand.

SOLUTION

A ball dropped from height h_a will have $v_i = \sqrt{2gh_a}$, where g is the local gravitational acceleration. Because $v_f = \sqrt{2gh_b}$, $e = \sqrt{h_b/h_a}$. So the variables are h_a, h_b, v_i, v_f, e, and g; h_a is an independent variable; h_b, v_i, v_f, and e are dependent variables; g is a parameter; and h_a and h_b are measurands.

Often, however, it is difficult to identify and control all of the variables that can influence an experimental result. Experiments involving biological systems often fall into this category. In these situations, repeated measurements are performed to arrive at statistical estimates of the measured variables, such as their means and standard deviations. **Repetition** implies that a set of measurements are repeated under the same, fixed operating conditions. This yields direct quantification of the variations that occur in the measured variables for the same experiment under fixed operating conditions. Often, however, the same experiment may be run under the same operating conditions at different times or places using the same or comparable equipment and facilities. Because uncontrollable changes may occur in the interim between running the experiments, additional variations in the measured variables may be introduced. These variations can be quantified by the **replication** (duplication) of the experiment. A **control experiment** is an experiment that is as nearly identical to the subject experiment as possible. Control experiments typically are performed to reconfirm a subject experiment's results or to verify a new experimental setup's performance. Finally, experiments can be categorized broadly into **timewise** and **sample-to-sample** experiments [5]. Values of a measurand are recorded in a continuous manner over a period of time in timewise experiments. Values are obtained for multiple samples of a measurand in sample-to-sample experiments. Both types of experiments can be considered the same when values of a measurand are acquired at discrete times. Here, what distinguishes between the two categories is the time interval between samples.

In the end, performing a good experiment involves identifying and controlling as many variables as possible and making accurate and precise measurements. It always should be performed with an eye out for discovery. To quote Sir Peter Medawar [6], "The merit of an experiment lies principally in its design and in the critical spirit in which it is carried out."

1.5   EXPERIMENTAL APPROACH

Park [7] remarks that "science is the systematic enterprise of gathering knowledge about the world and organizing and condensing that knowledge into testable laws and theories." Experiments play a pivotal role in this process. The general purpose of any experiment is to gain a better understanding about the process under investigation and, ultimately, to advance science. Many issues need to be addressed in the phases preceding, during, and following an experiment. These can be categorized as planning, design, construction, debugging, execution, data analysis, and reporting of results [5].

Prior to performing the experiment, a clear approach must be developed. The objective of the experiment must be defined along with its relation to the theory of the process. What are the assumptions made in the experiment? What are those made in the theory? Special attention should be given to ensuring that the experiment correctly reflects the theory. The process should be observed with minimal intervention, keeping in mind that the experiment itself may affect the process. All of the variables involved in the process should be identified. Which can be varied? Which can be controlled? Which will be recorded and how? Next, what results are expected? Does the experimental setup perform as anticipated? Then, after all of this has been considered, the experiment is performed.

Following the experiment, the results should be reviewed. Is there agreement between the experimental results and the theory? If the answer is yes, the results should be reconfirmed. If the answer is no, *both* the experiment and the theory should be examined carefully. Any measured differences should be explained in light of the uncertainties that are present in the experiment and in the theory.

Finally, the new results should be summarized. They should be presented within the context of uncertainty and the limitations of the theory and experiment. All this information should be presented such that another investigator can follow what was described and repeat what was done.

1.6   CLASSIFICATION OF EXPERIMENTS

There are many ways to classify experiments. One way is according to the intent or purpose of the experiment. Following this approach, most experiments can be classified as **variational**, **validational**, **pedagogical**, or **explorational**.

The goal of variational experiments is to establish (quantify) the mathematical relationships between the experiment's variables. This is accomplished by varying one or more of the variables and recording the results. Ideal variational experiments are those in which *all* the variables are identified and controlled. Imperfect variational experiments are those in which *some* of the variables are either identified or controlled. Experiments involving the determination of material properties or component or system behavior are variational. Standard testing also is variational.

Validational experiments are conducted to validate a specific hypothesis. They serve to evaluate or improve existing theoretical models. A critical, validational experiment, or Galilean experiment, is designed to refute a null hypothesis. An example would be an experiment designed to show that pressure does not remain constant when an ideal gas under constant volume is subjected to an increase in temperature.

Pedagogical experiments are designed to teach the novice or to demonstrate something that is already known. These are also known as Aristotelian experiments. Many experiments performed in primary and secondary schools are this type, such as the classic physics lab exercise designed to determine the local gravitational constant by measuring the time it takes a ball to fall a certain distance.

Explorational experiments are conducted to explore an idea or possible theory. These usually are based on some initial observations or a simple theory. All of the variables may not be identified or controlled. The experimenter usually is looking for trends in the data in hope of developing a relationship between the variables. Richard Feynman [8] aptly summarizes the role of experiments in developing a new theory, "In general we look for a new law by the following process. First we guess it. Then we compute the consequences of the guess to see what would be implied if this law that we guessed is right. Then we compare the result of the computation to nature, with experiment or experience, compare it directly with observation, to see if it works. If it disagrees with experiment it is wrong. In that simple statement is the key to science. It does not make any difference how beautiful your guess is. It does not make any difference how smart you are, who made the guess or what his name is—if it disagrees with experiment it is wrong. That is all there is to it."

An additional fifth category involves experiments that are far less common, which lead to discovery. Discovery can be either anticipated by theory (an analytic discovery), such as the discovery of the quark, or serendipitous (a synthetic discovery), such as the discovery of bacterial repression by penicillin. There also are thought (*gedanken* or Kantian) experiments that are posed to examine what would follow from a conjecture. Thought experiments, according to our formal definition, are *not* experiments because they do not involve any *physical* change in the process.

REFERENCES

[1] A. Gregory. 2001. *Eureka! The Birth of Science*. Duxford, Cambridge, UK: Icon Books.

[2] R. Harré. 1984. *Great Scientific Experiments*. New York: Oxford University Press.

[3] D. J. Boorstin. 1985. *The Discoverers*. New York: Vintage Books.

[4] G. Gale. 1979. *The Theory of Science*. New York: McGraw-Hill.

[5] H. W. Coleman and W. G. Steele. 1999. *Experimentation and Uncertainty Analysis for Engineers*, 2nd ed. New York: Wiley Interscience.

[6] P. Medawar. 1979. *Advice to a Young Scientist*. New York: Harper and Row.

[7] R. L. Park. 2000. *Voodoo Science: The Road from Foolishness to Fraud*. New York: Oxford University Press.

[8] R. Feynman. 1994. *The Character of Physical Law*, Modern Library edition. New York: Random House, 150.

REVIEW PROBLEMS

1. Variables manipulated by an experimenter are
  (a) independent, (b) dependent, (c) extraneous,
  (d) parameters, (e) presumed

2. Immediately following the announcement by the University of Utah on March 23, 1989, that Stanley Pons and Martin Fleischmann had "discovered" cold fusion, scientists throughout the world rushed to perform an experiment that typically would be classified as
  (a) variational, (b) validational, (c) pedagogical,
  (d) explorational, (e) serendipitous

3. If you were trying to perform a validational experiment to determine the base unit of mass, the gram, which of the following fluid conditions would be most desirable?

(a) a beaker of ice water, (b) a pot of boiling water, (c) a graduated cylinder of water at room temperature, (d) a thermometer filled with mercury

4. Match the following with the most appropriate type of variable:

independent	measured during the experiment
dependent	fixed throughout the experiment
extraneous	not controlled during the experiment
parameter	affected by a change made by the experimenter
measurand	changed by the experimenter

HOMEWORK PROBLEMS

1. Give one historical example of an inductivistic, a fallibilistic, and a conventionalistic experiment. State each of their significant findings.

2. Write a brief description of an experiment that you have performed or are familiar with, noting the specific objective of the experiment. List and define all of the independent and dependent variables, parameters, and measurands. Also provide any equation(s) that involve the variables and define each term.

3. Give one historical example of an experiment falling into each of the four categories of experimental purpose. Describe each experiment briefly.

4. Write a brief description of the very first experiment that you ever performed. What was its purpose? What were its variables?

5. What do you consider to be the greatest experiment ever performed? Explain your choice. You may want to read about the 10 "most beautiful experiments of all time" voted by physicists as reported by George Johnson in the *New York Times* on September 24, 2002, in an article titled "Here They Are, Science's 10 Most Beautiful Experiments." Also see R. P. Crease. 2003. *The Prism and the Pendulum: The Ten Most Beautiful Experiments in Science*. New York: Random House.

CHAPTER REVIEW

ACROSS

1. an experiment designed to teach
2. Plato's school of thought
8. Aristotle's school of thought
10. a variable affected by the experiment

DOWN

1. variable fixed throughout the experiment
3. process by which laws are arrived at from experiments
4. repeating experiments under the same, fixed operating conditions
5. physical quantities of the process under investigation
6. duplication of an experiment
7. a thought experiment
9. variable that is measured

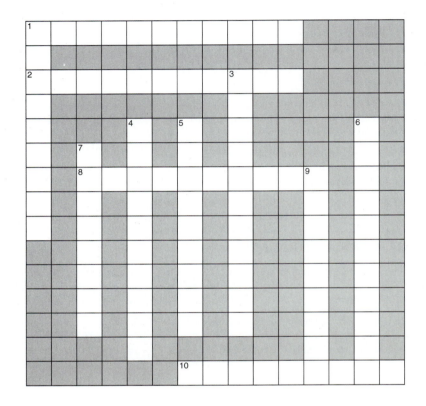

Former Weight Standards

Credit: Science Museum / Science & Society Picture Library, London.

Systems of standard weights first developed around the world out of a need to weigh precious metals accurately. This group includes a Siamese dragon, a Chinese jade, a Babylonian stone weight, a standard kilogram, a Burmese elephant, an Ashanti gold weight in the image of a warrior, and an Assyrian lion. Before such standardization, as summarized aptly by Bishop Fleetwood in 1745, "An acre is not an acre, a bushel not a bushel if you travel ten miles. A pound is not a pound if you go from a goldsmith to a grocer, nor a gallon a gallon if you go from the alehouse to the tavern." (From an exhibit on Weights and Measures at the Science Museum, London.)

2

UNITS AND SIGNIFICANT FIGURES

Units thus resemble sports officials: the only time you pay real attention to them is when something stupid happens.

Steve Minsky, "Measure for Measure," *Scientific American*, August 2000, 96.

Scientists lost a $125 million spacecraft as it approached Mars last week essentially because they confused feet and pounds with meters and kilograms, according to the National Aeronautics and Space Administration.

"A Little Metric Misstep Cost NASA $125 Million," *International Herald Tribune*, October 2, 1999.

Maximum Height: 11'3" 3.4290 metres

A sign on a bus in Edinburgh, Scotland, September, 1998.

CHAPTER OUTLINE

2.1   CHAPTER OVERVIEW

This chapter introduces two important topics: systems of units and significant figures. These topics often are considered too mundane to occupy valuable lecture time and therefore are left to students to learn on their own time. In fact, most students and their teachers spend hardly any time on these topics. Consequently, they often cannot identify the proper units of a particular dimension, convert its value from one system of units to another, and express it with the proper number of significant figures. It cannot be overemphasized that it is essential for a good scientist or engineer to have an excellent grasp of systems of units and significant figures.

2.2   LEARNING OBJECTIVES

You should be able to do the following after completing this chapter:

- Know the meaning of the following terms: dimension, unit, measure, consistent, base unit, supplementary unit, and derived unit
- Know the seven fundamental dimensions in the SI and Technical English systems
- Know the units for the following quantities in both the SI and Technical English systems: mass, length, time, velocity, acceleration, force, pressure, work, energy, torque, and power
- Convert between SI and Technical English systems
- Identify the least and most significant digits in a number
- Express calculated results with the correct number of significant figures

2.3   ENGLISH AND METRIC SYSTEMS

We live in a world in which we are constantly barraged by numbers and units. Examples include 56 kilobyte per second modems, 120 gigabyte hard drives, 720 megabyte CDs, a pint of Guinness drank in an English pub by an American tourist weighing 15 stones, 103 mile per hour fastballs, and over 2 meter-high aerial dunk shots. A visitor from the early 1900s would have no idea what we are talking about! We speak a foreign language that appears confusing to most. But units and measures are not meant to confuse. They were developed for us to communicate effectively, both commercially and technically. They are the structure behind our technical accomplishments. Without them, the Tower of Babel still would be under construction!

In the United States, two *languages* of systems currently are *spoken*. These loosely are referred to as the English and the metric systems. This *bilingual* situation can lead to some serious mistakes. A contemporary example of this is the loss of a $125 million

Mars Climate Orbiter on September 23, 1999, which was referred to in the headlines quoted at the beginning of this chapter. Basically, one group of scientists calculated the thrust of the orbiter's engines in units of pounds, but NASA assumed the values were in units of newtons. This led to approximately a 100 mile difference from its intended orbit, causing the spacecraft to burn up during orbital insertion! So, effective technical communication requires the abilities to *speak* the language of both systems of units and to *translate* between them. Before studying each system, however, it would be good to delve into a little of their history.

The English system of units evolved over centuries starting from the Babylonians, followed by the Egyptians, Greeks, Romans, Anglo-Saxons, and Norman-French. It was the Romans who introduced the base of 12 in the English system, where one Roman *pes* (foot) equaled 12 Roman *unciae* (inch). It was not until around the early 1500s that people began to consider quantifying and standardizing dimensions such as time and length. The yard, for example, has its origin with Saxon kings, whose *gird* was the circumference of the waist. It was King Edgar who, in an apparent attempt to provide a standard of measurement, declared that the yard should be the distance from the tip of his outstretched fingers to his nose. Other royal declarations, such as one made by Queen Elizabeth I defining the *statute* mile to be 5280 feet (8 furlongs at 220 yards per furlong) instead of the Roman mile (the distance of 1000 Roman soldier paces or 5000 feet), served to standardize what has become known as the English system of units.

The metric system, on the other hand, was not burdened with units of anthropometric origin, as was the English system. The metric system did not arise until near the end of the Period of Enlightenment, around the end of the 18th century. Thus, its development followed a more rational and scientific approach. Prior to its introduction, practically no single unit of measure was consistent. Footplates abounded, which marked the lengths of the most common *footmeasures* in Europe. In Rhineland, a foot was 31 centimeters, whereas in Gelderland it was 27 centimeters. The pound in Amsterdam was 494 grams. Slightly farther south in the Hague, it was 469 grams. This presented considerable confusion and impeded intercity commerce.

In 1670, a decimal system based on the length of 1 arc minute of the great circle of the earth was proposed by Gabriel Mounton. Jean Picard, in 1671, proposed that the length standard be defined as the length of a clock's pendulum whose period was a specified time. It was not until 1790 that a commission appointed by the French Academy of Sciences developed and formalized a decimal-based system defining length, mass, and volume. The unit of length, the meter, equaled 1 ten-millionth of the distance from the north pole to the equator along the meridian of the earth running from Dunkerque, France, through Paris to Barcelona, Spain. The unit of mass, the gram, was defined in terms of a liquid volume, where 1 gram equaled the mass of 1 cubic centimeter of water at its temperature of maximum density. The unit of volume, the liter, equaled 1 cubic decimeter. This approach established mass and volume as supplementary units in terms of a base unit (the meter), which was to a physical standard (the earth's circumference).

In 1866, the U.S. congress made it lawful to use the metric system in the United States in contracts, dealings, and court proceedings. Various metric units were defined in terms of their English counterparts. For example, the meter was defined as *exactly* 39.37 inches. In 1875, the United States signed with 16 other countries an international treaty called the Metric Conversion, which established a permanent international bureau of standards in Sèvres, France, and standards for length and mass. In 1893, the U.S. customary units (those based on the English system) were redefined in terms of their metric standards (which was opposite the approach taken in 1866). The yard became *exactly* 0.9144 meters (hence, the foot became *exactly* 0.3048 meters and the inch became *exactly* 0.0254 meters) and the pound *exactly* 0.453 592 4 kilograms. Since the treaty, almost all world countries have officially accepted this system. Over the years it was revised and simplified, eventually resulting in the *Le Système International d'Unités* (International System of Units). This system, abbreviated as SI, was adopted by the General Conference in 1960 and is what people today call the metric system.

The United States made a valiant attempt to adopt the metric system in 1975 with the Metric Conversion Act, which required federal agencies to use the metric system by 1992. Some signs along interstate highways showed distances to cities in both English and metric units. Soft drinks appeared in the market in liter bottles. A national chain of stores began selling metric tools. Beyond that, however, little happened with the general public and industry. Since then, most of these highway signs have disappeared. The liter-size plastic bottles and the metric tools remain as epitaphs. As of 2003, the United States, Liberia, and Myanmar are the only three countries out of the 192 countries in the world that have not formally adopted the metric system.

Today, the responsibility of maintaining the standards of measure in the United States rests with the National Institute of Standards and Technology (NIST). NIST provides a wealth of information on systems of units and their origin [1].

2.4 SYSTEMS OF UNITS

The measurement of a physical quantity involves the process of assigning a specific value with units to the physical quantity. The quantity has a **dimension** and its **unit** determines its **measure** or **magnitude**. For example, a sheet of European A4 paper is 210 millimeters wide by 297 millimeters long. The dimension is length, the unit is millimeters, and the measures are 210 and 297. If this information is expressed in another system of units, the dimension still is length but the unit and measures will be different. There are seven **fundamental dimensions**: length, mass, time, temperature, electrical current, amount of substance, and luminous intensity.

A **system of units** is necessary to provide a framework in which physical quantities can be expressed and also related to one another through physical laws. Five different systems of units are presented in Table 2.1. SI is the universally accepted system. Unfortunately, the English Engineering system (U.S. Standard Engineering or old English) and the Technical English system (U.S. Customary or British Gravitational) still are championed by U.S. industry (*vox clamantis in deserto*). Use of the other two systems, Absolute Metric and Absolute English, continue to appear in some publications.

Table 2.1 Five systems of units (adapted from [2] and [3])

Quantity	International System (SI)	Absolute Metric System (CGS)	English Engineering System (EE)	Absolute English System (AE)	Technical English System (TE)
Length	meter (m)	centimeter (cm)	foot (ft)	foot (ft)	foot (ft)
Time	second (s)	second (s)	second (s)	second (s)	second (s)
Mass	kilogram (kg)	gram (g)	pound-mass (lbm)	pound-mass (lbm)	*slug*
Force	*newton* (N)	*dyne*	pound-force (lbf)	*poundal*	pound-force (lbf)
g_c	$1 \dfrac{\text{kg} \cdot \text{m}}{\text{N} \cdot \text{s}^2}$	$1 \dfrac{\text{g} \cdot \text{cm}}{\text{dyne} \cdot \text{s}^2}$	$32.174 \dfrac{\text{lbm} \cdot \text{ft}}{\text{lbf} \cdot \text{s}^2}$	$1 \dfrac{\text{lbm} \cdot \text{ft}}{\text{poundal} \cdot \text{s}^2}$	$1 \dfrac{\text{slug} \cdot \text{ft}}{\text{lbf} \cdot \text{s}^2}$

Physical quantities of different dimensions are related to one another through equations in the form of definitions and natural laws. Consider, for example, Newton's second law: $F = ma$, where F denotes force, m is mass, and a is acceleration. In SI, a gravitational force of 9.81 newtons (N) (9.806 65 *exactly*) is required for a 1 kilogram (kg) mass to accelerate by 9.81 meters (m) per second (s) squared. The equation $F = ma$ gives two equations, a **numerical equation** that contains only the measures of the physical quantities, $9.81 = 1 \cdot 9.81$, and a **unit equation**, $\text{N} = 1 \cdot \text{kg} \cdot \text{m/s}^2$. The units within a system are **consistent** or **coherent** if no numerical factors other than 1 occur in all unit equations, as in this example.

A system of units is comprised of **base**, **supplementary**, and **derived** units. Base units are *dimensionally independent*. There is a base unit for every fundamental dimension contained in a particular system of units. Supplementary units are considered dimensionless, such as the radian, and do not represent a fundamental dimension. Derived units literally are derived from the base and supplemental units, and, therefore, are comprised of products, quotients, and powers of base and supplemental units. In the SI system, for example, the kilogram, meter, and second are base units. The newton is a derived unit because it represents a force, which is derived from the base units of mass times acceleration (as expressed by Newton's second law). Hence, a newton equals a kilogram times a meter divided by a second squared ($\text{N} = \text{kg} \cdot \text{m/s}^2$). The base units of the quantities listed in Table 2.1 are printed in normal font; those of the derived quantities are in italics.

STATEMENT | **Example 2.1**

Give the fundamental dimensions, unit, and measure of your weight in the Technical English system and in the International system.

SOLUTION

Assume an example weight of 172 pounds in the Technical English system. Weight is a force, which is mass times the local gravitational acceleration. So, for both the Technical English and International systems of units the fundamental dimensions of weight are mass, length, and time. The unit of force is lbf in the Technical English system and N in the International system. The measure is 172 in the Technical English system and 765, which equals 172/0.2248, in the International system.

Example 2.2	**STATEMENT**

In the English Engineering system 1 lbf is required to accelerate 1 lbm 32.174 ft/s^2. Is this system of units consistent?

SOLUTION

The unit equation for this circumstance is 1 lbf = 1 lbm × 32.174 ft/s^2. That is, 1 lbf equals 32.174 lbm · ft/s^2. A numerical factor other than 1 (here 32.174) appears in the unit equation, so the English Engineering system is *not* consistent.

Example 2.3	**STATEMENT**

What are the base units of mass and of force in the Technical English and English Engineering systems of units?

SOLUTION

In both the Technical English and English Engineering systems of units, force is a base unit and its unit is lbf. In the Technical English system, mass is a derived unit, the slug. Its base units are lbf, ft, and s (1 slug = 1 lbf/1 ft/s^2). In the English Engineering system, mass is a base unit and its unit is lbm.

For SI there are seven base units corresponding to the seven fundamental dimensions. These include the meter (m) for length, the kilogram (kg) for mass, the second (s) for time, the kelvin (K) for temperature, the ampere (A) for electric current, the mole (mol) for the amount of substance or quantity of matter, and the candela (cd) for luminous intensity. The corresponding fundamental symbols used for dimensional analysis are L for length, M for mass, T for time, Θ for temperature, $\mathcal{A}$ for electric current, $\mathcal{M}$ for the amount of substance or quantity of matter, and $\mathcal{K}$ for luminous intensity. There are two supplemental units, the radian (rad), which defines a plane angle, and the steradian (sr), which defines a solid angle. The seven base and two supplementary units of the SI system are discussed further in the following section. All other SI units are derived from these nine units, some having symbols and some not. For example, the volt, denoted by the symbol V, is a derived SI unit that represents the electric potential. Expressed in base units it equals kg · m^2/(s^3 · A).

There is a convention for the capitalization of unit abbreviations. Unit abbreviations named in honor of persons begin with a capital letter. For example, the pascal (Pa) is named after Blaise Pascal (1623–1662) and the hertz (Hz) is named after Heinrich Hertz (1857–1894). All other units, when spelled out, begin with lowercase letters, with a few exceptions. One is the SI abbreviation for volume, the liter (L). This abbreviation is capitalized to avoid confusion with the lowercase letter l and with the numeral 1.

A system of units can be created from a given number of base units. The MKSA system of units (actually a subsystem of the SI system) is a consistent system of units used for mechanics, electricity, and magnetism, that has the four base units m, kg, s, and A. A coherent system for mechanics is the MKS system having the three base units m, kg, and s. Many other systems abound such as the electrostatic CGS and the electromagnetic CGS systems. Their ubiquitous presence implicitly cautions us when converting from one system of units to another and further supports the use of one consistent system by everyone. Consider the beginning sentence in an article describing the construction of Washington, D.C.'s, Metro transit system that appeared in the July, 1976, *Newsletter of the National Safety Council*, "Mining recently has been completed for a 2200 foot-long (670 millimeters) twin tunnel section." Apparently this system was designed for either humans or insects, depending on which system of units is used!

Now return to the five systems of units presented in Table 2.1. There are basically four dimensions involved in each of these systems when used in mechanics: length, time, mass, and force. The English Engineering system is unique in that each of these four dimensions is defined to be independent. That is, for this particular system, the foot, second, pound-mass, and pound-force are base units. The unit for force is defined as the force with which the standard pound-mass is attracted to the earth at a location where the gravitational acceleration equals 32.1740 ft/s^2. The four dimensions are related through the equation $F = m \cdot a/g_c$, where F denotes the force in units of lbf, m is the mass in units of lbm, a is the gravitational acceleration (32.1740 ft/s^2), and g_c is a constant that relates the units of force, mass, length, and time. From this equation, for *only* the English Engineering system, $g_c = 32.1740$ lbm·ft/lbf·s^2. Thus, this system is *not* consistent, as was shown in Example 2.2.

STATEMENT   **Example 2.4**

A sounding rocket traveling at a constant velocity of 200 miles per hour in steady, level flight ejects 0.700 lbm/s of exhaust gas from its exit nozzle. Determine the rocket's thrust for both the English Engineering and International systems of units.

SOLUTION

Thrust is the equal and opposite reaction to the force that the exhaust gas exerts on the rocket nozzle. Because the rocket is traveling at a constant velocity, Newton's second law tells us that the thrust equals the velocity times the exhaust-gas mass flow rate.

In the English Engineering system:

$$\text{thrust} = V\frac{dm}{dt} = \underbrace{200\frac{\text{mile}}{\text{hour}} \times 5280\frac{\text{ft}}{\text{mile}} \times \frac{1}{3600}\frac{\text{hour}}{\text{s}}}_{293\,\frac{\text{ft}}{\text{s}}} \times 0.700\frac{\text{lbm}}{\text{s}} \times \underbrace{\frac{1}{32.174}\frac{\text{lbf} \times \text{s}^2}{\text{lbm} \times \text{ft}}}_{g_c} = 6.37\,\text{lbf}.$$

In the International system of units:

$$\text{thrust} = \underbrace{293\frac{\text{ft}}{\text{s}} \times 12\frac{\text{in.}}{\text{ft}} \times 0.0254\frac{\text{m}}{\text{in.}}}_{89.3\,\frac{\text{m}}{\text{s}}} \times \underbrace{0.700\frac{\text{lbm}}{\text{s}} \times \frac{1}{2.2046}\frac{\text{kg}}{\text{lbm}}}_{0.318\,\frac{\text{kg}}{\text{s}}} = 28.4\,\text{N}.$$

Each of the other four systems derive one of their four dimensions from the other three. The derived unit for each of these systems (force for three of these systems and mass for one) is given in italics in Table 2.1. For example, in the Absolute Metric system, the derived unit is the dyne, which, when expressed in terms of the base units, becomes g·cm/s^2. Note that each of these four systems is consistent as a consequence of this approach. This is indicated by the numerical factor of 1 for g_c.

2.5   SI STANDARDS

Next, let us examine the current definitions of the seven base and two supplementary units of the SI system.

The SI base unit of the dimension of time is the **second (s)**. It is defined as the duration of 9 192 631 770 cycles of the radiation associated with the transition between two hyperfine levels of the ground state of cesium-133. The conversion of this duration of cycles into time is accomplished by passing many cesium-133 atoms through a system of magnets and a resonant cavity driven by an oscillator into a detector (this device is called an atomic-beam spectrometer). Only those atoms that have undergone transition reach the detector. When 9 192 631 770 cycles of a detected atom in transition have occurred, the atomic clock advances 1 s. People certainly have come a long way since defining the meter in terms of a geophysical dimension that is changing constantly.

The SI base unit of length is the **meter (m)**. The meter was defined in 1983 to be the length that light travels in a vacuum during the interval of time equal to 1/299 792 458 s. Although it is related to the dimension of time, it is a base unit because it is not derived from other units. This definition uncoupled the meter from its 200-year-old terrestrial origin.

The SI base unit of mass is the **kilogram (kg)**. This is the only base unit still defined in terms of an artifact. The international standard is a cylinder of platinum-iridium alloy kept by the International Bureau of Weights and Measures in Sèvres, France. A copy of this cylinder, a secondary standard, is at the NIST in Gaithersburg, Maryland, where it serves as the primary standard in the United States. The kilogram is the only SI base unit linked to a unique physical object. This will end soon when the kilogram is redefined in terms of a more accurate, atom-based standard [10].

The **kelvin (K)** is the SI base unit of temperature. The kelvin is based on the triple point of pure water, where pure water coexists in solid, liquid, and vapor states. This occurs at 273.16 K and 0.0060 atmospheres of pressure. Thus, a kelvin is 1/273.16 of the thermodynamic temperature of the triple point of pure water. Absolute zero, at which all molecular motion ceases, is 0 K.

The SI base unit of electric current is the **ampere (A)**. The ampere is defined in terms of the force produced between two parallel, current carrying wires. Specifically, an ampere is the amount of current that must be maintained between two wires separated by 1 meter in free space in order to produce a force between the two wires equal to 2×10^{-7} N/m of wire length.

The **mole (mol)** is the SI base unit for the amount of substance. It is the amount of substance of a system that contains as many elementary entities as there are the number of atoms in 0.012 kg of carbon-12 ($6.022\ 142 \times 10^{23} = N_a$). That is, 1 mol contains N_a entities, where N_a is Avogadro's number. The entities can be either atoms, molecules, ions, electrons, or other particles or groups of such particles. The entities could even be golf balls! So, 1 mol of carbon-12 has a mass of 0.012 kg, 1 mol of monatomic oxygen has 0.016 kg, and 1 mol of diatomic oxygen has 0.032 kg. Each contain 6.022142×10^{23} entities, which would be atoms for carbon-12 and for monatomic oxygen and molecules for diatomic oxygen. The mass of 1 mole of a sub-stance is determined from its molecular (atomic) weight. Its SI units are kg/kg-mole. The atomic mass unit, typically designated by the symbol *amu*, *exactly* equals 1/12 the mass of one atom of the most abundant isotope of carbon, carbon-12, which is 1.6603×10^{-27} kg. This unit of mass is called a dalton unit.

The SI base unit of luminous intensity is the **candela (cd)**. One candela is the luminous intensity in a given direction of a source that emits radiation at only the specific wavelength of 540×10^{12} Hz and that has a radiant intensity in that direction of 1/683 watts per steradian. A 100-watt lightbulb has the luminous intensity of approximately 135 cd and a candle has approximately 1 cd.

There are two SI supplementary (dimensionless) units, the **radian (rad)** and the **steradian (sr)**. The radian is based on a circle and the steradian on a sphere. One radian is the plane angle with its vertex at the center of the circle that is subtended by an arc whose length is equal to the radius of the circle. Hence, there are 2π radians over the circumference of a circle. The steradian is the solid angle at the center of a sphere that subtends an area on the surface of the sphere equal to the *square* of the radius. Thus, there are 4π steradians over the surface of a sphere.

The base units of time, electric current, and amount of substance are the same in both the SI and English systems. The systems differ only in the units for the dimensions of length, mass, temperature, and luminous intensity. Presently, the level of accuracy for most base units is 1 part in 10 000 000 [10].

2.6 TECHNICAL ENGLISH AND SI CONVERSION FACTORS

People working in technical fields today must learn both the Technical English and SI systems and be proficient in converting between them. This is particularly true for the dimensions of mechanical, thermal, rotational, acoustical, photometric, electrical, magnetic, and chemical systems. The units used in the SI and Technical English systems for these dimensions are presented in tables in Appendix C that are based on information from [2, 5–9]. Often, the knowledge of one conversion factor for each dimension is sufficient to construct other conversion factors for that dimension. Table 2.2 lists some conversion factors between units in SI, English Engineering, and Technical English. The SI units for electrical and magnetic systems are presented in Chapter 4. There are many electronic work sheets available on the World Wide Web that automatically do unit conversions [11]. Also refer to the standards used for SI unit conversion [12].

2.6.1 LENGTH

For the dimension of length, 1 in. equals 2.54 cm *exactly*. Using this conversion, 1 ft = 0.3048 m *exactly*, 1 yd = 0.9144 m *exactly*, and 1 mi = 1.609 344 km *exactly*. A 10-km race is approximately 6.2 mi. Note that a period is used after the abbreviation

Table 2.2   Some useful conversion factors

Dimension	Units with Factors
Length	1 m = 3.2808 ft
	1 km = 0.621 mi
Volume	1 L = 0.001 m^3 = 61.02 in.^{3}
Mass	1 kg = 2.2046 lbm = 0.068 522 slug
Force	1 N = 0.2248 lbf = 7.233 pdl
Work, energy	1 kJ = 737.562 ft · lbf = 0.947 817 Btu
Power	1 kW = 1.341 02 hp = 3414.42 Btu/hr
Pressure, stress	1 atm = 14.696 psi = 101 325 Pa
	= 407.189 in. H$_2$O = 760.00 mm Hg = 1 bar
Density	1 slug/ft^3 = 512.38 kg/m^3
Temperature	K = °C + 273.15
	K = (5/9) °F + 255.38
	K = (5/9) °R
	°F = (9/5) °C + 32.0
	°F = °R − 459.69

for inch. This is the *only* unit abbreviation that is followed by a period, so as not to confuse it with "in," the preposition. All other unit abbreviations are *not* followed by a period.

2.6.2 AREA AND VOLUME

For area and volume, the square and the cube of the length dimension, respectively, are considered. The SI units of area and volume are m^2 and m^3. However, the liter (L), which equals 1 cubic decimeter or 1/1000 m^3, often is used. One L of liquid is approximately 1.06 quarts or 0.26 gallons. A 350-cubic-inch engine has a total cylinder displacement volume of approximately 5.7 L. Curiously, when the American tourist drinks an English pint of Guinness, he consumes 20 liquid UK ounces. A pint in his home country is 16 liquid ounces. In the United States, 1 liquid gallon (gal) = 4 liquid quarts (qt) = 8 liquid pints (pt) = 16 liquid cups (c). Further, 1 liquid cup (c) = 8 liquid ounces (oz) = 16 liquid tablespoons (Tbl) = 48 liquid teaspoons (tsp). The liquid (fluid) ounce is a unit of volume. The ounce when specified *without* the liquid prefix is a unit of mass, where 16 oz = 1 lbm.

2.6.3 DENSITY

The SI units of density are kg/m^3. Most gases have densities on the order of 1 kg/m^3 and most liquids and solids on the order of 1000 to 10 000 kg/m^3. For example, at 1 atm and 300 K air has a density of 1.161 kg/m^3, water 1000 kg/m^3, and steel

2.1 | MATLAB SIDEBAR

MATLAB is structured to handle large collections of numbers, known as arrays. Data files can be read into, operated on, and written from the MATLAB workspace. Array operations follow the rules of matrix algebra. An array is denoted by a single name that must begin with a letter and be fewer than 20 characters.

Operations involving scalars (single numbers, zero-dimensional arrays) are made using the standard operator notations of addition (+), subtraction (−), multiplication (*), right division (/), left division (\) and exponentiation (^). Here, for scalars a and b, a/b is $\frac{a}{b}$ and a\b is $\frac{b}{a}$.

The elements of row (horizontal) and column (vertical) vectors (one-dimensional arrays) are contained within square brackets, [ ]. In a row vector, the elements are separated by either spaces or commas; in a column vector by semicolons. The row vector A comprised of the elements 1, 3, and 5 would be set by the MATLAB command: A = [1 3 5] or A = [1,3,5]. The analogous column vector B would be defined by the MATLAB command: B = [1;3;5]. A row vector can be transformed into a column vector or vice versa by the transpose operator ('). Thus, the following statements are true: A = B' and B = A'.

A two-dimensional array consisting of m rows and n columns is known as an "m by n" or "$m \times n$" matrix. The number of rows always is stated first. A transpose operation on an $m \times n$ matrix will create an $n \times m$ matrix. The transpose operation gives the complex conjugate transpose if the matrix contains complex elements. A real-numbered transpose of a complex-element matrix is produced by using the dot transpose operator (.'). When the original matrix is real-numbered, then either the transpose or the dot transpose operators yield the same result.

The elements within an array are identified by referring to the row, column locations of the elements. The colon operator (:) is used to select an individual element or a group of elements. The statement A(3:8) refers to the third through eighth elements of the vector A. The statement A(2,3:8) refers to the elements in the third through eighth columns in the second row of the matrix A. The statement A(3:8,:) refers to the elements in the third through eighth rows in *all* columns of the matrix A.

7854 kg/m^3. The density of air can be determined over the temperature range from approximately 160 K to 2200 K using the equation of state for a perfect gas:

$$\rho = \frac{p}{R \cdot T} = \frac{p \cdot \mathcal{MW}}{\mathcal{R} \cdot T}, \qquad \textbf{[2.1]}$$

where ρ is the density, p is the pressure, T is the temperature, $\mathcal{R}$ is the universal gas constant equal to 8313.3 J/(kg-mole · K), $\mathcal{MW}$ is the molecular weight, and R is the gas constant, which equals $\mathcal{R}/\mathcal{MW}$. For air, $R = 287.04$ J/(kg · K), based on its molecular weight of 28.966 kg/kg-mole. The density of air at sea level is 1.225 0 kg/m^3. The density of water (±0.2 %) at 1 atm over the temperature range from 0 to 100 °C is given by the curve fit [4]:

$$\rho = 1000 - 0.0178|T - 4|^{1.7}, \qquad \textbf{[2.2]}$$

where the density is expressed in units of kg/m^3 and the temperature in units of °C.

2.6.4 MASS AND WEIGHT

The conversion of mass is straightforward. One lbm equals 0.453 592 4 kg *exactly*. So, 1 slug is approximately 14.59 kg. In terms of base units, 1 slug equals 1 lbf · s^2/ft.

2.2 | MATLAB SIDEBAR

There are several other MATLAB commands that are useful in array manipulation and identification. The MATLAB command linspace(x1,x2,n) creates a vector having *n* regularly spaced values between x1 and x2. The number of rows (*m*) and columns (*n*) in the matrix A can be identified by the MATLAB command [m n] = size(A). The MATLAB command cat(n,A,B,...) concatenates the arrays *A* and *B* and beyond together along the dimension *n*. So, if A = [1 3] and B = [2 4], then X = cat(1,A,B) creates the array X in which A and B are concatenated in their row dimension, where X = [1 3;2 4]. Likewise, cat(2,A,B) produces X = [1 3 2 4]. An m × n matrix consisting of either all zeros or all ones is created by the MATLAB commands zeros(m,n) and ones(m,n), respectively.

Be careful when performing matrix manipulation. Remember that two matrices can be multiplied together *only if* they are conformable. That is, the number of columns in the first matrix must equal the number of rows in the second matrix that multiplies the first one. Thus, if A is $m \times p$ and B is $p \times q$, then C = A*B is $m \times q$. However, B*A is not conformable, which demonstrates that matrix multiplication is not commutative. The associative, A(B + C) = AB + AC, and distributive, (AB)C = A(BC), properties hold for matrices. Element-by-element multiplication and division of matrices uses different operator sequences. The MATLAB command A.*B signifies that each element in A is multiplied by its corresponding element in B, provided that A and B have the same dimensions. Likewise, the MATLAB command A./B implies that each element in A is divided by its corresponding element in B. Further, the MATLAB command A.^3 raises each element to the third power in A.

Thus, the mass of the 15 stone American tourist in the English pub is 6.52 slugs in the Technical English system and 210 lbm in the English Engineering system (1 stone = 14 lbm).

Weight, which is a force, is the product of mass and acceleration. The tourist's weight is 210 lbf in both the Technical English and English Engineering systems. This seems confusing. The units and measures of the tourist's mass are different in these two systems. Yet, the weight units and measures are the same! Such system-conversion confusion usually arises when those speaking the English system do not specify what *dialect* they are using (Technical or Engineering). This invariably leads to the common question, "Should the mass be divided by 32.2 to compute the force or not?" The answer is yes if you are *speaking* English Engineering and no if you're *speaking* Technical English. Let us see why.

To avoid confusion in problems involving mass, acceleration, and force for the different systems, Newton's second law can be written as $F = ma/g_c$. This effectively keeps the measures of the dimensions correct for all systems. For consistent systems, the measure of g_c is unity. So $F = ma$ can be used directly. For example, in the Technical English system 1 lbf will accelerate 1 slug at 1 ft/s^2. For the inconsistent English Engineering system, g_c equals 32.174 lbm ft/lbf · s^2. So, $F = ma/g_c$ must be used. For example, 1 lbf will accelerate 32.174 lbm at 1 ft/s^2 or 1 lbm at 32.174 ft/s^2. By comparing the units of mass between the two English systems, 1 slug = 32.174 lbm. Such confusion usually compels unit-challenged individuals to learn the SI system for the sake of simplification.

Example 2.5

STATEMENT

Compute for both the Technical English and International systems of units the mass and weight of air at 300 K in a room with internal dimensions of 12 ft × 12 ft × 10 ft.

SOLUTION

The volume of the air in the room is 1440 ft^3. The density of air at 1 atm and 300 K is 1.16 kg/m^3 = 0.002 26 slug/ft^3. So, in Technical English, the mass of the air is 3.26 slugs and its weight is 3.26 slugs × 32.174 ft/s^2 = 105 lbf. In SI the mass is 47.6 kg and its weight is 47.6 kg × 9.81 m/s^2 = 467 N. Also note that in the English Engineering system, the density of air would be equal to 0.0727 lbm/ft^3. Thus, in English Engineering, the mass of the air is 105 lbm and its weight is 105 lbm × g/g_c = 105 lbm × 32.174 ft/s^2/32.174 lbm · ft/lbf · s^2 = 105 lbf. Note that the force in both the Technical English and English Engineering systems has the same measure but the mass does not.

Keep in mind that for an object of a given mass, its acceleration and weight change with distance from the center of the gravitational field of the body to which it is attracted. The weight w of a body is related to its mass m through Newton's law of gravitational attraction as

$$w(z) = mg_o \left(\frac{R}{R+z} \right)^2 = mg(z), \qquad \text{[2.3]}$$

where R is the radius of the body (R = 6.378 15 × 10^6 m for Earth), g_o the *local* gravitational acceleration (g_o equals 9.806 65 m/s^2 at sea level on Earth), and z is the distance away from the body (z = 0 at sea level).

Example 2.6

STATEMENT

Compute the gravitational acceleration in SI units at an altitude of 35 000 ft, where commercial jet airplanes fly.

SOLUTION

First the altitude is converted to the SI unit of meters. Here 35 000 ft/3.2808 ft/m = 10 668 m. Then, using the expression for $g(z)$ from Equation 2.3 yields g(10 668 m) = 0.996 66 · g_o. So, the change in the gravitational acceleration from that at sea level is very small, less than a half of a percent.

2.6.5 FORCE

The unit of force in SI is the newton (N), named after Sir Isaac Newton (1642–1727). A force of 1 N accelerates a 1 kg mass at 1 m/s^2. One N is approximately 0.225 lbf. Curiously, this is the approximate weight of an apple or, alternatively, the force felt by your hand when holding an apple. So, if a popular hamburger chain converted to metric, then its quarter pound hamburger would become a newton burger! In the English Engineering system there are pounds of force and pounds of mass, which are designated by lbf and lbm, respectively. In the Technical English system there is only one pound, the pound force. The unit of lbf (as opposed to lb) is used in Technical English to designate the pound force to avoid any ambiguity.

Force per unit area is pressure or stress. The SI unit for this is the pascal (Pa), which equals one N/m^2. One atmosphere, approximately 14.696 psia (pounds per square inch absolute) equals 101.325 kPa. The pressure at the center of the earth is 5.8×10^7 kPa and that of the best laboratory vacuum is 1.45×10^{-16} kPa [6].

Example 2.7

STATEMENT

At an altitude of 35 000 ft above sea level the atmospheric pressure is 205 mm Hg. Assuming that the typical area of an airplane's passenger window is 80 in.^{2}, determine the net force on the window during flight at that altitude.

SOLUTION

Assume that the pressure inside the airplane is 1 atm = 760 mm Hg. So, the force on the window will be outward because of the higher pressure inside the cabin. The pressure difference will be equal to $760 - 205 = 555$ mm Hg = 10.7 lbf/in.^{2}. Thus, the net outward force is (10.7 lbf/in.^{2}) (80 in.^{2}) = 859 lbf = 3.82 kN.

2.6.6   WORK AND ENERGY

The SI unit for work or energy is the joule (J), named after the British scientist James Joule (1818–1889). Joule is best known for the classic experiment in which he demonstrated the equivalence of energy and work. In fact, energy is *defined* as the ability to do work. One J is 0.2288 calories (cal), or approximately 0.738 ft · lbf, or approximately 9.48×10^{-4} Btus (British thermal units). One Calorie (with a capital C, abbreviated Cal) is 1000 calories = 1 kcal. Thus, 1 kJ = 0.2288 Cal. Most people count Calories when on a diet. It takes approximately 0.016 J of work to lift a teaspoonful of ice cream from the table to your mouth (a distance of approximately 1/3 m) to gain approximately 35 kJ of energy from the ice cream. That is not much caloric expenditure for a lot of caloric gain!

Example 2.8

STATEMENT

A person eats a cup of ice cream. How many miles would the person have to jog to expend the energy he just consumed? How much weight would he gain if he did not jog off the calories added by eating the ice cream?

SOLUTION

For energy to be conserved and the person not to change weight, the energy contained in the ice cream must equal the energy expended in jogging. Assume that 100 Cal are expended for each mile jogged. A cup of ice cream contains approximately 400 Cal. Thus, he would have to jog 4 mi. If he didn't jog, the 400 Cal would be converted into a mass of body fat whose weight is approximately 1/8 lbf on earth. This is because 1 g of fat produces 9 Cal of energy. So, 400 Cal is converted into 44.4 g of body fat. The weight of this mass on earth is 0.436 N, which is approximately 1/8 lbf.

2.6.7   POWER

Power is work or energy per unit time. The SI unit of power is the watt (W), which is a joule per second (J/s). This is named after the British engineer James Watt (1736–1819). Intensity usually refers to power per unit area, or in SI units, W/m². Flux often denotes intensity per unit time, or in SI units, $W/(m^2 \cdot s)$.

2.6.8   TEMPERATURE

The temperature scales are related as shown in Table 2.2. Water boils at approximately 212 °F, 100 °C, 373.15 K, and 671.67 °R, depending on the local pressure. Note that °C stands for degrees Celsius (not degrees Centigrade, which is no longer preferable) and K stands for kelvin (not degrees kelvin). Note the lack of the ° symbol here. The Kelvin and Rankine scales are *absolute* (thermodynamic) temperature scales. An absolute temperature is independent of the properties of a particular system and is based on the second law of thermodynamics. Temperatures of 0 K and 0 °R are absolute zero. In the Kelvin scale, a value of 273.16 is assigned to the triple point of water. An International Practical Temperature scale (IPTS-68) was adopted in 1968 by the International Committee of Weights and Measurements. This scale covers the temperature range from 13.81 K (the triple point of hydrogen) to 1377.58 K (the freezing point of gold at 1 atm). It specifies a series of 11 temperatures based on the triple, freezing, and boiling points of various substances, the temperature measurement instruments to be used for calibration purposes over a specified temperature range, and the equations for interpolating temperatures between the 11 fixed points.

STATEMENT   **Example 2.9**

Sir Isaac Newton developed his own temperature scale in 1701 where water was "just freezing" at 0 units and "boyles vehemently" at 34.4 units. Six units of his temperature scale corresponded to "air at midsummer." What was the temperature of the midsummer air in London in units of his contemporary Gabriel Fahrenheit's temperature scale?

SOLUTION

From the temperature difference between the boiling and freezing of water, it is known that 212 °F−32 °F = 180 °F, which corresponds to 34.4 units of Newton's scale. Thus, the conversion factor from Newton's to Fahrenheit's units is 5.23, *assuming* that both scales are linear in between the two temperatures. This implies that the midsummer's air is 6 × 5.23 °F/Newton unit + 32 °F = 63.4 °F.

2.6.9   OTHER PROPERTIES

The properties of gases, liquids, and solids can be expressed in terms of base and supplementary units. Absolute or dynamic viscosity μ is a fluid property that is related to the fluid's shear stress (force per unit area) τ and rate of shear strain (strain per unit time) $d\theta/dt$, by the expression $\tau = \mu d\theta/dt$. Thus, the fundamental dimensions of absolute viscosity are $ML^{-1}T^{-1}$, and the SI base units are $kg/(m \cdot s)$. Kinematic viscosity ν is the ratio of absolute viscosity to density μ/ρ and has the SI units of m^2/s. The absolute viscosity of air and water are affected weakly by pressure and strongly by temperature. The absolute viscosity for air can be determined using Sutherland's law [4]:

$$\mu = \mu_o \left(\frac{T}{T_o}\right)^{3/2}\left(\frac{T_o + S}{T + S}\right), \qquad \textbf{[2.4]}$$

where μ_o equals 1.71×10^{-5} kg/(m · s), $T_o = 273$ K, and $S = 110.4$ K for air. The absolute viscosity for water (± 1 %) can be determined by the curve fit [4]:

$$\mu = \mu_o \exp\left[-1.94 - 4.80\left(\frac{T_o}{T}\right) + 6.74\left(\frac{T_o}{T}\right)^2\right], \qquad \textbf{[2.5]}$$

where μ_o equals 1.792×10^{-3} kg/(m · s) and $T_o = 273.16$ K.

The units of most properties can be found using an expression, typically a physical law or definition, that relates the property to other terms in which the units are known. For example, the gas constant R is related to the speed of sound (distance per unit time), a, by the expression $a = \sqrt{\gamma R T}$, where T is the temperature, and γ equals the ratio of specific heats C_p/C_v. Thus, the base units of R are m^2/s^2 · K, or equivalently J/kg · K.

Note that C_p and C_v are functions of temperature. For air at 300 K, $C_p = 1.0035$ kJ/(kg · K) and $C_p = 0.7165$ kJ/(kg · K), which yields $\gamma_{air} = 1.4$. The temperature and pressure of the standard atmosphere of air at sea level are 288.15 K and 101 325 Pa, respectively. The speed of sound for air at these conditions is 340.43 m/s. Variations of atmospheric pressure, temperature, density, and speed of sound with altitude for the 1976 Standard Atmosphere are available in graphical and computational forms [13].

Finally, a word of caution is necessary. A unit balance always needs to be done when performing calculations involving unfamiliar quantities. Do this even when working within one system of units and using units that can be expressed directly in terms of base units. Conversion factors may be needed, especially when dealing with electrical and magnetic units. Example 2.10 serves to illustrate this point.

Example 2.10

STATEMENT

Determine the charge in units of coulombs of a 1 μm diameter oil droplet that is charged to the Rayleigh limit. Also express this in terms of the number of elementary charges. The Rayleigh limit charge q_{Ray} is given by the expression

$$q_{Ray} = \sqrt{2\pi \sigma_l d_p^3},$$

where $\sigma_l = 0.04$ N/m and d_p denotes the droplet diameter.

SOLUTION

Noting that $d_p = 1 \times 10^{-6}$ m and making substitutions into the expression for charge in terms of SI units yields

$$q_{\text{Ray}} = \sqrt{(2\pi)(0.04)(1 \times 10^{-18})} = 5.02 \times 10^{-10} \sqrt{\text{J} \cdot \text{m}}.$$

But is this the value of q_{Ray} in units of coulombs? In other words, is the unit C equal to the units $\sqrt{\text{J} \cdot \text{m}}$? The answer is no. A conversion factor of $\sqrt{4\pi\epsilon_o}$ that has units of $\text{C}/\sqrt{\text{J} \cdot \text{m}}$ is required, in which ϵ_o is the permittivity of free space that equals 8.85×10^{-12} F/m. This is because the units F/m equal the units $\text{C}^2/(\text{J} \cdot \text{m})$. The measure of the conversion factor is 5.02×10^{-10}. Another useful unit conversion is $(4\pi\epsilon_o)(\text{V}^2) = \text{N}$. Thus,

$$q_{\text{Ray}} = 5.02 \times 10^{-10} \sqrt{4\pi\epsilon_o} = (5.02 \times 10^{-10})(1.06 \times 10^{-5}) = 5.32 \times 10^{-15} \, \text{C}.$$

The number of elementary charges n_e equals $q_{\text{Ray}}/e = 5.32 \times 10^{-15}/1.60 \times 10^{-19} = 33\,000$.

2.7 PREFIXES

Often it is convenient to use **scientific notation** to avoid writing very large and very small numbers, as in Example 2.10. Positive and negative powers of 10 are used to shorten numbers by moving the decimal point. Examples are $1000 = 1 \times 10^3$, $0.001 = 1 \times 10^{-3}$, and, as seen earlier, $6.022\,137 \times 10^{23}$, which was used to represent 602 213 700 000 000 000 000 000. Sometimes the notation $\text{E}\pm n$ is used to replace $\times 10^n$, particularly with computer output where exponents cannot be generated. For example, $3.254 \times 10^8 = 3.254\,\text{E}+8$.

Table 2.3 lists the decimal prefixes. The SI preference is to express a number using a prefix such that its numerical value is between 0.1 and 1000. For example, the mass of the earth, which is 5.98×10^{24} kg, becomes 5.98 Ykg. Along these lines, there are approximately 10π Ms in 1 year.

In some situations, the magnitude of the number goes beyond where these prefixes can be used. Consider the estimate for the energy released in the Big Bang, 10^{68} J. When using the American system of numeration (see the Table of Numbers in your dictionary), this becomes 0.1 million vigintillion J! This system uses Latin prefixes for the "illion" in the unit and follows the simple formulation that $number = 10^{3(n+1)}$, where n specifies the name of the prefix. For example, the prefix "tri" means three, which corresponds to the number $10^{3(3+1)} = 10^{12}$ or one *trillion*. Likewise, one quattuordecillion ("qua" is 4 plus "dec," which is 10) equals $10^{3(4+10+1)} = 10^{45}$. For really large numbers, use the *googol*, which was "coined in the late 1930s by the nine-year-old nephew of the American mathematician Edward Kasner when he was asked to come up with the name for a very large number" [14]. One googol equals 10^{100}. Its cousin, one googolplex, equals $10^{10^{100}}$. Obviously, the use of scientific notation

Table 2.3   Prefixes for units

Factor	Prefix	Symbol
10^{24}	yotta	Y
10^{21}	zeta	Z
10^{18}	exa	E
10^{15}	peta	P
10^{12}	tera	T
10^{9}	giga	G
10^{6}	mega	M
10^{3}	kilo	k
10^{2}	hecto	h
10^{1}	deka	da
10^{-1}	deci	d
10^{-2}	centi	c
10^{-3}	milli	m
10^{-6}	micro	μ
10^{-9}	nano	n
10^{-12}	pico	p
10^{-15}	femto	f
10^{-18}	atto	a
10^{-21}	zepto	z
10^{-24}	yocto	y

usually is preferred in such cases unless you are out to impress your colleagues with your extensive vocabulary!

Example 2.11

STATEMENT

The American system of numeration differs from the British system of numeration, which also is used by most European countries. In the British system 10^6 is a "thousand thousands," 10^9 is a "milliard," and 10^{12} is a "billion." Beyond a British billion, the formulation is *number* $= 10^{6n}$, where n specifies the Latin name of the prefix of "illion." Determine if a British billionaire is richer than an American trillionaire.

SOLUTION

For numbers equal to and beyond 10^{12}, the formulations given for each of the American and British systems can be used. Thus, the ratio of a British number to an American number equal to and beyond 10^{12} will be given by the formulation $ratio = 10^{6n}/10^{3(n+1)} = 10^{3(n-1)}$. So, a British billionaire ($n = 2$) is actually a thousand times richer than an American billionaire, provided that the exchange rate between British pounds and American dollars is 1:1, which it is not. In fact, a British billionaire is approximately 1600 times richer than an American billionaire and 1.6 times richer than an American trillionaire!

2.8 SIGNIFICANT FIGURES

The ubiquitous use of calculators and computers has led to assignments and lab reports with far too many digits in every number! This situation begs for us to revisit the concept of **significant figures**. This is especially important because the number of digits present in a result implies the *precision* of the result. It goes without saying that the proper use of significant figures is an essential element in the presentation of both experimental and calculated results and their uncertainties.

How is the number of significant figures determined? The number of significant figures is the number of digits between and including the least and the most significant digits. The leftmost *nonzero* digit is called the **most significant digit**; the rightmost *nonzero* digit, the **least significant digit**. If there is a decimal point in the number, then the rightmost digit is the least significant digit even if it is a zero. These rules imply that the following numbers have five significant figures: 1.0000, 2734.2, 53 267., 428 970, 10 101, and 0.008 976 0.

But what happens when no decimal point is present, a zero is the rightmost digit, and it *is* significant? This situation is ambiguous and can be avoided by expressing the number in scientific notation, where 428 970 becomes $4.289\,70 \times 10^5$. The convention here is that *all* of the digits present in scientific notation are significant. In this case, there are six significant figures.

How is a number **rounded off** to drop its insignificant figures? Again, there are rules. To round off a number, the number first is truncated to its desired length. Then the *excess* digits are expressed as a decimal fraction. Depending on whether this fraction is less than, equal to, or greater than 1/2 determines the fate of the least significant digit in our truncated number. If it is greater than 1/2, we round up the least significant digit by one; if it is less than 1/2, it is left alone. If the fraction equals 1/2, the least significant digit is rounded up by one only if that digit is odd. This method reduces any systematic errors that can arise if that number actually resulted from a rounding at a previous step in the calculations. In this light, it is better *not* to round off numbers during the sequential analysis of data and only round off the final results. If numbers are rounded off every time during many sequential calculations, the results are skewed and a systematic error is introduced.

STATEMENT   **Example 2.12**

Round off the following numbers to three significant figures: 23 421, 16.024, 273.61, 5.6850×10^3, and 5.6750×10^3.

SOLUTION

The answers are 23 400, 16.0, 274, 5.68×10^3, and 5.68×10^3. Note that 16.024 when rounded off to three significant figures is 16.0, where the 0 is significant because it is to the right of the decimal point. Also note that the last two numbers when rounded off to three significant figures become the same. This is because of our rule for round off when the truncated fraction equals 1/2.

The misuse of significant figures occurs everywhere, from the laboratory reports of college students in Indiana to the buses in Edinburgh, Scotland. Let's examine the sign in the front of the Edinburgh bus that was presented at the beginning of this chapter. The maximum height was 11'3"or 135 in. There are three significant figures. So, the maximum height in meters should be written as 3.43 m, not 3.4290 m as shown having five significant figures. If the maximum height in meters was treated to have the correct number of significant figures, then the English system equivalent should have been written as 11'3.00".

Consider another example. The weight of a large steel cylinder is computed from measurements of its diameter and length. Let its length L be equal to 3.32 m (three significant figures) and its diameter d equal to 0.3605 m (four significant figures). Its volume V would be computed using the formula $V = \pi d^2 L/4$ and would be equal to 0.339 m^3. This results from rounding off the computed value of 0.338 874 1 . . . to the required number of three significant figures. This is because the number of significant figures in a computed result equals the minimum number of significant figures in any number used in the computation. Now, to convert from this volume to mass, suppose that the density of the steel ingot equals 7835 kg/m^3. This yields a mass of 2660 kg (rounded off from 2655.0793). Note that although the density has four significant figures, only three significant figures are retained in the result. Converting to weight gives 26 000 N (rounded off from 26 037.4), assuming a gravitational acceleration of 9.806 65 m/s^2. There are three significant figures in the final result, although it appears that there are only two. For this situation, the result should be expressed in scientific notation as 2.60×10^4 N, which implies three significant figures.

Example 2.13 | **STATEMENT**

Determine in the appropriate SI units the value with the correct number of significant figures of the work done by a 1.460×10^6 lbf force over a 2.3476 m distance.

SOLUTION

There are four significant figures for the force and five for the distance. Because work is the product of force and distance (assuming that the force is applied along the direction of motion), work will have four significant figures. The SI unit of work is the joule, where $J = N \cdot m$. Now 1.460×10^6 lbf equals 6.495×10^6 N. So, the work is 1.525×10^6 J or 1.525 MJ.

Finally, what happens to the number of significant figures when you are converting from one unit of a dimension to another, say from inches to feet? The number of significant figures does not change (assuming, which usually is the case, that the conversion factors are known *exactly*).

For example, consider a distance measurement with an uncertainty of 0.125 in. In units of feet with the correct number of significant figures, the uncertainty would be 0.0104 ft. Note there are three significant figures in both numbers, even though when converting to units of feet, inches are divided by 12 (which appears to have only two significant figures). The number of significant figures remains at three because the conversion from inches to feet is an *exact* conversion (it could be divided by 12.000 000 when converting from inches to feet).

STATEMENT   **Example 2.14**

Convert 100.0185 °C to temperature in K.

SOLUTION

The conversion factor from degrees Celsius to kelvin is K = 273.15 + °C. At first hand, it appears that there are only five significant figures in the conversion equation. But this is not so because the conversion equation is *exact*. So, 100.0185 °C = 373.1685 K, where both temperatures have seven significant figures.

Thus far, the rules of significant figures to numerical calculations appear straightforward. However, applying them directly to experimental results and their uncertainties sometimes leads to ambiguous situations that require common sense and good judgment to resolve.

When expressing a measured value with its associated uncertainty, the precision should be the same between the measured value and its uncertainty. This is an accepted convention in uncertainty analysis. In the previous example of determining the weight of the ingot, the two dimensions of length had different numbers of significant figures. This could result from using one type of instrument to measure the ingot's length and another to measure its diameter. The number of significant figures should correspond with the uncertainty in the measurement. For example, if the uncertainty in a measurement is ±0.05, then the measurement should be expressed with the same precision, say 1.23 ± 0.05.

Next consider the following apparent dilemma. A measured temperature of 54.0 °C is specified and its uncertainty is ±0.5 °C. Convert it into units of kelvin. Following the rules of significant figures the temperature becomes 327 K, where three significant figures are maintained in the conversion. Now our uncertainty of ±0.5 °C translates directly into an uncertainty of ±0.5 K because the conversion relation between °C and K is linear. Thus, a change in +0.5 °C or −0.5 °C from a temperature in °C is a change in +0.5 K or −0.5 K from the corresponding temperature in K. But look at what has happened. The measured temperature in K has lost the precision specified by the uncertainty. If the level of precision must be the same between a measured value and its uncertainty, then the converted, measured temperature needs to be 327.2 K. Following this convention, it appears that a significant figure was gained in the process!

In the end, there is no single, correct answer for this problem. It all depends on the purpose in performing the conversion. If temperature is only computed, then the converted, measured temperature is 327 K according to the rules of significant figures. If an experimental result is expressed with its uncertainty, then the converted, measured temperature is 327.2 K ± 0.5 K according to convention in uncertainty analysis.

The relation between significant figures and measurement uncertainty has been covered only briefly. This topic is important because many measured quantities are often reported with more significant figures than the uncertainty of the instruments used to measure them. This consequently leads to reporting the corresponding experimental results with more significant figures than the measurement uncertainty.

This approach is wrong and can be very misleading when interpreting experimental results. Measurement uncertainty will be discussed in Chapter 9.

REFERENCES

[1] http://physics.nist.gov/cuu/Units/current.html.

[2] T. G. Beckwith, R. D. Marangoni, and J. H. Leinhard V. 1993. *Mechanical Measurements*, 5th ed. New York: Addison-Wesley.

[3] G. J. Van Wylen and R. E. Sonntag. 1965. *Fundamentals of Classical Thermodynamics*. New York: John Wiley and Sons.

[4] F. M. White. 1999. *Fluid Mechanics*, 4th ed. New York: McGraw-Hill.

[5] D. Halliday and R. Resnick. 1966. *Physics*, Combined edition. Parts I and II. New York: John Wiley and Sons.

[6] S. Strauss. 1995. *The Sizesaurus*. New York: Kodansha International.

[7] C. F. Bohren. 1991. *What Light Through Yonder Window Breaks?* New York: John Wiley and Sons.

[8] 1976. *About Sound*, Washington D.C.: U.S. Environmental Protection Agency, Office of Noise Abatement and Control.

[9] L. Pauling. 1970. *General Chemistry*, 3rd ed. San Francisco: W.H. Freeman.

[10] T. Crump. 2001. *A Brief History of Science*. London: Robinson.

[11] http://www.megaconverter.com/Mega2/. This site also refers to many other conversion sites.

[12] 1976. *Standard for Metric Practice: ANSI/ASTM E 380-76*. New York: American Society for Testing and Materials.

[13] http://www.digitaldutch.com.

[14] J. Gullberg. 1997. *Mathematics from the Birth of Numbers*. New York: W.W. Norton and Company.

REVIEW PROBLEMS

1. What has the most mass?
  (**a**) 1 slug, (**b**) 1 kg, (**c**) 1 lbm, (**d**) 1 g, (**e**) 1 N

2. Four ounces weigh approximately how many newtons?

3. A scientist developing instruments for use in nanotechnology would be most interested in measuring which of the following lengths?
  (**a**) the mileage between San Francisco and New York,
  (**b**) the length of the leg of an ant,
  (**c**) the diagonal of a unit cell of iron,
  (**d**) the chord of an airfoil

4. Which of the following is *not* equivalent to the SI unit of energy?
  (**a**) $kg \cdot m^2/s^2$, (**b**) $Pa \cdot m^2$, (**c**) $N \cdot m$, (**d**) $W \cdot s$

5. The SI system has how many base units?
  (**a**) 2, (**b**) 3, (**c**) 4, (**d**) 7, (**e**) 250

6. What is the weight in newtons of a mass of 51 slugs with the correct number of significant figures?

7. An astronaut weighs 164 pounds on Earth (assume that Technical English is spoken). What is the astronaut's weight (in the appropriate SI unit and with the correct number of significant figures) on the surface of Mars where the gravitational acceleration is 12.2 feet per second squared?

8. An astronaut weighs 162 pounds on Earth (assume that Technical English is spoken). What is the astronaut's mass (expressed in the appropriate SI unit and with the correct number of significant figures) on the surface of Mars where the gravitational acceleration is 12.2 feet per second squared?

9. Which four base units are the same for both the SI and English Engineering systems?

10. A robotic manipulator weighs 393 lbf (Technical English) on Earth. What is the weight of the probe on the moon's surface in newtons (to the nearest tenth of a newton) if the lunar gravitational acceleration is 1/6 of that on Earth?

11. How much work is required to raise a 50 g ball 23 in. vertically upward? Express your answer in units of ft-lbf to the nearest one-thousandth of a ft-lbf.

12. How many molecules of water are there in 36 g of water?

13. If the Mars Rover weighs 742 N on Mars, where $g = 3.71$ m/s^2, what would be its mass (in slugs) on Jupiter, where $g = 23.12$ m/s^2, expressed with the correct number of significant figures?

14. How many significant figures does the number 001 001.0110 have?

15. The pressure acting on a 1.25 in.^{2} test specimen equals 15 MPa. What is the force (in N) acting on the specimen expressed with the correct number of significant figures?

16. If $w = (5.50/0.4) + 0.06$, what is the value of w with the correct number of significant figures?

17. The number 4 578.500 rounded off to four significant figures is
(a) 4580, (b) 4579.0, (c) 4579, (d) 4578, (e) 4570

18. The number 001 001.0110 has how many significant figures?
(a) 10, (b) 9, (c) 8, (d) 7, (e) 4

19. The number 11.285 00E12 has how many significant figures?
(a) 5, (b) 6, (c) 7, (d) 12, (e) 13

HOMEWORK PROBLEMS

1. The following presents the original definitions of some of the more customary English units. Try to guess the unit's name for each:
(a) the distance from the outstretched fingers to the tip of the nose of King Edgar,
(b) the distance covered by 36 barleycorns laid end to end,
(c) the width of the thumb of a king or three barleycorns laid end to end,
(d) the distance a Roman soldier traveled in a thousand paces,
(e) the length of a Viking's outstretched arms, and
(f) the amount of land that could be plowed with a yoke of oxen in a day.

2. Determine your
(a) mass in kilograms, (b) weight in newtons,
(c) height in meters, (d) volume in liters, and
(e) density in kilograms per cubic meter. Finally,
(f) compare your density to that of water at standard conditions.

3. Show that a quarter pound hamburger sold in a metric country would be (approximately) a "Newton Burger."

4. Show that 1 μcentury approximately equals 1 hr.

5. Compute how many seconds there are in 1 year and express this result in scientific notation and a familiar numerical constant.

6. On Earth an astronaut weighs 145 pounds (assume Technical English is spoken). Compute

(a) this astronaut's weight (in the appropriate SI unit) on the surface of Mars where the gravitational acceleration is 12.2 ft/s^2,
(b) her mass on Mars in the appropriate Technical English unit,
(c) her mass on Mars in the appropriate SI unit, and
(d) her mass on Earth in the appropriate SI unit.

7. The Reynolds number Re is a dimensionless number used in fluid mechanics and is defined as Re $= \rho V D/\mu$, where ρ is the fluid density, μ is the fluid absolute viscosity, V is the fluid velocity, and D is the characteristic length dimension of the body immersed in the moving fluid. Because this number has no units, it should be independent of the system of units chosen for ρ, μ, V, and D. In the International System of units, $\rho = 1.16$ kg/m^3, $\mu = 1.85 \times 10^{-5}$ N·s/m^2, $V = 5.0$ m/s, and $D = 0.254$ m. Using this information, compute
(a) values for ρ, μ, V, and D in the English Engineering system,
(b) Re based on the International System, and
(c) Re based on the English Engineering system.

8. The power coefficient C_P for a propeller is a nondimensional number that is defined as $C_P = P/\rho n^3 D^5$, where P is the power input to the propeller, ρ is the density of the fluid (usually air), n is the propeller's revolutions per second, and D is the propeller diameter. For $\rho = 0.002\ 11$ slug/ft^3, $n = 2400$ rpm, $D = 6.17$ ft, and $P = 139$ hp,
(a) express these four values in SI units and
(b) compute C_P based on the SI units.

9. The advance ratio J for a propeller is defined as $J = V/nD$, where V is the velocity, n is the propeller's revolutions per second, and D is the propeller diameter. For $V = 198$ ft/s, $n = 2400$ rpm, and $D = 6.17$ ft,

   (a) show that J is a nondimensional number by "balancing" the units and

   (b) compute the value of J.

10. An engineering student measures an ambient lab pressure and temperature of 405.35 in. H_2O and 70.5 °F, respectively, and a wind tunnel dynamic pressure (using a pitot-static tube) of 1.056 kN/m². Assume that $R_{air} = 287.04$ J/(kg · K). Determine with the correct number of significant figures

   (a) the room density using the perfect gas law in SI units (state the units with your answer) and

   (b) the wind tunnel velocity using Bernoulli's equation in units of ft/s. Bernoulli's equation states that for irrotational, incompressible flow, the dynamic pressure equals one-half the product of the density times the square of the velocity.

11. An engineer using a barometer measures the laboratory temperature and pressure to be 70.0 °F and 29.92 in. Hg, respectively. He then conducts a wind tunnel experiment using a pitot-static tube and an inclined manometer to determine the wind tunnel velocity through Bernoulli's equation. He measures a pressure difference of 3.22 in. H_2O. Determine the tunnel velocity and express it with the correct number of significant figures in units of m/s.

CHAPTER REVIEW

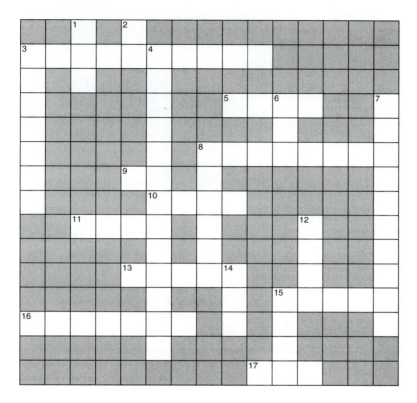

ACROSS

3. when no numerical factors other than 1 appear in all unit equations
5. prefix for 10 to the minus 18
8. SI base unit of mass
9. inconsistent system of units
10. SI base unit for the amount of substance
11. dimensionally independent units
13. 14 lbm
15. number of fundamental dimensions
16. a quantity's magnitude
17. the absolute metric system

DOWN

1. approximate density of air in SI units
2. the international system of units
3. SI base unit for luminous intensity
4. dimensionless units
6. approximate weight of two apples in SI units
7. length, mass, intensity, etc.
8. SI base unit of temperature
12. number of significant digits in 0.001 234 5
14. prefix for 10 to the plus 18
15. pest in a Technical English garden

The Cooke and Wheatstone
Morse Tape Perforator

In the middle of the 19th century communication made a significant leap forward by the development of the telegraph. In 1837 Charles Wheatstone and William Cooke produced the first practical telegraph, which was first conceived by Stephen Gray in 1727. The device continually was improved. That pictured here was designed originally around 1858 for recording transmitted code as holes in paper tape. Code also could be transmitted in an opposite manner by what became known as the Wheatstone transmitter. Wheatstone also is credited in applying paper tape for the data storage and transmission. This remained a major method of data retention for well over 100 years.

TECHNICAL COMMUNICATION

"The horror of the moment," the King went on, "I shall never forget." "You will, though," the Queen said, "if you don't make a memorandum of it."

Lewis Carroll. 1945.
Alice's Adventures in Wonderland. Racine: Whitman Publishing Company.

CHAPTER OUTLINE

3.1   CHAPTER OVERVIEW

The queen was right. All is lost if you do not make a memorandum of the results. This chapter describes the tools needed to help you prepare for technical communication. Suggested formats for technical memos and technical reports are presented. Guidelines for proper writing in general as well as specific to technical memoranda are given. Graphical presentation is discussed. Also, guidelines for oral technical presentations are summarized.

3.2   LEARNING OBJECTIVES

You should be able to do the following after completing this chapter:

- Know the standard formats of technical memos and reports
- Make clear, easy-to-understand figures with appropriate labeling
- Know the proper formats for presenting numbers and units
- Produce a high-quality technical memo or report following the appropriate formats
- Give an excellent oral technical presentation

3.3   GUIDELINES FOR WRITING

A short list of critical writing rules is presented in the following. These rules relate to neither the style nor the content of the writing. They account only for the most fundamental aspects of clearly written communication. All technical memoranda and reports must adhere to these rules. There are many good texts that present the styles for writing, including Strunk and White [1] and Baker [2].

3.3.1   WRITING IN GENERAL

- Words must be spelled properly.

   Incorrect: Mispellings should be avoided.
   Correct: Misspellings should be avoided.

- Sentence fragments must be avoided.

   Incorrect: First, a look behind the scenes.
   Correct: First, we will look behind the scenes.

- The subject and verb within the sentence must agree.

   Incorrect: A motion picture can improve upon a book, but they usually do not.

Correct: A motion picture can improve upon a book, but it usually does not.

- Avoid abrupt changes in tense; past tense is best.

  Incorrect: We weigh the sample …
  Correct: The investigators weighed the sample …

- Avoid abrupt changes in person; third person is best.

  Incorrect: We weigh the sample …
  Correct: The investigators weighed the sample …

- Avoid abrupt changes in voice; active voice is best.

  Incorrect: It was decided …
  Correct: The investigators decided …

- Contractions should be avoided.

  Incorrect: Don't use contractions.
  Correct: Do not use contractions.

- Avoid splitting infinitives.

  Incorrect: … to enable us to effectively plan our advertising …
  Correct: … to enable us to plan effective advertising …

- Avoid dangling participles.

  Incorrect: Going home, the walk was slippery.
  Correct: When I was going home, the walk was slippery.

- Modifiers must be hyphenated properly.

  Incorrect: … the red, hot flame …
  Correct: … the red-hot flame …

- A sentence should not end with a preposition.

  Incorrect: What did she write with?
  Correct: With what did she write?

- Proper end punctuation must be used.

  Incorrect: Be careful
  Correct: Be careful.

3.3.2 WRITING TECHNICAL MEMORANDA

Writing a good technical memorandum requires practice. Very few people have the innate ability to write memoranda well, especially technical memoranda. They learn to do so through experience. Some guidelines can be followed that relate to style. The following suggestions help to make a better document:

- Write technical memoranda in the third person.

- Use the past tense throughout technical memoranda.

- Limit the length of sentences. Break long sentences up into shorter ones. Scientists and engineers tend to write long sentences.

- Segment ideas into paragraphs such that the reader is led through the presentation in a smooth and effortless fashion.

- Type all memoranda. Choose a word processing software package and learn how to use it effectively. This will help to produce a professionally presented document, which usually includes text, figures, tables, and equations.

- Proofread and check for spelling errors. It often helps to have someone else do the proofreading.

- Provide a plausible explanation based on scientific principles whenever predictions and measurements differ. Uncertainties or mistakes are not acceptable reasons. Support any explanation by some simple calculations.

- Report the average value with its uncertainty whenever reporting results based on multiple measurements. Avoid simply listing all the measured values.

- Use correct English. Do not confuse commonly used words such as "its" and "it's." The former is possessive; the later is a contraction. Other examples include "affect" and "effect," "farther" and "further," "ensure" and "insure," "because" and "since," "approximately" and "about," and "decrease" and "drop."

One of the most frustrating experiences for a reader is to read a document having many mistakes and missing essential information. Some absolute rules can be established for writing technical memoranda. A memoranda that violates any one of these rules is incomplete.

1. Every variable and symbol, either measured or analytical, must be identified.

2. Every variable's units must be presented.

3. The proper number of significant figures must be used with all numbers.

4. Uncertainties must be given for every measured and predicted variable. Nominal values must be included. The assumed confidence level must be stated. Often it is easiest to present uncertainties in a table and include supporting calculations in an appendix.

5. The physical concepts behind a model must be explained. Do not just present the model's equations.

6. Do not use relative words, such as "good," "reasonable," "acceptable," "significant," and so forth, when describing an agreement between values. Quantitative statements must be made when making a comparison, such as "x differed from y by z %."

7. Each figure or table included must be referred to and discussed in the text. Do not say "calibration data is shown in Figure 1" or "results are presented in figures 1 through 6" and then never discuss what is shown in each figure.

8. Equations must be punctuated with commas or periods, as if they were part of a sentence. Do not let them dangle in space.

9. A "0" must be included in the front of the decimal point if no other number is present. The decimal point can be missed by the reader when the "0" is absent.

10. All pages must be numbered consecutively except the cover sheet.

3.3.3 NUMBER AND UNIT FORMATS

The presentation of numbers and units should follow specific formats [3]. A few of these guidelines that are very appropriate to presenting technical information include the following:

- Use SI units. Give the equivalent values in other units in parentheses following the SI unit values only if necessary.

- Avoid using unacceptable abbreviations such as sec for second, cc for cubic centimeter, l for liter, ppm for parts per million. For example, express 7 ppm as 7 μL/L.

- Include units for each number when using composite expressions, such as those involving areas, volumes, and ranges. Volume, for example, would be written as 2 m $\times$ 3 m $\times$ 5 m. The correct expression for a range of values would be 23 L/s to 45 L/s. Use the word "to" instead of a dash when expressing a range. For example, write 5 to 10 rather than 5–10.

- Use Arabic numerals and symbols for units, such as "the mass was 15 kg." Keep a space between the numeral and the symbol. This is even true for percentages, which should be expressed as x % and *not* as x%. Only an angle does not have a space between it and its symbol.

- Italicize quantity symbols, such as l, V, and t for length, volume, and time, respectively.

- Put unit symbols in Roman type. Subscripts and superscripts may be in either Roman or italic type.

- When there are more than four digits in a number on either side of the decimal marker, use spaces instead of commas to separate numbers into groups of three, counting in both directions from the decimal marker, such as 3.141 592 654.

- Express all logarithms using log, with their bases as subscripts, such as $\log_e(x)$ and $\log_{10}(x)$. Do not use $\ln(x)$ for the natural logarithm.

- Use decimal prefixes with a number's units, keeping its numerical value between 0.1 and 1000, such as 1.05 MJ instead of 1.05×10^6 J.

3.3.4 GRAPHICAL PRESENTATION

The proper presentation of quantitative information is essential to good technical communication. Most quantitative information is presented graphically in a variety of ways [4]. The types of plots commonly used for the graphical presentation of experimental data include Cartesian, semilogarithmic, logarithmic, stem, stair, and bar formats.

The Cartesian plot is the most common type of plot, in which the values of the dependent variable are plotted versus the values of the independent variable. Typically one or two dependent variables are included in one plot, having either two or three dimensions. An example of a Cartesian plot is shown in Figure 3.1.

Some physical systems respond in an exponential manner to external forcing. Many physical variables are related to one another through a power law, either linear, quadratic, or higher order. Possible relations can be ascertained by plotting the dependent variable versus the independent variable, such as versus time. Figure 3.2 shows three graphical representations of the same power-law relation. For $y = a \exp(bt)$, a semilog plot with the y-axis as the logarithmic axis gives a line of slope value $b / \log_e(10)$ and intercept value $\log_{10}(a)$. For $y = ax^b$, a log-log plot of y versus x will yield a line of slope value b and intercept value $\log_{10}(a)$.

Sometimes a series of values that were acquired sequentially needs to be examined. This usually is done to observe the trend of the values in time. Figure 3.3 displays three types of plots of the same data. A stem plot extends a line from the abscissa up to the ordinate value that is designated by a marker. A stairs plot connects

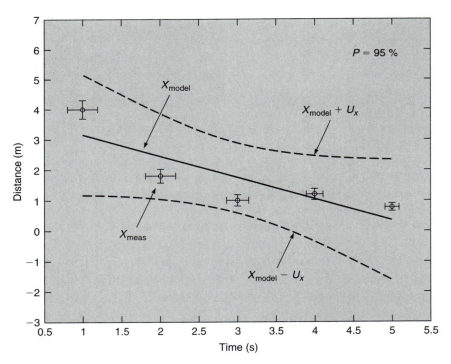

Figure 3.1   Graphical presentation of a comparison between predictive and experimental results.

3.1 | MATLAB SIDEBAR

MATLAB has numerous built-in commands for plotting. The MATLAB command `plot(x,y)` constructs a Cartesian plot of the values of y along the ordinate versus the values of x along the abscissa. The x,y pairs can be plotted with or without markers, connecting lines, or color. The markers, line styles, and colors are specified as arguments of the `plot` command. The symbols for markers are . (point), o (circle), x (x-mark), + (plus), and * (star). Those for line styles are - (solid line), : (dotted line), -. (dash-dot line), and – (dashed line). Those for colors are y (yellow), m (magenta), c (cyan), g (green), b (blue), w (white), and k (black). The following MATLAB command will plot y versus x with x,y pairs as red circles and z versus x with x,z pairs as green stars with a dashed line connecting the x,z pairs:

```
plot(x,y,'ro',z,x,'g--')
```

Grids, axes labels, and a title can be added to the plot using the MATLAB commands `grid`, which turns on grid lines, `xlabel('put the x-label here')`, `ylabel('put the y-label here')` and `title('put the title here')`. Maximum and minimum values of the axes can be specified by the MATLAB command `axis([xmin xmax ymin ymax])`. Text can be added to the plot at a specific x,y location (xloc, yloc) by the MATLAB command `text(xloc,yloc,'put text here')`. Most of these tasks can be accomplished by enabling the plot editor under the *tools* menu above the figure, once the figure is plotted.

3.2 | MATLAB SIDEBAR

There are two ways that multiple plots can be made on the same graph. If each dependent variable has the same number of values then they all can be plotted versus the independent variable using one MATLAB command. For example, if y1, y2, and y3 each are measured at a time, t, in an experiment, a plot of y1, y2, and y3 versus t is achieved by the MATLAB command:

```
plot(t,y1,t,y2,t,y3)
```

This command will *not* work when any of the dependent variables has a different number of values. A plot similar to that achieved by the above command can be gotten for this situation using the MATLAB `hold on` command, which would be

```
plot(t,y1)
hold on
plot(t,y2)
hold on
plot(t,y3)
```

Multiple, separate plots can be made using the MATLAB command `figure(n)` before each of n `plot(x,y)` commands. For convenience, begin with n equal to 1. Each figure appears in a separate window.

lines from each ordinate value to the next, mimicking a stairway. The bar plot gives a rectangular bar of a fixed width for each ordinate value. The type of presentation determines which plot is best.

Often when the amounts that contribute to a whole need to be displayed, the pie chart is used. This graphically is in the shape of a circular pie, in which each contributing amount is displayed as a sector of the pie. Each sector's area is its proportional contribution to the total area. Usually, each sector has a different color.

3.3 | MATLAB Sidebar

MATLAB has a number of commands that can be used to obtain different types of plots. In some instances, the independent variable may be related to the dependent variable through an exponent or power law. Such relations can be assessed readily using MATLAB plotting commands, such as `semilogx`, `semilogy`, and `loglog`. For example, if the underlying relation is given by $y = 2x^3$, then a plot of y versus x using the MATLAB `plot` command will appear as a hyperbola. Likewise, a plot of y versus x using the MATLAB `semilogx` command will show exponential behavior and a plot of y versus x using the MATLAB `loglog` command will be a line. These three cases are displayed in Figure 3.2. This figure was obtained using the following MATLAB command sequence:

```
subplot(3,1,1)
fplot('2*x^3',[0,10],'k')
subplot(3,1,2)
xx=0:0.01:10;
y=2*xx.^3;
semilogx(xx,y,'k')
subplot(3,1,3)
loglog(xx,y,'k')
```

If $y = \exp(x)$, then what MATLAB plotting command would yield a line? Note that in MATLAB, the command `log` represents the natural logarithm and the command `log10` stands for the common (base 10) logarithm.

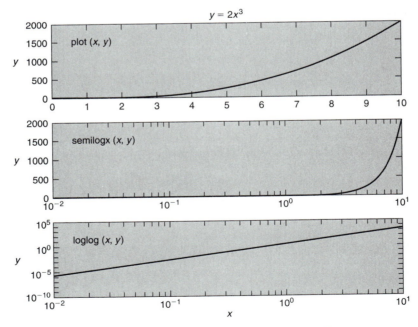

Figure 3.2   The same function obtained using the MATLAB `plot` (top), `semilogx` (middle), and `loglog` (bottom) commands.

3.4 | **MATLAB SIDEBAR**

MATLAB has plotting commands that can be used to display data in different ways. The commands stem, stairs, and bar can be used to examine the serial behavior of data. A series of ten data values are plotted in Figure 3.3 using these commands. Each plot presents an alternative approach to presenting the same information.

```
subplot(3,1,1)
stem(t,d,'k')
subplot(3,1,2)
stairs(t,d,'k')
subplot(3,1,3)
bar(t,d,'k')
```

3.5 | **MATLAB SIDEBAR**

The MATLAB plotting command pie(x) can be used to make a pie chart. The command automatically normalizes the each value of x by the sum of all of the x values. In this manner, each area in the pie is proportional to the total area of the pie. Adding a matrix to the argument of pie explodes user-designated areas of the pie. This is accomplished by setting values of 1 for each corresponding area to be exploded and keeping all other designates equal to 0. The following sequence of commands was used to construct Figure 3.4, which plots both the unexploded and exploded version of the pie.

```
x=[1,3,4,3,7,2,5];
ex=[0,0,1,0,1,0,1];
subplot(1,2,1)
pie(x)
subplot(1,2,2)
pie(x,ex)
```

Some data may follow an angular dependence, such as those representing acoustic radiation or the surface pressure distribution around an object. A polar plot, such as that shown in Figure 3.5, can be constructed for this purpose. The magnitude of the variable for a given θ is plotted as a distance from the origin. Data that represent cyclic processes will pass more than once through the origin.

The following guidelines should be followed when constructing plots (refer to Figure 3.1 as an example):

- A title or caption must be present.
- Both the abscissa and ordinate must be labeled with the name of the quantity plotted and its units in brackets or parentheses.
- Tick marks should be used on each side of both axes. Internal tick marks are preferable.
- All curves must be labeled either on the plot using arrow indicators or in the plot's legend when more than one curve is plotted.
- Analytical results must be plotted as a solid curve. Do not use symbols.
- Numerical results must be plotted as a dashed or dotted curve. Do not use symbols.

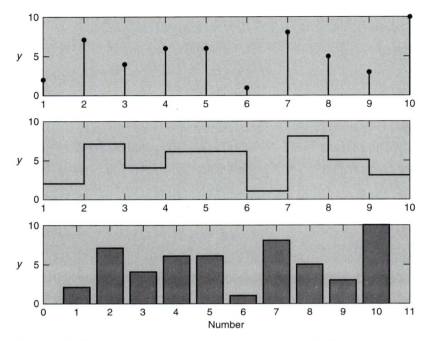

Figure 3.3   Ten sequential values plotted using the MATLAB stem (top), stairs (middle), and bar (bottom) commands.

- Experimental results must be denoted with symbols and error bars, using the same symbol for a given data set.
- Any curve representing an estimate must be presented with $\pm$ confidence limit curves evaluated at P % confidence.

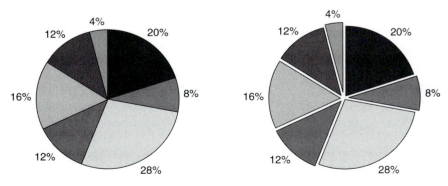

Figure 3.4   A standard pie chart (left) and a pie chart with "exploded" areas (right).

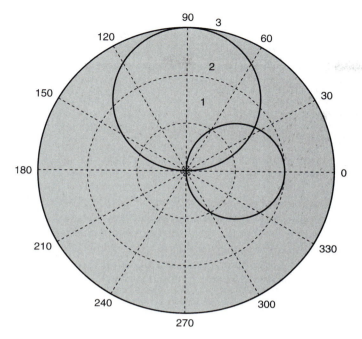

Figure 3.5   A polar plot of $3 \sin(\theta)$ centered along the $\theta = 90$ axis and $2 \cos(\theta)$ centered along the $\theta = 0$ axis.

3.6 | MATLAB SIDEBAR

The MATLAB `polar` command can be used to plot data that may follow some angular dependence. The following sequence of commands was used to construct Figure 3.5, which plots sine and cosine functional dependencies.

```
theta=0:0.01:2*pi;
x=3*sin(theta);
y=2*cos(theta);
```

```
polar(theta,x)
hold on
polar(theta,y)
```

Note that $3 \sin(\theta)$ is represented by a circle that extends to a value of 3 and is centered along the $\theta = 90$ axis. Likewise, $2 \cos(\theta)$ is shown as a circle that extends to a value of 2 and is centered along the $\theta = 0$ axis.

3.4   THE TECHNICAL MEMO

The format of a technical memo is similar to that commonly used in industry and at national laboratories. It is a concise, formal presentation of your findings on a particular technical issue. It is *not* a comprehensive explanation of the theoretical or experimental methods that you have used, but rather a summary highlighting the results that you have obtained. Typically, the body of a technical memo should not exceed two to three pages in length, excluding any supporting material that usually is placed in appendices. The suggested format is described next.

Date: 1/18/05

To: Professor P. F. Dunn

From: I. M. A. Student

Subject: Rocket Thrust Measurement

Summary: This should be one paragraph that summarizes the important results and states the significant conclusions. When writing this section, assume that this may be the only part of the technical memo that actually is read. Thus, it needs to be self-contained and not refer to any other written section, graphs, and tables that are contained in the body of the memo. Key results must be presented. Values of important experimental parameter ranges must be included. If theory also is presented, a quantitative statement about agreement or disagreement with the experimental results needs to be made.

Findings: This part covers in more detail what was summarized above. Enough information must be provided such that an engineer at your level could critically evaluate the approach and methods, and understand how you arrived at the results and conclusions, without your providing any information beyond what is written. Only the most important figures and tables need to be included here. Supporting material such as additional plots, program listings, or flow charts can be included as attachments (appendices) to the memo. All figures and tables must be numbered and referred to properly in the text. Include only the material that you consider necessary to support the conclusions. Never present results, especially in figures and tables, that you do not specifically discuss in this section. Do not attach volumes of computer-generated output without any explanation. The reader is impressed not by the volume of data collected but by the value of it.

References: References must be numbered in consecutive order. Do *not* include any references that are not cited in the memo. The following reference format conforms to that specified by *The Chicago Manual of Style* [5]:

Journal article references: 1. N. O. D. Student, and S. A. M. College. 2010. Measurements through a Hard Semester. *J. Heat Mass Transfer* 11:548–556.

This includes the last names and first and middle initials of all of the authors, the year of publication, the title of the article, the abbreviated title of the journal (italicized), the volume number, and inclusive page numbers of the article.

Book references: 2. M. A. Saad. 1992. *Compressible Fluid Flow*. 2nd ed. New Jersey: Prentice-Hall Inc.

The book title is italicized.

Appendices: These addenda contain supplementary material, such as detailed derivations or calculations. They are not meant to contain a record of everything that was done on the memo's subject. Include only what is needed to support material presented in the body of the memo. Do not place the results here and then refer to them from the memo.

3.5 THE TECHNICAL REPORT

This section describes the format that typically is required for laboratory reports. This format essentially parallels that of many journal publications. Each of the following sections need to be included in the report. As stated for the technical memo, write the report as if you were writing to another engineer at your level. The suggested format is described next.

1. **Title Sheet**: List the class and its number, report title and number, your name, all group members' names, and the date. This is the cover sheet of the report. It is not numbered.

2. **Abstract**: The primary purpose of the abstract is to provide the reader with a brief and sufficient summary of the project and its results. It is to be short (no more than approximately 100 words) and informative. It must indicate clearly the nature and range of the results contained in the report. The abstract must stand alone. No citing of numbered references, symbols, and so forth, must be made unless they are understood without any reference to the report. The easiest procedure is to write the abstract summarizing the entire body of the report *after* the report has been written.

3. **Table of Contents**: Each of the subsequent sections should be listed with their corresponding pages in the report.

4. **List of Symbols (Nomenclature)**: English symbols are first listed in alphabetical order, then Greek symbols in alphabetical order. Be sure to describe adequately your nomenclature, for example, not just "viscosity" but either "absolute (dynamic) viscosity" or "kinematic viscosity." Also note that in some cases the mere descriptive name of the symbol is not sufficient. For example, when listing coordinates be sure to specify the coordinates' directions with respect to some reference point. Also, when describing nondimensional numbers, specify their definitions in terms of the other symbols listed. The best procedure in gathering the nomenclature is to construct the list of symbols *after* the body of the report has been written.

5. **Introduction**: This section introduces the reader to the nature of the problem under investigation. It explains the history and relevance of such an experiment and its application. Previously published papers relevant to the experiment should be cited here. The general objectives of the experiment should be stated. Do not simply summarize the experimental objectives. Provide a guide for the reader as to what will follow in the report.

6. **Approach**: This section sometimes is referred to as *methods* or *procedure*. It needs to describe briefly the experimental, analytical, and numerical methods used to arrive at the results. There must be sufficient detail to permit a critical evaluation of the methods used and replication of the results by another party. It is not necessary,

however, to give full descriptions of all of the methods that are described in detail elsewhere, for example, how a particular numerical integration scheme works step-by-step. Uncertainty estimates for all parameters and procedures used to arrive at the results must be provided. Usually, it is preferable to present these estimates in a table. A block diagram or flow chart of the steps in the approach can be very helpful to the reader. Alternatively, a step-by-step approach can be put in narrative form. A flow chart should be included for each computer program used. A listing of each program should be presented in an appendix and documented with sufficient comments such that it can be followed easily by the reader.

7. **Results:**   The results of your experiment are presented here, usually facilitated by graphs, figures, and tables. The findings of the experiment are presented but neither discussed nor evaluated in this section. Keep in mind that you want to be concise when presenting the results. Put results in graphical form whenever possible. Sample calculations can be put in an appendix. When the results cover several aspects of the project, subdivide this section such that each part deals with one major aspect. The results of an uncertainty analysis must be provided. Detailed, supporting calculations should be presented in an appendix. Mention specifically what is contained in *each* figure and table. Do *not* say, for example, that "the results are shown in figures 1 through 6," and then fail to explain what is presented in each figure.

8. **Discussion:**   This section should include a discussion and evaluation of the results obtained and their relation to other pertinent studies, if any. The findings of your experiment are interpreted in this section. Express your scientifically justified opinion in this section about the facts that were presented in the previous section. Remember the distinction between fact and opinion. Point out the limitations of how you approached the experiment and how you would improve on your approach. Be constructively critical. Describe what you have learned from the experiment.

9. **Conclusions:**   Briefly conclude the major findings of your experiment. Do not introduce anything new or continue to discuss the results.

10. **References:**   These follow the same guidelines as for the technical memo.

11. **Appendices:**   These follow the same guidelines as for the technical memo.

3.6   THE ORAL TECHNICAL PRESENTATION

Many technical societies now have web pages that provide instructions for speakers [6]. The mode of oral presentation has changed considerably over the past several years. Most professional presentations now are made using standard software packages. The resulting, user-designed slides are projected digitally. Thus, the visual format of the presentation becomes very important.

The success of an excellent presentation lies in its organization, preparation, and delivery. The amount of time spent in preparing an oral presentation should be equivalent to that spent in writing a technical paper. Sometimes preparing an oral presentation is even more challenging. The listener is not as attentive as a reader. Presentation time is limited. The presenter must speak with confidence and enthusiasm for the talk to be effective. Adequate time for preparation and practice must be spent to produce a professional and well-received presentation. Practice also allows timing to be rehearsed.

Most presentations have two implicit goals: to deliver information and to have the audience understand it. This requires that information be presented clearly and concisely such that it is easily understood. The presentation must be substantive, including a statement of the problem, followed by a description of its solution and the results.

Some guidelines for a good oral presentation follow:

- Start with a title slide, followed immediately by a slide that outlines the talk.

- Break the body of the talk into sections, each making a specific point.

- Guide the listeners through the presentation by referring back to the outline at appropriate times.

- Conclude with a slide that summarizes the main points.

- Minimize the number of words and information presented on a slide. Keep it simple. Going into unnecessary detail will only lose the audience.

- Use an appropriate font size and type that is supported by a contrasting, simple, and pleasing background. Place an 8 in. by 10 in. copy of a slide on the floor. You should be able to read it clearly while standing directly over it. Refer to Figure 3.6 as an example.

- Follow the same rules for figures and equations that would be done for a written document. Label the axes. Provide units. Define all variables.

- Use your notes as a guide. Do not read directly from them or your paper.

- Stay focused on the topic. Avoid rambling beyond what was planned.

- Watch the time. Pace yourself. Do not exceed the time allotment.

- Speak enthusiastically. Do not speak in a monotone. Vary the rate and pitch.

- Try to stand near one place. Avoid walking aimlessly about.

- Avoid unconscious gestures, especially with your hands.

- Use a pointer to focus attention on an area of a slide. Be careful with unconsciously waving the pointer around, especially with a laser pointer, as it is very distracting.

- Look directly at the audience. Do not look over peoples' heads or stare at the floor.

- Avoid using vocal pauses, such as "you know" and "ah." Have someone listen to a practice presentation and note the number of vocal pauses made. You may be surprised at how many you say. Learn to break the habit.

- If you make a mistake, correct it and go on. Avoid joking about it or making excuses for it. Be confident!

- Finally, relax and enjoy giving the presentation. It represents your hard work and interest.

Figure 3.6   Example slide with suggested font size and layout.

REFERENCES

[1] W. Strunk, Jr., and E. B. White. 1979. *The Elements of Style*, 3rd ed. New York: MacMillan.

[2] S. Baker. 1984. *The Complete Stylist and Handbook*. New York: Harper and Row.

[3] http://physics.nist.gov/Pubs/SP811/cover.html.

[4] E. R. Tufte. 1998. *The Visual Display of Quantitative Information*, 2nd ed. Cheshire: Graphics Press.

[5] 2003. *The Chicago Manual of Style*, 15th ed. Chicago: The University of Chicago Press.

[6] http://www.aiaa.org/education/index.hfm?edu=13#j.

REVIEW PROBLEMS

1. A technical memo should contain only the following sections:

    (a) abstract, introduction, results;
    (b) abstract, results, conclusions;
    (c) introduction, results, conclusions;
    (d) summary, list of symbols, results, references;
    (e) summary, findings, references

2. Technical memoranda are presented customarily in what tense?

    (a) future, **(b)** passive, **(c)** past, **(d)** present, **(e)** subjunctive

CHAPTER REVIEW

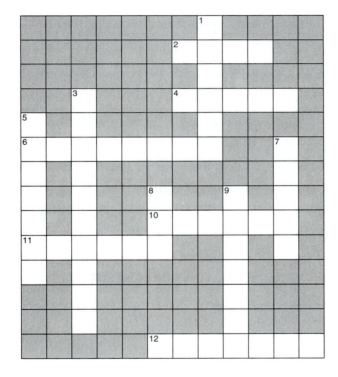

ACROSS

2. short form of written technical communication
4. type of curve for analytical results
6. a brief summary of a project and its results
10. to take out an insurance policy
11. to guarantee
12. wrote about Alice

DOWN

1. long form of written technical communication
3. section in which results are evaluated
5. longer in distance
7. the natural logarithm of x
8. a type of chart that is good to eat
9. longer in time

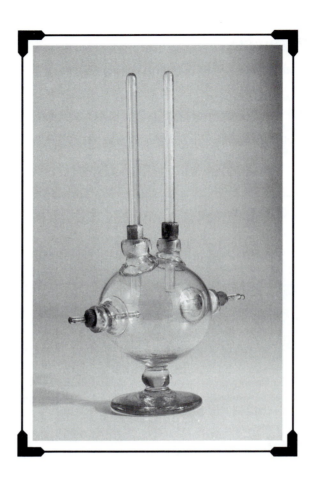

The Voltameter

Credit: Science Museum / Science & Society Picture Library, London.

In 1834, Michael Faraday (1791–1867), a British electrochemist, introduced the *volta-electrometer*. This term was shortened to *voltameter*, and then later replaced by the term *coulometer*, which was introduced by Theodore William Richards in 1902. This particular voltameter was used during the 19th century in the electrolysis of water. By applying a potential difference from a *voltaic pile* between two electrodes, a current was generated that electrolyzed the water.

4

BASIC ELECTRONICS

Nothing is too wonderful to be true, if it be consistent with the laws of nature and in such things as these, experiment is the best test of such consistency.

Michael Faraday (1791–1867) on March 19, 1849.
From a display at the Royal Institution, London.

. . . the language of experiment is more authoritative than any reasoning: facts can destroy our ratiocination— not vice versa.

Alessandro Volta (1745–1827),
quoted in *The Ambiguous Frog: The Galvani-Volta Controversy on Animal Electricity,* M. Pera, 1992.

CHAPTER OUTLINE

4.1   CHAPTER OVERVIEW

We live in a world full of electronic devices. Stop for a minute and think of all the ones you encounter each day. The clock radio usually is the first. This electronic marvel contains a digital display, a microprocessor, an AM/FM radio, and even a piezoelectric buzzer whose annoying sound beckons us to get out of bed. Before we even leave for work we have used electric lights, shavers, toothbrushes, blow dryers, coffee pots, toasters, microwave ovens, refrigerators, and televisions, to name a few. At the heart of all these devices are electrical circuits. For us to become competent experimentalists, we need to understand the basics of the electrical circuits that form the backbone of most instruments. In this chapter we will review some of the basics of electrical circuits. Then we will examine several more detailed circuits that comprise some common measurement systems.

4.2   LEARNING OBJECTIVES

You should be able to do the following after completing this chapter:

- Determine the equivalent resistance, capacitance, and inductance of series and parallel combinations of resistors, capacitors, and inductors.
- Know Kirchhoff's laws.
- Apply Kirchhoff's laws to elementary DC and AC circuit analysis.
- Know the relationship between input and output voltages and the resistances of a Wheatstone bridge.
- Know the relationship between resistances for a balanced Wheatstone bridge.
- Determine the Thévenin equivalent voltage and equivalent impedance for an elementary circuit.
- Know the rules for proper impedance matching and minimizing loading error.

4.3   CONCEPTS AND DEFINITIONS

Before proceeding to examine the basic electronics behind a measurement system's components, a brief review of some fundamentals is in order. This review includes the definitions of the more common quantities involved in electrical circuits, such as electric charge, electric current, electric field, electric potential, resistance, capacitance, and inductance. The SI dimensions and units for electrical and magnetic systems are summarized in Table C.6 of Appendix C. The origins of these and many other quantities involved in electromagnetism date back to a period rich in the ascent of science, the 17th through mid-19th centuries.

4.3.1   ELECTRIC CHARGE

Electric **charge**, q, previously called *electrical vertue* [1], has the SI unit of coulomb (C) named after the French scientist Charles Coulomb (1736–1806). The effect of charge was observed in early years when two similar materials were rubbed together and then found to repel each other. Conversely, when two dissimilar materials were rubbed together, they became attracted to each other. Amber, for example, when rubbed, would attract small pieces of feathers or straw. In fact, *electron* is the Greek word for amber.

It was Benjamin Franklin (1706–1790) who argued that there was only one form of electricity and coined the relative terms *positive* and *negative* charge. He stated that charge is neither created nor destroyed, rather it is conserved, and that it is only transferred between objects. Prior to Franklin's declarations, two forms of electricity were thought to exist: *vitreous* from glass or crystal, and *resinous*, from rubbing material like amber [1]. It now is known that positive charge indicates a deficiency of electrons and negative charge indicates an excess of electrons. Charge is not produced by objects, rather it is transferred between objects.

4.3.2   ELECTRIC CURRENT

The amount of charge that moves per unit time through or between materials is electric **current**, I. This has the SI unit of an ampere (A), named after the French scientist Andre Ampere (1775–1836). An ampere is a coulomb per second. This can be written as

$$I = \frac{dq}{dt}.$$ **[4.1]**

Current is a measure of the flow of electrons, where the charge of one electron is $1.602\,177\,33 \times 10^{-19}$ C. Materials that have many free electrons are called conductors (previously known as *nonelectrics* because they easily would loose their charge [1]). Those with no free electrons are known as insulators or dielectrics (previously known as *electrics* because they could remain charged) [1]. In between these two extremes lie the semiconductors, which have only a few free electrons. By convention, current is considered to flow from the **anode** (the positively charged terminal that loses electrons) to the **cathode** (the negatively charged terminal that gains electrons) even though the *actual* electron flow is in the *opposite* direction. Current moving from anode to cathode often is referred to as **conventional current**. This convention originated in the early 1800s when it was assumed that positive charge flowed in a wire. **Direct current** (DC) is constant in time and **alternating current** (AC) varies cyclically in time, as depicted in Figure 4.1. When current is alternating, the electrons do not flow in one direction through a circuit, but rather back and forth in both directions. The symbol for a current source in an electrical circuit is given in Figure 4.2.

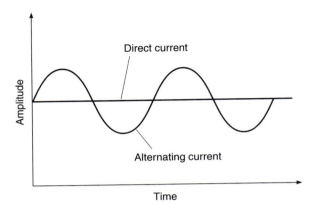

Figure 4.1   Direct and alternating currents.

4.3.3 ELECTRIC FORCE

When electrically charged bodies attract or repel each other, they do so because there is an **electric force** acting between the charges on the bodies. Coulomb's law relates the charges of the two bodies, q_1 and q_2, and the distance between them, R, to the electric force, F_e, by the relation

$$F_e = \frac{Kq_1q_2}{R^2},$$

[4.2]

where $K = 1/(4\pi\epsilon_o)$, with the permittivity of free space $\epsilon_o = 8.854\,187\,817 \times 10^{-12}$ F/m. The SI unit of force is the newton (N).

4.3.4 ELECTRIC FIELD

The **electric field**, E, is defined as the electric force acting on a positive charge divided by the magnitude of the charge. Hence, the electric field has the SI unit of newtons per coulomb. This leads to an equivalent expression $F_e = qE$. So, the work required to move a charge of 1 C through a unit electric field of 1 N/C a distance of 1 m is 1 N · m or 1 J. The SI unit of work is the joule (J).

4.3.5 ELECTRIC POTENTIAL

The **electric potential**, Φ, is the electric field potential energy per unit charge, which is the energy required to bring a charge from infinity to an arbitrary reference point in space. Often it is better to refer to the **potential difference**, $\Delta\Phi$, between two electric potentials. It follows that the SI unit for electric potential is joules per coulomb. This is known as the volt (V), named after Alessandro Volta (1745–1827). Volta invented the voltaic pile, originally made of pairs of copper and zinc plates separated by wet

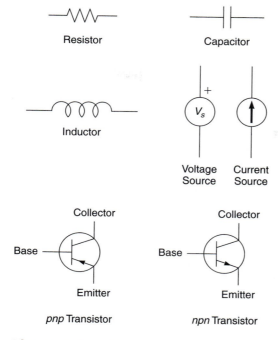

Figure 4.2   Basic circuit element symbols.

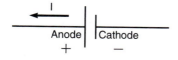

Figure 4.3   The battery.

paper, which was the world's first modern battery. In electrical circuits, a battery is indicated by a longer, solid line (the anode) separated over a small distance by a shorter, solid line (the cathode), as shown in Figure 4.3. The symbol for a voltage source is presented in Figure 4.2.

4.3.6   ELECTRIC RESISTANCE AND RESISTIVITY

When a voltage is applied across the ends of a conductor, the amount of current passing through it is linearly proportional to the applied voltage. The constant of proportionality is the electric **resistance**, R. The SI unit of resistance is the ohm (Ω), named after Georg Ohm (1787–1854).

Electric resistance can be related to electric **resistivity**, ρ, for a wire of cross-sectional area A and length L as

$$R = \frac{\rho L}{A}.$$ [4.3]

The SI unit of resistivity is $\Omega \cdot$ m. Conductors have low resistivity values (for example, Ag: $1.5 \times 10^{-8}\ \Omega \cdot$ m), insulators have high resistivity values (for example, quartz: $5 \times 10^{7}\ \Omega \cdot$ m), and semiconductors have intermediate resistivity values (for example, Si: $2\ \Omega \cdot$ m).

Resistivity is a property of a material and is related to the temperature of the material by the relation

$$\rho = \rho_0[1 + \alpha(T - T_0)],$$ [4.4]

where ρ_0 denotes the reference resistivity at reference temperature T_0 and α, the coefficient of thermal expansion of the material. For conductors, α ranges from approximately 0.0002/°C to 0.007/°C. Thus, for a wire

$$R = R_0[1 + \alpha(T - T_0)].$$ [4.5]

4.3.7 ELECTRIC POWER

Electric power is electric energy transferred per unit time, $P(t) = I(t)V(t)$. Using Ohm's law, it can be written also as $P(t) = I^2(t)R$. This implies that the SI unit for electric power is J/s or watt (W).

4.3.8 ELECTRIC CAPACITANCE

When a voltage is applied across two conducting plates separated by an insulated gap, a charge will accumulate on each plate. One plate becomes charged positively $(+q)$ and the other charged equally and negatively $(-q)$. The amount of charge acquired is linearly proportional to the applied voltage. The constant of proportionality is the **capacitance**, C. Thus, $q = CV$. The SI unit of capacitance is coulombs per volt (C/V). The symbol for capacitance, C, should not be confused with the unit of coulombs, C. The SI unit of capacitance is the farad (F), named after the British scientist Michael Faraday (1791–1867).

4.3.9 ELECTRIC INDUCTANCE

When a wire is wound as coil and current passed through it by applying a voltage, a magnetic field is generated that surrounds the coil. As the current changes in time, a changing magnetic flux is produced inside the coil, which in turn induces a back **electromotive force** (emf). This back emf opposes the original current, leading to either an increase or a decrease in the current, depending on the direction of the original current. The resulting magnetic flux, ϕ, is linearly proportional to the current. The constant of proportionality is called the electric **inductance**, denoted by L. The SI unit of inductance is the henry (H), named after the American Joseph Henry (1797–1878). One henry equals one weber per ampere or $L = \phi/I$.

Example 4.1

STATEMENT

0.3 A of current passes through an electrical wire when the voltage difference between its ends is 0.6 V. Determine [a] the wire resistance, R, [b] the total amount of charge that moves through the wire in 2 minutes, q_{total}, and [c] the electric power, P.

SOLUTION

[a] Application of Ohm's law gives $R = 0.6$ V/0.3 A $= 2 \, \Omega$. [b] Integration of Equation 4.1 gives $q(t) = \int_{t_1}^{t_2} I(t)dt$. Because $I(t)$ is constant, $q_{total} = (0.3 \text{ A})(120 \text{ s}) = 36$ C. [c] The power is the product of current and voltage. So, $P = (0.3 \text{ A})(0.6 \text{ V}) = 0.18$ W $= 0.2$ W, with the correct number of significant figures.

4.4 CIRCUIT ELEMENTS

At the heart of all electrical circuits are some basic circuit elements. These include the resistor, capacitor, inductor, transistor, ideal voltage source, and ideal current source. The symbols for these elements that are used in circuit diagrams are presented in Figure 4.2. These elements form the basis for more complicated devices such as operational amplifiers, sample-and-hold circuits, and analog-to-digital conversion boards, to name only a few (see [2]).

The resistor, capacitor, and inductor are **linear devices** because the complex amplitude of their output waveform is *linearly* proportional to the amplitude of their input waveform. A device is linear if (a) the response to $x_1(t) + x_2(t)$ is $y_1(t) + y_2(t)$, and (b) the response to $ax_1(t)$ is $ay_1(t)$, where a is any complex constant [4]. Thus, if the input waveform of a circuit comprised only of linear devices, known as a linear circuit, is a sine wave of a given frequency, its output will be a sine wave of the *same* frequency. Usually, however, its output amplitude will be different than its input amplitude and its output waveform will lag the input waveform in time. If the lag is between one-half to one cycle, the output waveform appears to lead the input waveform, although it always lags the input waveform. The response behavior of linear systems to various input waveforms is presented in Chapter 5. The current-voltage relations for the resistor, capacitor, and inductor are summarized in Table 4.1.

Table 4.1 Resistor, capacitor, and inductor current and voltage relations.

Element	Unit Symbol	$I(t)$	$V(t)$	$V_{I=\text{const}}$
Resistor	R	$V(t)/R$	$RI(t)$	RI
Capacitor	C	$CdV(t)/dt$	$(1/C)\int_0^t I(\tau)d\tau$	It/C
Inductor	L	$(1/L)\int_0^t V(\tau)d\tau$	LdI/dt	0

4.4.1 The Resistor

The basic circuit element used more than any others is the resistor. Its current-voltage relation is defined through Ohm's law,

$$R = \frac{V}{I}.$$ [4.6]

Thus, the current in a resistor is related *linearly* to the voltage difference across it, or vice versa. The resistor is made out of a conducting material, such as carbon, carbon film, or metal film. Typical resistances range from a few ohms to more than 10^7 Ω.

4.4.2 The Capacitor

The current flowing through a capacitor is related to the product of its capacitance and the time rate of change of the voltage difference, where

$$I = \frac{dq}{dt} = C\frac{dV}{dt}.$$ [4.7]

For example, 1 μA of current flowing through a 1-μF capacitor signifies that the voltage difference across the capacitor is changing at a rate of 1 V/s. If the voltage is not changing in time, there is no current flowing through the capacitor. The capacitor is used in circuits where the voltage varies in time. In a DC circuit, a capacitor acts as an open circuit. Typical capacitances are in the μF to pF range.

4.4.3 The Inductor

Faraday's law of induction states that the change in an inductor's magnetic flux, ϕ, with respect to time equals the applied voltage, $d\phi/dt = V(t)$. Because $\phi = LI$,

$$V(t) = L\frac{dI}{dt}.$$ [4.8]

Thus, the voltage across an inductor is related *linearly* to the product of its inductance and the time rate of change of the current. The inductor is used in circuits in which the current varies in time. The simplest inductor is a wire wound in the form of a coil around a nonconducting core. Most inductors have negligible resistance when measured directly. When used in an AC circuit, the inductor's back emf controls the current. Larger inductances impede the current flow more. This implies that an inductor in an AC circuit acts like a resistor. In a DC circuit, an inductor acts as a short circuit. Typical inductances are in the mH to μH range.

4.4.4 The Transistor

The transistor was developed in 1948 by William Shockley, John Bardeen, and Walter Brattain at Bell Telephone Laboratories. The common transistor consists of two types of semiconductor materials, *n*-type and *p*-type. The *n*-type semiconductor material has an excess of free electrons and the *p*-type material a deficiency. By using just two materials to form a *pn* junction, one can construct a device that allows current to flow in only one direction. This can be used as a **rectifier** to change alternating current to direct current. Simple junction transistors are basically three sections of semiconductor material sandwiched together, forming either *pnp* or *npn* transistors. Each section has its own wire lead. The center section is called the **base**, one end section the **emitter**, and the other the **collector**. In a *pnp* transistor, current flow is into the emitter. In an *npn* transistor, current flow is out of the emitter. In both cases, the emitter-base junction is said to be forward-biased or conducting (current flows forward from *p* to *n*). The opposite is true for the collector-base junction. It is always reverse-biased or nonconducting. Thus, for a *pnp* transistor, the emitter would be connected to the positive terminal of a voltage source and the collector to the negative terminal through a resistor. The base would also be connected to the negative terminal through another resistor. In such a configuration, current would flow into the emitter and out of both the base *and* the collector. The voltage difference between the emitter and the collector causing this current flow is termed the base bias voltage. The ratio of the collector-to-base current is the (current) gain of the transistor. Typical gains are up to approximately 200. The characteristic curves of a transistor display collector current versus the base bias voltage for various base currents. Using these curves, the gain of the transistor can be determined for various operating conditions. Thus, transistors can serve many different functions in an electrical circuit, such as current amplification, voltage amplification, detection, and switching.

4.4.5 THE VOLTAGE SOURCE

An *ideal* voltage source, shown in Figure 4.4, with $R_{out} = 0$ maintains a fixed voltage difference between its terminals, independent of the resistance of the load connected to it. It has a zero output impedance and can supply infinite current. An *actual* voltage source has some internal resistance. So the voltage supplied by it is limited and equal to the product of the source's current and its internal resistance, as dictated by Ohm's law. A good voltage source has a very low output impedance, typically less than 1 Ω. If the voltage source is a battery, it has a finite lifetime of current supply, as specified by its capacity. Capacity is expressed in units of current times lifetime (which equals

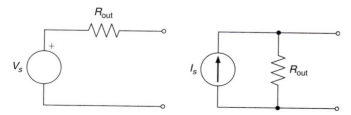

Figure 4.4   Voltage and current sources.

its total charge). For example, a 1200-mA-hour battery pack is capable of supplying 1200 mA of current for 1 hour or 200 mA for 6 hours. This corresponds to a total charge of 4320 C (0.2 A $\times$ 21 600 s).

4.4.6   THE CURRENT SOURCE

An *ideal* current source, depicted in Figure 4.4, with $R_{out} = \infty$ maintains a fixed current between its terminals, independent of the resistance of the load connected to it. It has an infinite output impedance and can supply infinite voltage. An *actual* current source has an internal resistance less than infinite. So the current supplied by it is limited and equal to the ratio of the source's voltage difference to its internal resistance. A good current source has a very high output impedance, typically greater than 1 MΩ. Actual voltage and current sources differ from their ideal counterparts only in that the actual impedances are neither zero nor infinite, but finite.

4.5   RLC COMBINATIONS

Linear circuits typically involve resistors, capacitors, and inductors connected in various *series* and *parallel* combinations. Using the current-voltage relations of the circuit elements and examining the potential difference between two points on a circuit, some simple rules for various combinations of resistors, capacitors, and inductors can be developed.

First, examine Figure 4.5 in which the *series* combinations of two resistors, two capacitors, and two inductors are shown. The potential difference across an ith resistor is IR_i, across an ith capacitor is q/C_i, and across an ith inductor is $L_i \, dI/dt$. Likewise, the total potential difference, V_T, for the series resistors' combination is $V_T = IR_T$, for the series capacitors' combination is $V_T = q/C_T$, and for the series inductors' combination is $V_T = L_T \, dI/dt$. Because the potential differences across resistors,

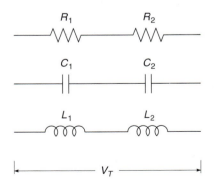

Figure 4.5   Series R, C, and L circuit configurations.

capacitors, and inductors in series add, $V_T = V_1 + V_2$. Hence, for the resistors' series combination, $V_T = IR_1 + IR_2 = IR_T$, which yields

$$R_T = R_1 + R_2. \qquad \textbf{[4.9]}$$

For the capacitors' series combination, $V_T = q/C_1 + q/C_2 = q/C_T$, which implies

$$\frac{1}{C_T} = \frac{1}{C_1} + \frac{1}{C_2}. \qquad \textbf{[4.10]}$$

For the inductors' series combination, $V_T = L_1\, dI/dt + L_2\, dI/dt = L_T\, dI/dt$, which gives

$$L_T = L_1 + L_2. \qquad \textbf{[4.11]}$$

Thus, when in series, resistances and inductances add, and the reciprocals of capacitances add.

Next, look at Figure 4.6 in which the *parallel* combinations of two resistors, two capacitors, and two inductors are displayed. The same expressions for the ith and total potential differences hold as before. Hence, for the resistors' parallel combination, $I_T = I_1 + I_2$, which leads to

$$\frac{1}{R_T} = \frac{1}{R_1} + \frac{1}{R_2}. \qquad \textbf{[4.12]}$$

For the capacitors' parallel combination, $q_T = q_1 + q_2$, which leads to

$$C_T = C_1 + C_2. \qquad \textbf{[4.13]}$$

For the inductors' parallel combination, $I_T = I_1 + I_2$, which gives

$$\frac{1}{L_T} = \frac{1}{L_1} + \frac{1}{L_2}. \qquad \textbf{[4.14]}$$

Thus, when in parallel, capacitances add and the reciprocals of resistances and inductances add.

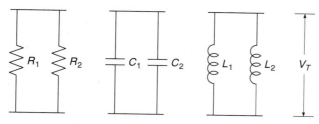

Figure 4.6   Parallel R, C, and L circuit configurations.

Example 4.2

STATEMENT

Determine the total equivalent resistance, R_T, and total equivalent capacitance, C_T, for the respective resistance and capacitance circuits shown in Figure 4.7.

SOLUTION

For the two resistors in parallel, the equivalent resistance, R_a, is

$$\frac{1}{R_a} = \frac{1}{4} + \frac{1}{4} = 2 \ \Omega.$$

The two other resistors are in series with R_a, so $R_T = R_a + 2 + 6 = 10 \ \Omega.$

For the two capacitors in parallel, the equivalent capacitance, C_b, is $C_b = 3 + 3 = 6 \ \mu F.$ This is in series with the two other capacitors, which implies that

$$\frac{1}{C_T} = \frac{1}{2} + \frac{1}{3} + \frac{1}{C_b} = \frac{1}{2} + \frac{1}{3} + \frac{1}{6} = 1.$$

So, $C_T = 1 \ \mu F.$

To aid further with circuit analysis, two laws developed by G. R. Kirchhoff (1824–1887) can be used. **Kirchhoff's current (or first) law,** which is conservation of charge, states that at any junction (node) in a circuit, the current flowing into the junction must equal the current flowing out of it, which implies that

$$\sum_{\text{node}} I_{\text{in}} = \sum_{\text{node}} I_{\text{out}}. \qquad \textbf{[4.15]}$$

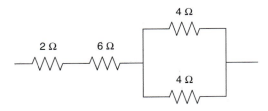

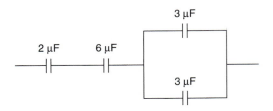

Figure 4.7   Resistor and capacitor circuits.

A **node** in a circuit is a point where two or more circuit elements meet. **Kirchhoff's voltage (or second) law**, which is conservation of energy, says that around any loop in a circuit, the sum of the potential differences equals zero, which gives

$$\sum_{i,\text{cl.loop}} V_i = 0. \qquad \textbf{[4.16]}$$

A **loop** is a closed path that goes from one node in a circuit back to itself without passing through any intermediate node more than once. Any *consistent* sign convention will work when applying Kirchhoff's laws to a circuit. Armed with this information, some important DC circuits can be examined now.

4.6   ELEMENTARY DC CIRCUIT ANALYSIS

In DC circuits, current is steady in time. Thus, there is no inductance, even if an inductor is present. An actual inductor, however, has resistance. This typically is on the order of 10 Ω. Often, an inductor's resistance in a DC circuit is neglected.

The first elementary DC circuit to analyze is that of a DC electric motor in series with a battery, as shown in Figure 4.8. Examine the battery first. It has an internal resistance, R_{batt}, and an open circuit voltage (potential difference), V_{oc}. R_{batt} is the resistance and V_{oc} is the potential difference that would be measured across the terminals of the battery if it were isolated from the circuit by not being connected to it. However, when the battery is placed in the circuit and the circuit switch is closed such that current, I_a, flows around the circuit, the situation for the battery changes. The measured potential difference across the battery now is less because current flows through the battery, effectively leading to a potential difference across R_{batt}. This yields

$$V_{\text{batt}} = V_{\text{oc}} - I_a R_{\text{batt}}, \qquad \textbf{[4.17]}$$

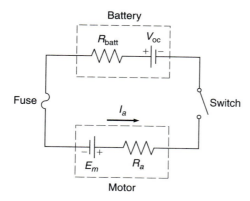

Figure 4.8   A battery and electric motor circuit.

in which V_{batt} represents the closed-circuit potential difference across the battery. Similarly, the DC motor has an internal resistance, R_a, which is mainly across its armature. It also has a potential difference, E_m, across its terminals when operating with the battery connected to it. To summarize, V_{oc} is measured across the battery terminals when the switch is open, and V_{batt} is measured when the switch is closed.

Now what is E_m in terms of the known quantities? To answer this, apply Kirchhoff's second law around the circuit loop when the switch is closed. Starting from the battery's anode and moving in the direction of the current around the loop, this gives

$$V_{oc} - I_a R_{batt} - E_m - I_a R_a = 0, \qquad\qquad \text{[4.18]}$$

which immediately leads to

$$E_m = V_{oc} - I_a(R_{batt} + R_a). \qquad\qquad \text{[4.19]}$$

This equation reveals a simple fact: the relatively high battery and motor internal resistances lead to a decrease in the motor's maximum potential difference. This consequently results in a decrease in the motor's power output to a device such as the propeller of a remotely piloted aircraft.

Example 4.3 | **STATEMENT**

For the electrical circuit shown in Figure 4.9, determine [a] the magnitude of the current in the branch between nodes A and B, and [b] the direction of that current.

SOLUTION

Application of Kirchhoff's second law to the left loop gives $5\,\text{V} - (1\,\Omega)(I_1\,\text{A}) - 2\,\text{V} - (1\,\Omega)(I_3\,\text{A}) = 0$. Similar application to the right loop yields $2\,\text{V} - (2\,\Omega)(I_2\,\text{A}) + (1\,\Omega)(I_3\,\text{A}) = 0$. At node A, application of Kirchhoff's first law implies $I_1 - I_2 - I_3 = 0$. These three expressions can be solved to yield $I_1 = 2.2\,\text{A}$, $I_2 = 1.4\,\text{A}$, and $I_3 = 0.8\,\text{A}$. Because I_3 is positive, the direction shown in Figure 4.9, from node A to node B, is correct.

The second elementary direct-current circuit is a **Wheatstone bridge**. The Wheatstone bridge is used in a variety of common instruments such as pressure transducers and hot-wire anemometers. Its circuit, shown in Figure 4.10, consists of four resistors (R_1 through R_4), each two comprising a pair (R_1 and R_2; R_3 and R_4) in series that is

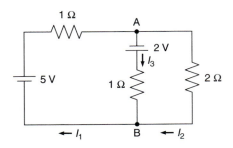

Figure 4.9   An electrical circuit.

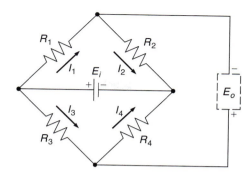

Figure 4.10   The Wheatstone bridge configuration.

connected to the other pair in parallel, and a voltage source, E_i, connected between the R_1–R_3 and the R_2–R_4 nodes. The voltage output of the bridge, E_o, is measured between the R_1–R_2 and the R_3–R_4 nodes. E_o is measured by an ideal voltmeter with an infinite input impedance such that no current flows through the voltmeter.

An expression needs to be developed that relates the bridge's output voltage to its input voltage and the four resistances. There are four unknowns, I_1 through I_4. This implies that four equations are needed to reach the desired solution. Examination of the circuit reveals that there are four closed loops for which four equations can be written by applying Kirchhoff's second law. The resulting four equations are as follows:

$$E_i = I_1 R_1 + I_2 R_2, \qquad \text{[4.20]}$$

$$E_i = I_3 R_3 + I_4 R_4, \qquad \text{[4.21]}$$

$$E_o = I_4 R_4 - I_2 R_2, \qquad \text{[4.22]}$$

and

$$E_o = -I_3 R_3 + I_1 R_1. \qquad \text{[4.23]}$$

Kirchhoff's first law leads to $I_1 = I_2$ and $I_3 = I_4$, assuming no current flows through the voltmeter. These two current relations can be used in Equations 4.20 and 4.21 to give

$$I_1 = \frac{E_i}{R_1 + R_2} \qquad \text{[4.24]}$$

and

$$I_3 = \frac{E_i}{R_3 + R_4}. \qquad \text{[4.25]}$$

These two expressions can be substituted into Equation 4.23, yielding the desired result

$$E_o = E_i \left[\frac{R_1}{R_1 + R_2} - \frac{R_3}{R_3 + R_4} \right].$$ **[4.26]**

Equation 4.26 leads to some interesting features of the Wheatstone bridge. When there is no voltage output from the bridge, the bridge is considered to be balanced even if there is an input voltage present. This immediately yields the balanced bridge equation

$$\frac{R_1}{R_2} = \frac{R_3}{R_4}.$$ **[4.27]**

This condition can be exploited to use the bridge to determine an unknown resistance, say R_1, by having two other resistances fixed, say R_2 and R_3, and varying R_4 until the balanced bridge condition is achieved. This is called the **null method**. This method is used to determine the resistance of a sensor that usually is located remotely from the remainder of the bridge. An example is the hot-wire sensor of an anemometry system used in the constant-current mode to measure local fluid temperature.

The bridge can be used also in the **deflection method** to provide an output voltage that is proportional to a *change* in resistance. Assume that resistance R_1 is the resistance of a sensor, such as a fine wire or a strain gage. The sensor is located remotely from the remainder of the bridge circuit in an environment in which the temperature increases from some initial state. Its resistance will change by an amount δR from R_1 to R_1'. Application of Equation 4.26 yields

$$E_o = E_i \left[\frac{R_1'}{R_1' + R_2} - \frac{R_3}{R_3 + R_4} \right].$$ **[4.28]**

Further, if all the resistances are initially the same, where $R_1 = R_2 = R_3 = R_4 = R$, then Equation 4.28 becomes

$$E_o = E_i \left[\frac{\delta R/R}{4 + 2\delta R/R} \right] = E_i \cdot f\left(\frac{\delta R}{R} \right).$$ **[4.29]**

Thus, when using the null and deflection methods, the Wheatstone bridge can be utilized to determine a resistance or a change in resistance.

One practical use of the Wheatstone bridge is in a force measurement system. This system is comprised of a cantilever beam, rigidly supported on one end, that is instrumented with four strain gages, two on top and two on bottom, as shown in Figure 4.11. The strain gage is discussed further in Chapter 6. A typical strain gage is shown in Figure 6.2 of that chapter. The electrical configuration is called a four-arm bridge. The system operates by applying a known force, F, near the end of the beam along its centerline and then measuring the output of the Wheatstone bridge formed by the four strain gages. When a load is applied in the direction shown in Figure 4.11, the beam will deflect downward, giving rise to a tensile strain, ϵ_L, on the top of the beam and a compressive strain, $-\epsilon_L$, on the bottom of the beam. Because a strain gage's resistance increases with strain, $\delta R \sim \epsilon_L$, the resistances of the two tensile strain gages will increase and those of the two compressive strain gages will decrease. In general, following the notation in the Figure 4.11, for the applied load condition

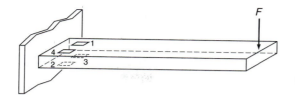

Figure 4.11   Cantilever beam with four strain gages.

$$R_1' = R_1 + \delta R_1; \; R_4' = R_4 + \delta R_4; \; R_2' = R_2 - \delta R_2; \; R_3' = R_3 - \delta R_3. \qquad \textbf{[4.30]}$$

If all four gages are identical, where they are of the same pattern with $R_1 = R_2 = R_3 = R_4 = R$, the two tensile resistances will *increase* by δR and the two compressive ones will *decrease* by δR. For this case, Equation 4.28 simplifies to

$$E_o = E_i \cdot \left(\frac{\delta R}{R}\right). \qquad \textbf{[4.31]}$$

For a cantilever beam shown in Figure 4.11, the strain along the length of the beam on its top side is proportional to the force applied at its end, F. Thus, $\epsilon_L \sim F$. If strain gages are aligned with this axis of strain, then $\delta R \sim \epsilon_L$, as discussed in Section 6.4.2. Thus, the voltage output of this system, E_o, is *linearly* proportional to the applied force, F. Further, with this strain gage configuration, variational temperature and torsional effects are compensated for automatically. This is an inexpensive, simple yet elegant measurement system that can be calibrated and used to determine unknown forces. This configuration is the basis of most force balances used for aerodynamic and mechanical force measurements.

STATEMENT   **Example 4.4**

Referring to Figure 4.10, if $R_1 = 1\,\Omega$, $R_2 = 3\,\Omega$, and $R_3 = 2\,\Omega$, determine [a] the value of R_4 such that the Wheatstone bridge is balanced, and [b] the bridge's output voltage under this condition.

SOLUTION

Equation 4.27 specifies the relationship between resistances when the bridge is balanced. Thus, $R_4 = R_2 R_3 / R_1 = (3)(2)/1 = 6\,\Omega$. Because the bridge is balanced, its output voltage is zero. This can be verified by substituting the four resistance values into Equation 4.26.

4.7   ELEMENTARY AC CIRCUIT ANALYSIS

As shown in Table 4.1, the expressions for the voltages and currents of AC circuits containing resistors, capacitors, and inductors involve differentials and integrals with respect to time. Expressions for $V(t)$ and $I(t)$ of AC circuits can be obtained directly by solving the first-order and second-order ordinary differential equations that govern their behavior. The differential-equation solutions for RC and RLC circuits subjected to step and sinusoidal inputs are presented in Chapter 5. At this point, a working knowledge of AC circuits can be gained through some elementary considerations.

When capacitors and inductors are exposed to time-varying voltages in AC circuits, they each create a **reactance** to the voltage. Reactance plus resistance equals **impedance**. Symbolically, $X + R = Z$. Often an RLC component is described by its impedance because it encompasses both resistance and reactance. Impedance typically is considered *generalized resistance* [2]. For DC circuit analysis, impedance is resistance because there is no reactance. For this case, $Z = R$.

Voltages and currents in AC circuits usually do not vary simultaneously in the same manner. An increase in voltage with time, for example, can be followed by a corresponding increase in current at some later time. Such changes of voltage and current in time are characterized best by using complex number notation. This notation is described in more detail in Chapter 11.

Assume that the voltage, $V(t)$, and the current, $I(t)$, are represented by the complex numbers $V_o e^{i\phi}$ and $I_o e^{i\phi}$, respectively. Here $e^{i\phi}$ is given by Euler's formula,

$$e^{i\phi} = \cos\phi + i\,\sin\phi, \qquad\qquad \textbf{[4.32]}$$

where $i = \sqrt{-1}$. In electrical engineering texts, j is used to denote $\sqrt{-1}$ because i is used for the current. Throughout this text, i symbolizes the imaginary number. The real voltage and real current are obtained by multiplying each by the complex number representation $e^{i\omega t}$ and then taking the real part, Re, of the resulting number. That is,

$$V(t) = \text{Re}(Ve^{i\omega t}) = \text{Re}(V)\cos\omega t - \text{Im}(V)\sin\omega t \qquad \textbf{[4.33]}$$

and

$$I(t) = \text{Re}(Ie^{i\omega t}) = \text{Re}(I)\cos\omega t - \text{Im}(I)\sin\omega t, \qquad \textbf{[4.34]}$$

in which ω is the frequency in rad/s. The frequency in cycles/s is f, where $2\pi f = \omega$.

Expressions for capacitive and inductive reactances can be derived using the voltage and current expressions given in Equations 4.33 and 4.34 [2]. For a capacitor, $I(t) = C\,dV(t)/dt$. Differentiating Equation 4.33 with respect to time yields the current across the capacitor,

$$I(t) = -V_o C\omega\,\sin\omega t = \text{Re}\left(V_o \frac{e^{i/\omega C}}{1/i\omega C}\right). \qquad \textbf{[4.35]}$$

The denominator of the real component is the capacitive reactance,

$$X_C = \frac{1}{i\omega C}. \qquad\qquad \textbf{[4.36]}$$

In other words, $V(t) = I(t)X_C$.

For the inductor, $V(t) = L\,dI(t)/dt$. Differentiating Equation 4.34 with respect to time yields the voltage across the inductor,

$$V(t) = -I_o L\omega\,\sin\omega t = \text{Re}(i\omega L I_o e^{i/\omega C}). \qquad \textbf{[4.37]}$$

The numerator of the real component is the inductive reactance,

$$X_L = i\omega L. \qquad \textbf{[4.38]}$$

Simply put, $V(t) = I(t)X_L$.

Because the resistances of capacitors and inductors are effectively zero, their impedances equal the reactances. Further, the resistor has no reactance, so its impedance is its resistance. Thus, $Z_R = R$, $Z_C = 1/i\omega C$, and $Z_L = i\omega L$.

Ohm's law is still valid for AC circuits. It now can be written as $V = ZI$. Also the rules for adding resistances apply to adding impedances, where, for impedances in series

$$Z_T = \sum Z_i, \qquad \textbf{[4.39]}$$

and for impedances in parallel

$$Z_T = \frac{1}{\sum(1/Z_i)}. \qquad \textbf{[4.40]}$$

STATEMENT

Example 4.5

An electrical circuit loop is comprised of a resistor, R, a voltage source, E_o, and a 1-μF capacitor, C, that can be added into the loop by a switch. When the switch is closed, all three electrical components are in series. When the switch is open, the loop has only the resistor and the voltage source in series. $E_o = 3$ V and $R = 2$ Ω. Determine the expression for the current as a function of time, $I(t)$, immediately after the switch is closed.

SOLUTION

Applying Kirchhoff's second law to the loop gives

$$R I(t) + \frac{1}{C} \int I(t)dt = E_o.$$

This equation can be differentiated with respect to time to yield

$$R\frac{dI(t)}{dt} + \frac{I(t)}{C} = \frac{dE_o}{dt} = 0,$$

because E_o is a constant 3 V. This equation can be integrated to obtain the result $I(t) = C_1 e^{-t/RC}$, where C_1 is a constant. Now at $t = 0$ s, current flows through the loop and equals E_o/R. This implies that $C_1 = 3/2 = 1.5$ A. Thus, for the given values of R and C, $I(t) = 1.5e^{-t/(2\times10^{-6})}$. This means that the current in the loop becomes almost zero in approximately 10 μs after the switch is closed.

4.8 EQUIVALENT CIRCUITS

Thévenin's equivalent circuit theorem states that any two-terminal network of linear impedances, such as resistors and inductors, and voltage sources can be replaced by

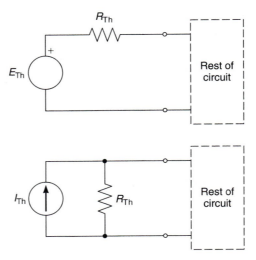

Figure 4.12   Thévenin and Norton equivalent circuits.

an equivalent circuit consisting of an ideal voltage source, E_{Th}, in series with an impedance, Z_{Th}. This circuit is shown in the top of Figure 4.12.

Norton's equivalent circuit theorem states that any two-terminal network of linear impedances and current sources can be replaced by an equivalent circuit consisting of an ideal current source, I_{Th}, in parallel with an impedance, Z_{Th}. This circuit is illustrated in the bottom of Figure 4.12.

The voltage of the Thévenin equivalent circuit is the current of the Norton equivalent circuit times the equivalent impedance. Obtaining the equivalent impedance sometimes can be tedious, but it is very useful in understanding circuits, especially the more complex ones.

The **Thévenin equivalent voltage** and equivalent impedance can be determined by examining the open-circuit voltage and the short-circuit current. The Thévenin equivalent voltage, E_{Th}, is the open-circuit voltage, which is the potential difference that exists between the circuit's two terminals when nothing is connected to the circuit. This would be the voltage measured using an ideal voltmeter. Simply put, $E_{Th} = E_{oc}$, where the subscript oc denotes open circuit. The **Thévenin equivalent impedance**, Z_{Th}, is E_{Th} divided by the short-circuit current, I_{sc}, where the subscript sc denotes short circuit. The short-circuit current is the current that would pass through an ideal ammeter connected across the circuit's two terminals.

An example diagram showing an actual circuit and its Thévenin equivalent is presented in Figure 4.13. In this figure, Z_{Th} is represented by R_{Th} because the only impedances in the circuit are resistances. R_m denotes the meter's resistance, which would be infinite for an ideal voltmeter and zero for an ideal ammeter. The Thévenin equivalents can be found in the following manner.

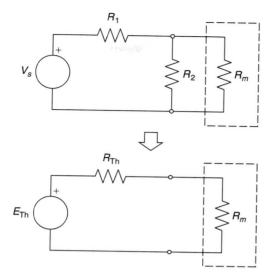

Figure 4.13   A circuit and its Thévenin
equivalent.

For the actual circuit, Kirchhoff's voltage law implies

$$E_i = I(R_1 + R_2).$$ **[4.41]**

Also, the voltage measured by the ideal voltmeter, E_m, using Ohm's law and noting that $R_2 \ll R_m$, is

$$E_m = E_{Th} = IR_2.$$ **[4.42]**

Combining Equations 4.41 and 4.42 yields the open-circuit or Thévenin equivalent voltage,

$$E_{Th} = E_i \frac{R_2}{R_1 + R_2}.$$ **[4.43]**

Further, the short-circuit current would be

$$I_{sc} = \frac{E_i}{R_1}.$$ **[4.44]**

So, the Thévenin equivalent resistance is

$$R_{Th} \equiv \frac{E_{Th}}{I_{sc}} = \frac{R_1 R_2}{R_1 + R_2}.$$ **[4.45]**

The resulting Thévenin equivalent circuit is shown in the bottom of Figure 4.13.

An alternative approach to determining the Thévenin impedance is to replace all voltage sources in the circuit by their internal impedances and then find the circuit's output impedance. Usually the voltage sources' internal impedances are negligible and can be assumed to be equal to zero, effectively replacing all voltage sources by short circuits. For the circuit shown in Figure 4.13, this approach would lead to having the resistances R_1 and R_2 in parallel to ground, leading directly to Equation 4.45.

This alternative approach can be applied also when determining the Thévenin equivalent resistance, which is the output impedance, of the Wheatstone bridge circuit shown in Figure 4.10. Assuming a negligible internal impedance for the voltage source E_i, R_{Th} is equivalent to the parallel combination of R_1 and R_2 in series with the parallel combination of R_3 and R_4. That is,

$$R_{Th} = \frac{R_1 R_2}{R_1 + R_2} + \frac{R_3 R_4}{R_3 + R_4}.$$ [4.46]

Example 4.6

STATEMENT

For the circuit shown in the top of Figure 4.13, determine the Thévenin equivalent resistance and the Thévenin equivalent voltage, assuming that $V_s = 20$ V, $R_1 = 6\ \Omega$, $R_2 = 3\ \Omega$, and $R_m = 3$ MΩ.

SOLUTION

Because $R_m \gg R_2$, the Thévenin equivalent voltage is given by Equation 4.43 and the Thévenin equivalent resistance by Equation 4.45. Substitution of the given values for $V_s = E_i$, R_1 and R_2 into these equations yields $E_{Th} = (20)[3/(6+3)] = 6.67\ V$ and $R_{Th} = [(6)(3)]/(6+3) = 2\ \Omega$.

4.9 METERS

All voltage and current meters can be represented by Thévenin and Norton equivalent circuits, as shown in Figure 4.14. These meters are characterized by their input impedances. An *ideal* **voltmeter** has an infinite input impedance such that no current flows through it. An *ideal* **ammeter** has zero input impedance such that all the connected circuit's current flows through it. The actual devices differ from their ideal counterparts only in that the actual impedances are neither zero nor infinite, but finite. A voltmeter is attached in parallel to the point of interest in the circuit. An ammeter is attached in series with the point of interest in the circuit. A good voltmeter has a very

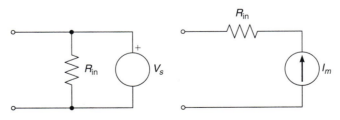

Figure 4.14   Voltage and current meters.

high input impedance, typically greater than 1 MΩ. Because of this, a good voltmeter connected to a circuit draws negligible current from the circuit and, therefore, has no additional voltage difference present between the voltmeter's terminals. Likewise, because a good ammeter has a very low input impedance, typically less than 1 Ω, almost all of the attached circuit's current flows through the ammeter.

Resistance measurements typically are made using an **ohmmeter**. The resistance actually is determined by passing a known current through the test leads of a meter and the unknown resistance and then measuring the total voltage difference across them. This is called the *two-wire method*. This approach is valid provided that the unknown resistance is much larger than the resistances of the test leads. In practice, this problem is circumvented by using a multimeter and the *four-wire method*. This method requires the use of two additional test leads. Two of the leads carry a known current through the unknown resistance and then back to the meter, while the other two leads measure the resulting voltage difference across the unknown resistance. The meter determines the resistance by Ohm's law and then displays it.

4.10 IMPEDANCE MATCHING AND LOADING ERROR

When the output of one electronic component is connected to the input of another, the output signal may be altered, depending on the component impedances. Each measurement circumstance requires a certain relation between the output component's output impedance and the input component's input impedance to avoid signal alteration. If this impedance relation is not maintained, then the output component's signal will be altered upon connection to the input component. A common example of impedance mismatch is when an audio amplifier is connected to a speaker with a high input impedance. This leads to a significant reduction in the power transmitted to the speaker, which results in a low volume from the speaker.

A **loading error** can be introduced whenever one circuit is attached to another. Loading error, e_{load}, is defined in terms of the difference between the true output impedance, R_{true}, the impedance that would be measured across the circuit's output terminals by an ideal voltmeter, and the impedance measured by an actual voltmeter, R_{meas}. Expressed on a percentage basis, the % loading error is

$$e_{\text{load}} = 100 \left[\frac{R_{\text{true}} - R_{\text{meas}}}{R_{\text{true}}} \right]. \qquad \textbf{[4.47]}$$

Loading errors that occur when measuring voltages, resistances, or current can be avoided by following two simple rules. These rules, summarized at the end of this section, can be derived by considering two circuits, one in which an actual voltage source circuit is connected to an actual voltmeter, and the other in which an actual current source circuit is connected to an actual ammeter. These circuits are shown in Figure 4.15.

For the voltage circuit, Kirchhoff's voltage law applied around the outer circuit loop gives

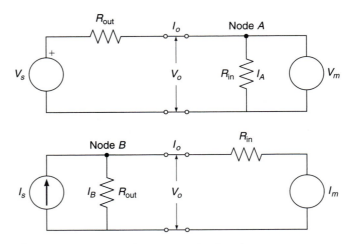

Figure 4.15   Voltage circuit (top) and current circuit (bottom) illustrating loading error.

$$V_m = V_s - I_o R_{out}. \tag{4.48}$$

Kirchhoff's current law applied at node A yields

$$I_o = I_A = \frac{V_m}{R_{in}}, \tag{4.49}$$

where all of the current flows through the voltmeter's R_{in}. Substituting Equation 4.49 into Equation 4.48 results in

$$V_m = V_s \left[\frac{1}{1 + (R_{out}/R_{in})} \right] = V_s \left[\frac{R_{in}}{R_{in} + R_{out}} \right]. \tag{4.50}$$

When $R_{in} \gg R_{out}$, $V_m = V_s$. Noting for this voltage-measurement case that $R_{true} = R_{out}$ and $R_{meas} = (R_{in} R_{out})/(R_{in} + R_{out})$, the loading error becomes

$$e_{load,V} = \left[\frac{R_{out}}{R_{in} + R_{out}} \right]. \tag{4.51}$$

For the current circuit, Kirchhoff's current law applied at node B yields

$$I_s = I_B + I_o. \tag{4.52}$$

Kirchhoff's voltage law applied around the circuit loop containing R_{in} and R_{out} gives

$$I_m R_{in} = I_B R_{out}. \tag{4.53}$$

Substituting Equation 4.53 into Equation 4.52 results in

$$I_m = I_s \left[\frac{1}{1 + (R_{in}/R_{out})} \right] = I_s \left[\frac{R_{out}}{R_{in} + R_{out}} \right]. \tag{4.54}$$

When $R_{in} \ll R_{out}$, $I_m = I_s$. Noting for the current-measurement case that $R_{true} = R_{in}$ and $R_{meas} = (R_{in}R_{out})/(R_{in} + R_{out})$, the loading error becomes

$$e_{load,I} = \left[\frac{R_{in}}{R_{in} + R_{out}} \right]. \qquad \textbf{[4.55]}$$

Loading errors can be avoided between two circuits by connecting them via a buffer that has near-infinite input and near-zero output impedances. This is one of the many uses of operational amplifiers, which are presented in Chapter 6.

STATEMENT   **Example 4.7**

Determine the minimum input impedance, R_{min}, of a voltage measurement circuit that would have less than 0.5 % loading error when connected to a circuit having an output impedance of 50 Ω.

SOLUTION

Direct application of Equation 4.51 implies

$$\frac{0.5}{100} = \frac{50 \; \Omega}{50 \; \Omega + R_{min}}.$$

Solving for the minimum input impedance gives $R_{min} = 9950 \; \Omega$, or approximately 10 kΩ. This condition can be met by using a unity-gain operational amplifier in the noninverting configuration at the input of the voltage-measurement circuit (see Chapter 6).

The impedance relation for optimum power transmission between an output source and an input circuit can be determined [3]. For the voltage circuit in Figure 4.15, noting that the power received, P_{in}, equals V_{in}^2/R_{in}, Equation 4.50 becomes

$$P_{in} = V_s^2 \left[\frac{R_{in}}{(R_{in} + R_{out})^2} \right]. \qquad \textbf{[4.56]}$$

Differentiating Equation 4.56 with respect to R_{in}, setting the result equal to zero, and solving for R_{in} gives

$$R_{in} = R_{out}. \qquad \textbf{[4.57]}$$

Substitution of Equation 4.57 into the derivative equation shows that this condition ensures a maximum transmission of power. Equation 4.57 represents true **impedance matching**, where the two impedances have the *same* value.

STATEMENT   **Example 4.8**

Determine the power that is transmitted, P_t, between two connected circuits if the output-circuit impedance is 6.0 Ω, the input-circuit impedance is 4.0 Ω, and the source voltage is 12 V.

SOLUTION
Substitution of the given values into Equation 4.50 gives $V_m = 12[4/(6+4)] = 4.8$ V. Now, the power transmitted is given by $P_t = V_{in}^2/R_{in} = 4.8^2/4 = 5.8$ W, with the correct number of significant figures.

Impedance matching also is critical when an output circuit that generates waveforms is connected by a cable to a receiving circuit. In this situation, the high-frequency components of the output circuit can reflect back from the receiving circuit. This essentially produces an input wave to the receiving circuit that is different from that intended. When a cable with characteristic impedance, R_{cable}, is connected to a receiving circuit of load impedance, R_{in}, and these impedances are matched, then the input wave will not be reflected. The reflected wave amplitude, A_r, is related to the incident wave amplitude, A_i, by

$$A_r = A_i \left[\frac{R_{cable} - R_{in}}{R_{cable} + R_{in}} \right]. \qquad \textbf{[4.58]}$$

When $R_{cable} < R_{in}$, the reflected wave is inverted. When $R_{cable} > R_{in}$, the reflected wave is not inverted [2].

The rules for impedance matching and for loading-error minimization, as specified by Equations 4.50, 4.54, 4.57, and 4.58, are as follows:

- **Rule 1—for loading-error minimization**:   When measuring a voltage, the input impedance of the measuring device must be much greater than the equivalent circuit's output impedance.

- **Rule 2—for loading-error minimization**:   When measuring a current, the input impedance of the measuring device must be much less than the equivalent circuit's output impedance.

- **Rule 3—for impedance matching**:   When transmitting power to a load, the output impedance of the transmission circuit must equal the input impedance of the load for maximum power transmission.

- **Rule 4—for impedance matching**:   When transmitting high-frequency signals through a cable, the cable impedance must equal the load impedance of the receiving circuit.

4.11   ELECTRICAL NOISE

Electrical noise is defined as anything that obscures a signal [2]. Noise is characterized by its amplitude distribution, frequency spectrum, and the physical mechanism responsible for its generation. Noise can be subdivided into **intrinsic noise** and **interference noise**. Intrinsic noise is random and primarily the result of thermally induced molecular motion in any resistive element (Johnson noise), current fluctuations in a material (shot noise), and local property variations in a material ($1/f$ or pink noise).

The first two are intrinsic and cannot be eliminated. The latter can be reduced through quality control of the material that is used.

Noise caused by another signal is called interference noise. Interference noise depends on the amplitude and frequency of the noise source. Common noise sources include AC-line power (50 Hz to 60 Hz), overhead fluorescent lighting (100 Hz to 120 Hz), and sources of radio-frequency (RF) and electromagnetically induced (EMI) interference, such as televisions, radios, and high-voltage transformers.

The causes of electrical interference include local electric fields, magnetic fields, and ground loops. These noticeably affect analog voltage signals with amplitudes less than 1 V. A surface at another electric potential that is near a signal-carrying wire will establish an undesirable capacitance between the surface and the wire. A local magnetic field near a signal-carrying wire will induce an additional current in the wire. A current flowing through one ground point in a circuit will generate a signal in another part of the circuit that is connected to a different ground point.

Most interference noise can be attenuated to acceptable levels by proper shielding, filtering, and amplification. For example, signal wires can be shielded by a layer of conductor that is separated from the signal wire by an insulator. The electric potential of the shield can be driven at the same potential as the signal through the use of operational amplifiers and feedback, thereby obviating any undesirable capacitance [5]. Pairs of insulated wires carrying similar signals can be twisted together to produce signals with the same mode of noise. These signals subsequently can be conditioned using common-mode rejection techniques. Use of a single electrical ground point for a circuit almost always will minimize ground-loop effects. Signal amplification and filtering also can be used. In the end though, it is better to eliminate the sources of noise than to try to cover them up.

The magnitude of the noise is characterized through the signal-to-noise ratio (SNR). This is defined as

$$\text{SNR} \equiv 10 \, \log_{10}\left[\frac{V_s^2}{V_n^2}\right], \qquad \textbf{[4.59]}$$

where V_s and V_n denote the source and noise voltages, respectively. The voltage values usually are rms values (see Chapter 11). Also, a center frequency and range of frequencies are specified when the SNR is given.

REFERENCES

[1] P. Fara. 2002. *An Entertainment for Angels: Electricity in the Enlightenment.* Duxford: Icon Books.

[2] P. Horowitz and W. Hill. 1989. *The Art of Electronics*, 2nd ed. Cambridge: Cambridge University Press.

[3] M. Histand and D. Alciatore. 2003. *Introduction to Mechatronics and Measurement Systems*, 2nd ed. New York: McGraw-Hill.

[4] A. V. Oppenheim and A. S. Willsky. 1997. *Signals and Systems*, 2nd ed. New York: Prentice Hall.

[5] P. F. Dunn and W. A. Wilson. 1977. Development of the Single Microelectrode Current and Voltage Clamp for Central Nervous System Neurons. *Electroencephalography and Clinical Neurophysiology* 43:752–756.

REVIEW PROBLEMS

1. Three 11.9-µF capacitors are placed in series in an electrical circuit. Compute the total capacitance in µF to one decimal place.

2. Which of the following combination of units is equivalent to 1 J?
(a) $1\,C \cdot A \cdot W$, (b) $1\,W \cdot s/C$, (c) $1\,N/C$,
(d) $1\,C \cdot V$

3. For the electrical circuit depicted in Figure 4.16, given $R_1 = 16\,\Omega$, $R_3 = 68\,\Omega$, $I_1 = 0.9\,A$, $I_3 = 0.2\,A$, and $R_2 = R_4$, find the voltage potential, E, to the nearest whole volt.

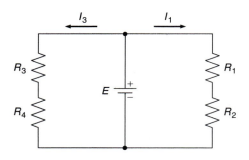

Figure 4.16   Electrical circuit.

4. The ends of a wire 1.17 m in length are suspended securely between two insulating plates. The diameter of the wire is 0.000 05 m. Given that the electrical resistivity of the wire is $1.673 \times 10^{-6}\,W \cdot m$ at 20.00 °C and that its coefficient of thermal expansion is $56.56 \times 10^{-5}\,°C$, compute the internal resistance in the wire at 24.8 °C to the nearest whole Ω.

5. A wire with the same material properties given in the previous problem is used as the R_1 arm of the Wheatstone bridge as shown in Figure 4.10. The Wheatstone bridge is designed to be used in deflection method mode and to act as a transducer in a system used to determine the ambient temperature in the laboratory. The length of the copper wire is fixed at 1.00 m and the diameter is 0.0500 mm. $R_2 = R_3 = R_4 = 154\,\Omega$ and $E_i = 10.0\,V$. For a temperature of 25.8 °C, compute the output voltage, E_o, in V to the nearest hundredth.

6. Which of the following effects would most likely NOT result from routing an AC signal across an inductor?
(a) a change in the frequency of the output alternating current
(b) a back electromagnetic force on the input current
(c) a phase lag in the output AC signal
(d) a reduction in the amplitude of the AC signal

7. Match each of the following terms related to electrical circuits with the person for whom the quantity's unit is named.

current	James Joule
charge	Charles Coulomb
electrical field work	Georg Ohm
electric potential	James Watt
resistance	André Ampère
power	Michael Faraday
inductance	Joseph Henry
capacitance	Alessandro Volta

8. Given the electrical circuit in Figure 4.17, where $R_1 = 37\,\Omega$, $R_2 = 65\,\Omega$, $R_3 = 147\,\Omega$, $R_4 = 126\,\Omega$, and $R_5 = 25\,\Omega$, find the total current drawn by all of the resistors to the nearest tenth A.

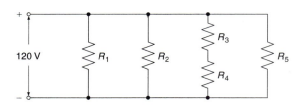

Figure 4.17   Resistor circuit.

9. This and the following four questions pertain to the electrical circuit diagram given in Figure 4.18. A Wheatstone bridge is used as a transducer for a resistance temperature device (RTD), which forms the R_1 leg of the bridge. The coefficient of thermal expansion for the RTD is 0.0005 °C. The reference resistance of the device is 25 Ω at a reference temperature of 20.0 °C. Compute the resistance of the RTD at 67 °C to the nearest tenth of an Ω. You will use the procedure to arrive at this answer in the next problem.

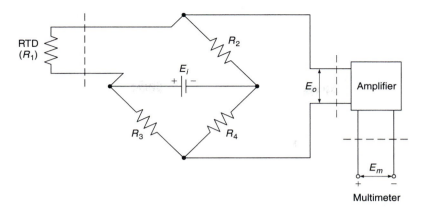

Figure 4.18   Temperature measurement system.

10. For the Wheatstone bridge shown, $R_2 = R_3 = R_4 = 25\ \Omega$ and $E_i = 5\,\mathrm{V}$. The maximum temperature to be sensed by the RTD is 78 °C. Find the maximum output voltage from the Wheatstone bridge to the nearest thousandth volt. The answer to this question will be used in the following problem. (Hint: Your answer should be between 0.034 and 0.049.)

11. A constant gain amplifier, with gain factor G, conditions the output voltage from the Wheatstone bridge. The multimeter used to process the output voltage from the amplifier, E_m, has a full-scale output of +10 V. To the hundreds place (for example, 1100), determine the maximum gain factor possible. The answer to this question will be used in the following problem.

12. The RTD senses a temperature of 60 °C. Compute the voltage output to the multimeter, E_m, to the nearest hundredth V.

13. In this RTD measurement system, what methodology classifies the use of the Wheatstone bridge?
(**a**) deflection method,
(**b**) null method,
(**c**) strain gage method,
(**d**) resistance-temperature method

14. Which of a following is a consequence of the conservation of energy?
(**a**) Ohm's law,
(**b**) Kirchhoff's first law,
(**c**) potential differences around a closed loop sum to zero,
(**d**) reciprocals of parallel resistances add

15. The measurement system shown in Figure 4.19 is a cantilever beam, fixed at one end, with a pair of strain gages on each the top and the bottom of the beam. The four gages, having equal unstrained resistance, are arranged in a Wheatstone bridge configuration. A point force is applied to the free end of the beam. Which of the following statements is *not* true about the depicted measurement system?
(**a**) The change in resistance in each gage is proportional to the applied force.
(**b**) Temperature and torsional effects are compensated.
(**c**) Strain in the beam is proportional to the bridge output voltage.
(**d**) Compression on a lower side gage causes an increase in its resistance.

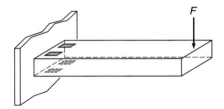

Figure 4.19   Cantilever beam measurement system.

16. An initially balanced Wheatstone bridge has $R_1 = R_2 = R_3 = R_4 = 120\ \Omega$. If R_1 increases by 20 Ω, what is the ratio of the bridge's output voltage to its excitation voltage?

17. A Wheatstone bridge can be used to determine unknown resistances using the null method. The electrical circuit shown in Figure 4.20 (with no applied potential) forms the R_1 arm of the Wheatstone bridge. If $R_2 = R_3 = 31 \, \Omega$ and $R_c = 259 \, \Omega$, find the necessary resistance of arm R_4 to balance the bridge. Resistances R_1, R_2, R_3, and R_4 refer to the resistances in the standard Wheatstone bridge configuration. Use the standard Wheatstone bridge. Respond in Ω to the nearest Ω.

18. A Wheatstone bridge has resistances $R_2 = 10 \, \Omega$, $R_3 = 14 \, \Omega$, and $R_4 = 3 \, \Omega$. Determine the value of R_1 in Ω when the bridge is used in the null method. Round off your answer to the nearest whole number of Ω.

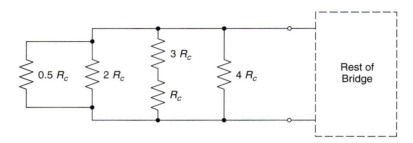

Figure 4.20   Wheatstone bridge circuit.

HOMEWORK PROBLEMS

1. Consider the pressure measurement system shown in Figure 4.21. The Wheatstone bridge of the pressure transducer is initially balanced at $p = p_{atm}$. Determine [a] the value of R_x (in Ω) required to achieve this balanced condition, and [b] E_o (in V) at this balanced condition. Finally, determine [c] the value of E_i (in V) required to achieve $E_o = 50.5$ mV when the pressure is changed to $p = 111.3$ kPa. Note that $R_s (\Omega) = 100[1 + 0.2(p - p_{atm})]$, with p in kPa.

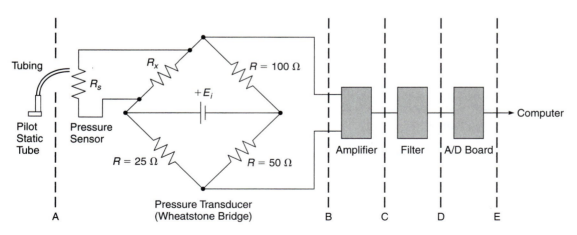

Figure 4.21   An example pressure measurement system configuration.

2. Consider the temperature measurement system shown in Figure 4.22. At station B determine [a] E_o (in V) when $T = T_o$, [b] E_o (in V) when $T = 72$ °F, and [c] the bridge's output impedance (in Ω) at $T = 72$ °F. Note that the sensor resistance is given by $R_s = R_o[1 + \alpha(T - T_o)]$, with $\alpha = 0.004/°F$ and $R_o = 25$ Ω at $T_o = 32$ °F. Also $E_i = 5$ V.

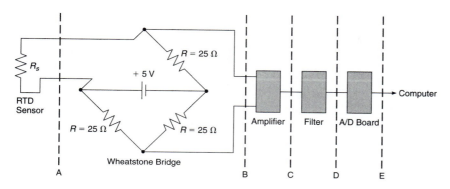

Figure 4.22   An example temperature measurement system configuration.

3. Consider the Wheatstone bridge that is shown in Figure 4.10. Assume that the resistor R_1 is actually a thermistor whose resistance, R, varies with the temperature, T, according to the equation

$$R = R_o \cdot \exp\left[\beta\left(\frac{1}{T} - \frac{1}{T_o}\right)\right], \qquad \textbf{[4.60]}$$

where $R_o = 1000$ Ω at $T_o = 26.85$ °C $= 300$ K (absolute) and $\beta = 3500$. Both T and T_o must be expressed in absolute temperatures in Equation 4.60. (Recall that the absolute temperature scales are either K or °R). Assume that $R_2 = R_3 = R_4 = R_o$. [a] Determine the normalized bridge output, E_o/E_i, when $T = 400$ °C. [b] Write a program to compute and plot the normalized bridge output from $T = T_o$ to $T = 500$ °C. [c] Is there a range of temperatures over which the normalized output is linear? [d] Over what temperature range is the normalized output very insensitive to temperature change?

4. For the test circuit shown in Figure 4.23, derive an expression for the output voltage, E_o, as a function of the input voltage, E_i, and the resistances shown for [a] the ideal case of the perfect voltmeter having $R_m = \infty$, and

[b] the nonideal voltmeter case when R_m is finite. Show mathematically that the solution for case [b] becomes that for case [a] when $R_m \to \infty$.

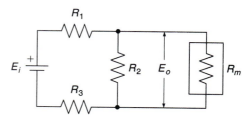

Figure 4.23   Test circuit.

5. An inexpensive voltmeter is used to measure the voltage to within 1 % across the power terminals of a stereo system. Such a system typically has an output impedance of 500 Ω and a voltage of 120 V at its power terminals. Assuming that the voltmeter is 100 % accurate such that the instrument and zero-order uncertainties are negligible, determine the minimum input impedance (in Ω) that this voltmeter must have to meet the 1 % criterion.

6. A voltage divider circuit is shown in Figure 4.24. The common circuit is used to supply an output voltage E_o that is less than a source voltage E_i. [a] Derive the expression for the output voltage, E_o, measured by the meter, as a function of E_i, R_x, R_y, and R_M, assuming that R_M is *not* negligible with respect to R_x and R_y. Then, [b] show that the expression derived in part [a] reduces to $E_o = E_i(R_x/R_T)$ when R_M becomes infinite.

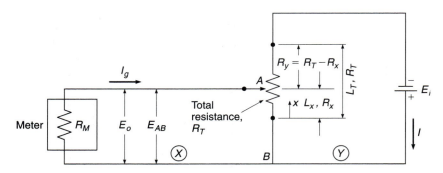

Figure 4.24   The voltage divider circuit.

CHAPTER REVIEW

ACROSS

4. the unit for voltage
6. the integral of current in time
10. a passive circuit element
11. coined the terms *positive* and *negative* for charge
12. measures resistance
14. the unit for inductance

DOWN

1. its voltage is proportional to *di/dt*
2. another possible arrangement of circuit elements
3. measures voltage
5. one possible arrangement of circuit elements
7. measures current
8. its current is proportional to *dV/dt*
9. the unit for charge
11. the unit for capacitance
13. the unit for resistance

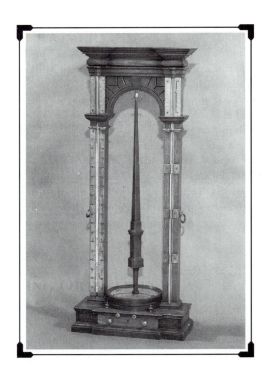

A Pressure-Temperature-Humidity Measurement System

The barometer, thermometer, and hygrometer each became standard instruments in the early 1700s to measure pressure, temperature, and humidity. By the mid-1700s they were being combined, forming a local-environment measurement system. This particular three-part table instrument was made in 1739 by George Graham (1673–1751), a leading clockmaker in his day. Humidity was measured by the stretching and shrinking of a piece of weighted whipcord, suspended inside the hygrometer's tapered wooden spindle. Each instrument had a different time response.

chapter

5

CALIBRATION AND RESPONSE

A single number has more genuine and permanent value than an expensive library full of hypotheses.

Robert J. Mayer, c. 1840.

Measures are more than a creation of society, they **create** *society.*

Ken Alder. 2002.
The Measure of All Things. London: Little, Brown.

It is easier to get two philosophers to agree than two clocks.

Lucius Annaeus Seneca, c. 40.

CHAPTER OUTLINE

5.1   CHAPTER OVERVIEW

In this chapter, the performance of a measurement system is investigated. Calibration methods are presented that ensure recorded values are accurate indicators of the variables sensed. Both the static and the dynamic response characteristics of linear measurement systems are examined. First-order and second-order systems are considered in detail, including how their output can lag temporally changes that occur in the experiment's environment. With this information, approaches to data acquisition and signal processing, which are the subjects of subsequent chapters, then can be considered.

5.2   LEARNING OBJECTIVES

You should be able to do the following after completing this chapter:

- Know the meaning of static calibration, dynamic calibration, static sensitivity, sequential and random calibration, calibration accuracy, and calibration precision
- Know the difference between quasi-static and dynamic processes
- Know the general formulation equations of a dynamic system and the corresponding expressions for zero-, first-, and second-order systems
- Calculate the response of a first-order system to either a step or a sinusoidal input given the initial and final conditions and the time constant of the system
- Know the definitions of the magnitude ratio, phase shift, and dynamic error for a first-order system
- Determine the magnitude and phase response of a second-order system given a damping coefficient value
- Know the definitions of ringing frequency, rise, and settling times for a second-order system
- Know the difference between underdamped, critically damped, and overdamped second-order systems

5.3   STATIC RESPONSE CHARACTERIZATION

Measurement systems and their instruments are used in experiments to obtain measurand values that usually are either steady or varying in time. For both situations, errors arise in the measurand values simply because the instruments are not perfect; their outputs do not precisely follow their inputs. These errors can be quantified through the process of **calibration**.

In a calibration, a known input value (called the **standard**) is applied to the system and then its output is measured. Calibrations can either be **static** (not a function of time) or **dynamic** (both the **magnitude** and the **frequency** of the input signal can be a function of time). Calibrations can be performed in either **sequential** or **random** steps. In a sequential calibration, the input is increased systematically and then decreased. Usually this is done by starting at the lowest input value and calibrating at every other input value up to the highest input value. Then the calibration is continued back down to the lowest input value by covering the alternate input values that were skipped during the upscale calibration. This helps to identify any unanticipated variations that could be present during calibration. In a random calibration, the input is changed from one value to another in no particular order.

From a calibration experiment, a **calibration curve** is established. A generic static calibration curve is shown in Figure 5.1. This curve has several characteristics. The **static sensitivity** refers to the slope of the calibration curve at a particular input value, x_1. This is denoted by K, where $K = K(x_1) = (dy/dx)_{x=x_1}$. Unless the curve is linear, K will not be a constant. More generally, sensitivity refers to the smallest change in a quantity that an instrument can detect, which can be determined knowing the value of K and the smallest indicated output of the instrument. There are two **ranges** of the calibration, the input range, $x_{max} - x_{min}$, and the output range, $y_{max} - y_{min}$.

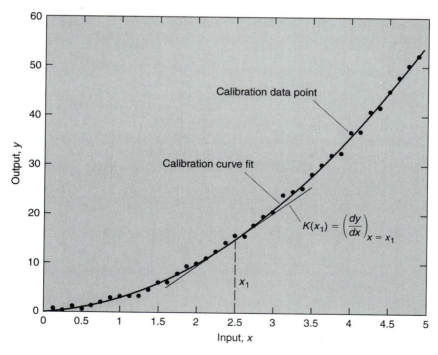

Figure 5.1   Typical static calibration curve.

Calibration accuracy refers to how close the measured value of a calibration is to the **true value**. Typically, this is quantified through the **absolute error**, e_{abs}, where

$$e_{abs} = |\text{true value} - \text{indicated value}|. \qquad \textbf{[5.1]}$$

The **relative error**, e_{rel}, is

$$e_{rel} = \frac{e_{abs}}{|\text{true value}|}. \qquad \textbf{[5.2]}$$

The **accuracy** of the calibration, a_{cal}, is related to the absolute error by

$$a_{cal} = 1 - e_{rel}. \qquad \textbf{[5.3]}$$

Calibration precision refers to how well a particular value is indicated upon repeated but independent applications of a specific input value. An expression for the precision in a measurement and the uncertainties that arise during calibration are presented in Chapter 9.

Example 5.1

STATEMENT

A hot-wire anemometer system was calibrated in a wind tunnel using a pitot-static tube. The data obtained is presented in Table 5.1. Using this data, a linear calibration curve-fit was made, which yielded

$$E^2 = 10.207 + 3.284\sqrt{U}.$$

Determine the following for the curve-fit: [a] the sensitivity, [b] the maximum absolute error, [c] the maximum relative error, and [d] the accuracy at the point of maximum relative error.

SOLUTION

Because the curve-fit is linear, the sensitivity of the curve-fit is its slope, which equals 3.284 $V^2/\sqrt{m/s}$. The calculated voltages, E_c, from the curve-fit expression are given in Table 5.1. Inspection of the results reveals that the maximum difference between the measured and calculated voltages is 0.02 V, which occurs at a velocity of 6.10 m/s. Thus, the maximum absolute error, e_{abs}, is 0.02 V, as defined by Equation 5.1. The relative error, e_{rel}, is defined by Equation 5.2. This also occurs at a velocity of 6.10 m/s, although maximum relative error does not always occur at the same calibration point as the maximum absolute error. Here, $e_{rel} = 0.02/4.30 = 0.004\,67$. Consequently, by Equation 5.3, the accuracy at the point of maximum relative error is $1 - 0.0046\,7 = 0.9953\,3$, or 99.5 %.

Table 5.1   Hot-wire anemometer system calibration data.

Velocity, U (m/s)	Measured Voltage, E_m (V)	Calculated Voltage, E_c (V)
0.00	3.19	3.19
3.05	3.99	3.99
6.10	4.30	4.28
9.14	4.48	4.49
12.20	4.65	4.66

5.1 | MATLAB SIDEBAR

For a space-delimited array, for example Y.dat, that is contained within the same working directory, the MATLAB command load Y.dat will bring the array into the workspace. This subsequently is referred to as Y in the workspace. By typing Y, all the elements of that array will be displayed. An M-file can be written to prompt the user to enter the array's file name. This is accomplished with the MATLAB command sequence

```
filename = input('enter the file name
            with its extension: ','s')
eval(['load ',filename])
```

Note that a space is required after the word *load* in the second line. The element in the second row and third column of that file, if it existed, would be identified subsequently by the command filename(2,3).

If a data file is located in a directory other than the current path directory, then the following command sequence will change to the other directory (here its path is C:\otherdir) containing the file (here called some.dat), load the file into the workspace, and then return to the original directory (here identified as pdir):

```
pdir = pwd;
cd C:\otherdir
load some.dat
```

```
eval(['cd ',pdir])
```

Note the required space after cd in the last line. This command sequence is quite useful when dealing with many files located in different directories.

Also, the MATLAB command fscanf can be used to read formatted data from a file. Here, the file OUT.dat would be opened or created, written to, and closed using the commands:

```
fid = fopen('OUT.dat')
myformat = '%g %g %g'
A = fscanf(fid,myformat,[3 inf])
A = A'
fclose(fid)
```

The fscanf command reads data from the file specified by an integer file identifier (fid) established by the command fopen. Each column of data is read into the matrix A *as a row* according to the specified format (%g allows either decimal or scientific formatted data to be read). If the specified format does not match the data, the file reading stops. The [3 inf] term specifies the size of A, here three rows of arbitrary (inf) length. The matrix A then is transposed to mirror the original column structure of Y.dat. After these operations, the file is closed.

5.4   DYNAMIC RESPONSE CHARACTERIZATION

In reality, almost every measurement system does not respond instantaneously to an input that varies in time. Often there is a time delay and amplitude difference between the system's input and output signals. This obviously creates a measurement problem. If these effects are not accounted for, dynamic errors will be introduced into the results.

To properly assess and quantify these effects, an understanding of how measurement systems respond to transient input signals must be gained. The ultimate goal would be to determine the output (response) of a measurement system for all conceivable inputs. The dynamic error in the measurement can be related to the difference between the input and output at a given time. In this chapter, only the

---

5.2 | **MATLAB SIDEBAR**

The MATLAB M-file `plotxy.m` plots the (x,y) data pairs of a user-specified data file in which the variable values are listed in columns. The data values in any two user-specified columns can be plotted as pairs. The M-file plots the data pairs as circles with lines in between the pairs.

5.3 | **MATLAB SIDEBAR**

For situations either when a file contains some text in rows or columns that is not needed within the workspace or when a file needs to be created without text from one that does have text, the MATLAB command `textread` can be used. For example, assume that the file WTEXT.dat contains four columns consisting of text in columns 1 and 3 and decimal numbers in columns 2 and 4. The following command sequence reads WTEXT.dat and then stores the decimal numbers as vectors x and y for columns 2 and 4, respectively, in the workspace:

```
myformat = '\%*s \%f \%*s \%f'
```

```
[x,y] = textread('WTEXT.dat',myformat)
```

Note that if the input file is constructed with text in the first *n* rows and then numbers, say three numbers, separated by spaces in subsequent rows, the sequence of commands would be:

```
myformat = '\%f \%f \%f'
[x,y,z] = textread('WTEXT.dat',
               myformat,'headerlines',n)
```

Remember that the format must be consistent with the variables represented by numbers in each row.

---

basics of this subject will be covered. The response characteristics of several specific systems (zero-, first-, and second-order) to specific transient inputs (step and sinusoidal) will be studied. Hopefully, this brief foray into dynamic system response will give an appreciation for the problems that can arise when measuring time-varying phenomena.

First, examine the general formulation of the problem. The output signal, $y(t)$, in response to the input forcing function of a *linear* system, $F(t)$, can be modeled by a *linear* ordinary differential equation with *constant* coefficients ($a_0, \ldots, a_n$) of the form

$$a_n \frac{d^n y}{dt^n} + a_{n-1} \frac{d^{n-1} y}{dt^{n-1}} + \cdots + a_1 \frac{dy}{dt} + a_0 y = F(t). \qquad \textbf{[5.4]}$$

In this equation, n represents the **order** of the system. The input forcing function can be written as

$$F(t) = b_m \frac{d^m x}{dt^m} + b_{m-1} \frac{d^{m-1} x}{dt^{m-1}} + \cdots + b_0 x, \ \ m \le n, \qquad \textbf{[5.5]}$$

where $b_0, \ldots, b_n$ are *constant* coefficients, $x = x(t)$ is the forcing function, and m represents its order, where m must always be less than or equal to n to avoid having an overdeterministic system. By writing $F(t)$ as a polynomial, the ability to describe almost any shape of forcing function is retained.

5.4 | MATLAB SIDEBAR

An experimentalist wishes to load a data file into the MATLAB workspace, convert some of its information and then store that converted information in another data file. Assume that the input data file called IN.dat consists of four rows of unknown length. The first row is time, the third distance, and the fourth force. Information in the second row is not needed. The desired output file will be called OUT.dat and consists of two columns, the first time and the second work. What are the MATLAB commands needed to accomplish this?

The first task is to load the input file into the MATLAB workspace, as described before. Then each row of data would be given its name by

```
time = filename(1,:);
distance = filename(3,:);
force = filename(4,:);
```

Next, a matrix called A is created that would have time as its first column and work, which equals force times distance, as its second column. This is done by the commands

```
work = force.*distance
A = [time;work]
```

Finally, the MATLAB command `fprintf` can be used. Here, the file OUT.dat is opened or created, written to, and finally closed using the commands

```
fid = fopen('OUT.dat','wt')
myformat = '%12.6f %12.6f\n'
fprintf(fid,myformat,A)
fclose(fid)
```

The letters wt in the second line create a file in the text mode (t) to which the data can be written (w). The \n in the third line starts the next row. The %12.6f sets the number's format to decimal format and stores the number as 12 digits, with 6 digits to the right of the decimal point. Alternatively, scientific format (%e) can be used. Specifying the format %g allows MATLAB to choose either decimal or scientific format, whichever is shorter.

The output response, $y(t)$, actually represents a physical variable followed in time. For example, it could be the displacement of the mass of an accelerometer positioned on a fluttering aircraft wing or the temperature of a thermocouple positioned in the wake of a heat exchanger. The exact ordinary differential equation governing each circumstance is derived from a conservation law, for example, from Newton's second law for the accelerometer or from the first law of thermodynamics for the thermocouple.

To solve for the output response, the exact form of the input forcing function, $F(t)$, must be specified. This is done by choosing values for the $b_0, \ldots, b_n$ coefficients and m. Then Equation 5.4 must be integrated subject to the initial conditions.

In this chapter, two types of input forcing functions, step and sinusoidal, are considered for linear, first-order, and second-order systems. There are analytical solutions for these situations. Further, as will be shown in Chapter 11, almost all types of functions can be described through Fourier analysis in terms of the sums of sine and cosine functions. So, if a linear system's response for sinusoidal-input forcing is determined, then its response to more complicated input forcing can be described. This is done by linearly superimposing the outputs determined for each of the sinusoidal-input forcing components that were identified by Fourier analysis. Finally, note that many measurement systems are linear, but not all. In either case, the response of the system almost always can be determined numerically. Numerical solution methods for such a model will be discussed in Section 5.9.

Now consider some particular systems by first specifying the order of the systems. This is done by substituting a particular value for n into Equation 5.4.

- For $n = 0$, a zero-order system is specified by

$$a_0 y = F(t). \qquad \textbf{[5.6]}$$

Instruments that behave like zero-order systems are those whose output is directly coupled to its input. An electrical-resistance strain gage in itself is an excellent example of a zero-order system, where an input strain directly causes a change in the gage resistance. However, dynamic effects can occur when a strain gage is attached to a flexible structure. In this case, the strain gage response must be modeled as a higher-order system.

- For $n = 1$, a first-order system is given by

$$a_1 \dot{y} + a_0 y = F(t). \qquad \textbf{[5.7]}$$

Instruments whose responses fall into the category of first-order systems include thermometers, thermocouples, and other similar simple systems that produce a time lag between the input and output due to the *capacity* of the instrument. For thermal devices, the heat transfer between the environment and the instrument coupled with the thermal capacitance of the instrument produces this time lag.

- For $n = 2$, a second-order system is specified by

$$a_2 \ddot{y} + a_1 \dot{y} + a_0 y = F(t). \qquad \textbf{[5.8]}$$

Examples of second-order instruments include diaphragm-type pressure transducers, U-tube manometers, and accelerometers. This type of system is characterized by its *inertia*. In the U-tube manometer, for example, the fluid having inertia is moved by a pressure difference.

The responses of each of these systems is examined in the following sections.

5.5   ZERO-ORDER SYSTEM DYNAMIC RESPONSE

For a zero-order system the equation is

$$y = \left(\frac{1}{a_0}\right) F(t) = K F(t), \qquad \textbf{[5.9]}$$

where K is called the static sensitivity or steady-state gain. It can be seen that the output, $y(t)$, exactly follows the input forcing function, $F(t)$, in time and that $y(t)$ is amplified by a factor, K. Hence, for a zero-order system, a plot of the output signal values (on the ordinate) versus the input signal values (on the abscissa) should yield a straight line of slope, K. In fact, instrument manufacturers often provide values for the steady-state gains of their instruments. These values are obtained by performing *static* calibration experiments.

5.6   FIRST-ORDER SYSTEM DYNAMIC RESPONSE

First-order systems are slightly more complicated. Their governing equation is

$$\tau \dot{y} + y = K F(t), \qquad \qquad [5.10]$$

where τ is the **time constant** of the system $= a_1/a_0$. When the time constant is small, the derivative term in Equation 5.10 becomes negligible and the equation reduces to that of a zero-order system. That is, the smaller the time constant, the more instantaneous is the response of the system.

Now digress for a moment and examine the origin of a first-order-system equation. Consider a standard glass bulb thermometer initially at room temperature that is immersed into hot water. The thermometer takes a while to read the correct temperature. However, it is necessary to obtain an equation of the thermometer's temperature as a function of time after it is immersed in the hot water in order to be more specific.

Start by considering the liquid inside the bulb as a fixed mass across whose surfaces heat can be transferred. When the thermometer is immersed into the hot water, heat (a form of energy) will be transferred from the hotter body (the water) to the cooler body (the thermometer's liquid). This leads to an increase in the total energy of the liquid. This energy transfer is governed by the first law of thermodynamics (conservation of energy), which is

$$\frac{dE}{dt} = \frac{dQ}{dt}, \qquad \qquad [5.11]$$

in which E is the total energy of the thermometer's liquid, Q is the heat transferred from the hot water to the thermometer's liquid, and t is time. The rate at which the heat is transferred into the thermometer's liquid depends on the physical characteristics of the interface between the outside of the thermometer and the hot water. The heat is transferred convectively to the glass from the hot water and is described by

$$\frac{dQ}{dt} = h A (T_{hw} - T), \qquad \qquad [5.12]$$

where h is the convective heat transfer coefficient, A is the surface area over which the heat is transferred, and T_{hw} is the temperature of the hot water. Here it is assumed implicitly that there are no conductive heat transfer losses in the glass. All of the heat transferred from the hot water through the glass reaches the thermometer's liquid. Now as the energy is stored within the liquid, its temperature increases. For energy within the liquid to be conserved, it must be that

$$\frac{dE}{dt} = m C_v \frac{dT}{dt}, \qquad \qquad [5.13]$$

where T is the liquid's temperature, m is its mass, and C_v is its specific heat at constant volume.

Thus, upon substitution of Equations 5.12 and 5.13 into Equation 5.11

$$mC_v\frac{dT}{dt} = hA(T_{hw} - T).$$ [5.14]

Rearranging,

$$\frac{mC_v}{hA}\frac{dT}{dt} + T = T_{hw}.$$ [5.15]

Comparing this equation to Equation 5.10, it can be seen that $y = T$, $\tau = mC_v/hA$, and $T_{hw} = F(t)$ with $K = 1$. This is the linear, first-order differential equation with constant coefficients that relates the time rate of change in the thermometer's liquid temperature to its temperature at any instance of time and the conditions of the situation. Equation 5.15 must be integrated to obtain the desired equation of the thermometer's temperature as a function of time after it is immersed in the hot water.

Another example of a first-order system is an electrical circuit comprised of a resistor of resistance, R, and a capacitor of capacitance, C, both in series with a voltage source with voltage, $E_i(t)$. The voltage differences, ΔV, across each component in the circuit are $\Delta V = RI$ for the resistor and $\Delta V = Q/C$ for the capacitor, where the current, I, is related to the charge, Q, by $I = dQ/dt$. Application of Kirchhoff's voltage law to the circuit gives

$$RC\frac{dV}{dt} + V = E_i(t).$$ [5.16]

Comparing this equation to Equation 5.10, $\tau = RC$ and $K = 1$.

Now proceed to solve a first-order system equation to determine the response of the system subject to either a step change in conditions (by assuming a step-input forcing function) or a periodic change in conditions (by assuming a sinusoidal-input forcing function). The former, for example, could be the temperature of a thermometer as a function of time after it is exposed to a sudden change in temperature, as was examined above. The latter, for example, could be the temperature of a thermocouple in the wake behind a heated cylinder.

5.6.1   RESPONSE TO STEP-INPUT FORCING

Start by considering the governing equation for a first-order system

$$\tau\dot{y} + y = KF(t),$$ [5.17]

where the step-input forcing function, $F(t)$, is defined as A for $t > 0$ and the initial condition $y(0) = y_0$. Equation 5.17 is a linear, first-order ordinary differential equation. Its general solution (see [1]) is of the form

$$y(t) = c_0 + c_1 e^{-t/\tau}.$$ [5.18]

Substitution of this expression for y and the expression for its derivative $\dot{y}$ into Equation 5.17 yields $c_0 = KA$. Subsequently, applying the initial condition to Equation 5.18 gives $c_1 = y_0 - KA$. Thus, the specific solution can be written as

$$y(t) = KA + (y_0 - KA)\, e^{-t/\tau}. \qquad \textbf{[5.19]}$$

Now examine this equation. When the time equals zero, the exponential term is unity, which gives $y(0) = y_0$. Also, when time becomes very large with respect to τ, the exponential term tends to zero, which gives an output equal to KA. Hence, the output rises exponentially from its initial value of y_0 at $t = 0$ to its final value of KA at $t \gg \tau$. This is what is seen in the solution, as shown in the left graph of Figure 5.2. Note that at the dimensionless time $t/\tau = 1$, the value the signal reaches is approximately two-thirds (actually $1 - 1/e$ or 0.6321) of its final value. The time that it takes the system to reach 90 % of its final value (which occurs at $t/\tau = 2.303$) is called the **rise time** of a first-order system. At $t/\tau = 5$ the signal has reached greater than 99 % of its final value.

The term y_0 can be subtracted from both sides of Equation 5.19 and then rearranged to yield

$$M(t) \equiv \frac{y(t) - y_0}{y_\infty - y_0} = 1 - e^{-t/\tau}, \qquad \textbf{[5.20]}$$

noting that $y_\infty = KA$. Note $M(t)$ is called the **magnitude ratio** and is a dimensionless variable that represents the change in y at any time t from its initial value divided

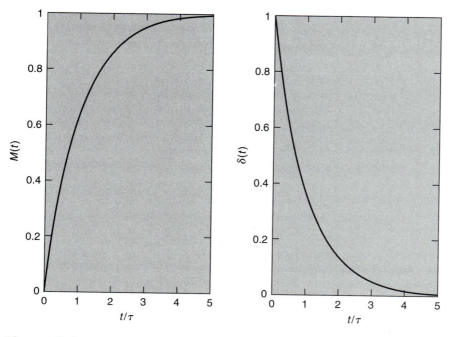

Figure 5.2   Response of a first-order system to step-input forcing.

5.5 | MATLAB SIDEBAR

The MATLAB M-file `fordstepmd.m` was used to generate Figure 5.2. It determines both the magnitude ratio and the dynamic error of a first-order system's response to a step-input forcing as a function of the dimensionless time.

by its maximum possible change. When y reaches its final value, $M(t)$ is unity. The right-hand side of Equation 5.20 is a dimensionless time, t/τ. Because Equation 5.20 is dimensionless, it is valid for *all* first-order systems responding to step-input forcing.

Alternatively, Equation 5.19 can be rearranged directly to give

$$\frac{y(t) - y_\infty}{y_0 - y_\infty} = e^{-t/\tau} \equiv \delta(t). \qquad \textbf{[5.21]}$$

In this equation, $\delta(t)$ represents the fractional difference of y from its final value. This can be interpreted as the fractional **dynamic error** in y. From Equations 5.20 and 5.21,

$$\delta(t) = 1 - M(t). \qquad \textbf{[5.22]}$$

This result is plotted in the right graph of Figure 5.2. At the dimensionless time $t/\tau = 1$, δ equals $0.3678 = 1/e$. Further, at $t/\tau = 5$, the dynamic error is essentially zero (0.007). That means for a first-order system subjected to a step change in input it takes approximately five time constants for the output to reach the input value. For perfect measurement system, there would be no dynamic error [$\delta(t) = 0$] and the output would always follow the input [$M(t) = 1$].

5.6.2 RESPONSE TO SINUSOIDAL-INPUT FORCING

Now consider a first-order system that is subjected to an input that varies sinusoidally in time. The governing equation is

$$\tau \dot{y} + y = KF(t) = KA \sin(\omega t), \qquad \textbf{[5.23]}$$

where K and A are arbitrary constants. The units of K would be those of y divided by those of A. The general solution is

$$y(t) = y_h + y_p = c_0 e^{-t/\tau} + c_1 + c_2 \sin(\omega t) + c_3 \cos(\omega t), \qquad \textbf{[5.24]}$$

in which c_0 through c_3 are constants, where the first term on the right-hand side of this equation is the homogeneous solution, y_h, and the remaining terms constitute the particular solution, y_p.

The constants c_1 through c_3 can be found by substituting the expression for y_p and the expression for its derivative into Equation 5.23. By comparing like terms in the resulting equation,

$$c_1 = 0, \qquad \textbf{[5.25]}$$

$$c_2 = \frac{KA}{\omega^2 \tau^2 + 1},$$ [5.26]

and

$$c_3 = -\omega \tau C_2 = \frac{-\omega \tau KA}{\omega^2 \tau^2 + 1}.$$ [5.27]

The constant c_0 can be found by applying the initial condition $y(0) = y_0$ to Equation 5.24

$$c_0 = y + 0 - c_3 = \frac{\omega \tau KA}{\omega^2 \tau^2 + 1}.$$ [5.28]

Thus, the final solution becomes

$$y(t) = y_0 + \omega \tau D e^{-t/\tau} + D \sin(\omega t) - \omega \tau D \cos(\omega t),$$ [5.29]

where

$$D = \frac{KA}{\omega^2 \tau^2 + 1}.$$ [5.30]

Now Equation 5.29 can be simplified further. The sine and cosine terms can be combined in Equation 5.29 into a single sine term using the trigonometric identity

$$\alpha \cos(\omega t) + \beta \sin(\omega t) = \sqrt{\alpha^2 + \beta^2} \sin(\omega t + \phi),$$ [5.31]

where

$$\phi = \tan^{-1}\left(\frac{\alpha}{\beta}\right).$$ [5.32]

Equating this expression with the sine and cosine terms in Equation 5.29 gives $\alpha = -\omega \tau D$ and $\beta = D$. Thus,

$$D \sin(\omega t) - \omega \tau D \cos(\omega t) = D\sqrt{\omega^2 \tau^2 + 1} = \frac{KA}{\sqrt{\omega^2 \tau^2 + 1}}$$ [5.33]

and

$$\phi = \tan^{-1}(-\omega \tau) = -\tan^{-1}(\omega \tau)$$ [5.34]

or, in units of degrees,

$$\phi^\circ = -\left(\frac{180}{\pi}\right) \tan^{-1}(\omega \tau).$$ [5.35]

The minus sign is present in Equations 5.34 and 5.35 by convention to denote that the output lags behind the input.

The final solution is

$$y(t) = y_0 + \frac{\omega \tau KA}{\omega^2 \tau^2 + 1} e^{-t/\tau} + \frac{KA}{\sqrt{\omega^2 \tau^2 + 1}} \sin(\omega t + \phi).$$ [5.36]

The second term on the right-hand side represents the **transient response** while the third term is the **steady-state response**. For $\omega \tau \ll 1$, the transient term becomes very small and the output follows the input. For $\omega \tau \gg 1$, the output is *attenuated* and its phase is *shifted* from the input by ϕ radians. The **phase lag** in seconds (lag time), β, is given by

$$\beta = \frac{\phi}{\omega}. \qquad \textbf{[5.37]}$$

Examine this response further in a dimensionless sense. The magnitude ratio for this input-forcing situation is the ratio of the magnitude of the steady-state output to that of the input. Thus,

$$M(\omega) = \frac{KA/\sqrt{\omega^2\tau^2 + 1}}{KA} = \frac{1}{\sqrt{\omega^2\tau^2 + 1}}. \qquad \textbf{[5.38]}$$

The dynamic error, using its definition in Equation 5.22 and Equation 5.38, becomes

$$\delta(\omega) = 1 - \frac{1}{\sqrt{\omega^2\tau^2 + 1}}. \qquad \textbf{[5.39]}$$

Shown in Figures 5.3 and 5.4, respectively, are the magnitude ratio and the phase shift plotted versus the product $\omega\tau$. First examine Figure 5.3. For values of $\tau\omega$ less than approximately 0.1, the magnitude ratio is very close to unity. This implies that the system's output closely follows its input in this range. At $\omega\tau$ equal to unity, the

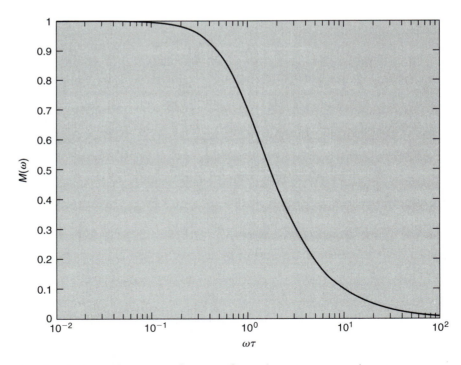

Figure 5.3　　The magnitude ratio a first-order system responding to sinusoidal-input forcing.

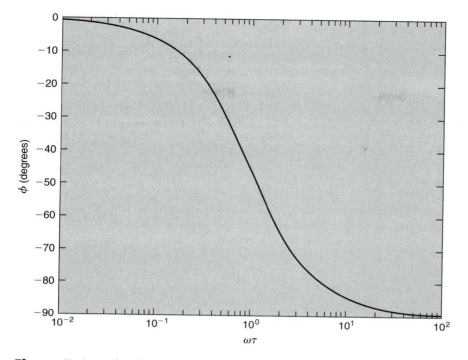

Figure 5.4 The phase shift a first-order system responding to sinusoidal-input forcing.

magnitude ratio equals 0.707, that is, the output amplitude is approximately 71 % of its input. Here, the dynamic error would be $1 - 0.707 = 0.293$ or approximately 29 %. Now look at Figure 5.4. When $\omega\tau$ is unity, the phase shift equals $-45°$. That is, the output signal lags the input signal by 45° or one-eighth of a cycle.

The magnitude ratio often is expressed in units of decibels, abbreviated as dB. The decibel's origin began with the introduction of the Bel, defined in terms of the ratio of output power, P_2, to the input power, P_1, as

$$\text{Bel} = \log_{10}\left(\frac{P_2}{P_1}\right).$$

[5.40]

5.6 | **MATLAB SIDEBAR**

The MATLAB M-files `fordsineM.m` and `fordsinep.m` plot the steady-state first-order system's magnitude ratio and phase lag, respectively, in response to sinusoidal-input forcing. These are plotted versus the dimensionless parameter $\omega\tau$. Figures 5.3 and 5.4 were made using `fordsineM.m` and `fordsinep.m`, respectively.

The decibel, equal to 10 Bels, was defined as

$$\text{Decibel} = 10 \log_{10}\left(\frac{P_2}{P_1}\right).$$ **[5.41]**

Equation 5.41 is used to express sound intensity levels, where P_2 corresponds to the sound intensity and P_1 to the reference intensity, 10^{-12} W/m^2, which is the lowest intensity that humans can hear. The Saturn V on launch has a sound intensity of 172 dB; human hearing pain occurs at 130 dB; a soft whisper at a distance of 5 m is 30 dB.

There is one further refinement in this expression. Power is a squared quantity, $P_2 = Q_2^2$ and $P_1 = Q_1^2$, where Q_2 and Q_1 are the base measurands, such as volts for an electrical system. With this in mind, Equation 5.41 becomes

$$\text{Decibel} = 10 \log_{10}\left(\frac{Q_2}{Q_1}\right)^2 = 20 \log_{10}\left(\frac{Q_2}{Q_1}\right).$$ **[5.42]**

Equation 5.42 is the basic definition of the decibel as used in measurement engineering. Finally, Equation 5.42 can be written in terms of the magnitude ratio

$$\text{dB} = 20 \log_{10} M(\omega).$$ **[5.43]**

The point $M(\omega) = 0.707$, which is a decrease in the system's amplitude by a factor of $1/\sqrt{2}$, corresponds to an attenuation of the system's input by -3 dB. Sometimes, this is called the *half-power point* because, at this point, the power is one-half the original power.

Example 5.2

STATEMENT

Convert the sound intensity level of 30 dB to $\log_e M(\omega)$.

SOLUTION

The relationship between logarithms of bases a and b is

$$\log_b x = \frac{\log_a x}{\log_a b}.$$

For this problem the bases are e and 10. So,

$$\log_e M(\omega) = \frac{\log_{10} M(\omega)}{\log_{10} e}.$$

Now, $\log_{10} e = 0.434\,294$. Also, $\log_e 10 = 2.302\,585$, and $e = 2.718\,282$. Using Equation 5.43, $\log_{10} M(\omega)$ at 30 dB equals 1.500. Thus

$$\log_e M(\omega) = \frac{1.500}{0.434} = 3.456.$$

The relationship between logarithms of two bases is used often when converting back and forth between base 10 and base e systems.

Systems often are characterized by their bandwidth and center frequency. **Bandwidth** is the range of frequencies over which the output amplitude of a system remains above 70.7 % of its input amplitude. Over this range, $M(\omega) \geq 0.707$ or -3 dB. The lower frequency at which $M(\omega) < 0.707$ is called the **low cutoff frequency**. The higher frequency at which $M(\omega) < 0.707$ is called the **high cutoff frequency**. The **center frequency** is the frequency equal to one-half the sum of the low and high cutoff frequencies. Thus, the bandwidth is the difference between the high and low cutoff frequencies. Sometimes bandwidth is defined as the range of frequencies that contain most of the system's energy or which the system's gain is almost constant. However, the preceding quantitative definition is preferred and used most frequently.

STATEMENT | **Example 5.3**

Determine the low and high cutoff frequencies, center frequency, and the bandwidth in units of Hz of a first-order system having a time constant of 0.1 s that is subjected to sinusoidal-input forcing.

SOLUTION

For a first-order system, $M(\omega) \geq 0.707$ from $\omega\tau = 0$ to $\omega\tau = 1$. Thus, the low cutoff frequency is 0 Hz and the high cutoff frequency is 1 rad/[(0.1 s)(2π rad/cycle)] $= 5/\pi$ Hz. The bandwidth equals $5/\pi$ Hz $- 0$ Hz $= 5/\pi$ Hz. The center frequency is $5/2\pi$.

Example 5.4 illustrates how the time constant of a thermocouple affects its output.

STATEMENT | **Example 5.4**

Consider an experiment in which a thermocouple that is immersed in a fluid and connected to a reference junction/linearizer/amplifier microchip with a static sensitivity of 5 mV/°C. Its output is $E(t)$ in mV. The fluid temperature varies sinusoidally in °C as $115 + 12\sin(2t)$. The time constant τ of the thermocouple is 0.15 s. Determine $E(t)$, the dynamic error $\delta(\omega)$, and the time delay $\beta(\omega)$ for $\omega = 2$. Assume that this system behaves as a first-order system.

SOLUTION

It is known that

$$\tau \dot{E} + E = KF(t).$$

Substitution of the given values yields

$$0.15\dot{E} + E = 5[115 + 12\sin 2t] \qquad \qquad \textbf{[5.44]}$$

with the initial condition of $E(0) = (5 \text{ mV/°C})(115 \text{ °C}) = 575$ mV.

To solve this linear, first-order differential equation with constant coefficients, a solution of the form $E(t) = E_h + E_p$ is assumed, where $E_h = C_0 e^{-t/\tau}$ and $E_p = c_1 + c_2 \sin 2t + c_3 \cos 2t$. Substitution of this expression for $E(t)$ into the left-hand side and grouping like terms gives

$$c_1 = 575, \quad c_2 = 55.1, \ \text{and} \ c_3 = -16.5.$$

Equation 5.44 then can be rewritten as

$$E(t) = k_0 e^{-t/0.15} + 575 + 55.1 \sin 2t - 16.5 \cos 2t.$$

Using the initial condition,
$$c_0 = 16.5.$$

Thus, the final solution for $E(t)$ is

$$E(t) = 575 + 16.5e^{-t/0.15} + 55.1 \sin 2t - 16.5 \cos 2t$$

or, in units of °C temperature

$$T(t) = 115 + 3.3e^{-t/0.15} + 11.0 \sin 2t - 3.3 \cos 2t.$$

The output (measured) temperature is plotted in Figure 5.5 along with the input (actual) temperature. A careful comparison of the two signals reveals that the output lags the input in time and has a slightly attenuated amplitude. At $t = 2$ s, the actual temperature is ~ 106 °C, which is less than the measured temperature of ~ 109 °C. Whereas, at $t = 3$ s, the actual temperature is ~ 112 °C, which is greater than the measured temperature of ~ 109 °C. So, for this type of forcing, the measured temperature can be greater or less than the actual temperature, depending on the time at which the measurement is made.

The time lag and the percent reduction in magnitude can be found as follows. The dynamic error is

$$\delta(\omega = 2) = 1 - M(\omega = 2) = 1 - \frac{1}{[1 + (2 \times 0.15)^2]^{1/2}} = 0.04,$$

which is a 4 % reduction in magnitude. The time lag is

$$\beta(\omega = 2) = \frac{\phi(\omega = 2)}{\omega} = \frac{-\tan^{-1} \omega\tau}{\omega} = \frac{(-16.7°)(\pi \ \text{rad}/180°)}{2 \ \text{rad/s}} = -0.15 \ \text{s},$$

which implies that the output signal lags the input signal by 0.15 s. The last two terms in the temperature expression can be combined using a trigonometric identity (see Chapter 11), as

$$11.0 \sin 2t - 3.3 \cos 2t = 11.48 \sin(2t - 0.29), \qquad \textbf{[5.45]}$$

where 0.29 rad = 16.7° is the phase lag found before.

5.7 SECOND-ORDER SYSTEM DYNAMIC RESPONSE

The response behavior of second-order systems is more complex than first-order systems. Their behavior is governed by the equation

$$\frac{1}{\omega_n^2}\ddot{y} + \frac{2\zeta}{\omega_n}\dot{y} + y = KF(t), \qquad \textbf{[5.46]}$$

where $\omega_n = \sqrt{a_0/a_2}$ denotes the natural frequency, and $\zeta = a_1/2\sqrt{a_0 a_2}$ is the damping ratio of the system. Note that when $2\zeta \gg 1/\omega_n$, the second derivative term in Equation 5.46 becomes negligible with respect to the other terms and the system

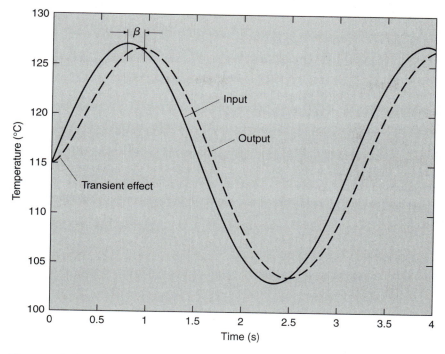

Figure 5.5   The time history of the thermocouple system.

behavior approaches that of a first-order system with a system time constant equal to $2\zeta/\omega_n$.

Equation 5.46 could represent, among other things, either a mechanical spring-mass-damper system or an electrical capacitor-inductor-resistor circuit, both with forcing. The solution to this type of equation is rather lengthy and is described in detail in many applied mathematics texts (see [1]). Now examine where such an equation would come from by considering an example.

A familiar situation occurs when a bump in the road is encountered with a car. If the car has a good suspension system it will absorb the effect of the bump. The bump hardly will be felt. On the other end, if the suspension system is old, an up-and-down motion is present that may take several seconds to die out. This is the response of a linear, second-order system (the car with its suspension system) to an input forcing (the bump).

The car with its suspension system can be modeled as a mass (the body of the car and its passengers) supported by a spring (the suspension coil) and a damper (the shock absorber) in parallel (usually there are four sets of spring-dampers, one for each wheel). Newton's second law can be applied, which states that the mass times the acceleration of a system is equal to the sum of the forces acting on the system.

This becomes

$$m\frac{d^2y}{dt^2} = \sum_i F_i = F_g + F_s(t) + F_d(t) + F(t), \qquad \textbf{[5.47]}$$

in which y is the vertical displacement; F_g is the gravitational force ($= mg$); $F_s(t)$ is the spring force ($= -k[L^* + y]$), where k is the spring constant and L^* is the initial compressed length of the spring; $F_d(t)$ is the damping force ($= -\gamma dy/dt$), where γ is the damping coefficient; and $F(t)$ the forcing function. Note that the spring and damping forces are negative because they act *opposite* to the direction of motion. The height of the bump as a function of time as dictated by the speed of the car would determine the exact shape of $F(t)$. Now when there is no vertical displacement, which is the case just before the bump is encountered, the system is in equilibrium and y does not change in time. Equation 5.47 reduces to

$$0 = mg - kL^*. \qquad \textbf{[5.48]}$$

This equation can be used to replace L^* in Equation 5.47 to arrive at

$$\frac{m}{k}\frac{d^2y}{dt^2} + \frac{\gamma}{k}\frac{dy}{dt} + y = \frac{1}{k}F(t). \qquad \textbf{[5.49]}$$

Comparing this equation to Equation 5.46, $\omega_n = \sqrt{k/m}$, $\zeta = \gamma/\sqrt{4km}$, and $K = 1/k$.

Another example of a second-order system is an electrical circuit comprised of a resistor, R, a capacitor, C, and an inductor, L, in series with a voltage source with voltage, $E_i(t)$, that completes a closed circuit. The voltage differences, ΔV, across each component in the circuit are $\Delta V = RI$ for the resistor, $\Delta V = L dI/dt$ for the inductor, and $\Delta V = Q/C$ for the capacitor, where the current I is related to the charge, Q, by $I = dQ/dt$. Application of Kirchhoff's voltage law to the circuit's closed loop gives

$$LC\frac{d^2I}{dt^2} + RC\frac{dI}{dt} + I = C\frac{dE_i(t)}{dt}. \qquad \textbf{[5.50]}$$

Comparing this equation to Equation 5.46, $\omega_n = \sqrt{1/LC}$, $\zeta = R/\sqrt{4L/C}$, and $K = C$.

The approach to solving a nonhomogeneous, linear, second-order, ordinary differential equation with constant coefficients of the form of Equation 5.46 involves finding the homogeneous, $y_h(t)$, and particular, $y_p(t)$, solutions and then linearly superimposing them to form the complete solution, $y(t) = y_h(t) + y_p(t)$. The values of the arbitrary coefficients in the $y_h(t)$ solution are determined by applying the specified initial conditions, which are of the form $y(0) = y_o$ and $\dot{y}(0) = \dot{y}_o$. The values of the arbitrary coefficients in the $y_p(t)$ solution are found through substitution of the general form of the $y_p(t)$ solution into the differential equation and then equating like terms.

The form of the homogeneous solution to Equation 5.46 depends on roots of its corresponding characteristic equation

$$\frac{1}{\omega_n^2} r^2 + \frac{2\zeta}{\omega_n} r + 1 = 0,$$ [5.51]

which are

$$r_{1,2} = -\zeta\omega_n \pm \omega_n \sqrt{\zeta^2 - 1}.$$ [5.52]

Depending on the value of the discriminant $\sqrt{\zeta^2 - 1}$, there are three possible families of solutions (see Appendix I):

- $\zeta^2 - 1 > 0$: the roots are real, negative, and distinct. The general form of the solution is

$$y_h(t) = c_1 e^{r_1 t} + c_2 e^{r_2 t}.$$ [5.53]

- $\zeta^2 - 1 = 0$: the roots are real, negative, and equal to $-\omega_n$. The general form of the solution is

$$y_h(t) = c_1 e^{rt} + c_2 t e^{rt}.$$ [5.54]

- $\zeta^2 - 1 < 0$: the roots are complex and distinct. The general form of the solution is

$$y_h(t) = c_1 e^{r_1 t} + c_2 e^{r_2 t} = e^{\lambda t}(c_1 \cos \mu t + c_2 \sin \mu t),$$ [5.55]

using Euler's formula $e^{it} = \cos t + i \sin t$ and noting that

$$r_{1,2} = \lambda \pm i\mu$$ [5.56]

with $\lambda = -\zeta\omega_n$ and $\mu = \omega_n \sqrt{1 - \zeta^2}$.

All three general forms of solutions have exponential terms that are negative in time. Thus, as time increases, all homogeneous solutions tend toward a value of zero. Such solutions often are termed **transient solutions**. When $0 < \zeta < 1$ (when $\sqrt{\zeta^2 - 1} < 0$) the system is called **underdamped**; when $\zeta = 1$ (when $\sqrt{\zeta^2 - 1} = 0$) it is called **critically damped**; when $\zeta > 1$ (when $\sqrt{\zeta^2 - 1} > 0$) it is called **overdamped**. The reasons for these names will be obvious later. Now examine how a second-order system responds to step and sinusoidal inputs.

5.7.1 RESPONSE TO STEP-INPUT FORCING

The responses of a second-order system to a step input having $F(t) = A$ for $t > 0$ with the initial conditions $y(0) = 0$ and $\dot{y}(0) = 0$ are as follows:

- For the underdamped case ($0 < \zeta < 1$)

$$y(t) = KA \left\{ 1 - e^{-\zeta\omega_n t} \left[\frac{1}{\sqrt{1 - \zeta^2}} \sin(\omega_n t \sqrt{1 - \zeta^2} + \phi) \right] \right\},$$ [5.57]

where

$$\phi = \sin^{-1}(\sqrt{1 - \zeta^2}).$$ **[5.58]**

As shown by Equation 5.57, the output initially overshoots the input, lags it in time, and is oscillatory. As time goes on the oscillations damp out and the output approaches and eventually reaches the input value. A special situation arises for the no-damping case when $\zeta = 0$. For this situation the output lags the input and repeatedly overshoots and undershoots it forever.

- For the critically damped case ($\zeta = 1$)

$$y(t) = KA\left[1 - e^{-\omega_n t}(1 + \omega_n t)\right].$$ **[5.59]**

No oscillation is present in the output. Rather, the output slowly and monotonically approaches the input, eventually reaching it.

- For the overdamped case ($\zeta > 1$)

$$y(t) = KA\left\{1 - e^{-\zeta\omega_n t}\left[\cosh(\omega_n t\sqrt{\zeta^2 - 1}) + \frac{\zeta}{\sqrt{\zeta^2 - 1}}\sinh(\omega_n t\sqrt{\zeta^2 - 1})\right]\right\}.$$
[5.60]

The behavior is similar to the $\zeta = 1$ case. Here the larger the value of ζ, the longer time it takes for the output to reach the input signal's value.

Note that in the equations of all three cases the quantity $\zeta\omega_n$ in the exponential terms multiplies the time. Hence, the quantity $1/\zeta\omega_n$ represents the time constant of the system. The larger the value of the time constant, the longer it takes the response to approach steady-state. Further, because the magnitude of the step-input forcing equals KA, the magnitude ratio, $M(t)$, for all three cases is obtained simply by dividing the right-hand sides of Equations 5.57, 5.59, and 5.60 by KA.

Equations 5.57 through 5.60 appear rather intimidating. It is helpful to plot these equations rewritten in terms of their magnitude ratios and examine their form. The system responses to step-input forcing is shown in Figure 5.6 for various values of ζ. The quickest response to steady-state is when $\zeta = 0$ (that is when the time constant $1/\zeta\omega_n$ is minimum). However, such a value of ζ clearly is not optimum for

---

5.7 | **MATLAB SIDEBAR**

The MATLAB M-file secordstep.m plots second-order system response to step-input forcing. It accepts a user-specified value of ζ. This M-file was used to create Figures 5.6 and 5.7. For convenience, the natural frequency is set equal to 1 in the M-file, but can be changed.

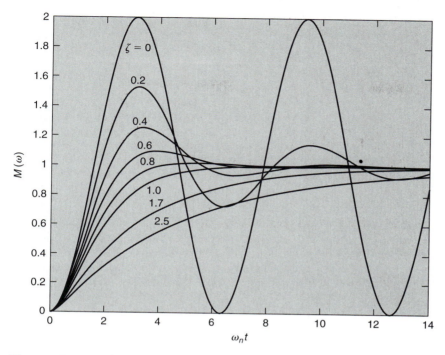

Figure 5.6   The magnitude ratio of a second-order system responding to step-input forcing.

a measurement system because the amplitude ratio overshoots, then undershoots, and continues to oscillate about a value of $M(\omega) = 1$ forever. The oscillatory behavior is known as *ringing* and occurs for all values of $\zeta < 1$.

Shown in Figure 5.7 is the response of a second-order system having a value of $\zeta = 0.2$ to step-input forcing. Note the oscillation in the response about an amplitude ratio of unity. In general, this oscillation is characterized by a period T_d, where $T_d = 2\pi/\omega_d$ with the **ringing frequency** $\omega_d = \omega_n\sqrt{1 - \zeta^2}$. The **rise time** for a second-order system is the time required for the system to initially reach 90 % of its steady-state value. The **settling time** is the time beyond which the response remains within ± 10 % of its steady-state value.

A value of $\zeta = 0.707$ quickly achieves a steady-state response. Most second-order instruments are designed for this value of ζ. When $\zeta = 0.707$ the response overshoot is within 5 % of $M(t) = 1$ within about one-half of the time required for a $\zeta = 1$ system to achieve steady-state. For values of $\zeta > 1$, the system eventually reaches a steady-state value, taking longer times for larger values of ζ.

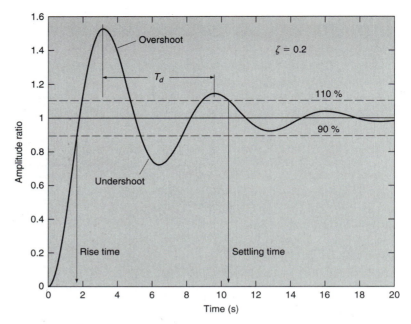

Figure 5.7   The temporal response of a second-order system with $\zeta = 0.2$ to step-input forcing.

5.8 | **MATLAB SIDEBAR**

When examining the response of a system to an input forcing, often one is interested in finding the time it takes for the system finally to reach a steady-state value to within some percentage tolerance. The MATLAB M-file sstol.m accomplishes this task. This M-file uses the MATLAB command break in the conditional loop

```
if abs((x(i)-meanx)>delxtol)
    tsave=t(i+1);
    j=i;
    break
end
```

which causes the program to exit the conditional loop and save the time at which the difference between x and its mean value exceeds a certain tolerance.

Figure 5.8 was produced by sstol.m for the case of a second-order system with a damping coefficient equal to 0.3 in response to step-input forcing. This M-file is constructed to receive a user-specified input data file that consists of two columns, time and amplitude. The percent plus-minus tolerance also is user-specified. The M-file also indicates the time at which the signal reaches and stays within the tolerance limit. The figure label lists values of the steady-state mean amplitude, the time at which the signal stays within the specified tolerance and the tolerance percentage. The time to reach steady-state actually is determined by examining the data file in reverse order. This way, it is simple to determine the tolerance time by finding when the amplitude first exceeds the tolerance level.

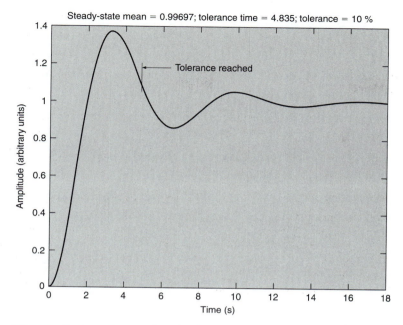

Figure 5.8   Output plot of M-file `sstol.m` used to determine steady-state time.

5.7.2   RESPONSE TO SINUSOIDAL-INPUT FORCING

The response of a second-order system to a sinusoidal input having $F(t) = KA\sin(\omega t)$ with the initial conditions $y(0) = 0$ and $\dot{y}(0) = 0$ is

$$y_p(t) = \frac{KA\sin[\omega t + \phi(\omega)]}{\{[1 - (\omega/\omega_n)^2]^2 + [2\zeta\omega/\omega_n]^2\}^{1/2}},$$ **[5.61]**

where the phase lag in units of radians is

$$\phi(\omega) = -\tan^{-1}\frac{2\zeta\omega/\omega_n}{1 - (\omega/\omega_n)^2} \quad \text{for} \quad \frac{\omega}{\omega_n} \leq 1,$$ **[5.62]**

or

$$\phi(\omega) = -\pi - \tan^{-1}\frac{2\zeta\omega/\omega_n}{1 - (\omega/\omega_n)^2} \quad \text{for} \quad \frac{\omega}{\omega_n} > 1.$$ **[5.63]**

Note that Equation 5.61 is the particular solution, which also is the steady-state solution. This is because the homogeneous solutions for all ζ are transient and tend toward a value of zero as time increases. Hence, the steady-state magnitude ratio based on the input $KA\sin(\omega t)$, Equation 5.61 becomes

$$M(\omega) = \frac{1}{\{[1 - (\omega/\omega_n)^2]^2 + [2\zeta\omega/\omega_n]^2\}^{1/2}}.$$ **[5.64]**

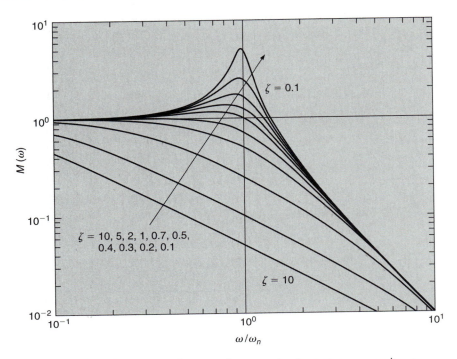

Figure 5.9   The magnitude ratio of a second-order system responding to sinusoidal-input forcing.

These equations show that the system response will contain both magnitude and phase errors. The magnitude and phase responses for different values of ζ are shown in Figures 5.9 and 5.10, respectively. Note that the magnitude ratio is a function of frequency, ω, for the sinusoidal-input forcing case, whereas it is a function of time, t, for the step-input forcing case.

First examine the magnitude response shown in Figure 5.9. For low values of ζ, approximately 0.6 or less, and $\omega/\omega_n \leq 1$, the magnitude ratio exceeds unity. The maximum magnitude ratio occurs at the value of $\omega/\omega_n = \sqrt{1 - 2\zeta^2}$. For $\omega/\omega_n \geq$ ~1.5, the magnitude ratio is less than unity and decreases with increasing values of ω/ω_n.

Typically, magnitude attenuation is given in units of dB/decade or dB/octave. A **decade** is defined as a 10-fold increase in frequency (any 10:1 frequency range). An **octave** is defined as a doubling in frequency (any 2:1 frequency range). For example, using the information in Figure 5.9, there would be an attenuation of approximately -8 dB/octave [$= 20\log(0.2) - 20\log(0.5)$] in the frequency range $1 \leq \omega/\omega_n \leq 2$ when $\zeta = 1$.

Now examine the phase response shown in Figure 5.10. As ω/ω_n increases, the phase response becomes more negative. That is, the output signal begins to lag the

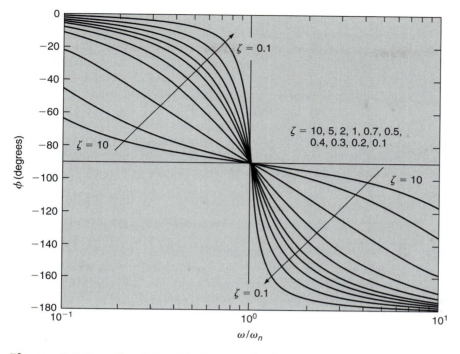

Figure 5.10   The phase shift of a second-order system responding to a sinusoidal-input forcing.

5.9 | MATLAB SIDEBAR

The MATLAB M-files `secordsineM.m` and `secordsinep.m` plot the steady-state second-order system's magnitude ratio and phase lag, respectively, in response to sinusoidal-input forcing. Both M-files accept user-specified values of ζ. Figures 5.9 and 5.10 were made using `secordsineM.m` and `secordsinep.m`, respectively.

---

input signal in time, with this lag time increasing with ω/ω_n. For values of $\omega/\omega_n < 1$, this lag is greater for greater values of ζ. At $\omega/\omega_n = 1$, all second-order systems having any value of ζ have a phase lag of $-90°$ or $1/4$ of a cycle. For $\omega/\omega_n > 1$, the increase in lag is less for systems with greater values of ζ.

5.8 HIGHER-ORDER SYSTEM DYNAMIC RESPONSE

As seen in this chapter, the responses of linear, first- and second-order systems to simple step and sinusoidal inputs are rather complex. Most experiments involve more than one instrument. Thus, the response of most experimental measurement systems will be even more complex than the simple cases examined here.

When each instrument in a measurement system is linear, as described in Chapter 4, the total measurement system response can be calculated easily. For the overall system [a] the static sensitivity is the *product* of all of the static sensitivities, [b] the magnitude ratio is the *product* of all of the magnitude ratios, and [c] the phase shift is the *sum* of all of the phase shifts.

In the end, the most appropriate way to determine the dynamic response characteristics of a measurement system is through a dynamic calibration. This can be accomplished by subjecting the system to a range of either step or sinusoidal inputs of amplitudes and frequencies that span the entire range of those that would be encountered in an actual experiment. With this approach, the system's dynamic errors can be quantified accurately.

Example 5.5

STATEMENT

A pressure transducer is connected through flexible tubing to a static pressure port on the surface of a cylinder that is mounted inside a wind tunnel. The structure of the flow local to the port is such that the static pressure, $p(t)$, varies as

$$p(t) = 15 \sin 2t,$$

in which t is time. Both the tubing and the pressure transducer behave as second-order systems.

The natural frequencies of the transducer, $\omega_{n,\text{trans}}$, and the tubing, $\omega_{n,\text{tube}}$, are 2000 rad/s and 4 rad/s, respectively. Their damping ratios are $\zeta_{\text{trans}} = 0.7$ and $\zeta_{\text{tube}} = 0.2$, respectively. Find the magnitude attenuation and phase lag of the pressure signal, as determined from the output of the pressure transducer, and then write the expression for this signal.

SOLUTION

Because this measurement system is linear, the system's magnitude ratio, $M_s(\omega)$, is the product of the components' magnitude ratios and the phase lag, $\phi_s(\omega)$, is the sum of the components' phase lags, where ω the circular frequency of the pressure. Thus,

$$M_s(\omega) = M_{\text{tube}}(\omega) \cdot M_{\text{trans}}(\omega)$$

and

$$\phi_s(\omega) = \phi_{\text{tube}}(\omega) + \phi_{\text{trans}}(\omega).$$

Also, $\omega/\omega_{\text{tube}} = 2/4 = 0.5$ and $\omega/\omega_{\text{trans}} = 2/2000 = 0.001$. Application of Equations 5.62 and 5.64, noting $\zeta_{\text{trans}} = 0.7$ and $\zeta_{\text{tube}} = 0.2$, yields $\phi_{\text{tube}} = -21.8°$, $\phi_{\text{trans}} = -0.1°$, $M_{\text{tube}} = 1.86$, and $M_{\text{trans}} = 1.00$. Thus, $\phi_s(2) = -21.8° + -0.1° = -21.9°$ and $M_s(2) = (1.86)(1.00) = 1.86$. The pressure signal, as determined from the output of the transducer, is $p_s(t) = (15)(1.86)\sin[2t - (21.9)(\pi/180)] = 27.9 \sin(2t - 0.38)$. Thus, the magnitude of the pressure signal at the output of the measurement system will appear 186 % greater than the actual pressure signal and be delayed in time by 0.19 s $[(0.38 \text{ rad})/(2 \text{ rad/s})]$.

5.9 NUMERICAL SOLUTION METHODS

Differential equations governing a system's response to input forcing may be nonlinear and not have exact solutions. Fortunately, methods are available to numerically integrate most ordinary differential equations and obtain the system response [1]. The basic solution approach is to reduce any higher-order ordinary differential equations to a system of coupled, first-order ordinary differential equations. Then, the first-order equations are solved using finite-difference methods. For example, the second-order ordinary differential equation

$$\frac{d^2 y(t)}{dt^2} - cy(t) = d \qquad \textbf{[5.65]}$$

can be reduced to two first-order ordinary differential equations by using the substitution $dy/dt = z(t)$, which yields the system of equations

$$\frac{dy(t)}{dt} = z(t)$$

$$\frac{dz(t)}{dt} = cy(t) + d. \qquad \textbf{[5.66]}$$

Two initial conditions are needed to obtain a specific solution.

The numerical solution of a first-order ordinary differential equation can be obtained using various finite-difference methods [2]. The exact differential, $dy(t)/dt = f(y, t)$, is approximated by a finite difference. There are many ways to approximate $dy(t)/dt$. The choice depends on the required accuracy and computation time. A straightforward finite-difference approximation for $dy(t)/dt$ is the forward Euler expression

$$f(y_n, t_n) \approx \frac{y_{n+1} - y_n}{\Delta t}, \qquad \textbf{[5.67]}$$

where n and $n+1$ denote the nth and $n+1$-th points. Equation 5.67 leads directly to

$$y_{n+1} = y_n + \Delta t \; f(y_n, t_n). \qquad \textbf{[5.68]}$$

The expression for $f(y, t)$ is obtained from the governing first-order ordinary differential equation. An initial condition, $y(0)$, also is specified. This permits the value of y_{n+1} to be computed for a fixed Δt from Equation 5.68. This algorithm is applied successively up to the desired final time.

Other methods can be used to determine an expression analogous to Equation 5.67. All these methods are easy to implement. The following more commonly used methods replace $f(y_n, t_n)$ in Equation 5.68 by

$$f(y_{n+1}, t_{n+1}) \qquad \textbf{[5.69]}$$

for the backward Euler method, and

$$\frac{f(y_n, t_n) + f(y_{n+1}, t_{n+1})}{2} \qquad \textbf{[5.70]}$$

for the improved Euler method. The improved Euler method is more accurate than the forward and backward Euler methods. The fourth-order Runge-Kutta method replacement for Equation 5.67 is

$$\frac{k_1 + 2k_2 + 2k_3 + k_4}{6},$$ **[5.71]**

where

$$k_1 = f(y_n, t_n),$$

$$k_2 = f\left(y_n + \frac{\Delta t}{2}k_1, t_n + \frac{\Delta t}{2}\right),$$

$$k_3 = f\left(y_n + \frac{\Delta t}{2}k_2, t_n + \frac{\Delta t}{2}\right), \text{ and}$$

$$k_4 = f(y_n + \Delta t\, k_3, t_n + \Delta t).$$ **[5.72]**

The fourth-order Runge-Kutta method is more accurate than any Euler method. It is the most frequently used method, which is quite sufficient for most numerical integrations [3].

Example 5.6 | **STATEMENT**

A first-order system is described by the equation

$$\frac{dy(t)}{dt} - 2y(t) = F(t) = 0.5 - t,$$

with the initial condition $y(0) = 1$. Solve this differential equation numerically and analytically. Use four numerical methods: [1] forward Euler, [2] backward Euler, [3] improved Euler, and [4] fourth-order Runge-Kutta. Use a step size of 0.05 s. Plot all the results for comparison.

SOLUTION

The MATLAB M-file odeint.m can be used for this purpose. The results are presented in Figure 5.11. The fourth-order Runge-Kutta and improved Euler solutions follow the exact solution closely. The backward Euler method underestimates the response. The forward Euler method overestimates the response.

5.10 | **MATLAB SIDEBAR**

MATLAB can be used to numerically integrate first-order ordinary differential equations. The M-file odeint.m numerically integrates and plots a user-specified first-order ordinary differential equation over a user-specified range of the independent variable. The results of four methods (forward Euler, backward Euler, improved Euler, and fourth-order Runge-Kutta) are plotted and compared with the exact result. MATLAB also has the built-in functions ode23 and ode45 that perform numerical integration using mixed-order Runge-Kutta methods.

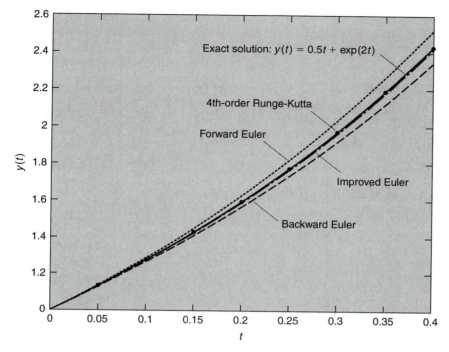

Figure 5.11 The response of the system $dy(t)/dt - 2y(t) = F(t) = 0.5 - t$ to forcing as determined by odeint.m.

REFERENCES

[1] W. E. Boyce and R. C. Di Prima. 1997. *Elementary Differential Equations and Boundary Value Problems*, 6th ed. New York: John Wiley and Sons.

[2] S. Nakamura. 1995. *Numerical Analysis and Graphic Visualization with MATLAB*. New York: Prentice-Hall.

[3] W. H. Press, S. A. Teukolsy, W. T. Vetterling, and B. P. Flannery. 1992. *Numerical Recipes*, 2nd ed. New York: Cambridge University Press.

REVIEW PROBLEMS

1. Does a smaller-diameter thermocouple or a larger-diameter thermocouple have the larger time constant?

2. The dynamic error in a temperature measurement using a thermocouple is 70 % at 3 s after an input-step change in temperature. Determine the magnitude ratio of the thermo-couple's response at 1 s. Report your answer to the nearest hundredth.

3. Determine the % dynamic error of a measurement system that has an output of $3\sin(200t)$ for an input of $4\sin(200t)$.

4. Determine the attenuation (reduction) in units of dB/decade for a measurement system that has an output of $3\sin(200t)$ for an input of $4\sin(200t)$ and an output of $\sin(2000t)$ for an input of $4\sin(2000t)$.

5. Is a strain gage in itself classified as a zero-, first-, second-, or higher-order system?

6. Determine the damping ratio of an RLC circuit with $LC = 1$ that has a magnitude ratio of 8 when subjected to a sine wave input with a frequency of 1 rad/s. Report your answer to the nearest hundredth.

7. Determine the phase lag in degrees for a simple RC filter (one made of only a resistor and a capacitor) with $RC = 5$ s when its input signal has a frequency of $1/\pi$ Hz.

8. A first-order system is subjected to a step input of magnitude B. The time constant in terms of B equals [a] $0.707B$, [b] $0.5B$, [c] $\left(1 - \frac{1}{e}\right)B$, [d] B/e.

9. A second-order system with $\zeta = 0.5$ and $\omega_n = 2$ rad/s is subjected to a step input of magnitude B. The system's time constant equals [a] 0.707 s, [b] 1.0 s, [c] $\left(1 - \frac{1}{e}\right)$ s, [d] not enough information.

10. A second-order system with $\zeta = 0.5$ and $\omega_n = 2$ rad/s is subjected to a sinusoidal input of magnitude $B\sin(4t)$. The phase lag of the output signal in units of degrees is [a] −3, [b] −146, [c] −34, [d] −180.

11. A first-order system is subjected to an input of $B\sin(10t)$. The system's time constant is 1 s. The amplitude of the system's output is approximately [a] $0.707B$, [b] $0.98B$, [c] $\left(1 - \frac{1}{e}\right)B$, [d] $0.1B$.

12. A first-order system is subjected to an input of $B\sin(10t)$. The system's time constant is 1 s. The time lag of the system's output is [a] −0.15 s, [b] −0.632 s, [c] $-\pi$ s, [d] −84.3 s.

HOMEWORK PROBLEMS

1. A first-order system has $M(f = 200$ Hz$) = 0.707$. Determine [a] its time constant (in ms), and [b] its phase shift (in °).

2. A thermometer held in room-temperature air is suddenly immersed into a beaker of cold water. Its temperature as a function of time is recorded. Determine the thermocouple's time constant by performing a least-squares *linear* regression analysis, after transforming the temperatures into their appropriate nondimensional variables. Assume first-order system behavior. The data is listed in Table 5.2.

3. A first-order system with a time constant equal to 10 ms is subjected to a sinusoidal forcing with an input amplitude equal to 8.00 V. When the input forcing frequency equals 100 rad/s, the output amplitude is 5.66 V; when the input forcing frequency equals 1000 rad/s, the output amplitude is 0.80 V. Determine [a] the magnitude ratio for the 100 rad/s forcing case, [b] the roll-off slope (in units of dB/decade) for the $\omega\tau = 1$ to $\omega\tau = 10$ decade, and [c] the phase lag (in °) for the 100 rad/s forcing case.

4. The dynamic error in a temperature measurement using a thermometer is 70 % at 3 s after an input step change in temperature. Determine [a] the magnitude ratio at 3 s, [b] the thermometer's time constant (in s), and [c] the magnitude ratio at 1 s.

5. A thermocouple is immersed in a liquid to monitor its temperature fluctuations. Assume the thermocouple acts as a first-order system. The temperature fluctuations (in °C) vary in time as $T(t) = 50 + 25\cos(4t)$. The output of the thermocouple transducer system (in V) is linearly proportional to temperature and has a static sensitivity of 2 mV/°C. A step-input calibration of the system reveals that its rise time is 4.6 s. Determine the system's [a] time constant (in s), [b] output $E(t)$ (in mV), and [c] time lag (in s) at $\omega = 0.2$ rad/s.

6. A knowledgeable aerospace student selects a pressure transducer (with $\omega_n = 6284$ rad/s and $\zeta = 2.0$) to investigate the pressure fluctuations within a laminar separation bubble on the suction side of an airfoil. Assume that the transducer behaves as an overdamped second-order system. If the experiment requires that the transducer response has $M(\omega) \geq 0.707$ and $|\phi(\omega)| \leq 20°$, determine the maximum frequency (in Hz) that the transducer can follow and accurately meet the two criteria.

7. A strain gage system is mounted on an airplane wing to measure wing oscillations and strain during wind gusts. The system is second order, having a 90 % rise time of 100 ms, a *ringing* frequency of 1200 Hz, and a damping ratio of 0.8. Determine [a] the dynamic error when subjected to a 1 Hz oscillation, and [b] the time lag (in s).

Table 5.2   Thermocouple response data

Time (ms)	Temperature (°C)
0	24.8
40	22.4
120	19.1
200	15.5
240	13.1
400	9.76
520	8.15
800	6.95
970	6.55
1100	6.15
1400	5.75
1800	5.30
2000	5.20
2200	5.00
3000	4.95
4000	4.95
5000	4.95
6000	4.95
7000	4.95

8. In a planned experiment a thermocouple is to be exposed to a step change in temperature. The response characteristics of the thermocouple must be such that the thermocouple's output reaches 98 % of the final temperature within 5 s. Assume that the thermocouple's bead (its sensing element) is spherical with a density equal to 8000 kg/m^3, a specific heat at constant volume equal to 380 J/(kg · K) and a convective heat transfer coefficient equal to 210 W/(m^2· K). Determine the *maximum* diameter that the thermocouple can have and still meet the desired response characteristics.

9. Determine by calculation the damping ratio value of a second-order system that would be required to achieve a magnitude ratio of unity when the sinusoidal input forcing frequency equals the natural frequency of the system.

10. The pressure tap on the surface of a heat exchanger tube is connected via flexible tubing to a pressure transducer. *Both* the tubing and the transducer behave as second-order systems. The natural frequencies are 30 rad/s for the tubing and 6280 rad/s for the transducer. The damping ratios are 0.45 for the tubing and 0.70 for the transducer. Determine the magnitude ratio and the phase lag for the system when subjected to a sinusoidal forcing having a 100-Hz frequency. What, if anything, is the problem with this system for this application?

11. Determine the percent dynamic error in the temperature measured by a thermocouple having a 3-ms time constant when subjected to a temperature that varies sinusoidally in time with a frequency of 531 Hz.

12. The output of an underdamped second-order system with $\zeta = 0.1$ subjected to step-input forcing initially oscillates with a period equal to 1 s until the oscillation dissipates. The same system then is subjected to sinusoidal-input forcing with a frequency equal to 12.62 rad/s. Determine the phase lag (in °) at this frequency.

CHAPTER REVIEW

ACROSS

1. known calibration input
5. when ζ is less than one
8. $\log_{10}(P_2/P_1)$
12. fixed in time
13. a lag in time
14. process of quantifying errors

DOWN

1. time for a system to stay within 10 % of its final value
2. refers to indicating the true value exactly
3. type of equation that governs linear first- and second-order systems
4. when the input is increased systematically
6. refers to indicating a particular value upon repeated measurement
7. a 10-fold increase in frequency
9. variable in time
10. time for a system to reach 90 % of its final value
11. time constant of a first-order *RC* circuit

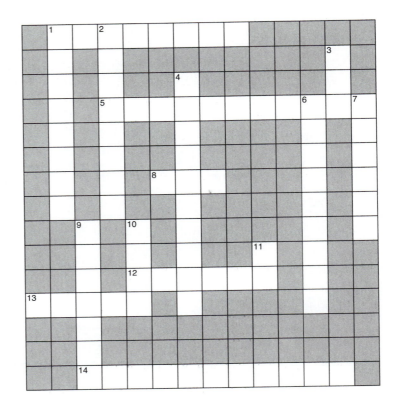

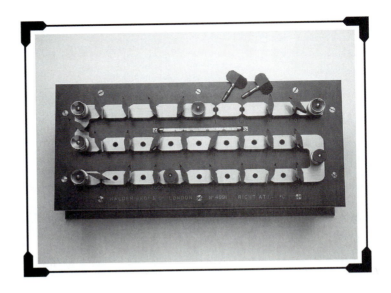

The Cooke and Wheatstone Bridge

Credit: Science Museum / Science & Society Picture Library, London.

This measurement device was introduced to the Royal Society in 1843 in a paper *An Account of Several New Processes for Determining the Constants of a Voltaic Circuit* by Charles Wheatstone (1802–1875), an English physicist and Professor of Experimental Philosophy at King's College, London. In that paper he also introduced a unit of resistance based on a 1-ft length of copper weighing 100 grains. The device, which was known in the 19th century as the Cooke and Wheatstone bridge, produced by the entrepreneur William Cooke and Wheatstone. Samuel Christie actually was the first person to describe the bridge in 1833. Now it is known simply as the Wheatstone bridge. The bridge employs two voltage dividers connected to a voltage source to measure resistance. One divider is made up of one known and one unknown resistor. The other is made up of two more known resistors. Using this configuration, the value of the unknown resistor can be determined.

chapter

6

MEASUREMENT SYSTEMS

Measure what can be measured and make measurable what cannot be measured.

Galileo Galilei, c. 1600.

If you can measure that of which you speak, and can express it by a number, you know something of your subject; but if you cannot measure it, your knowledge is of a meagre and unsatisfactory kind.

Lord Kelvin, 1891

I profess to be a scientific man, and was exceedingly anxious to obtain accurate measurements of her shape; but ... I did not know a word of Hottentot ... Of a sudden my eye fell upon my sextant ... I took a series of observations upon her figure in every direction, up and down, crossways, diagonally, and so forth ... and thus having obtained both base and angles, I worked out the results by trigonometry and logarithms.

Sir Francis Galton,
Narrative of an Explorer in Tropical South Africa, 1853.

CHAPTER OUTLINE

6.1   CHAPTER OVERVIEW

The workhorse of an experiment is its measurement system. This is the equipment used from sensing an experiment's environment to recording the results. This chapter begins by identifying the main elements of a measurement system. The basic electronics behind most of these elements is covered in Chapter 4. The sensor and the transducer, the first two elements of a measurement system, will be examined first. Several sensor/transducer examples will be presented and discussed. Then the essentials of amplifiers will be covered, which include operational amplifiers that are the basic, active elements of all circuit boards today. Finally, filters and contemporary analog-to-digital processing methods will be considered. The chapter is concluded by examining three typical measurement systems.

6.2   LEARNING OBJECTIVES

You should be able to do the following after completing this chapter:

- Know each element and its primary function in a measurement system
- Differentiate between a sensor and a transducer
- Describe how resistive and capacitive sensors each work
- Give an example of a common sensor/transducer system and describe how it works
- Give examples of several sensor/transducers, their characteristics, and the physical principles on which they are based
- Determine the gain of an amplifier
- Describe the major attributes of an operational amplifier
- Calculate the output voltages for the six common operational amplifier configurations
- Describe the various types of filters
- Know the electrical components and configuration for simple RC low-pass and high-pass filters
- Know the definitions of a simple RC filter's time constant, cut-off frequency, magnitude ratio, and phase lag
- Determine the magnitude ratio and phase lag of a simple RC filter given its input signal frequency
- Know the definitions of a decibel, decade, and octave, and determine the magnitude ratio attenuation in dB per decade or octave for a given system
- Describe how a digital filter works
- Describe how the two most common types of analog-to-digital converters work

- Determine the number of possible digital values, resolution, and absolute quantization error for an M-bit A/D converter
- Identify each of the elements of a typical measurement system
- Determine the specific characteristics of a measurement system's elements given its design constraints

6.3   MEASUREMENT SYSTEM ELEMENTS

A measurement system is comprised of the equipment used to sense an experiment's environment, to change and condition what is sensed into a recordable form, and to record its values. Formally, the elements of a measurement system include the sensor, the transducer, the signal conditioner, and the signal processor. These elements, acting in concert, sense the physical variable, provide a response in the form of a signal, condition and process the signal, and store its value.

A measurement system's main purpose is to produce an accurate numerical value of the measurand. Ideally, the recorded value should be the exact value of the physical variable sensed by the measurement system. In practice, the perfect measurement system neither exists nor is needed. A result needs to have only a certain accuracy that is achieved using the most simple equipment and measurement strategy. This can be accomplished provided there is a good understanding of the system's response characteristics.

To accomplish the task of measurement, the system must perform several functions in series. These are illustrated schematically in Figure 6.1. First, the physical variable must be sensed by the system. The variable's stimulus determines a specific state of the sensor's properties. Any detectable physical property of the sensor can serve as the sensor's **signal**. When this signal changes rapidly in time, it is referred to as an **impulse**. So, by definition, the **sensor** is a device that senses a physical stimulus and converts it into a signal. This signal usually is electrical, mechanical, or optical.

For example, as depicted by the words in the shaded box in Figure 6.1, the temperature of a gas (the physical stimulus) results in an electrical resistance (the signal) of a resistance temperature device (RTD — a temperature sensor) that is located in the gas. This is because the resistance of the RTD sensor (typically a fine platinum wire) is proportional to the change in temperature from a reference temperature. Thus, by measuring the RTD's resistance, the local temperature can be determined. In some situations, however, the signal may not be amenable to direct measurement. This requires that the signal be changed into a more appropriate form, which, in almost all circumstances, is electrical. Most of the sensors in our bodies have electrical outputs.

The device that changes (transduces) the signal into the desired quantity (be it electrical, mechanical, optical, or another form) is the **transducer**. In the most general sense, a transducer transforms energy from one form to another. Usually, the transducer's output is an electrical signal, such as a voltage or current. For the RTD example, this would be accomplished by having the RTD's sensor serve as one resistor

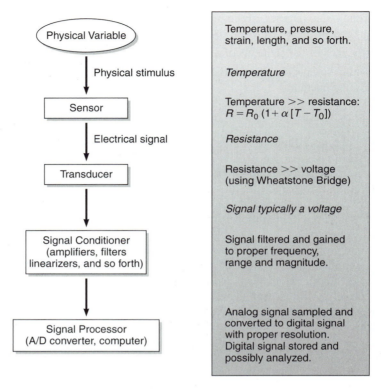

Figure 6.1   The general measurement system configuration.

in an electrical circuit (a Wheatstone bridge) that yields an output voltage proportional to the sensor's resistance. Often, either the word sensor or the word transducer is used to describe the combination of the actual sensor and transducer. A transducer also can change an input into an output providing motion. In this case, the transducer is called an **actuator**. Sometimes, the term transducer is considered to encompass both sensors and actuators [2]. So, it is important to clarify what someone specifically means when referring to a transducer.

Often after the signal has been transduced, its magnitude still may be too small or may contain unwanted electrical noise. In this case, the signal must be conditioned before it can be processed and recorded. In the signal-conditioning stage, an amplifier may be used to increase the signal's amplitude or a filter may be used to remove the electrical noise or some unwanted frequency content in the signal. The **signal conditioner**, in essence, puts the signal in its final form to be processed and recorded.

In most situations, the conditioner's output signal is analog (continuous in time) and the **signal processor** output is digital (discrete in time). So, in the signal processing stage, the signal must be converted from analog to digital. This is accomplished by adding an analog-to-digital (A/D) converter, which usually is contained within the computer that is used to record and store data. That computer also can be used to analyze the resulting data or to pass this information to another computer.

A standard glass-bulb thermometer contains all the elements of a measurement system. The sensor is actually the liquid within the bulb. As the temperature changes, the liquid volume changes, either expanding with an increase in temperature or contracting with a decrease in temperature. The transducer is the bulb of the thermometer. A change in the volume of the liquid inside the bulb leads to a mechanical displacement of the liquid because of the bulb's fixed volume. The stem of the thermometer is a signal conditioner that physically amplifies the liquid's displacement and the scale on the stem is a signal processor that provides a recordable output.

Thus, a measurement system performs many different tasks. It senses the physical variable, transforms it into a signal, transduces and conditions the signal, and then records and stores a corresponding numerical value. How each of these elements function is considered in the following.

6.4 SENSORS AND TRANSDUCERS

A sensor senses the process variable through its contact with the physical environment, and the transducer transduces the sensed information into a different form, yielding a detectable output. Contact need not be physical. The sensor, for example, could be an optical pyrometer located outside of the environment under investigation. This is a **noninvasive** sensor. An **invasive** or *in situ* sensor is located within the environment. Ideally, invasive sensors should not disturb the environment.

Usually the signals between the sensor and transducer and the detectable output are electrical, mechanical, or optical. Electrically based sensors and transducers can be **active** or **passive**. Passive elements require no external power supply. Active elements require an external power supply to produce a voltage or current output. Mechanically based sensors and transducers usually use a secondary sensing element that provides an electrical output. Often the sensor and transducer are combined physically into one device.

Sensors and transducers can be found everywhere. The sensor/transducer system in a house thermostat basically consists of a metallic coil (the sensor) with a small glass capsule (the transducer) fixed to its top end. Inside the capsule is a small amount of mercury and two electrical contacts (one at the bottom of the capsule and one at its top). When the thermostat's set temperature equals the desired room temperature, the mercury is at the bottom of the capsule such that no connection is made via the electrically conducting mercury and the two contacts. The furnace and its blower are off. As the room temperature decreases, the metallic coil contracts, thereby tilting the capsule and causing the mercury to close the connection between the two contacts. The capsule, in effect, transduces the length change in the coil into a digital (on/off) signal.

Another type of sensor/transducer system is in a telephone mouthpiece. This consists of a diaphragm with coils housed inside a small magnet. There is a separate system for both the mouth piece and the ear piece. The diaphragm is the sensor and its coils within the magnet's field are the transducer. Talking into the mouth piece generates pressure waves that cause the diaphragm with its coils to move within the magnetic field. This induces a current in the coil, which is transmitted (after some modification) to another telephone. When the current arrives at the ear piece, it flows through the coils of the ear piece's diaphragm inside the magnetic field and causes the diaphragm to move. This sets up pressure waves that strike a person's eardrum as sound. Newer phones use piezosensor/transducers that generate an electric current when subjected to pressure waves and alternatively pressure waves from an applied current. Today, most signals are digitally encoded for transmission either in optical pulses through fibers or in electromagnetic waves to and from satellites. Yet even with this new technology, the sensor still is a surface that moves and the transducer still converts this movement into an electrical current.

6.4.1 SENSOR PRINCIPLES

Sensors are available today that sense almost anything imaginable. New ones are being developed constantly. Sensors can be categorized into **domains**, according to the type of physical variables that they sense [1], [2]. These domains and the sensed variables are

- Chemical: chemical concentration, composition, and reaction rate;
- Electrical: current, voltage, resistance, capacitance, inductance, and charge;
- Magnetic: magnetic field intensity, flux density, and magnetization;
- Mechanical: displacement or strain, level, position, velocity, acceleration, force, torque, pressure, and flow rate;
- Radiant: electromagnetic wave intensity, wavelength, polarization, and phase; and
- Thermal: temperature, heat, and heat flux.

The first step in understanding sensor functioning is to gain a sound knowledge about the basic principles behind sensor design and operation. This is especially true today because sensor designs change almost daily. Once these basic principles are understood, then any standard measurement textbook, for example [3], [4], [5], [6], and [7], can be consulted to obtain descriptions of innumerable devices based on these principles. For the most current information, many sensor and transducer manufacturers now provide information via the Internet that describes their product's mode of operation and specific operational characteristics.

Sensors always are based on some physical principle or law [8]. These principles and laws are presented in Appendix E. The choice of either designing or selecting a particular sensor starts with identifying the physical variable to be sensed and the physical principle or law associated with that variable. Then, the sensor's input/output characteristics must be identified. These include, but are not limited to, the sensor's

- Operational bandwidth;
- Magnitude and frequency response over that bandwidth;

- Sensitivity;
- Accuracy;
- Voltage or current supply requirements;
- Physical dimensions, weight, and materials;
- Environmental operating conditions (pressure, temperature, relative humidity, air purity, radiation);
- Type of output (electrical, mechanical);
- Further signal conditioning requirements;
- Operational complexity; and
- Cost.

The final choice of sensor can involve some or all of these considerations. Example 6.1 illustrates how the design of a sensor can be a process that often involves reconsideration of the design constraints before arriving at the final design.

STATEMENT

Example 6.1

A design engineer intends to scale down a pressure sensor to fit inside an ultraminiature robotic device. The pressure sensor consists of a circular diaphragm that is instrumented with a strain gage. The diaphragm is deflected by a pressure difference that is sensed by the gage and transduced by a Wheatstone bridge. The diaphragm of the full-scale device has a 1-cm radius, is 1-mm thick, and is made of stainless steel. The designer plans to make the miniature diaphragm out of silicon. The miniature diaphragm is to have a 600-μm radius, operate over the same pressure difference range, and have the same deflection. The diaphragm deflection, δ, at its center is

$$\delta = \frac{3(1 - v^2)r^4 \Delta p}{16Eh},$$

in which v is Poisson's ratio, E is Young's modulus, r is the diaphragm radius, h is the diaphragm thickness, and Δp is the pressure difference. Determine the required diaphragm thickness to meet these criteria and comment on the feasibility of the new design.

SOLUTION

Assuming that Δp remains the same, the new thickness is

$$h_n = h_o \left[\frac{(1 - v_n^2)r_n^4 E_o}{(1 - v_o^2)r_o^4 E_n} \right].$$

The properties for stainless steel are $v_o = 0.29$ and $E_o = 203$ GPa. Those for silicon are $v_n = 0.25$ and $E_n = 190$ GPa. Substitution of these and the aforementioned values into the expression yields $h_n = 1.41 \times 10^{-8}$ m $= 14$ nm. This thickness is too small to be practical. An increase in h_n by a factor of 10 will increase the Δp range likewise. Recall that this design required a similar deflection. A new design would be feasible if the required deflection for the same transducer output could be reduced by a factor of 1000, such as by the use of a piezoresistor on the surface of the diaphragm. This would increase h_n to 14 μm, which is reasonable using current microfabrication techniques. Almost all designs are based on many factors, which usually require compromises to be made.

6.4.2 EXAMPLES OF SENSORS

Sensor/transducers can be developed for different measurands and be based on the same physical principle or law. Likewise, sensor/transducers can be developed for the same measurand and be based on different physical principles or laws. A thin wire sensor's resistance inherently changes with strain. This wire can be mounted on various structures and used with a Wheatstone bridge to measure strain, force, pressure, or acceleration. A thin wire's resistance also inherently changes with temperature. This, as well as other sensors, such as a thermocouple, a thermistor, and a constant-current anemometer, can be used to measure temperature. Table 6.1 lists a number of sensor/transducers, their measurands, and other characteristics. Each type of sensor listed is considered next.

Fine Wire or Strain Gage Sensor   A sensor based on the principle that a change in resistance can be produced by a change in a physical variable is, perhaps, the most common type of sensor. A resistance sensor can be used to measure displacement, strain, force, pressure, acceleration, flow velocity, temperature, and heat or light flux.

One simple sensor of this type is a pure metal wire or strip whose resistance changes with temperature. The resistance of a *resistance temperature device* (RTD) is related to temperature by

$$R = R_o[1 + \alpha(T - T_o) + \beta(T - T_o)^2 + \gamma(T - T_o)^3 + \cdots], \qquad \textbf{[6.1]}$$

Table 6.1   Example sensors (L, length; R, resistance; δ, deflection; C, capacitance; U, velocity; T, temperature; f_D, Doppler frequency difference; V, voltage; RH, relative humidity; ϵ, dielectric constant.)

Measurand	Sensor	Transducer	Domain	Principle
strain	fine wire or strain gage	none	mechanical	$\Delta L \to \Delta R$
force	strain gage on structure	Wheatstone bridge	mechanical	$\delta \to \Delta L \to \Delta R$
pressure	strain gage on structure	Wheatstone bridge	mechanical	$\delta \to \Delta L \to \Delta R$
acceleration	strain gage on structure	Wheatstone bridge	mechanical	$\delta \to \Delta L \to \Delta R$
acceleration	capacitance sensor on structure	Wheatstone bridge	mechanical	$\delta \to \Delta L \to \Delta C$
velocity	fine wire	constant-T anemometer	mechanical	$U \to \Delta T \to \Delta R$
velocity	microparticles in laser beams	photodetector	mechanical	microparticle $U \to \Delta f_D$
temperature	dissimilar-wire junction	reference junction	thermal	$\Delta T \to \Delta V$
temperature	fine wire	Wheatstone bridge	thermal	$\Delta T \to \Delta R$
relative humidity	capacitance sensor	Wheatstone bridge	electrical	$\Delta RH \to \Delta \epsilon \to \Delta C$

where α, β, and γ are coefficients of thermal expansion and R_o is the resistance at the reference temperature T_o. A wire with a diameter on the order of 25 µm can be used to measure local velocity in a fluid flow. The fine wire is connected to a Wheatstone bridge-feedback amplifier circuit that is used to maintain the wire at a constant resistance, hence at a constant temperature above the fluid's temperature. As the wire is exposed to different velocities, the power required to maintain the wire at the constant temperature changes because of the changing heat transfer to the environment. The power is proportional to the square root of the fluid velocity. This system is called a *hot-wire anemometer*. Examples involving the hot-wire anemometer are presented in Chapters 5 and 9.

If a semiconductor is used instead of a conductor, a greater change in resistance with temperature can be achieved. This is a *thermistor*. Its resistance changes exponentially with temperature as

$$R = R_o \exp\left[\eta\left(\frac{1}{T} - \frac{1}{T_o}\right)\right], \qquad \textbf{[6.2]}$$

where η is a material constant. Thus, a thermistor usually gives better resolution over a small temperature range, whereas the RTD covers a wider temperature range. For both sensors, a transducer such as a Wheatstone bridge circuit typically is used to convert resistance to voltage.

The *strain gage* is the most frequently used resistive sensor. A typical strain gage is shown in Figure 6.2. The gage consists of a very fine wire of length L. When the wire is stretched, its length increases by ΔL, yielding a longitudinal strain of $\epsilon_L \equiv \Delta L / L$. This produces a change in resistance. Its width decreases by $\Delta d / d$, where d is the wire diameter. This defines the transverse strain $\epsilon_T \equiv \Delta d / d$. Poisson's ratio, ν, is defined as the negative of the ratio of transverse to longitudinal *local* strains, $-\epsilon_T / \epsilon_L$. The

Backing

Encapsulation

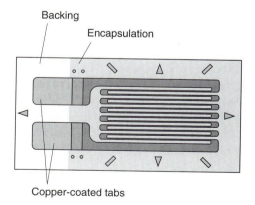

Copper-coated tabs

Figure 6.2 A strain gage with a typical sensing area of 5 mm × 10 mm.

negative sign compensates for the *decrease* in transverse strain that accompanies an *increase* in longitudinal strain, thereby yielding positive values for v. Poisson's ratio is a material property that couples these strains.

For a wire, the resistance R can be written as

$$R = \rho \frac{L}{A},$$ [6.3]

where ρ is the resistivity, L is the length, and A is the cross-sectional area. Taking the total derivative of Equation 6.3 yields

$$dR = \frac{\rho}{A} dL + \frac{L}{A} d\rho - \frac{\rho L}{A^2} dA.$$ [6.4]

Equation 6.4 can be divided by Equation 6.3 to give the relative change in resistance,

$$\frac{dR}{R} = (1 + 2v)\epsilon_L + \frac{d\rho}{\rho}.$$ [6.5]

Equation 6.5 shows that the relative resistance change in a wire depends on the strain of the wire and the resistivity change.

A **local gage factor**, G_l, can be defined as the ratio of the relative resistance change to the relative length change,

$$G_l = \frac{dR/R}{dL/L}.$$ [6.6]

This expression relates *differential* changes in resistance and length and describes a factor that is valid only over a very small range of strain.

An **engineering gage factor**, G_e, also can be defined as

$$G_e = \frac{\Delta R/R}{\Delta L/L}.$$ [6.7]

This expression is based on small, finite changes in resistance and length. This gage factor is the slope based on the total resistance change throughout the region of strain investigated. The local gage factor is the instantaneous slope of a plot of $\Delta R/R$ versus $\Delta L/L$. Because it is very difficult to measure local changes in length and resistance, the engineering gage factor typically is used more frequently. Equation 6.5 can be rewritten in terms of the engineering gage factor as

$$G_e = 1 + 2v + \left(\frac{\Delta \rho}{\rho} \cdot \frac{1}{\epsilon_L} \right).$$ [6.8]

For most metals, $v \approx 0.3$. The last term in brackets represents the strain-induced changes in the resistivity, which is a piezoresistive effect. This term is constant for typical strain gages and equals approximately 0.4. Thus, the value of the engineering gage factor is approximately 2 for most metallic strain gages.

An alternative expression for the relative change in resistance can be derived using statistical mechanics, in which

$$\frac{dR}{R} = 2\epsilon_L + \frac{dv_0}{v_0} - \frac{d\lambda}{\lambda} - \frac{dN_0}{N_0}. \qquad \textbf{[6.9]}$$

Here v_0 is the average number of electrons in the material in motion between ions, λ is the average distance traveled by an electron between collisions, and N_0 is the total number of conduction electrons. Equation 6.9 implies that the differential resistance change, and, thus, the gage factor, is independent of the material properties of the conductor. This also implies that the change in resistance only will be proportional to the strain when the sum of the changes on the right-hand side of Equation 6.9 is either zero or directly proportional to the strain. Fortunately, most strain gage materials have this behavior. So, when a strain gage is used in a circuit such as a Wheatstone bridge, strain can be converted into a voltage. This system can be used as a *displacement sensor*.

Strain gages also can be mounted on a number of different flexures to yield various types of sensor systems. One example is four strain gages mounted on a beam to determine its deflection, as described in Chapter 4. As force is applied to the beam, it deflects, producing a strain. This strain is converted into a change in the resistance of a strain gage mounted on the beam. This is called a *force transducer*. Another example involves one or more strain gages mounted on the surface of a diaphragm that separates two chambers exposed to different pressures. As the diaphragm is deflected because of a pressure difference between the two chambers, a strain is produced. The resultant resistance usually is converted into a voltage using a Wheatstone bridge. This system is called a *pressure transducer*, although it actually contains both a sensor (the strain gage) and a transducer (the Wheatstone bridge). A schematic of a miniature, integrated-silicon pressure sensor is shown in Figure 6.3. The calibration and use of this type of pressure sensor in a model rocket's onboard measurement system is presented in Section 6.8.

An *accelerometer* uses a strain gage flexure arrangement. An accelerometer in the 1970s typically contained a small mass that was moved against a spring as the device containing them was accelerated. The displacement of the mass was calibrated against a known force. This information then was used to determine the acceleration from the

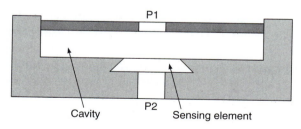

Figure 6.3   Schematic of an integrated-silicon pressure sensor.

displacement using Newton's second law. Newer designs then used strain gages or piezoelectric transducers instead of a spring, although the size did not change much. Now microaccelerometers are available [10]. These contain a very small mass, attached to a silicon cantilever beam that is instrumented with a piezoresistor. As the device is accelerated, the beam deflects, the piezoresistor is deformed and its resistance changes. The piezoresistor is incorporated into an onboard Wheatstone bridge circuit providing a voltage output that is linearly proportional to acceleration. The entire microaccelerometer and associated circuitry is several millimeters in dimension. The calibration and use of this type of accelerometer in a model rocket's onboard measurement system is presented in Section 6.8.

Capacitive Sensor   A *capacitive sensor* consists of two small conducting plates, each of area, A, separated by a distance, d, with a dielectric material in between. The capacitance between the two plates is

$$C = \frac{\epsilon_o \epsilon A}{d},$$ [6.10]

where ϵ_o is the permittivity of free space and ϵ is the relative permittivity. When used, for example, to measure pressure, the dielectric is air and one plate is held fixed. As the other plate moves because of the forces acting on it, the capacitance of the sensor changes. The change in capacitance is proportional to the difference in pressure from the reference pressure measured at zero plate deflection. When used in a capacitive Wheatstone bridge circuit, the pressure difference is converted into a voltage. This system forms a capacitive *pressure transducer*. A central plate fixed to a small mass can be used instead of air between the two capacitor plates. As the mass and its attached central plate are accelerated, the change in capacitance with respect to time is sensed. This is converted into a voltage that is proportional to the acceleration. This system constitutes an *accelerometer*. Another use of this type of sensor is to expose the dielectric material to moist air while keeping the distance between the two plates fixed. The permittivity of the dielectric material changes with relative humidity, which leads to a change in the sensor's capacitance. This type of sensor can be used as a *relative humidity sensor*.

Optically Based Sensor   An optically based measurement system can be designed to measure noninvasively the velocity and velocity fluctuations of a transparent fluid over the velocity range from ∼1 cm/s to ∼500 m/s with ∼1 % accuracy. This system is known as the *laser Doppler velocimeter* (LDV) and operates on the principle of the Doppler effect [9]. A coherent beam of laser light of a given frequency is directed into the moving fluid containing microparticles (∼1 μm diameter), which ideally follow the flow. Because these microparticles are moving with respect to the beam, the frequency of light as received by the microparticles is Doppler shifted. A photodetector in the same reference frame as the laser receives the light that is scattered from the microparticles. This scattered light is frequency shifted once again at the receiver. The frequency of the scattered light, however, is too high to be detected using conventional detectors.

This limitation can be overcome by using two beams of equal frequency, intensity, and diameter, and crossing them inside the flow. This produces an ellipsoidal measurement volume with submillimeter dimensions. This method is called the dual-beam or Doppler-frequency-difference method. The crossed beams produce an ellipsoidal measurement volume, on the order of 0.5 to 1 mm in length and 0.1 to 0.3 mm in diameter. The velocity component, U, of the flow perpendicular to the bisector of the incident beams separated by an angle, θ, is related to the Doppler frequency difference, f_D, by

$$U = \frac{\lambda f_D}{2 \sin(\theta/2)},$$ **[6.11]**

where λ is the wavelength of the incident laser light. The Doppler frequency difference is the difference between the frequencies of the two scattered light beams, as received in the laboratory reference frame. Further modifications can be made by adding other beams of different frequencies in different Cartesian coordinate directions to yield all three components of the velocity. Also, frequency shifting usually is employed to compensate for insensitivity of Equation 6.11 to flow direction. If two additional, equal-spaced detectors are added, the phase lag between the signals of the three detectors is related to the diameter of the microparticle passing through the measurement volume. This system is called a *phase-Doppler anemometer*.

6.4.3   SENSOR SCALING

Sensors have evolved considerably since the beginning of scientific instruments [12]. Marked changes have occurred in the last 300 years. The temperature sensor serves as a good example. Daniel Gabriel Fahrenheit (1686–1736) produced the first mercury-in-glass thermometer in 1714 with a calibrated scale based on the freezing point of a certain ice/salt mixture, the freezing point of water, and body temperature. This device was accurate to within several degrees and was approximately the length scale of 10 cm. In 1821, Thomas Johann Seebeck (1770–1831) found that by joining two dissimilar metals at both ends to form a circuit, with each of the two junctions held at a different temperature, a magnetic field was present around the circuit. This eventually led to the development of the thermocouple. Until very recently, the typical thermocouple circuit consisted of two dissimilar metals joined at each end, with one junction held at a fixed temperature (usually the freezing point of distilled water contained within a thermally insulated flask) and the other at the unknown temperature. A potentiometer was used to measure the mV-level emf. Presently, because of the advance in microcircuit design, the entire reference temperature junction is replaced by an electronic one and contained with an amplifier and linearizer on one small chip. Such chips even are being integrated with other microelectronics and thermocouples such that they can be located in a remote environment and have the temperature signal transmitted digitally with very low noise to a receiving station. The simple temperature sensor has come a long way since 1700.

Sensor development has advanced rapidly since 1990 because of MEMS (microelectromechanical system) sensors technology [2]. The basic nature of sensors has not changed, although their size and applications have changed. Sensors, however, simply cannot be scaled down in size and still operate effectively. Scaling laws for microdevices, such as those proposed by W. S. N. Trimmer in 1987, must be followed in their design [10]. As sensor sizes are reduced to millimeter and micrometer dimensions, their sensitivities to physical parameters can change. This is because some effects scale with the sensor's physical dimension. For example, the surface-to-volume ratio of a transducer with a characteristic dimension, L, scales as L^{-1}. So, surface area-active microsensors become more advantageous to use as their size is decreased. On the other hand, the power loss-to-onboard power scales as L^{-2}. So, as an actuator that carries its own power supply becomes smaller, power losses dominate and the actuator becomes ineffective. Further, as sensors are made with smaller and smaller amounts of material, the properties of the material may not be isotropic. A sensor having an output that is related to its property values may be less accurate as its size is reduced. For example, the temperature determined from the change in resistance of a miniature resistive element is related to the coefficients of thermal expansion of the material. If property values change with size reduction, further error will be introduced if macroscale coefficient values are used.

The scaling of most sensor design variables with length are summarized in Table 6.2. This can be used to examine the scaling of some conventional sensors. Consider the laminar flow element, which is used to determine a liquid flow rate. The element basically consists of many parallel tubes through which the bulk flow is subdivided to achieve laminar flow through each tube. The flow rate, Q, is related to the pressure difference, Δp, measured between two stations separated by a distance, L, as

$$Q = C_o \frac{\pi D^4 \Delta p}{128 \mu L}, \qquad \text{[6.12]}$$

where D is the internal diameter of the pipe containing the flow tubes, μ is the absolute viscosity of the fluid, and C_o is the flow coefficient of the element. What happens if this device is reduced in size by a factor of 10 in both length and diameter? According to Equation 6.12, assuming C_o is constant, for the same Q, a Δp 1000 times greater is required! Likewise, to maintain the same Δp, Q must be reduced by a factor of 1000. The latter is most likely the case. Thus, a MEMS-scale laminar flow element can handle only much smaller flow rates as compared to the conventional laminar flow element.

Example 6.2

STATEMENT

Equation 6.12 is valid for a single tube when $C_o = 1$, where it reduces to the Hagen-Poiseuille law. How does the pressure gradient scale with a reduction in the tube's diameter if the same velocity is maintained?

Table 6.2   Variable scaling with length.

Variable	Equivalent	Dimensions	Scaling
displacement	distance	L	L
strain	length change/length	$\Delta L/L$	L^0
strain rate or shear rate	strain change/time	$L^0 T^{-1}$	L^0
velocity	distance/time	LT^{-1}	L
surface	width × length	L^2	L^2
volume	width × length × height	L^3	L^3
force	mass × acceleration	$L^3 L T^{-2}$	L^4
line force	force/length	$L^3 T^{-2}$	L^3
surface force	force/area	$L^3 L^{-1} T$	L^2
body force	force/volume	$L^3 L^{-2} T$	L
work, energy	force × distance	$L^3 L^2 T^{-2}$	L^5
power	energy/time	$L^3 L^2 T^{-3}$	L^5
power density	power/volume	$L^3 L^{-1} T^{-3}$	L^2
electric current	charge/time	QT^{-1}	L^0
electric resistance	resistivity × length/cross-sectional area	ρL^{-1}	L^{-1}
electric field potential	voltage	V	L^0
electric field strength	voltage/length	VL^{-1}	L^{-1}
electric field energy	permittivity × electric field strength ^{2}	$V^2 L^{-2}$	L^{-2}
resistive power loss	voltage^2/resistance	$V^2 L$	L
electric capacitance	permittivity × plate area/plate spacing	$L^2 L^{-1}$	L
electric inductance	voltage/change of current in time	$VT^2 Q^{-1}$	L^0
electric potential energy	capacitance × voltage^2	LV^2	L
electrostatic potential energy	capacitance × voltage^2 with $V \sim L$	LV^2	L^3
electrostatic force	electrostatic potential energy change/distance	$L^3 L^{-1}$	L^2
electromagnetic force	electromagnetic potential energy change/distance	$L^2 L^2$	L^4
flow rate	velocity × cross-sectional area	LL^2	L^3
pressure gradient	surface force/length	$L^2 L^{-1}$	L

SOLUTION

The velocity, U, is the flow rate divided by the tube's cross-sectional area, $U = 4Q/(\pi d^2)$, where d is the tube diameter. Thus, Equation 6.12 can be written $\Delta p/L = 32\mu U d^2$. This implies that the pressure gradient increases by a factor of 100 as the tube diameter is reduced by a factor of 10. Clearly, this presents a problem in sensors using microcapillaries under these conditions. This situation necessitates the development of other means to move liquids in microscale sensors, such as piezoelectric and electrophoretic methods.

Decisions on the choice of a microsensor or microactuator are not based exclusively on length-scaling arguments. Other factors may be more appropriate. This is illustrated by Example 6.3.

Example 6.3	**STATEMENT**

STATEMENT

Most conventional actuators use electromagnetic forces. Are either electromagnetic or electrostatic actuators better for microactuators based on force-scaling arguments?

SOLUTION

Using Table 6.2, the electrostatic force scales as L^2 and the electromagnetic force as L^4. So, a reduction in L by a factor of 100 leads to a reduction in the electrostatic force by a factor of 1×10^4 and in the electromagnetic force by a factor of 1×10^8! If these forces are comparable at the conventional scale, then the electrostatic force is 10 000 times larger than the electromagnetic force at this reduced scale.

The final choice of which type of micro-actuator to use, however, may be based on other considerations. For example, Madou [11] argues that energy density also could be the factor on which to scale. Energy densities several orders of magnitude higher can be achieved using electromagnetics as compared to electrostatics, primarily because of limitations in electrostatic energy density. This could yield higher forces using electromagnetics as compared to electrostatics for comparable microvolumes.

6.5   AMPLIFIERS

An amplifier is an electronic component that scales the magnitude of an input analog signal, $E_i(t)$, producing an output analog signal, $E_o(t)$. In general, $E_o(t) = f\{E_i(t)\}$. For a linear amplifier $f\{E_i(t)\} = GE_i(t)$; for a logarithmic amplifier $f\{E_i(t)\} = G \cdot \log_x [E_i(t)]$, where G is the gain of the amplifier. Amplifiers are often used to increase the output signal of a transducer to a level that utilizes the full-scale range of an A/D converter located between the transducer and the board. This minimizes errors that arise when converting a signal from analog to digital format.

The **common-mode rejection ratio** (CMRR) is another characteristic of amplifiers. It is defined as

$$CMRR = 20 \log_{10} \frac{G_d}{G_c}, \qquad [6.13]$$

in which G_d is the gain when different voltages are applied across the amplifier's positive and negative input terminals, and G_c when the same voltages are applied. Ideally, when two signals of the same voltage containing similar levels of noise are applied to the inputs of an amplifier, its output should be zero. Realistically, however, the amplifier's output for this case is not zero, but rather is some finite value. This implies that the amplifier effectively has gained the signal difference by a factor of G_c, when, ideally, it should have been zero. Thus, the lower G_c is, the better it is. Typically, CMRR values greater than 100 are considered high and desirable for most applications.

Today, almost all amplifiers used in common measurement systems are operational amplifiers. An *op amp* actually is comprised of many transistors, resistors, and capacitors, presented in the form of an integrated circuit. For example, the LM124 series op amp, whose schematic diagram is shown in Figure 6.4, consists of 13 transistors, 2 resistors, 1 capacitor, and 4 current sources.

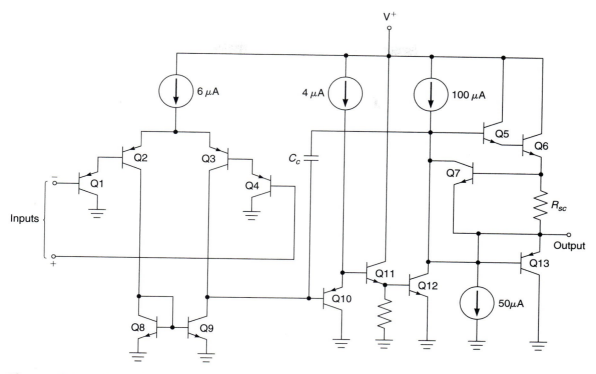

Figure 6.4 Internal layout of a low-cost field-effect-transistor (FET) operational amplifier (National Semiconductor Corporation LM124 series).

When used in an open-loop configuration, as shown in Figure 6.5, the output is not connected *externally* to the input. It is, of course, connected through the internal components of the op amp. For the open-loop configuration, $E_o(t) = A[E_{i2}(t) - E_{i1}(t) - V_o]$, where V_o is the op amp's offset voltage, which typically is zero; E_{i1} is called the *inverting* input and E_{i2} the *noninverting* input. Because A is so large, this configuration is used primarily in situations to measure very small differences between the two inputs, when $E_{i2}(t) \cong E_{i1}(t)$.

The op amp's major attributes are as follows:

- Very high input impedance ($> 10^7 \ \Omega$)
- Very low output impedance ($< 100 \ \Omega$)
- High internal open-loop gain ($\sim 10^5$ to 10^6)

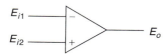

Figure 6.5 An operational amplifier in an open-loop configuration.

These attributes make the op amp an ideal amplifier. Because the input impedance is very high, very little current is drawn from the input circuits. Also, negligible current flows between the inputs. The high internal open-loop gain ensures that the voltage difference between the inputs is zero. The very low output impedance implies that the output voltage is independent of the output current.

When used in the closed-loop configuration, as depicted in Figure 6.6, the output is connected externally to the input. That is, a feedback loop is established between the output and the input. The exact relation between $E_o(t)$ and $E_{il}(t)$ and $E_{i2}(t)$ depends on the specific feedback configuration.

Op amps typically can be treated as black boxes when incorporating them into a measurement system. Many circuit design handbooks provide equations relating an op amp's output to its input for a specified task. This can be a simple task such as inverting and gaining the input signal (the inverting configuration), not inverting but gaining the input signal (the noninverting configuration), or simply passing the signal through it with unity gain (the voltage-follower configuration). An op amp used in the voltage-follower configuration serves as an impedance converter. When connected to the output of a device, the op amp effectively provides a very low output impedance to the device–op amp system. This approach minimizes the loading errors introduced by impedance mismatching that are described in Chapter 4. Op amps also can be used to add or subtract two inputs or to integrate or differentiate an input with respect to time, as well as many more complex tasks. The six most common op amp configurations and their input-output relations are presented in Figure 6.7.

Example 6.4

STATEMENT

Derive the expression given for the input-output relation of the differential amplifier shown in Figure 6.7.

SOLUTION

Let node A denote that which connects R_1 and R_2 at the op amp's positive input and node B that which connects R_1 and R_2 at the op amp's negative input. Essentially no current passes through the op amp because of its very high input impedance. Application of Kirchhoff's

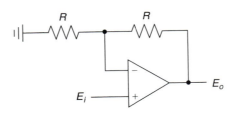

Figure 6.6   An operational amplifier in a closed-loop configuration.

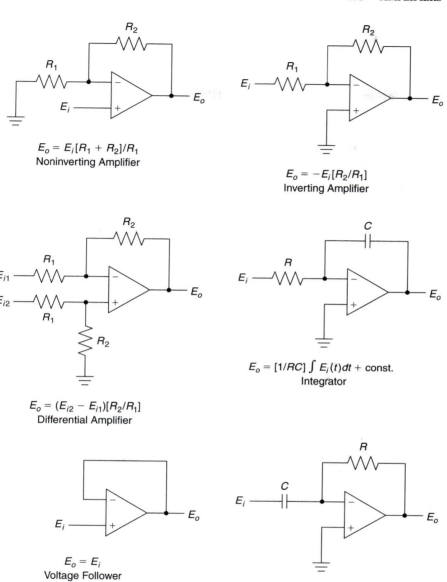

$E_o = E_i[R_1 + R_2]/R_1$
Noninverting Amplifier

$E_o = -E_i[R_2/R_1]$
Inverting Amplifier

$E_o = (E_{i2} - E_{i1})[R_2/R_1]$
Differential Amplifier

$E_o = [1/RC] \int E_i(t)dt + \text{const.}$
Integrator

$E_o = E_i$
Voltage Follower

$E_o = -RC[dE_i(t)/dt]$
Differentiator

Figure 6.7   Other operational-amplifier configurations.

first law at node A gives

$$\frac{E_{i2} - E_A}{R_1} = \frac{E_A - 0}{R_2}.$$

This implies

$$E_A = \left[\frac{R_2}{R_1 + R_2}\right]E_{i2}.$$

Application of Kirchhoff's first law at node B yields

$$\frac{E_{i1} - E_B}{R_1} = \frac{E_B - E_o}{R_2}.$$

This gives

$$E_B = \left[\frac{R_1 R_2}{R_1 + R_2}\right]\left[\frac{E_{i1}}{R_1} + \frac{E_o}{R_2}\right].$$

Now $E_A = E_B$ because of the op amp's high internal open-loop gain. Equating the expressions for E_A and E_B gives the desired result, $E_o = (E_{i2} - E_{i1})(R_2/R_1)$.

In fact, op amps are the foundations of many signal-conditioning circuits. One example is the use of an op amp in a simple sample-and-hold circuit, as shown in Figure 6.8. In this circuit, the output of the op amp is held at a constant value ($= G \cdot E_i$) for a period of time (usually several μs) after the normally closed (NC) switch is held open using a computer's logic control. Sample-and-hold circuits are common features of A/D converters, which are covered later in this chapter. They provide the capability to simultaneously acquire the values of several signals. These values are then held by the circuit for a sufficient period of time until all of them can be stored in the computer's memory.

Quite often in measurement systems, a differential signal, such as that across the output terminals of a Wheatstone bridge, has a small (on the order of tens of millivolts), DC-biased (on the order of volts) voltage. When this is the case, it is best to use an *instrumentation amplifier*. An instrumentation amplifier is a high-gain, DC-coupled differential amplifier with a single output, high-input impedance, and high CMRR [13]. This configuration ensures that the millivolt-level differential signal is amplified sufficiently and that the DC-bias and interference-noise voltages are rejected.

6.6 FILTERS

Another measurement system component is the filter. Its primary purpose is to remove signal content at unwanted frequencies. Filters can be passive or active. Passive filters are comprised of resistors, capacitors, and inductors that require no external power

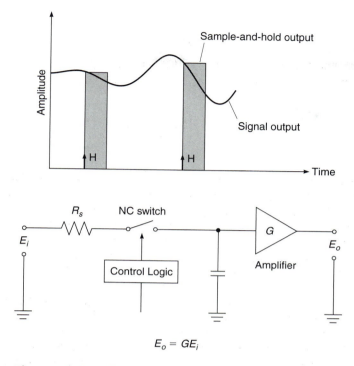

Figure 6.8 The sample-and-hold circuit.

supply. Active filters use resistors and capacitors with operational amplifiers, which require power. Digital filtering also is possible, where the signal is filtered *after* it is digitized.

The most common types of *ideal* filters are presented in Figure 6.9. The term *ideal* implies that the magnitude of the signal passing through the filter is not attenuated over the desired passband of frequencies. The term *band* refers to a range of frequencies and the term *pass* denotes the unaltered passing. The range of frequencies over which the signal is attenuated is called the stopband. The **low-pass filter** passes lower signal frequency content up to the cut-off frequency, f_c, and the **high-pass filter** passes content above f_c. Low- and high-pass filters can be combined to form either a **band-pass filter** or a *notch* filter, each having two cut-off frequencies, f_{cL} and f_{cH}. Actual filters do not have perfect step changes in amplitude at their cut-off frequencies. Rather, they experience a more gradual change, which is characterized by the roll-off at f_c, specified in terms of the ratio of amplitude change to frequency change.

The simplest filter can be made using one resistor and one capacitor. This is known as a **simple *RC* filter**, as shown in Figure 6.10. Referring to the top of that figure, if E_o is measured across the capacitor to ground, it serves as a low-pass filter. Lower-frequency signal content is passed through the filter, whereas high-frequency

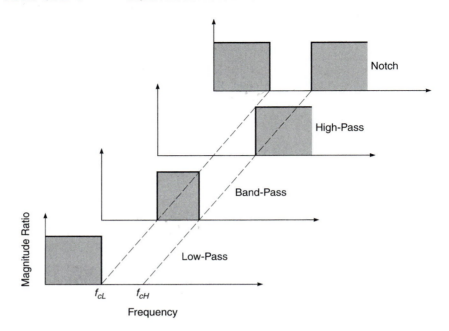

Figure 6.9   Ideal filter characteristics.

content is not. Conversely, if E_o is measured across the resistor to ground, it serves as a high-pass filter, as shown in the bottom of the figure. Here, higher-frequency content is passed through the filter, whereas lower-frequency content is not. For both filters, because they are not ideal, some fraction of intermediate frequency content is passed through the filter. The time constant of the simple RC filter, τ, equals $R \cdot C$. A unit balance shows that the units of $R \cdot C$ are $(V/A) \cdot (C/V)$ or s. An actual filter differs from an ideal filter in that an actual filter alters both the *magnitude* and the *phase* of the signal, but does *not* change its frequency.

    Actual filter behavior can be understood by first examining the case of a simple sinusoidal input signal to a filter. This is displayed in Figure 6.11. The filter's input signal (denoted by A in the figure) has a peak-to-peak amplitude of E_i, with a one-cycle period of T seconds. That is, the signal's input frequency, f, is $1/T$ cycles/s or Hz. Sometimes the input frequency is represented by the circular frequency, ω, which has units of rad/s. So, $\omega = 2\pi f$. If the filter only attenuated the input signal's amplitude, it would appear as signal B at the filter's output, having a peak-to-peak amplitude equal to E_0. In fact, however, an actual filter also delays the signal in time by Δt between the filter's input and output, as depicted by signal C in the figure. The output signal is said to lag the input signal by Δt. This **time lag** can be converted into a phase lag or phase shift by noting that $\Delta t / T = \phi/360°$, which implies that $\phi = 360°(\Delta t / T)$. By convention, the phase lag equals $-\phi$. The magnitude ratio, $M(f)$, of the filter equals $E_o(f)/E_i(f)$. For different input signal frequencies, both M and ϕ will have different values.

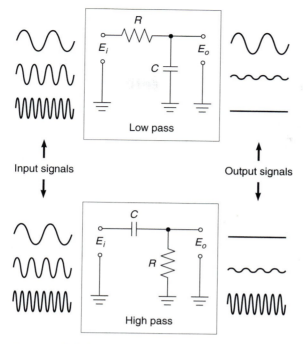

Figure 6.10 Simple RC low-pass and high-pass filters.

Analytical relationships for $M(f)$ and $\phi(f)$ can be developed for simple filters. Typically, M and ϕ are plotted each versus $\omega\tau$ or f/f_c, both which are dimensionless, as shown in Figures 5.3 and 5.4 of Chapter 5. The **cutoff frequency**, ω_c, is defined as the frequency at which the power is one-half of its maximum. This occurs at $M = 0.707$, which corresponds to $\omega\tau = 1$ for first-order systems, such as simple filters [14]. Thus, for simple filters, $\omega_c = 1/(RC)$ or $f_c = 1/(2\pi RC)$. In fact, for a simple low-pass RC filter,

$$M(\omega) = \frac{1}{\sqrt{1 + (\omega\tau)^2}} \qquad \text{[6.14]}$$

and

$$\phi = -\tan^{-1}(\omega\tau). \qquad \text{[6.15]}$$

Using these equations, $M(\omega = 1/\tau) = 0.707$ and $\phi = -45°$. That is, at an input frequency equal to the cut-off frequency of an actual RC low-pass filter, the output signal's amplitude is 70.7 % of the signal's input amplitude and it lags the input signal by 45°. For an RC high-pass filter, the phase lag equation is given by Equation 6.15 and the magnitude ratio is

$$M(\omega) = \frac{\omega\tau}{\sqrt{1 + (\omega\tau)^2}}. \qquad \text{[6.16]}$$

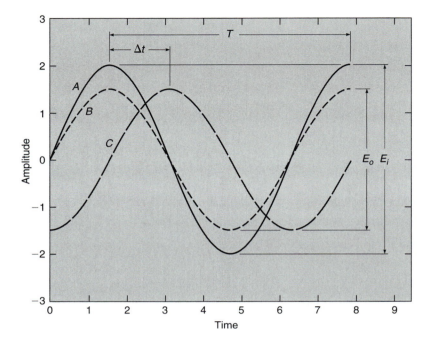

Figure 6.11   Generic filter input/output response characteristics.

These equations are derived in Chapter 5.

An active low-pass Butterworth filter configuration is shown in Figure 6.12. Its time constant equals R_2C_2 and its magnitude ratio and phase lag are given

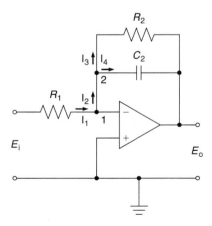

Figure 6.12   The active low-pass Butterworth filter.

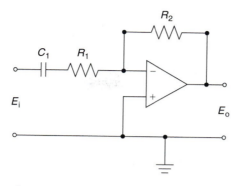

Figure 6.13 The active high-pass Butterworth filter.

by Equations 6.14 and 6.15, respectively. An active high-pass Butterworth filter configuration is displayed in Figure 6.13. Its time constant equals $R_1 C_1$ and its phase lag is given by Equation 6.15. Its magnitude ratio is

$$M(\omega) = \left[\frac{R_2}{R_1}\right] \cdot \left[\frac{\omega\tau}{\sqrt{1 + (\omega\tau)^2}}\right]. \qquad \textbf{[6.17]}$$

Other classes of filters have different response characteristics. Refer to [13] for detailed descriptions or [4] for an overview.

STATEMENT

Example 6.5

For the circuit depicted in Figure 6.12, determine the equation relating the output voltage, E_o, to the input voltage, E_i.

SOLUTION

The op amp's major attributes ensure that no current flows into the op amp and that the voltage difference between the two input terminals is zero. Assigning currents and nodes as shown in Figure 6.12 and applying Kirchhoff's current law and Ohm's law to node 1 gives

$$I_1 = I_2$$
$$\frac{E_1}{R_1} = I_2.$$

Applying Kirchhoff's current law and Ohm's law at node 2 results in

$$I_2 = I_3 + I_4$$
$$\frac{E_1}{R_1} = -\frac{E_0}{R_2} - C_2\dot{E}_0.$$

Dividing the above equation through by C_2 and rearranging terms yields

$$\frac{E_1}{C_2 R_1} = -\frac{E_0}{C_2 R_2} - \dot{E}_0,$$
$$\dot{E}_0 + \frac{1}{C_2 R_2}E_0 = -\frac{1}{C_2 R_1}E_1.$$

This is a first-order, ordinary differential equation whose method of solution is presented in Chapter 5.

Digital filters operate on a digitally converted signal. The filter's cutoff frequency adjusts automatically with sampling frequency and can be as low as a fraction of a Hz [13]. An advantage that digital filters have over their analog counterparts is that digital filtering can be done *after* data has been acquired. This approach allows the original, unfiltered signal content to be maintained. Digital filters operate by successively weighting each input signal value that is discretized at equal-spaced times, x_i, with k number of weights, h_k. The resulting filtered values, y_i, are given by

$$y_i = \sum_{k=-\infty}^{\infty} h_k x_{i-k}.$$ [6.18]

The values of k are finite for real digital filters. When the values of h_k are zero except for $k \geq 0$, the digital filter corresponds to a real analog filter. Symmetrical digital filters have $h_{-k} = h_k$, which yield phase shifts of $0°$ or $180°$.

Digital filters can use their output value for the ith value to serve as an additional input for the $(i + 1)$-th output value. This is known as a **recursive** digital filter. When there is no feedback of previous output values, the filter is a **nonrecursive** filter. A low-pass, digital recursive filter [13] can have the response

$$y_i = ay_{i-1} + (1 - a)x_i,$$ [6.19]

where $a = \exp(-t_s/\tau)$. Here, t_s denotes the time in between samples and τ the filter time constant, which equals RC. For this filter to operate effectively, $\tau \gg t_s$. Or, in other words, the filter's cutoff frequency must be much less than the Nyquist frequency. The latter is covered extensively in Chapter 12.

An example of this digital filtering algorithm is shown in Figure 6.14. The input signal of $\sin(0.01t)$ is sampled 10 times per second. Three output cases are plotted, corresponding to the cases of $\tau = 10$, 100, and 1000. Because of the relatively high sample rate used, both the input and output signals appear as analog signals, although both actually are discrete. When $\omega\tau$ is less than one, there is little attenuation in the signal's amplitude. In fact, the filtered amplitude is 99 % of the original signal's amplitude. Also, the filtered signal lags the original signal by only $5°$. At $\omega\tau = 1$, the amplitude attenuation factor is 0.707 and the phase lag is $45°$. When $\omega\tau = 10$, the attenuation factor is 0.90 and the phase lag is $85°$. This response mirrors that of an analog filter, as depicted in Figure 6.11 and described further in Chapter 5.

Almost all signals are comprised of multiple frequencies. At first, this appears to complicate filter performance analysis. However, almost any input signal can be decomposed into the sum of many sinusoidal signals of different amplitudes and frequencies. This is the essence of Fourier analysis, which is examined in Chapter 11. For a linear, time-invariant system such as a simple filter, the output signal can be reconstructed from its Fourier component responses.

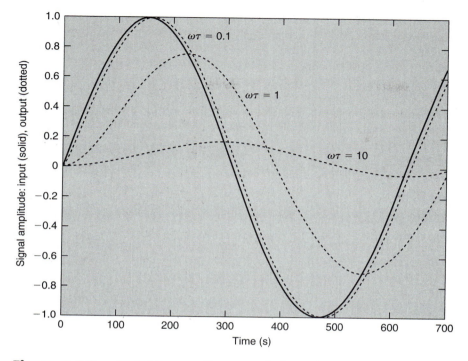

Figure 6.14   Digital low-pass filtering applied to a discretized sine wave.

6.7   ANALOG-TO-DIGITAL CONVERTERS

The last measurement system element encountered typically is the A/D converter. This component serves to translate analog signal information into the digital format that is used by a computer. In the computer's binary world, numbers are represented by 0's and 1's in units called bits. A bit value of either 0 or 1 is stored physically in a computer's memory cell using a transistor in series with a capacitor. An uncharged or charged capacitor represents the value of 0 or 1, respectively. Similarly, logic gates comprised of *on-off* transistors perform the computer's calculations.

Decimal numbers are translated into binary numbers using a decimal-to-binary conversion scheme. This is presented in Table 6.3 for a 3-bit scheme. A series of locations, which are particular addresses, are assigned to a series of bits that represent decimal values corresponding from right to left to increasing powers of 2. The least significant (rightmost) bit (LSB) represents a value of 2^0, whereas the most significant (leftmost) bit (MSB) of an M-bit scheme represents a value of $2^M - 1$. For example, for the 3-bit scheme shown in Table 6.3, when the LSB and MSB are *on* and the intermediate bit is *off*, the binary equivalent 101 of the decimal number 5 is stored.

Example 6.6	**STATEMENT**

STATEMENT

Convert the following decimal numbers into binary numbers: [a] 5, [b] 8, and [c] 13.

SOLUTION

An easy way to do this type of conversion is to note that the power of 2 in a decimal number is equal to the number of zeros in the binary number. [a] $5 = 4 + 1 = 2^2 + 1$. Therefore, the binary equivalent of 5 is $100 + 1 = 101$. [b] $8 = 2^3$. Therefore, the binary equivalent of 8 is 1000. [c] $13 = 8 + 4 + 1 = 2^3 + 2^2 + 1$. Therefore, the binary equivalent of 13 is $1000 + 100 + 1 = 1101$.

There are several different methods to perform analog-to-digital conversion electronically. The two most common ones are the successive-approximation and ramp-conversion methods. The **successive-approximation method** utilizes a digital-to-analog (D/A) converter and a differential op amp that subtracts the analog input signal from the D/A converter's output signal. The conversion process begins when the D/A converter's signal is incremented in voltage steps from 0 V using digital logic. When the D/A converter's signal rises to within ϵ V of the analog input signal, the differential op amp's output, now equal to ϵ V, causes the logic control to stop incrementing the D/A converter and tells the computer to store the converter's digital value. The **ramp-conversion method** follows a similar approach by increasing a voltage and comparing it to the analog input signal's voltage. The increasing signal is produced using an integrating op amp configuration, in which the op amp configuration is turned on through

Table 6.3   Binary-to-decimal conversion.

	4	2	1	Conversion	Decimal Equivalent
On Decimal Value					
Off Decimal Value	0	0	0	⇓	⇓
Binary Representation	0	0	0	$0 \cdot 4 + 0 \cdot 2 + 0 \cdot 1$	0
	0	0	1	$0 \cdot 4 + 0 \cdot 2 + 1 \cdot 1$	1
	0	1	0	$0 \cdot 4 + 1 \cdot 2 + 0 \cdot 1$	2
	0	1	1	$0 \cdot 4 + 1 \cdot 2 + 1 \cdot 1$	3
	1	0	0	$1 \cdot 4 + 0 \cdot 2 + 0 \cdot 1$	4
	1	0	1	$1 \cdot 4 + 0 \cdot 2 + 1 \cdot 1$	5
	1	1	0	$1 \cdot 4 + 1 \cdot 2 + 0 \cdot 1$	6
	1	1	1	$1 \cdot 4 + 1 \cdot 2 + 1 \cdot 1$	7

a switch controlled by the computer. In parallel, the computer starts a binary counter when the op amp configuration is turned on. When the analog input and op amp configuration signals are equal, the computer stops the binary counter and stores its values.

The terminology used for an M-bit A/D converter is summarized in Table 6.4. The values listed in the table for the LSB and MSB are when the bit is *on*. The bit equals 0 when it is *off*. The minimum decimal value that can be represented by the converter equals 0. The maximum value equals $2^M - 1$. Thus, 2^M possible values can be represented. The weight of a bit is defined as the value of the bit divided by the number of possible values. The resolution and absolute quantization error are based on an M-bit, unipolar A/D converter with a full-scale range (FSR) equal to 10.00 V, where $Q_o = 1000$ mV/bit. Most A/D converters used today are 12 or 16 bit, providing signal resolutions of 2.44 and 0.153 mV/bit, respectively.

An analog signal is continuous in time and therefore comprised of an infinite number of values. An M-bit A/D converter, however, can only represent the signal's amplitude by a finite set of 2^M values. This presents a signal resolution problem. Consider the analog signal represented by the solid curve shown in Figure 6.15. If the signal is sampled discretely at δt time increments, it will be represented by the values indicated by the solid circles. With **discrete** *sampling*, only the signal values between the sample times are lost, but the signal's exact amplitude values are maintained at each sample time. Yet, if this information is stored using the **digital** *sampling* scheme of the A/D converter, the signal's *exact* amplitude values also are lost. In fact, for the 12-bit A/D converter used to sample the signal shown in Figure 6.15, the particular signal is represented by only four possible values (0 mV, 2.44 mV, 4.88 mV, and 7.32 mV), as indicated by the ×'s in the figure. Thus, a signal whose amplitude is within the range of $\pm Q/2$ of a particular bit's value will be assigned the bit's value. This error is termed the **absolute quantization error** of an A/D converter.

Quite often, if a signal's amplitude range is on the order of the A/D converter's resolution, an amplifier will be used before the A/D converter to gain the signal's amplitude and, therefore, reduce the absolute quantization error to an acceptable

Table 6.4 *M*-bit terminology.

Term	Formula	$M = 8$	$M = 12$
MSB value	2^{M-1}	128	2048
LSB value	2^0	1	1
Maximum possible value	$2^M - 1$	255	4095
Minimum possible value	0	0	0
Number of possible values	2^M	256	4096
MSB weight	2^{-1}	1/2	1/2
LSB weight	2^{-M}	1/256	1/4096
Resolution, Q (mV/bit) for $E_{FSR} = 10$ V	$E_{FSR}/2^M$	39.10	2.44
Dynamic range (dB)	$20 \cdot \log_{10}(Q/Q_o)$	-28	-52
Absolute quantization error (mV)	$\pm Q/2$	± 19.60	± 1.22

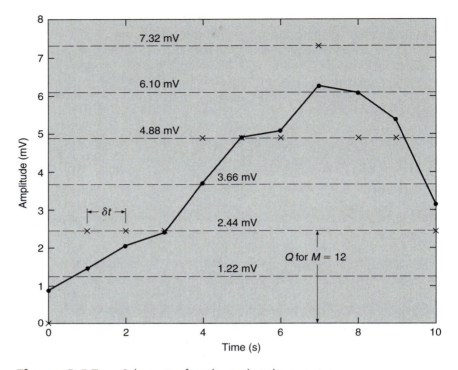

Figure 6.15   Schematic of analog-to-digital conversion.

level. An alternative approach is to use an A/D board with better resolution, but, almost always, this is a more expensive approach.

6.8   EXAMPLE MEASUREMENT SYSTEMS

The designs of three actual measurement systems are presented in this section to illustrate the different choices that can be made. The final design of each system involves many trade-offs between the accuracy and the cost of components. The elements of each of these systems are summarized in Table 6.5. Several major differences can be noted. The most significant are the choices of the sensor/transducer and of the signal processing system. These are dictated primarily by the environments in which each is designed to operate. System no. 1 is developed to be located near the test section of a subsonic wind tunnel, to be placed on a small table, and to use an existing personal computer. System no. 2 is designed to be located on one cart that can be moved to a remote location where the rocket motor is tested. A digital oscilloscope is chosen for its convenience of remote operation and its triggering and signal storage capabilities. System no. 3 is developed to be placed inside of a 2-in.-internal-diameter model rocket fuselage and then launched over 100 m into the air with accelerations and

Table 6.5   The elements of three measurement systems.

System No.	Variable (result)	Sensor/ Transducer	Signal Conditioner	Signal Processor
1	pressure (→velocity)	strain gage, Wheatstone bridge	amplifier, filter	A/D converter, computer
2	force (→thrust)	strain gage, Wheatstone bridge	amplifiers, filter	digital oscilloscope
3a	pressure (→velocity)	piezoresistive element	amplifier	microcontroller system
3b	acceleration (acceleration)	differential-capacitive structure	amplifier	microcontroller system

velocities as high as 60 m/s^2 (~6 g) and 50 m/s, respectively. These conditions constrain the size and weight of the measurement system package. A small, battery-powered, microcontroller-based data acquisition system with an A/D converter, amplifier, and memory is designed specifically for this purpose [15].

Measurement system no. 1 is designed to measure the velocity of air flowing in a wind tunnel, as shown in Figure 6.16. Its pitot-static tube is located within a wind tunnel. The pitot-static tube and tubing are passive and simply transmit the total and static pressures to a sensor located outside of the wind tunnel. The actual sensor is a strain gage mounted on a flexible diaphragm inside a pressure transducer housing. The static pressure port is connected to one side of the pressure transducer's diaphragm chamber; the total pressure port to the other side. This arrangement produces a flexure of the diaphragm proportional to the dynamic pressure (the physical stimulus), which strains the gage, changing its resistance (the electrical impulse). This resistance change imbalances a Wheatstone bridge operated in the deflection mode, producing a voltage

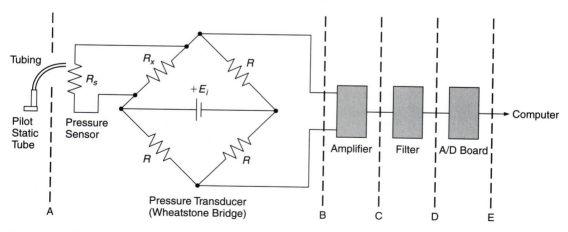

Figure 6.16   An example pressure measurement system.

at station B. Beyond station B, the signal is amplified, filtered, converted into its digital format, and finally stored by the computer. Another example measurement system, used to measure temperature, is considered in this chapter's homework problems.

The velocity measurement system is to be designed such that the input voltage to the A/D converter, E_D, is 10 V when the wind tunnel velocity is 100 m/s. The design also is subject to the additional conditions specified in Table 6.6. Given these constraints, the desired input/output characteristics of each measurement system element can be determined for stations A through E, as denoted in Figure 6.16. Determination of the performance characteristics for each stage is as follows:

- Station A: The velocity, V, of 100 m/s yields a dynamic pressure, Δp, of 5700 N/m^2 using Bernoulli's equation, $\Delta p = 0.5\rho V^2$. The density, ρ, equals 1.14 kg/m^3, as determined using Equation 2.1.

- Station B: The dynamic pressure produces a force, F, on the diaphragm, which has an area, A, equal to 1 cm^2. The resulting force is 0.57 N, noting that the force equals the pressure difference across the diaphragm times its area. A longitudinal strain on the diaphragm, ϵ_L, is produced by F, where $\epsilon_L = C_o \cdot F$. The resulting strain is 5.7×10^{-4}. According to Equation 6.7, this gives $\delta R / R = 1.14 \times 10^{-3}$. The Wheatstone bridge is operated in the deflection method mode with all resistances equal to 120 Ω at 294 K and $V = 0$ m/s. The output voltage, $E_o = E_B$, is determined using Equation 4.29 and equals 1.42 mV.

- Station C: The relatively low output voltage from the Wheatstone bridge needs to be amplified to achieve the A/D input voltage, E_D, of 10 V. Assuming that the filter's magnitude ratio is unity, the gain of the amplifier equals E_D/E_B, which is 10/0.142 or 70.4. An op amp in the noninverting configuration is used. Its input-output voltage relation is given in Figure 6.7. $E_o/E_i = 70.4$ and $R_1 = 1$ MΩ implies that R_2 equals 69.4 MΩ.

- Station D: The measurement system operates at steady state. The voltages are DC, having zero frequency. Thus, the filter's magnitude ratio is unity. Therefore, $E_D = E_C$.

- Station E: If the A/D converter has a full scale input voltage, E_{FSR}, of 10 V, then the converter is at its maximum input voltage when $V = 100$ m/s. The

Table 6.6   Velocity measurement system conditions.

Component	Conditions	Unknown
Environment	$T = 294$ K, $p = 1$ atm	$\rho, \Delta p, F, \epsilon_L, \delta R$
Diaphragm	$\epsilon_L = 0.001 \times F$, $A = 1$ cm^2, $C_o = 0.001$	
Strain gage	$R = 120$ Ω at 294 K	
Wheatstone bridge	All $R = 120$ Ω at 294 K, $E_i = 5$ V	E_o
Amplifier	Noninverting op amp, $R_1 = 1$ MΩ	R_2
Filter	Low-pass with $R = 1$ MΩ, $C = 1$ μF	
A/D converter	$E_{FSR} = 10$ V, $Q < 1$ mV/bit	M

relationship between the A/D converter's E_{FSR}, Q, and the number of converter bits, M, is presented in Table 6.4. Choosing $M = 12$ does not meet the constraint. The next choice is $M = 16$. This yields $Q = 0.153$ mV/bit, which satisfies the constraint.

Many choices can be made in designing this system. For example, the supply voltage to the Wheatstone bridge could be increased from 5 V to 10 V or 12 V, which are common supply voltages. This would increase the output voltage of the bridge and, therefore, require less amplification to meet the 10-V constraint. Other resistances can be used in the bridge. A different strain gage can be used on the diaphragm. If the system will be used for nonsteady velocity measurements, then the time responses of the tubing, the diaphragm, and the filter need to be considered. Each can affect the magnitude and the phase of the signal. The final choice of specific components truly is an engineering decision.

Next, examine measurement system no. 2 that is designed to acquire thrust as a function of time of a model rocket motor. The first element of the measurement system consists of an aluminum, cantilevered beam with four 120-Ω strain gages, similar to that shown schematically in Figure 4.11. These strain gages comprise the legs of a Wheatstone bridge. The maximum output of the bridge is approximately 50 mV. So, the output of the bridge is connected to an instrumentation amplifier with a gain of 100 and then to a variable-gain, operational amplifier in the inverting configuration. This allows the signal's amplitude to be adjusted for optimum display and storage by the digital oscilloscope. A two-pole, low-pass Sallen-and-Key filter [13] receives the second amplifier's output, filters it, and then passes it to the digital oscilloscope. The filter's schematic is shown in Figure 6.17. Typical filter parameter values are $R = 200$ kΩ, $R_f = 200$ kΩ, $C = 0.1$ μF, and $K = 1.586$. A low-pass filter is used to eliminate the ~30-Hz component that is the natural frequency of the cantilevered beam. The beam's natural frequency was determined by a subsidiary experiment, in which a force was applied impulsively to the end of the beam (see the signal and its frequency spectrum in Figure 11.14). The experimental results of the original and filtered rocket motor thrust as a function of time are shown in Figure 6.18. The effect of the low-pass filter is clearly visible. Additional details about the experiment can be found in Appendix H.

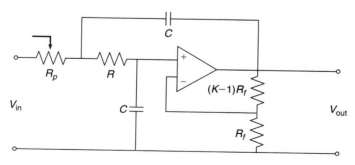

Figure 6.17 Two-pole, low-pass Sallen-and-Key filter.

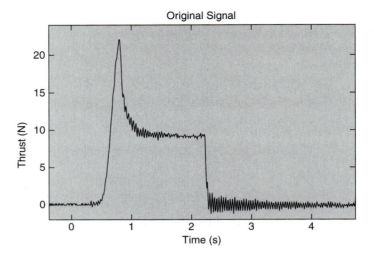

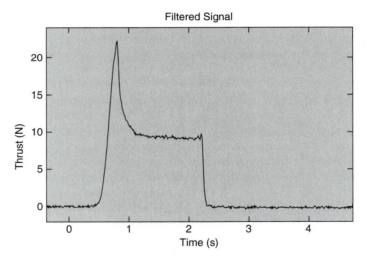

Figure 6.18   Original and filtered rocket motor thrust
signal.

Finally, consider the design of a measurement system no. 3, to be used remotely
in a model rocket to acquire the rocket's acceleration and velocity data during
ascent. The measurement system hardware consists of two sensor/transducers, one
for pressure and the other for acceleration, and a board containing a microcontroller-
based data acquisition system. The pressure transducer includes an integrated silicon
pressure sensor that is signal conditioned, temperature compensated, and calibrated
onchip. A single piezoresistive element is located on a flexible diaphragm. Total
and static pressure ports on the rocket's nose cone are connected with short

tubing to each side of the flexible diaphragm that is contained inside the transducer's housing. The difference in pressure causes the diaphragm to deflect, producing an output voltage that is directly proportional to the differential pressure, which, for this case, is the dynamic pressure. The single-avis ±5-g accelerometer contains a polysilicon surface sensor. A differential capacitor structure attached to the surface deflects under acceleration, causing an imbalance in the capacitor circuit. This produces an output voltage that is linearly proportional to the acceleration. The accelerometer calibration and pressure transducer curves are shown in Figures 6.19 and 6.20, respectively. Both sensor/transducer outputs are each routed into an amplifier with a gain of 16 and then to the inputs of a 12-bit A/D converter. The output digital signals are stored directly into memory (256 kB). The measurement system board has a mass of 33 g and dimensions of 4.1 cm by 10.2 cm. The onboard, 3.3-V, 720-mAh Li battery that powers the entire system has a mass of 39 g. All of the onboard data is retrieved after capture and downloaded into a laptop computer. A sample of the reconstructed data is displayed in Figure 6.21. The rocket's velocity in time can be determined from the pressure transducer's output. This is compared to the time integral of the rocket's acceleration in Figure 6.22. Finally, this information can be used with that on the rocket's drag to determine the maximum altitude of the rocket.

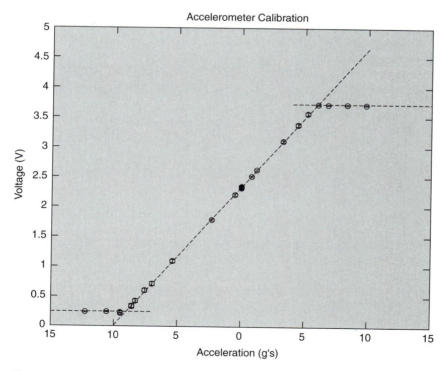

Figure 6.19 Calibration of the accelerometer.

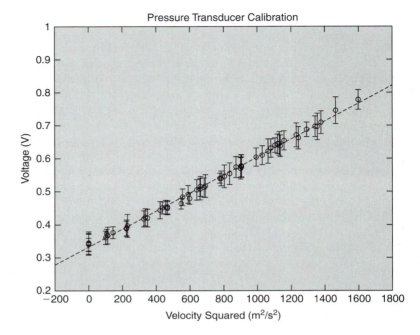

Figure 6.20   Calibration of the pressure transducer.

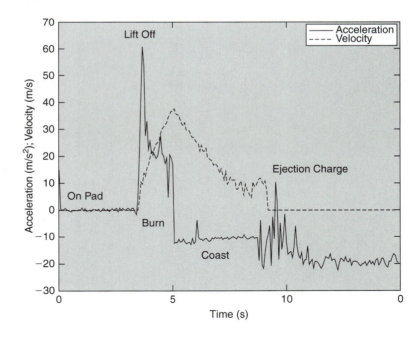

Figure 6.21   Example rocket velocity and acceleration data.

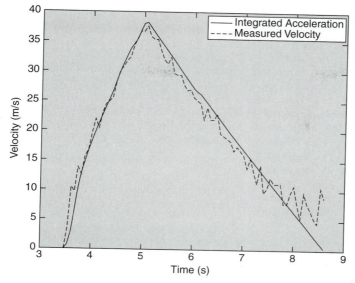

Figure 6.22 Integrated acceleration and velocity comparison.

REFERENCES

[1] R. M. White. 1987. A Sensor Classification Scheme. *IEEE Transactions on Ultrasonics, Ferroelectrics and Frequency Control* UFFC-34: 124–126.

[2] G. T. A. Kovacs. 1998. *Micromachined Transducers Sourcebook.* New York: McGraw-Hill.

[3] J. P. Bentley. 1988. *Principles of Measurement Systems*, 2nd ed. New York: John Wiley and Sons.

[4] A. J. Wheeler, and A. R. Ganji. 2004. *Introduction to Engineering Experimentation*, 2nd ed. New York: Prentice Hall.

[5] E. Doebelin. 2003. *Measurement Systems: Application and Design*, 5th ed. New York: McGraw-Hill.

[6] T. G. Beckwith, R. D. Marangoni, and J. H. Leinhard V. 1993. *Mechanical Measurements*, 5th ed. New York: Addison-Wesley.

[7] R. Figliola and D. Beasley. 2000. *Theory and Design for Mechanical Measurements*, 3rd ed. New York: John Wiley and Sons.

[8] D. G. Alciatore and M. B. Histand. 2003. *Introduction to Mechatronics and Measurement Systems*, 2nd ed. New York: McGraw-Hill.

[9] C. Crowe, M. Sommerfeld, and Y. Tsuji. 1998. *Multiphase Flows with Droplets and Particles.* New York: CRC Press.

[10] T. R. Hsu. 2002. *MEMS & Microsystems: Design and Manufacture.* New York: McGraw-Hill.

[11] M. Madou. 1997. *Fundamentals of Microfabrication.* New York: CRC Press.

[12] T. Crump. 2001. *A Brief History of Science as Seen Through the Development of Scientific Instruments.* London: Robinson.

[13] P. Horowitz and W. Hill. 1989. *The Art of Electronics*, 2nd ed. Cambridge: Cambridge University Press.

[14] A. V. Oppenheim and A. S. Willsky. 1997. *Signals and Systems*, 2nd ed. New York: Prentice Hall.

[15] T. S. Szarek. 2003. On the Use of Microcontrollers for Data Acquisition in an Introductory Measurements Course. *M.S. Thesis.* Department of Aerospace and Mechanical Engineering. Indiana: University of Notre Dame.

REVIEW PROBLEMS

1. Modern automobiles are equipped with a system to measure the temperature of the radiator fluid and output this temperature to a computer monitoring system. A thermistor is manufactured into the car radiator. A conducting cable leads from the thermistor and connects the thermistor to one arm of a Wheatstone bridge. The voltage output from the Wheatstone bridge is input into the car computer that digitally samples the signal ten times each second. If the radiator fluid temperature exceeds an acceptable limit, the computer sends a signal to light a warning indicator to alert the driver. Match the following components of the fluid temperature measurement system with their function in terms of a generalized measurement system.

radiator fluid temperature   sensor
thermistor   physical variable
Wheatstone bridge   transducer
car computer   signal processor

2. Which of the following instruments is used to interface analog systems to digital ones?
   (a) A/C converter, **(b)** D/C converter,
   (c) A/D converter, **(d)** AC/DC converter

3. An electronic manometer is calibrated using a fluid-based manometer as the calibration standard. The resulting calibration curve fit is given by the equation $V = 1.1897P - 0.0002$, where the unit of P is in. H_2O and V is volts. Identify the static sensitivity (in V/in. H_2O) from the choices [a] 0.0002, [b] $1.1897P^2 - 0.0002P$, [c] 1.1897, [d] −0.0002.

4. A metallic wire embedded in a strain gage is 4.2 cm long with a diameter of 0.07 mm. The gage is mounted on the upper surface of a cantilever beam to sense strain. Before strain is applied, the initial resistance of the wire is 64 Ω. Strain is applied to the beam, stretching the wire 0.1 mm, and changing its electrical resistivity by 2×10^{-8} $\Omega \cdot m$. If Poisson's ratio for the wire is 0.342, find the change in resistance in the wire due to the strain to the nearest hundredth ohm.

5. Determine the static sensitivity at $x = 2.00$ for a calibration curve having $y = 0.8 + 33.72x + 3.9086x^2$. Express the result with the correct number of significant figures.

6. A student records a small sample of three voltage measurements: 1.000 V, 2.000 V, and 3.000 V. Determine the uncertainty estimated with 50 % confidence in the population's true mean value of the voltage. Express your answer with the correct number of significant figures.

7. What is the time constant (in seconds) of a single-pole, low-pass, passive filter having a resistance of 2 kΩ and a capacitance of 30 µF?

8. A single-stage, low-pass RC filter with a resistance of 93 Ω is designed to have a cutoff frequency of 50 Hz. Determine the capacitance of the filter in units of µF.

9. Two resistors, R_A and R_B, arranged in parallel, serve as the resistance, R_1, in the leg of a Wheatstone bridge where $R_2 = R_3 = R_4 = 200$ Ω and the excitation voltage is 5.0 V. If $R_A = 1000$ Ω, what value of R_B is required to give a bridge output of 1.0 V?

HOMEWORK PROBLEMS

1. Consider the amplifier between stations B and C of the temperature-measurement system shown in Figure 6.23. [a] Determine the minimum input impedance of the amplifier (in Ω) required to keep the amplifier's voltage measurement loading error, e_V, less than 1 mV for the case when the bridge's output impedance equals 30 Ω and its output voltage equals 0.2 V. [b] Based on your answer in part [a], if an operational amplifier were used, would it satisfy the requirement of $e_V < 1$ mV? Answer yes or no and explain why or why not. [c] What would be the gain, G, required to have the amplifier's output equal to 9 V when $T = 72$ °F?

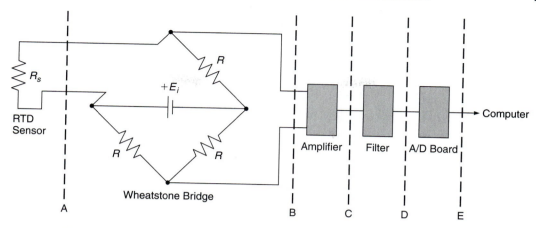

Figure 6.23   An example temperature-measurement system configuration.

2. Consider the A/D board between stations D and E of the temperature measurement system shown in Figure 6.23. Determine how many bits ($M = 4, 8, 12$, or 16) would be required to have less than ± 0.5 % quantization error for the input voltage of 9 V with $E_{FSR} = 10$ V.

3. The voltage from a 0-kg to 5-kg strain gage balance scale has a corresponding output voltage range of 0 to 3.50 mV. The signal is recorded using a new 16-bit A/D converter having a unipolar range of 0 V to 10 V, with the resulting weight displayed on a computer screen. An intelligent aerospace engineering student decides to place an amplifier in between the strain gage balance output and the A/D converter such that 1 % of the balance's full scale output will be equal to the resolution of 1 bit of the converter. Determine [a] the resolution (in mV/bit) of the converter and [b] the gain of the amplifier.

4. The operational amplifier shown in Figure 6.24 has an *open-loop* gain of 10^5 and an output resistance of 50 Ω.

Determine the *effective* output resistance (in Ω) of the op amp for the given configuration.

5. A single-stage, passive, low-pass (*RC*) filter is designed to have a cutoff frequency, f_c, of 100 Hz. Its resistance equals 100 Ω. Determine the filter's [a] magnitude ratio at $f = 1$ kHz, [b] time constant (in ms), and [c] capacitance (in μF).

6. A voltage-sensitive Wheatstone bridge (refer to Figure 6.25) is used in conjunction with a hot-wire sensor to measure the temperature within a jet of hot gas. The resistance of the sensor (in Ω) is $R_1 = R_o[1 + \alpha(T - T_o)]$, where $R_o = 50$ Ω is the resistance at $T_o = 0$ °C and $\alpha = 0.00395/°C$. For $E_i = 10$ V and $R_3 = R_4 = 500$ Ω, determine [a] the value of R_2 (in Ω) required to balance the bridge at $T = 0$ °C. Using this as a fixed R_2 resistance, further determine [b] the value of R_1 (in Ω) at $T = 50$ °C, and [c] the value of E_o (in V) at $T = 50$ °C. Next, a voltmeter having an input impedance of 1000 Ω is connected

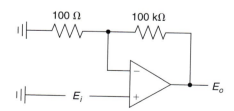

Figure 6.24   A closed-loop operational amplifier configuration.

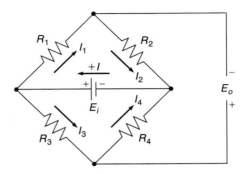

Figure 6.25   The Wheatstone bridge configuration.

across the bridge to measure E_o. Determine [d] the percentage loading error in the measured bridge output voltage. Finally, [e] state what other electrical component and in what specific configuration could be added in between the bridge and the voltmeter to reduce the loading error to a negligible value.

7. An engineer is asked to specify several components of a temperature measurement system. The output voltages from a type J thermocouple referenced to 0 °C vary lin-

early from 2.585 mV to 3.649 mV over the temperature range from 50 °C to 70 °C. The thermocouple output is to be connected directly to an A/D converter having a range from −5 V to +5 V. For both a 12-bit and a 16-bit A/D converter determine [a] the quantization error (in mV), [b] the percentage error at $T = 50$ °C, and [c] the percentage error at $T = 70$ °C. Now if an amplifier is installed in between the thermocouple and the A/D converter, determine [d] the amplifier's gain to yield a quantization error of 5 % or less.

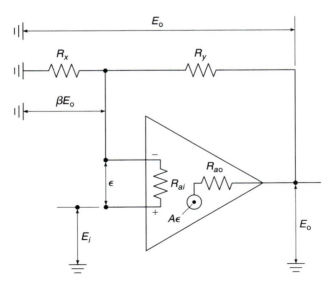

Figure 6.26   The operational amplifier in the voltage-follower configuration.

8. Consider the filter between stations C and D of the temperature measurement system shown in Figure 6.23. Assume that the temperature varies in time with frequencies as high as 15 Hz. For this condition, determine [a] the filter's cutoff frequency (in Hz), and [b] the filter's time constant (in ms). Next, find [c] the filter's output voltage (peak-to-peak) when the amplifier's output voltage (peak-to-peak) is 8 V and the temperature varies with a frequency of 10 Hz, and [d] the signal's phase lag through the filter (in ms) for this condition.

9. An op amp in the negative-feedback, voltage-follower configuration is shown in Figure 6.26. In this configuration, a voltage difference, ϵ, between the op amp's positive and negative inputs results in a voltage output of $A\epsilon$, where A is the open-loop gain. The op amp's input and output impedances are R_{ai} and R_{ao}, respectively. E_i is its input voltage and E_o is its output voltage. Assuming that there is negligible current flow into the negative input, determine [a] the value of β, and [b] the closed-loop gain, G, in terms of β and A. Finally, recognizing that A is very large ($\sim 10^5$ to 10^6), [c] derive an expression for E_o as a function of E_i, R_x, and R_y.

10. Refer to the information given for Problem 9. When E_i is applied to the op amp's positive input, a current I_{in}, flows through the input resistance, R_{ai}. The op amp's effective input resistance, R_{ci}, which is the resistance that would be measured between the op amp's positive and negative inputs by an ideal ohmmeter, is defined as E_i/I_{in}. [a] Derive an expression for R_{ci} as a function of R_{ai}, β, and A. Using this expression, [b] show that this is a very high value.

11. Refer to the information given for Problem 10. The op amp's output voltage for this configuration is $E_o = A(E_i - \beta E_o)$. Now assume that there is a load connected to the op amp's output that results in a current flow, I_{out}, across the op amp's output resistance, R_{ao}. This effectively reduces the op amp's output voltage by $I_{out} R_{ao}$. For the equivalent circuit, the Thévenin output voltage is E_o as given in the preceding expression and the Thévenin output impedance is R_{co}. [a] Derive an expression for R_{co} as a function of R_{ao}, β, and A. Using this expression, [b] show that this is a very low value.

CHAPTER REVIEW

ACROSS

1. removes unwanted signal content at certain frequencies
4. signal that changes rapidly in time
7. a sensor that provides motion
8. today's amplifier of choice
10. requiring external power
12. senses the physical stimulus and converts it into a signal
13. gage factor related to the *local* resistance change
14. gage factor related to the *total* resistance change

DOWN

2. input impedance of an ideal op amp
3. when the output is used as part of the input
4. located within the experiment's environment
5. output impedance of an ideal op amp
6. changes the signal into the desired quantity
9. requiring no external power
11. acronym for a very small sensor system

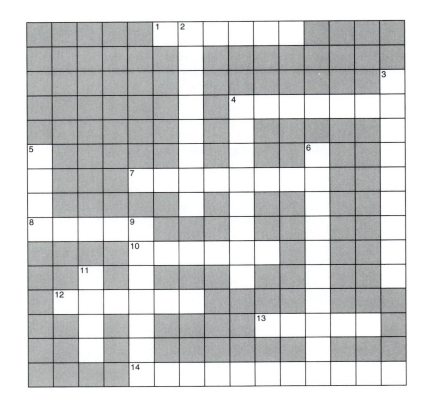

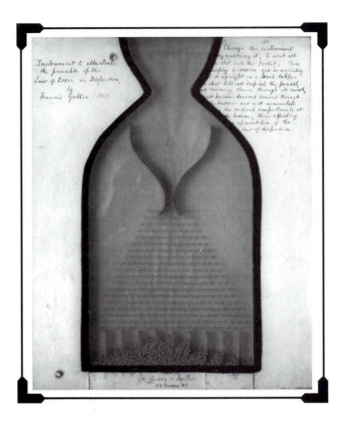

The Quincunx

Credit: Author's photograph, with thanks to The Galton Collection, University College London.

This instrument was conceived by Francis Galton (1822–1911), a British scientist, to illustrate the principle of the normal distribution, known then as the law of error or dispersion. This particular instrument, the first of its kind, was made by Tissley and Spiller, 172 Brompton Road, London. The instrument was named a quincunx because of the staggered arrangement of the harrow of pegs. Galton describes on the front face of the instrument how to operate it as follows: "Charge the instrument by reversing it to send all the shot into the pocket. Then sharply re-reverse and immediately put it upright on a level table. The shot will all drop into the funnel, and running thence through its mouth, will pursue devious courses through the harrow and will accumulate in the vertical compartments at the bottom, there affording a representation of the law of dispersion."

When you deal in large numbers, probabilities are the same as certainties. I wouldn't bet my life on the toss of a single coin, but I would, with great confidence, bet on heads appearing between 49 % and 51 % of the throws of a coin if the number of tosses was 1 billion.

Brian L. Silver. 1998.
The Ascent of Science. Oxford: Oxford University Press.

It is the nature of probability that improbable things will happen.

Aristotle, c. 350 B.C.

Chance favours only the prepared mind.

Louis Pasteur, 1879.

CHAPTER OUTLINE

7.1   CHAPTER OVERVIEW

Probability underlies all of our lives. How often has one heard that the chance of rain tomorrow will be 50 % or that there is a good chance to be a winner in today's lottery? It is hard to avoid its mention in an electronically connected society. But what science is behind such statements? Similar questions can be asked in relation to experiments, such as the probability that a pressure will exceed a certain limit.

In this chapter we will study some of the tools of probability. We will start by reviewing some basic concepts in probability and examining the differences between a population and a sample when using statistics. Next we will find out how to calculate and present the statistical information about a population or a sample. Finally, we will explore the concept of the probability density function and its integral, the probability distribution function. After finishing with this chapter, you will have studied most of the basic concepts of probability. This will prepare you to begin the study of statistics, which is the subject of Chapter 8.

7.2   LEARNING OBJECTIVES

You should be able to do the following after completing this chapter:

- Know the definitions for and differences between an event, an independent event, a dependent event, and an outcome
- Know the definitions for null and mutually exclusive sets
- Determine the intersection and union of multiple sets
- Apply Bayes' rule to determine the probability of an event using a test that has a specified accuracy
- Determine the number of permutations and combinations of an event
- Know the meaning of and differences between a population and a sample
- Know the meaning of and differences between the true mean and variance and the sample mean and variance
- Know the meaning of and differences between analog, discrete, and digital signal representations
- Know the definitions for a histogram, frequency distribution, and frequency density distribution and be able to plot each from discrete data
- Know the definitions for the first through fourth central moments of a specified probability density function and be able to calculate them

7.3   RELATION TO MEASUREMENTS

Probability and statistics are two distinct but closely related fields of science. Probability deals with the likelihood of events. The mathematics of probability shows us how to calculate the likelihood or chance of an event based on theoretical populations. Statistics involves the collection, presentation, and interpretation of data, usually for the purpose of making inferences about the behavior of an underlying population or for testing a theory. Both fields can be used to answer many practical questions that arise when performing an experiment, such as the following:

- How frequently does this event occur?
- What are the chances of rejecting a correct theory?
- How repeatable are the results?
- What confidence is there in the results?
- How can the fluctuations and drift in the data be characterized?
- How much data is necessary for an adequate sample?

Armed with a good grasp of probability and statistics, all of these questions can be answered quantitatively.

7.4 BASIC CONCEPTS IN PROBABILITY

The concept of the probability of an occurrence or outcome is intuitive. Consider the toss of a fair die. The probability of getting any one of the six possible numbers is 1/6. Formally this is written as $Pr[A] = 1/6$ or approximately 17 %, where the A denotes the occurrence of any one specific number. The **probability** of an occurrence can be defined as the number of times of the occurrence divided by the total number of times considered (the times of the occurrence plus the times of no occurrence). If the probabilities of getting 1 or 2 or 3 or 4 or 5 or 6 on a single toss are added, the result is $Pr[1] + Pr[2] + Pr[3] + Pr[4] + Pr[5] + Pr[6] = 6(1/6) = 1$. That is, the sum of all of the possible probabilities is unity. A probability of 1 implies absolute certainty; a probability of 0 implies absolute uncertainty or impossibility.

Now consider several tosses of a die and, based on these results, determine the probability of getting a specific number. Each toss results in an **outcome**. The tosses when the specific number occurred comprise the **set** of occurrences for the **event** of getting that specific number. The tosses in which the specific number did *not* occur comprise the **null set** or **complement** of that event. Remember, the event for this situation is not a single die toss, but rather all the tosses in which the specific number occurred. Suppose, for example, the die was tossed eight times, obtaining eight outcomes: 1, 3, 1, 5, 5, 4, 6, and 2. The probability of getting a 5 based on these outcomes would be $Pr[5] = 2/8 = 1/4$. That is, two of the eight possible outcomes comprise the set of events where 5 is obtained. The probability of the event of getting a 3 would be $Pr[3] = 1/8$. These results do not imply necessarily that the die is unfair, rather, that the die has not been tossed enough times to assess its fairness. This subject is considered later in this chapter. It relates directly to the question of how many measurements need to be taken to achieve a certain level of confidence in an experiment.

7.4.1   UNION AND INTERSECTION OF SETS

Computing probabilities in the manner just described is correct provided the one event that is considered has nothing in common with the other events. Continuing to use the previous example of eight die tosses, determine the probability that either an even number or the numbers 3 or 4 occur. There is Pr[even] = 3/8 (from 4, 6, and 2) and Pr[3] = 1/8 and Pr[4] = 1/8. Adding the probabilities, the sum equals 5/8. Inspection of the results, however, shows that the probability is 1/2 (from 3, 4, 6, and 2). Clearly the method of simply adding these probabilities for this type of situation is *not* correct.

To handle the more complex situation when events have members in common, the union of the various sets of events must be considered, as illustrated in Figure 7.1. The lined triangular region marks the set of events A and the circular region the set of events B. The complement of A is denoted by A'. The sample space is **exhaustive** because A and A' comprise the entire sample space. For two sets A and B, the **union** is the set of all members of A *or* B *or* both, as denoted by the region bordered by the dashed line. This is written as Pr[$A \cup B$].

If the sets of A and B are **mutually exclusive** where they do not share any common members, then

$$\Pr[A \cup B] = \Pr[A] + \Pr[B]. \qquad \textbf{[7.1]}$$

This would be the case if the sets A and B did not overlap in the figure (if the circular region was outside the triangular region). Thus, the probability of getting 3 or 4 in the eight-toss experiment is 1/4 (from 3 and 4).

If the sets do overlap and have common members, as shown by the cross-hatched region in the figure, then

$$\Pr[A \cup B] = \Pr[A] + \Pr[B] - \Pr[A \cap B], \qquad \textbf{[7.2]}$$

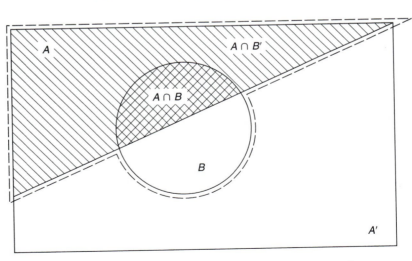

Figure 7.1   The union and the intersection of the sets A and B.

where $Pr[A \cap B]$ is the probability of the intersection of A and B. The **intersection** is the set of all members in *both A and B*. So, the correct way to compute the desired probability is $Pr[even \cup 3 \cup 4] = Pr[even] + Pr[3] + Pr[4] - Pr[even \cap 3 \cap 4] = 3/8 + 1/8 + 1/8 - 1/8 = 1/2$. $Pr[even \cap 3 \cap 4] = 1/8$, because only one common member, 4, occurred during the eight tosses.

7.4.2   CONDITIONAL PROBABILITY

The moment questions are asked such as—"What is the chance of getting a 4 on the second toss of a die given either a 1 or a 3 on the first toss?"—more thought is required before answering them. This is a problem in **conditional probability**, the probability of an event given that specified events have occurred in the past. This concept can be formalized by determining the probability that event B occurs given that event A occurred previously. This is written as $Pr[B \mid A]$, where

$$Pr[B \mid A] \equiv \frac{Pr[B \cap A]}{Pr[A]}. \qquad \textbf{[7.3]}$$

Rearranging this definition gives

$$Pr[B \cap A] = Pr[B \mid A]Pr[A]. \qquad \textbf{[7.4]}$$

This is known as the *multiplication rule* of conditional probability. Further, because $Pr[A \cap B] = Pr[B \cap A]$, Equation 7.4 implies that

$$Pr[B \mid A]Pr[A] = Pr[A \mid B]Pr[B]. \qquad \textbf{[7.5]}$$

Examination of Figure 7.1 further reveals that

$$Pr[A] = Pr[A \cap B] + Pr[A \cap B'], \qquad \textbf{[7.6]}$$

where $A \cap B'$ is shown as the lined region. Using the converse of Equation 7.4, Equation 7.6 becomes

$$Pr[A] = Pr[A \mid B]Pr[B] + Pr[A \mid B']Pr[B']. \qquad \textbf{[7.7]}$$

Equation 7.7 is known as the *total probability rule* of conditional probability. It can be extended to represent more than two mutually exclusive and exhaustive events [6].

When events A and B are mutually exclusive, then $Pr[B \mid A] = 0$, which leads to $Pr[B \cap A] = 0$. When event B is **independent** of event A, the outcome of event A has no influence on the outcome of event B. Then

$$Pr[B \mid A] = Pr[B]. \qquad \textbf{[7.8]}$$

Thus, for independent events, it follows from Equations 7.4 and 7.8 that

$$Pr[B \cap A] = Pr[B]Pr[A]. \qquad \textbf{[7.9]}$$

That is, the conditional probability of a series of independent events is the *product* of the individual probabilities of each of the events. Hence, to answer the question posed at the beginning of this section, $\Pr[B \cap A] = \Pr[A]\Pr[B] = (1/6)(2/6) = 2/36$. That is, there is approximately a 6 % chance that either an even number or the numbers 3 or 4 occurred.

The chance of winning the lottery can be determined easily using this information. The results suggest not to bet in lotteries. Assume that four balls are drawn from a bin containing white balls numbered 1 through 49, and then a fifth ball from another bin containing red balls with the same numbering scheme. Because each number selection is independent of the other number selections, the probability of guessing the numbers on the four white balls correctly would be $\Pr[1]\,\Pr[2]\,\Pr[3]\,\Pr[4] = (1/49)(1/48)(1/47)(1/46) = 1/(5\,085\,024)$. The probability of guessing the numbers on all five balls correctly would be $\Pr[1]\,\Pr[2]\,\Pr[3]\,\Pr[4]\,\Pr[5] = (1/(5\,085\,024))(1/49) = 1/(249\,166\,176)$ or about 1 chance in 250 million! That chance is about equivalent to tossing a coin and getting 28 heads in a row. Recognizing that the probability that an event will *not* occur equals one minus the probability that it will occur, the chance of *not* winning the lottery is 99.999 999 6 %. So, it is very close to impossible to win the lottery.

Other useful conditional probability relations can be developed. Using Equation 7.5 and its converse, noting that $\Pr[A \cap B] = \Pr[B \cap A]$,

$$\Pr[B \mid A] = \frac{\Pr[B]\Pr[A \mid B]}{\Pr[A]}. \qquad \textbf{[7.10]}$$

This relation enables us to determine the probability of event B occurring given that event A has occurred from the probability of event A occurring given that event B has occurred.

7.1 | MATLAB SIDEBAR

MATLAB works well with user-defined functions. In a user-defined function, input and output variables can be specified, as is done by the subroutine of a computer program. The general form of a MATLAB function is `function [output1, output2, ...] = functionname(input1, input2, ...)`. The function is saved as a MATLAB M-file and then referred to in another M-file or active workspace by its name with values given for its arguments. For example, the user-defined function called `triprob` (stored as `triprob.m` within the directory) is defined as

```
function [w] = triprob(x,y,z)
w = x*y*z
```

When the function is referred to by a statement such as "triprob(0.3,0.4,0.5);" that includes a semicolon, the output will be displayed with the variable name stated within the function's M-file. For this case, it would be "w = 0.06."

Example 7.1

STATEMENT

Determine the probability that the temperature, T, will exceed 100 °F in a storage tank whenever the pressure, p, exceeds 2 atmospheres. Assume the probability that the pressure exceeds 2 atmospheres whenever the temperature exceeds 100 °F is 0.30, the probability of the pressure exceeding 2 atmospheres is 0.10 and the probability of the temperature exceeding 100 °F is 0.25.

SOLUTION

Formally, $\Pr[p > 2 \mid T > 100] = 0.30$, $\Pr[p > 2] = 0.10$ and $\Pr[T > 100] = 0.25$. Using Equation 7.10, $\Pr[T > 100 \mid p > 2] = (0.25)(0.30)/(0.10) = 0.75$. That is, whenever the pressure exceeds 2 atmospheres there is a 75 % chance that the temperature will exceed 100 °F.

Equation 7.7 can be substituted into Equation 7.10 to yield

$$\Pr[B \mid A] = \frac{\Pr[A \mid B]\Pr[B]}{\Pr[A \mid B]\Pr[B] + \Pr[A \mid B']\Pr[B']}. \qquad \textbf{[7.11]}$$

This equation expresses what is known as *Bayes' rule*, which is valid for mutually exclusive and exhaustive events. It was discovered accidentally by the Reverend Thomas Bayes (1702–1761) while manipulating formulas for conditional probability [5]. Its power lies in the fact that it enables one to calculate probabilities inversely, to determine the probability of a *before* event conditional on an *after* event. This equation can be extended to represent more than two events [4]. Bayes' rule can be used to solve many practical problems in conditional probability. For example, the probability that a component identified as defective by a test of known accuracy actually is defective can be determined knowing the percentage of defective components in the population.

Before determining this probability specifically, examine these types of probabilities in a more general sense. Assume that the probability of an event occurring is p_1. A test can be performed to determine whether or not the event has occurred. This test has an accuracy of $100p_2$ %. That is, it is correct $100p_2$ % of the time and incorrect $100(1 - p_2)$ % of the time. What is the % probability that the test can predict an actual event? There are four possible situations that can arise: (1) the test indicates that the event has occurred *and* the event actually has occurred (a true positive), (2) the test indicates that the event has occurred *and* the event actually has not occurred (a false positive), (3) the test indicates that the event has not occurred *and* the event actually has occurred (a false negative), and (4) the test indicates that the event has not occurred *and* the event actually has not occurred (a true negative). Here, the terms positive and negative refer to the *indicated* occurrence of the event and the terms true and false refer to whether or not the indicated occurrence agrees with the *actual* occurrence of the event. So, the probabilities of the four possible combinations of events are $p_2 p_1$ for true positive, $(1 - p_2)(1 - p_1)$ for false positive, $(1 - p_2)p_1$ for false negative, and $p_2(1 - p_1)$ for true negative. Now the probability that an event actually occurred given that a test indicated that it had occurred would be the ratio of the actual probability of occurrence (the true positive probability) to the sum of all indicated positive occurrences (the true positive plus the false positive probabilities).

That is,

$$\Pr[A \mid IA] = \frac{p_2 p_1}{p_2 p_1 + (1 - p_2)(1 - p_1)}, \qquad \text{[7.12]}$$

where IA denotes the event of an *indicated* occurrence of A and A symbolizes the event of an *actual* occurrence.

Alternatively, Equation 7.12 can be derived directly using Bayes' rule. Here,

$$\Pr[A \mid IA] = \frac{\Pr[IA \mid A]\Pr[A]}{\Pr[IA \mid A]\Pr[A] + \Pr[IA \mid A']\Pr[A']}, \qquad \text{[7.13]}$$

which is identical to Equation 7.12 because $\Pr[IA \mid A] = p_2$, $\Pr[A] = p_1$, $\Pr[IA \mid A'] = (1 - p_2)$ and $\Pr[A'] = (1 - p_1)$.

Example 7.2

STATEMENT

An experimental technique is being developed to detect the removal of a microparticle from the surface of a wind tunnel wall as the result of a turbulent sweep event. Assume that a sweep event occurs 14 % of the time the detection scheme is operated. The experimental technique can detect a burst correctly 73 % of the time. [a] What is the probability that a sweep event will be detected during the time period of operation? [b] What is the probability that a sweep event will be detected if the experimental technique is correct 90 % of the time?

SOLUTION

The desired probability is the ratio of true positive identifications to true positive plus false positive identifications. Let p_1 be the probability of an actual sweep event occurrence during the time period of operation and p_2 be the experimental technique's reliability (the probability to identify correctly). For this problem, $p_1 = 0.14$ and $p_2 = 0.73$ for part [a] and $p_2 = 0.90$ for part [b]. Substitution of these values into Equation 7.12 yields $P = 0.31$ for part [a] and $P = 0.59$ for part [b]. First, note that the probability that a sweep event will be detected is only 31 % with a 73 % experimental technique reliability. This in part is because of the relatively low percentage of sweep event occurrences during the period of operation. Second, an increase in the technique's reliability from 73 % to 90 % or by 17 % increases the probability from 31 % to 59 % or by 28 %. A certain increase in technique reliability increases the probability of correct detection by relatively a much greater amount.

Example 7.3

STATEMENT

Suppose that 4 % of all transistors manufactured at a certain plant are defective. A test to identify a defective transistor is 97 % accurate. What is the probability that a transistor identified as defective actually is defective?

SOLUTION

Let event A denote that the transistor actually is defective and event B a positive indication of defectiveness. What is $\Pr[A \mid B]$? It is known that $\Pr[A] = 0.04$ and $\Pr[B \mid A] = 0.97$. It follows that $\Pr[A'] = 1 - \Pr[A] = 0.96$ and $\Pr[B \mid A'] = 1 - \Pr[B \mid A'] = 0.03$ because the set of all possible events are mutually exclusive and exhaustive. Direct application of Bayes' rule gives

$$\Pr[A \mid B] = \frac{(0.97)(0.04)}{(0.97)(0.04) + (0.03)(0.96)} = 0.57.$$

So, there is a 57 % chance that a transistor identified as defective actually is defective. At first glance, this percentage seems low. Intuitively, the value would be expected to be closer to the accuracy of the test (97 %). However, this is not the case. In fact, to achieve a 99 % chance of correctly identifying a defective transistor, the test would have to be 99.96 % accurate!

It is important to note that the way statistics are presented, whether in the form of probabilities, percentages, or absolute frequencies, makes a noticeable difference to some people in arriving at the correct result. Studies [6] have shown that when statistics are expressed as frequencies, a far greater number of people arrive at the correct result. Example 7.3 can be solved again by using an alternative approach [6].

Example 7.4

STATEMENT

Suppose that 4 % of all transistors manufactured at a certain plant are defective. A test to identify a defective transistor is 97 % accurate. What is the probability that a transistor identified as defective actually is defective?

SOLUTION

Step 1: Determine the base rate of the population, which here is what fraction of transistors are defective at the plant (0.04).

Step 2: Using the test's accuracy and the results of the first step, determine the fraction of defective transistors that are identified by the test to be defective ($0.04 \times 0.97 = 0.0388$).

Step 3: Using the fraction of good transistors in the population and the test's false-positive rate ($1 - 0.97 = 0.03$), determine the fraction of good transistors that are identified by the test to be defective ($0.96 \times 0.03 = 0.0288$).

Step 4: Determine the desired probability, which is 100 times the fraction in step 2 divided by the sum of the fractions in steps 2 and 3 ($0.0388 / [0.0388 + 0.0288] = 0.57$) or 57 %.

Which approach is easier to understand?

7.4.3 COINCIDENCES

Conditional probability can be used to explain what appear to be rare coincidences. My wife and I took a tour of Scotland in 1999 along with five other people whom we had never met before. When we boarded the tour bus in Scotland we were astounded to find out that five out of the seven of us lived in Indiana! Our reaction was common— what a rare coincidence, especially because only about one out of every 1000 people in the world (approximately 0.1 %) live in Indiana and about 3/4 of us on the bus were from Indiana. But then I started to think and ask questions. It turns out that the United Kingdom is a very popular vacation spot for people from the midwestern and eastern United States and that a commercial airline was having a special offer that included a flight and tour of Scotland for those flying out of Chicago's O'Hare and New York City's Kennedy airports (all of those on the bus lived near Chicago or New York City).

Granted these conditions do not explain why five people from Indiana versus another midwestern or eastern state were there, but they do make this coincidence much more probable and certainly not rare.

As remarked by Stewart [2], "Because we notice coincidences and ignore noncoincidences, we make coincidences seem more significant than they really are." In fact, even today many people still attribute the occurrence of apparently rarely occurring events to mysterious causes. Perhaps it is easier to believe in an inexplicable cause than to identify the conditions under which the event occurred. Most likely, such occurrences are not so rare after all.

7.4.4 PERMUTATIONS AND COMBINATIONS

The probability of an event can be determined knowing the number of occurrences of the event and the total number of occurrences of all possible events. Finding the number of all possible events sometimes can be confusing. Consider an experiment in which there are three possible occurrences denoted by a, b, and c, each of which can occur only once without replacement. What is the probability of getting c, then a, then b, which is $Pr[cab]$? To determine this probability, the total number of ways that a, b, and c can be arranged *respective* of their order must be known. That is, the number of **permutations** of a, b, and c must be determined. The number of **permutations** of n objects is

$$n! = n(n-1)(n-2)\cdots 1, \qquad \textbf{[7.14]}$$

where $n!$ is called **n factorial**. Sterling's formula is sometimes useful, where $n! \simeq \sqrt{2\pi n}n^n \exp(-n)$, which agrees with Equation 7.14 to within 1 % for $n > 9$. So, there are six possible ways (abc, acb, bca, bac, cab, and cba) to arrange a, b, and c. Thus, $Pr[cab] = 1/6$.

Now what if the experiment had four possible occurrences, a, b, c, and d, and $Pr[cab]$ needs to be determined? The number of permutations of n objects taken m at a time, P_m^n, is

$$P_m^n = \frac{n!}{(n-m)!} = n(n-1)(n-2)\cdots(n-m+1). \qquad \textbf{[7.15]}$$

So, there are $4!/(4-3)!$ or 24 possible ways to arrange three of the four possible occurrences. Thus, $Pr[cab] = 1/24$. The probability of getting c, then a, then b is reduced from 1/6 to 1/24 when the possibility of a fourth occurrence is introduced. Often it is easy to calculate the number of permutations using a spreadsheet program. For example, using Microsoft EXCEL, the value of Equation 7.15 is given by the command PERMUT(n,m).

Further consider the same experiment, but where the number of possible combinations of c, a, and b are determined *irrespective* of the order. That is, the number of **combinations** of n objects taken m at a time, C_m^n, given by

$$C_m^n = \frac{n!}{m!(n-m)!} = \frac{n(n-1)(n-2)\cdots(n-m+1)}{m!}, \qquad \textbf{[7.16]}$$

7.2 | **MATLAB SIDEBAR**

MATLAB can be used to determine the values of $n!$, P_m^n, C_m^n, and $C_{m(r)}^n$ using a simple "for" loop to determine the factorial of a number. This is accomplished by decrementing the loop from the specified value of n down to a value of 2 by -1 while forming the product of the decremented numbers. For example, to compute $n!$, it would be written in MATLAB script:

```
x = 1;
for i = n:-1:2
    x = i*x
```

```
end
nfac = x
```

The user-defined function `pandc.m` computes $n!$, P_m^n, C_m^n, and $C_{m(r)}^n$. Its arguments are n and m. With `pandc.m` in the MATLAB workspace, simply typing `pandc(n,m)` will list the desired four values. As an example, typing `pandc(4,3)` returns the values 24, 24, 4, and 20, corresponding to $4!$, P_3^4, C_3^4, and $C_{3(r)}^4$, respectively.

---

must be found. There are only $4!/(3!1!)$ or four possible combinations of three out of four possible occurrences (*abc, abd, cdb,* and *cda*). So, the $Pr[cab] = 1/4$ for this case. That is, there is a 10-time greater chance (25 % versus 2.5 %) of getting *a, b,* and *c* in any order versus getting the particular order of *c, a,* and *b*. Using Microsoft EXCEL, the value of Equation 7.16 is given by the command `COMBIN(n,m)`.

Finally, if there is repetition with replacement, then the number of possible combinations of n objects taken m at a time with **replacement**, $C_{m(r)}^n$, is

$$C_{m(r)}^n = \frac{(m + n - 1)!}{(m!)(n - 1)!}.$$

[7.17]

For our experiment, Equation 7.17 gives $6!/(3!3!)$ or 20 possible combinations of three out of four possible occurrences. This leads to $Pr[cab] = 1/20$. Clearly, when there is repetition, the number of possible combinations increases.

7.4.5 BIRTHDAY PROBLEMS

There are two classic birthday problems that challenge one's ability to compute the probability of an occurrence [1]. The first is to determine the probability of at least two out of n people having the same birth date of the year (same day and month but not necessarily the same year). The second is to determine the probability of at least two out of n people having a *specific* birth date, such as January 18. For simplicity, assume that there are 365 days per year and that the probability of having a birthday on any day of the year is the same (both assumptions are, in fact, not true).

Consider the first problem. Often it is easier first to compute the probability that an event will *not* occur and then subtract that probability from unity to obtain the probability that the event will occur. For the *second* person there is a probability of $364/365$ ($= [366 - n]/365$, where $n = 2$) of *not* having the same birth date of the year as the first person. For the third person it is $(364/365) \cdot (363/365)$ of *not* having the same birth date of the year as the first person *or* the second person. Here each event is independent of the other, so the joint probability is the product of the two. Continuing this logic, the probability, Q, of n people *not* having the same birth date of the year is

$$Q = \frac{364}{365} \cdot \frac{363}{365} \cdots \frac{(366 - n)}{365}. \qquad [7.18]$$

With a little algebra and using the definition of the factorial, Equation 7.18 can be rewritten as

$$Q = \frac{1}{365^n} \frac{365!}{(365 - n)!} = \frac{1}{365^n} P_n^{365}. \qquad [7.19]$$

Thus, the probability, P, of at least two of n people having the same birth date of the year is

$$P = 1 - Q = 1 - \frac{1}{365^n} P_n^{365}. \qquad [7.20]$$

For $n = 10$, Equation 7.20 gives $P = 11.69$ % and for $n = 50$, $P = 97.04$ %. How many people have to be in a room to have a greater than 50 % chance of at least two people having the same birth date of the year? The answer is at least 23 people ($P = 50.73$ %).

Now examine the second problem. This differs from the first problem in that a specific birth date is specified. The probability of the second of n people not having that specific birth date is $(364/365)$ and the probability of the third of n people not having that specific birth date is the same, and so on. Thus, the probability of n people *not* having a specific birth date is

$$Q = \left[\frac{364}{365} \right]^{n-1}. \qquad [7.21]$$

The probability of at least two of n people having a *specific* birth date of the year is

$$P = 1 - Q = 1 - \left[\frac{364}{365} \right]^{n-1}. \qquad [7.22]$$

For $n = 10$, Equation 7.22 gives $P = 2.44$ % and for $n = 50$, $P = 12.58$ %. So how many people have to be in a room to have a greater than 50 % chance of at least two people having a specific birth date of the year? The answer is at least 254 people ($P = 50.05$ %).

From these two examples it is clear to see that for equal probabilities, far fewer people are required to have at least two people with the same birth date of the year than to have at least two people with a specific birth date of the year.

7.5   SAMPLE VERSUS POPULATION

Quantitative information about a process or population usually is gathered through an experiment. From this information, certain characteristics of the process or population can be estimated. This approach is illustrated schematically in Figure 7.2.

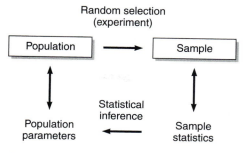

Figure 7.2   The finite sample versus the infinite population.

The **population** refers to the complete collection of all members relevant to a particular issue and the **sample** to a subset of that population. The sample is obtained through a process of random selection, where each member of the population has an equal probability of being chosen. **Statistics** of the sample, such as its **sample mean value**, $\bar{x}$, and its **sample variance**, S_x^2, can be computed. From these statistics, the population's **parameters**, which literally means *almost measurements*, such as its **true mean value**, x', and its **true variance**, σ^2, can be estimated using methods of *statistical inference*. The term *statistic* was defined by R. A. Fisher, the renowned statistician, as the number that is derived from observed measurements and that estimates a parameter of a distribution [5]. Other useful information also can be obtained using statistics, such as the probability that a future measurand will have a certain value. The interval within which the true mean and true variance are contained also can be ascertained assuming a certain level of confidence and the distribution of all possible values of the population.

The process of sampling implicitly involves random selection. When a measurement is made during an experiment, a value is selected randomly from an infinite number of possible values in the measurand population. That is, the process of selecting the next measurand value does not depend on any previously acquired measurand values. The specific value of the selected measurand is a **random variable**, which is a real number between $-\infty$ and $+\infty$ that can be associated with each possible measurand value. So, the term *random* refers to the selection process and *not* to the often-misinterpreted meaning that the acquired measurand values form a set of random numbers. If the selection process is not truly random, then erroneous conclusions about the population may be made from the sample.

7.6   PLOTTING STATISTICAL INFORMATION

Usually the first thing done after an experiment is to plot the data and to observe its trends. This data typically is a set of measurand values acquired with respect to time or space. The representation of the variation in a measurand's magnitude with respect to time or space is called its **signal**. This signal usually is one of three possible representations: **analog** (continuous in both magnitude and time or space), **discrete** (continuous in magnitude but at specific, fixed-interval values in time or space), and **digital** (having specific, fixed-interval values in both magnitude and time or space). These three representations of the same signal are illustrated in Figure 7.3 and discussed further in Chapter 11. In that figure, the analog signal is denoted by the solid curve, the digital signal by open circles, and the discrete signal by solid circles. The discrete representation sometimes is called a **scattergram**.

A cursory examination of the continuous signal displayed in Figure 7.3 shows that the signal has **variability** (its magnitude varies in time) and exhibits a **central tendency** (its magnitude varies about a mean value). How can this information be quantified? What other ways are there to view the sample such that more can be understood about the underlying physical process?

One way to view the central tendency of the signal and the frequency of occurrence of the signal's values is with a **histogram**, which literally is a *picture of cells*. Galileo may have used a frequency diagram to summarize some astronomical observations in 1632 [8]. John Graunt probably was the first to invent the histogram in 1662 in order to present the mortality rates of the black plague [9].

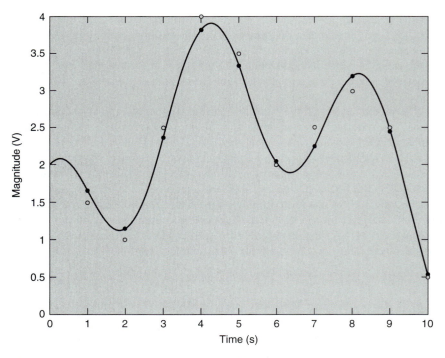

Figure 7.3   Various representations of the same signal.

7.3   │   **MATLAB SIDEBAR**

Figure 7.3 was constructed using the MATLAB M-file signals.m. The data for the solid curve was generated by creating an array of 1000 points from 0 to 10 and the corresponding magnitudes (y-values) by the commands:

```
t = 0:0.01:10;
ycont = 10*(1+0.2*sin(0.3*t)
        +0.1*cos(1.5*t))-9;
```

Strictly speaking, this results in a discrete signal. But because of the small increments in *t*, it will appear as a continuous signal in the figure. The discrete signal consists of 10 points found by the commands:

```
td = 0:1:10;
ydisc = 10*(1+0.2*sin(0.3*td)
        +0.1*cos(1.5*td))-9;
```

The digital signal was obtained by directly entering an array of magnitudes after rounding off the discrete values to the nearest 0.5:

```
ydig = [1.5,1.0,2.5,4.0,3.5,2.0,
        2.5,3.0,2.5,0.5];
```

These values also can be determined using MATLAB's round command.

The resulting plot was constructed as a series of three overlay plots because the magnitude arrays have different sizes. This was done using MATLAB's hold on command as follows:

```
plot(t,ycont,'k')
hold on
plot(td,ydisc,'kx')
hold on
plot(td,ydig,'ko')
```

If each of the *t* and *y* arrays were the same size, then the plot could be made using one command line:

```
plot(t,ycont,'k',td,ydisc,'kx',
     td,ydig,'ko')
```

Consider the *digital* representation of the signal shown in Figure 7.3. There are 10 values (1.5, 1.0, 2.5, 4.0, 3.5, 2.0, 2.5, 3.0, 2.5, and 0.5 V). The resolution of the digitization process for this case is 0.5 V. A histogram of this signal is formed by simply counting the number of times that each value occurs and then plotting the count for each value on the ordinate axis versus the value on the abscissa axis. The histogram is shown in Figure 7.4. Several features are immediately evident. The most frequently occurring value is 2.5 V. This value occurs 3 out of 10 ti   . So, it comprises 30 % of the signal's values. The range of values is from 0.5 V to 4.0 V. The average signal value appears to be between 2.0 V and 2.5 V (its actual value is 2.3 V).

What are the mechanics and rules behind constructing a histogram? In practice, there are two types of histograms. Equal-probability interval histograms have *class intervals* (bins) of variable width, each containing the same number of occurrences. Equal-width interval histograms have class intervals of fixed width, each possibly containing a different number of occurrences. The latter is used most frequently. It is more informative because it clearly shows both the frequency and the distribution of occurrences. The number of intervals, hence the interval width, and the interval origin must be determined first before constructing an equal-width interval histogram. There

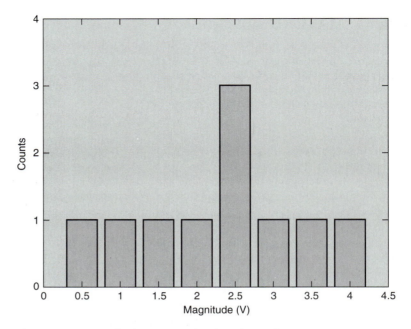

Figure 7.4   The histogram of a digital signal representation.

are many subtleties involved in choosing the optimum interval width and interval origin. The reader is referred to Scott [9] for a thorough presentation.

But how is the number of intervals chosen? Too few or too many intervals yield histograms that are not informative and do not reflect the distribution of the population. At one extreme, all the occurrences can be contained in one interval; at the other extreme, each occurrence can be in its own interval. Clearly, there must be an optimum number of intervals that yields the most representative histogram. An example of how the choice of the number of intervals affects the histogram's fidelity is presented in Figure 7.5. In that figure, the theoretical values of the population are represented by black dots in the left and center histograms and by a white curve in the right histogram. The data used for the histogram consisted of 5000 values drawn randomly from Student's t distribution [7] (this type of distribution is discussed in Chapter 8). In the left histogram, too few intervals are chosen and the population is overestimated in the left and right bins and underestimated in the center bin. In the right histogram, too many intervals are chosen and the population is consistently overestimated in almost all bins. In the middle histogram, the optimum number of class intervals is chosen and excellent agreement between the observed and expected values is achieved.

For equal-probability interval histograms, the intervals have different widths. The widths typically are determined such that the probability of an interval equals $1/K$, where K denotes the number of intervals. Bendat and Piersol [17] present a formula for K that was developed originally for continuous distributions by Mann and Wald [11] and modified by Williams [12]. It is valid strictly for $N \geq 450$ at the

7.4 | MATLAB SIDEBAR

Data that is stored in files can be read easily using MATLAB commands. The following series of MATLAB commands asks the user to identify a data file, which then is read into the MATLAB workspace by the commands:

```
filename = input('enter the
filename w/o its extension:','s');
ext = input('enter the files
extension, e.g., dat: ','s');
num = input('enter the column
number of the data to plot
(e.g., 1): ');
eval(['load ',filename,'.',ext]);
```

The number of rows and columns of the data array can be determined by the command:

```
[rows,cols] = size(eval
([filename,'(:,1)']));
```

Part or all of a particular column of the array can be named. For example, say that the data file was named `testone.dat`. Assume that the first two rows of each column are text identifiers that name the variable and its units. The third row and beyond is the actual data. If the first column were pressure data, then the actual data could be identified using the statement:

```
pressure = testone(3:length
                (testone),1);
```

The MATLAB `length` command determines the number of rows. The argument `3:length(testone)` identifies the third through last rows and the argument `1` the first column. The arguments `3:rows,1` would do the same for this case. Such commands can be embedded into the beginning of an M-file that is written to analyze and plot data.

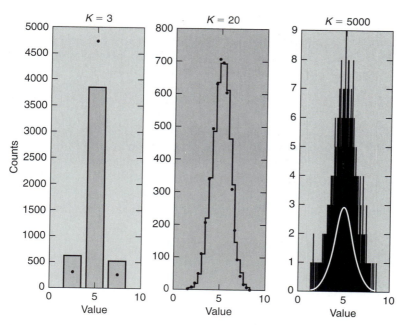

Figure 7.5 Histograms with different numbers of intervals for the same data.

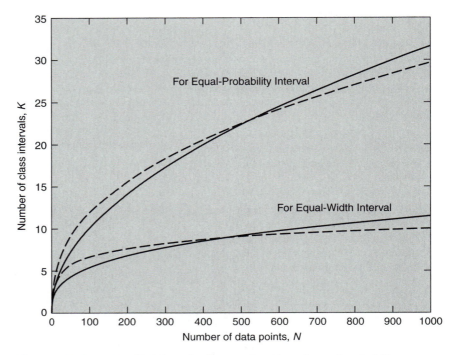

Figure 7.6 $K = f(N)$ formulas for equal-width and equal-probability histograms. Equal-probability interval symbols: modified Mann and Wald formula (dash) and square-root formula (solid). Equal-width interval symbols: Sturgis formula (dash) and Scott formula (solid).

95 % confidence level, although the Mann and Wald [11] state that it is probably valid for $N \geq 200$ or even lower N. The exact expression given by Mann and Wald is $K = 2[2(N-1)^2/c^2]^{0.2}$, where $c = 1.645$ for 95 % confidence. Various spreadsheets as well as Montgomery and Runger [4] suggest the formula $K = \sqrt{N}$, which agrees to within 10 % with the modified Mann and Wald formula up to approximately $N = 1000$.

7.5   MATLAB SIDEBAR

The M-file 3histos.m was used to construct Figure 7.5. It reads data from a user-specified file and then plots three histograms side by side. The number of intervals for each of the three histograms is user-specified. Theoretical values are generated by specifying the distribution of the population, which currently is set in the M-file to be the normal distribution.

7.6 | MATLAB SIDEBAR

The MATLAB M-file `mw.m` calculates the factor c and K of the Mann and Wald formula for user-specified values of the % level of significance (= 100 % − % confidence) and N. As an example, for $N = 250$ and 10 % level of significance, `mw.m` determines $c = 1.2816$ and $K = 18$.

For equal-width interval histograms, the interval width is constant and equal to the range of values (maximum minus minimum values) divided by the number of intervals. Sturgis's formula [10] determines K from the number of binomial coefficients needed to have a sum equal to N and can be used for values of N as low as approximately 30. Based on this formula, various authors, such as Rosenkrantz [3], suggest using values of K between 5 and 20 (this would cover between approximately $N = 2^5 = 32$ to $N = 2^{20} \simeq 10^6$). Scott's formula for K ([13] and [14]), valid for $N \geq 25$, was developed to minimize the integrated mean square error that yields the best fit between Gaussian data and its parent distribution. The exact expression is $\Delta x = 3.49\sigma N^{-1/3}$, where Δx is the interval width and σ is the standard deviation. Using this expression and assuming a range of values based on a certain % coverage, an expression for K can be derived.

The formulas for K for both types of histograms are presented in Table 7.1 and displayed in Figure 7.6. The number of intervals for equal-probability interval histograms at a given value of N is approximately two to three times greater than the corresponding number for equal-width interval histograms. More intervals are required for the former case because the center intervals must be narrow and numerous to maintain the same probability as those intervals on the tails of the distribution. Also, because of the low probabilities of occurrence at the tails of the distribution, some intervals for equal-width histograms may need to be combined to achieve greater than five occurrences in an interval. This condition (see [16]) is necessary for proper comparison between theoretical and experimental distributions. However, for small samples it may not be possible to meet this condition. Another very important condition that must be followed is that $\Delta x \geq u_x$, where u_x is the uncertainty of the measurement of x. That is, the interval width should *never* be smaller than the uncertainty of the measurand.

Table 7.1 Formulas for the number of histogram intervals with 95 % confidence.

Interval Type	Formula	Reference
Equal probability	$K = 1.87(N − 1)^{0.40}$	[11], [12], [17]
Equal probability	$K = \sqrt{N}$	[4], [15]
Equal width	$K = 3.322 \log N$	[10]
Equal width	$K = 1.15N^{1/3}$	[13], [14]

The MATLAB M-file `kforN.m` was used to construct Figure 7.6. The range of N can be varied by the user.

To construct equal-width histograms, these steps need to be followed:

1. Identify the minimum and maximum values of the measurand x, x_{min}, and x_{max}, thereby finding its range, $x_{range} = x_{min} - x_{max}$.

2. Determine the number of class intervals, K, using the appropriate formula for equal-width histograms. Preferably, this should be Scott's formula.

3. Calculate the width of each interval, where $\Delta x = x_{range}/K$.

4. Count the number of occurrences, n_j ($j = 1$ to K), in each Δx interval. Check that the sum of all the n_j's equals N, the total number of data points.

5. Be sure that the conditions for $n_j > 5$ and $\Delta x \geq u_x$ are met.

6. Plot n_j versus xm_j, where xm_j is discretized as the midpoint value of each interval.

Instead of examining the distribution of the *number* of occurrences of various magnitudes of a signal, the *frequency* of occurrences can be determined and plotted. The plot of $n_j/N = f_j$ versus xm_j is known as the **frequency distribution** (sometimes called the relative frequency distribution). The area bounded laterally by any two x-values in a frequency distribution equals the frequency of occurrence of that range of x-values or the probability that x will assume values in that range. Also, the sum of all the f_j's equals 1. The frequency distribution often is preferred over the histogram because it directly displays the probabilities of occurrence. Further, as the sample size becomes large, the sample's frequency distribution becomes similar to the distribution of the population's probabilities, which is called the probability density function.

The distribution of all of the values of the infinitely large population is given by its probability density function, $p(x)$. This will be defined in Section 7.7. Typically, $p(x)$ is normalized such that the integral of $p(x)$ over all x equals unity. This effectively sets the sum of the probabilities of all the values between $-\infty$ and $+\infty$ to be unity or 100 %. Similar to the frequency distribution, the area under the portion of the probability density function over a given measurand range equals the % probability that the measurands will have values in that range.

The M-file `histo.m` plots the equal-width interval histogram of a selected column from a user-specified data file. Scott's formula is used for K, although other formulas are provided. The MATLAB command `[cnts,binloc] = hist(data,k)` determines the number of counts `cnts` in each of `k` intervals from the array `data`. The abscissa position for each interval is specified by the array `binloc`. The command `bar(binloc,cnts)` plots the histogram.

7.9 | **MATLAB SIDEBAR**

The M-file `hf.m` plots the histogram and frequency distribution of a selected column of data. The frequency distribution is obtained by the MATLAB command `bar(binloc,cnts/N)` after determining the `binloc` and `cnts` arrays using MATLAB's `hist` command. An example of the resulting figure for a data array of 2564 measurements is shown in Figure 7.7.

To properly compare a frequency distribution with an assumed probability density function on the same graph, the frequency distribution first must be converted into a **frequency density distribution**. The frequency density is denoted by f_j^*, where $f_j^* = f_j/\Delta x$. This is because the probability density function is related to the frequency distribution by the following:

$$p(x) = \lim_{N\to\infty,\Delta x\to 0} \sum_{j=1}^{K} \frac{f_j}{\Delta x} = \lim_{N\to\infty,\Delta x\to 0} \sum_{j=1}^{K} f_j^*. \qquad \textbf{[7.23]}$$

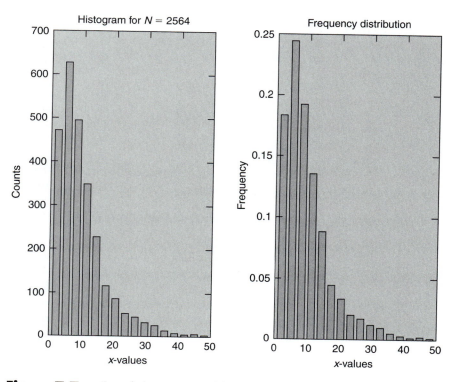

Figure 7.7   Sample histogram and frequency distributions of the same data.

The N required for this comparative limit to be attained within a certain confidence can be determined using the law of large numbers [3], which is

$$N \geq \frac{1}{4\epsilon^2(1 - P_o)}.$$ **[7.24]**

This law, derived by Jacob Bernoulli (1654–1705), considered the father of the quantification of uncertainty, was published posthumously in 1713 [5]. The law determines the N required to have a probability of at least P_o that $f_j^*(x)$ differs from $p(x)$ by less than ϵ.

Example 7.5

STATEMENT

Determine how many tosses would have to be made to assess whether or not a coin used in a coin toss is fair.

SOLUTION

Assume one wants to be at least 68 % confident ($P_o = 0.68$) that the coin's fairness is assessed to within 5 % ($\epsilon = 0.05$). Then, according to Equation 7.24, the coin must be tossed at least 310 times ($N \geq 310$).

Often the frequency distribution of a finite sample is used to identify the **probability density function** of its population. The probability density function's shape tells much about the physical process governing the population. Once the probability density function is identified, much more information about the process can be obtained.

7.7   THE PROBABILITY DENSITY FUNCTION

Consider the time history record of a random variable, $x(t)$. Its probability density function, $p(x)$, reveals how the values of $x(t)$ are distributed over the entire range of $x(t)$ values. The probability that $x(t)$ will lie between x^* and $x^* + \Delta x$ is given by

$$p(x) = \lim_{\Delta x \to 0} \frac{\Pr[x^* < x(t) \leq x^* + \Delta x]}{\Delta x}.$$ **[7.25]**

The probability that $x(t)$ is in the range x to $x + \Delta x$ over a total time period T also can be determined. Assume that T is large enough such that the statistical properties are truly representative of that time history. Further assume that a single time history is sufficient to fully characterize the underlying process. These assumptions are discussed further in Chapter 11. For the time history depicted in Figure 7.8, the total amount of time during the time period T that the signal is between x and $x + \Delta x$ is given by T_x, where

$$T_x = \sum_{j=1}^{m} \Delta t_j$$ **[7.26]**

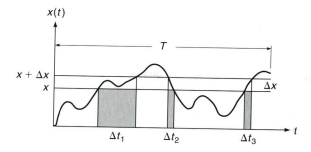

Figure 7.8   A time history record.

for m occurrences. In other words,

$$\Pr[x < x(t) \le x + \Delta x] = \lim_{T \to \infty} \left[\frac{T_x}{T} \right] = \lim_{T \to \infty} \frac{1}{T} \sum_{j=1}^{m} \Delta t_j. \qquad \textbf{[7.27]}$$

This implies that

$$p(x) = \lim_{\Delta x \to 0} \frac{1}{\Delta x} \left[\lim_{T \to \infty} \frac{1}{T} \sum_{j=1}^{m} \Delta t_j \right] = \lim_{\Delta x \to 0, T \to \infty} \left[\frac{T_x / T}{\Delta x} \right]. \qquad \textbf{[7.28]}$$

Likewise, x could be the number of occurrences of a variable with a Δx interval, n_j, where the total number of occurrences is N. Here N is like T and n_j is like T_x, so

$$p(x) = \lim_{\Delta x \to 0} \frac{1}{\Delta x} \left[\lim_{N \to \infty} \sum_{j=1}^{m} \frac{n_j}{N} \right] = \lim_{\Delta x \to 0, N \to \infty} \sum_{j=1}^{m} \left[\frac{n_j / N}{\Delta x} \right]. \qquad \textbf{[7.29]}$$

Equations 7.28 and 7.29 show that the limit of the frequency density distribution is the probability density function.

The probability density function of a signal that repeats itself in time can be found by applying the aforementioned concepts. To determine the probability density function of this type of signal, the signal only needs to be examined over one period T. Equation 7.28 then becomes

$$p(x) = \frac{1}{T} \lim_{\Delta x \to 0} \frac{1}{\Delta x} \sum_{j=1}^{m} \Delta t_j. \qquad \textbf{[7.30]}$$

Now as $\Delta x \to 0$, $\Delta t_j \to \Delta x \cdot |dt/dx|_j$. Thus, in the limit as $\Delta x \to 0$, Equation 7.30 becomes

$$p(x) = \frac{1}{T} \sum_{j=1}^{m} \left| \frac{dt}{dx} \right|_j, \qquad \textbf{[7.31]}$$

noting that m is the number of times the signal is between x and $x + \Delta x$.

Example 7.6

STATEMENT

Determine the probability density function of the periodic signal $x(t) = x_o \sin(\omega t)$ with $\omega = 2\pi/T$.

SOLUTION

Differentiation of the signal with respect to time yields

$$dx = x_o \omega \cos(\omega t) dt \quad \text{or} \quad dt = \frac{dx}{x_o \omega \cos(\omega t)}. \qquad \textbf{[7.32]}$$

Now, this particular signal resides two times during one period in the x to $x + \Delta x$ interval. Thus, for this signal, using Equations 7.31 and 7.32, the probability density function becomes

$$p(x) = \frac{\omega}{2\pi} 2 \left| \frac{1}{x_o \omega \cos(\omega t)} \right| = \left| \frac{1}{\pi x_o \cos(\omega t)} \right|. \qquad \textbf{[7.33]}$$

The probability density functions of other deterministic, continuous functions of time can be found using the same approach.

The probability density function also can be determined graphically by analyzing the time history of a signal in the following manner:

1. Given $x(t)$ and the sample period, T, choose an amplitude resolution Δx.

2. Determine T_x, then $T_x/(T \Delta x)$, noting also the midpoint value of x for each Δx interval.

3. Construct the probability density function by plotting $T_x/(T_x \Delta x)$ for each interval on the ordinate (y-axis) versus the midpoint value of x for that interval on the abscissa (x-axis).

Note that the same procedure can be applied to examining the time history of a signal that does not repeat itself in time. By graphically determining the probability density function of a known periodic signal, it is easy to observe the effect of the amplitude resolution Δx on the resulting probability density function in relation to the exact probability density function.

Example 7.7

STATEMENT

Consider the signal $x(t) = x_o \sin(2\pi t/T)$. For simplicity, let $x_o = 1$ and $T = 1$, so $x = \sin(2\pi t)$. Describe how the probability density function could be determined graphically.

SOLUTION

First choose $\Delta x = 0.10$. As illustrated in Figure 7.9, for the interval $0.60 < \sin(2\pi t) \le 0.70$, $T_x = 0.020 + 0.020 = 0.040$, which yields $T_x/(T_x \Delta x) = 0.40$ for the midpoint value of $x = 0.65$. Likewise, for the interval $0.90 < \sin(2\pi t) \le 1.00$, $T_x = 0.14$, which yields $T_x/(T \Delta x) = 1.40$ for the midpoint value of $x = 0.95$. Using this information gathered for all Δx intervals, an estimate of the probability density function can be made by plotting $T_x/(T/\Delta x)$ versus x.

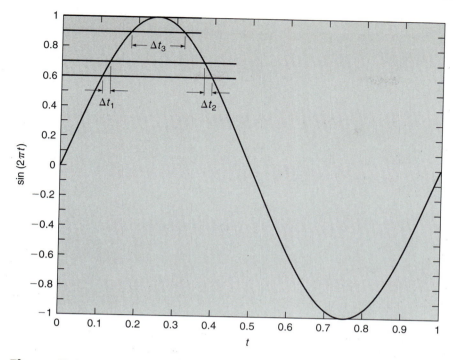

Figure 7.9   Constructing p(x) from the time history record.

7.10 | MATLAB SIDEBAR

The probability density function of a periodic signal can be estimated and plotted using MATLAB. The periodic signal $x(t) = x_o \sin(2\pi t/T)$ is shown in the left-most plot of Figure 7.10. Two approximations of the probability density function using $\Delta x = 0.40$ and 0.025 are shown in the two right-most plots. The exact probability density function expression for this signal is given by Equation 7.33. It is symmetric about the $x(t) = 0$ line, with a minimum value of $1/\pi = 0.3183$ at $x(t) = 0$. As Δx is reduced, the resulting probability density function more closely approximates the exact one. The three plots in Figure 7.10 were determined using the MATLAB m-file xpofx.m. This M-file accepts a user-specified $x(t)$ and two Δx resolutions.

One simple way of interpreting the underlying probability density function of a signal is to consider it as the projection of the density of the signal's amplitudes, as illustrated in Figure 7.11 for the signal $x = \sin(2\pi t)$. The more frequently occurring values appear more dense as viewed along the horizontal axis.

In some situations, particularly when examining the time history of a random signal, the aforementioned procedure of measuring the time spent at each Δx interval to determine $p(x)$ becomes quite laborious. Recall that if $p(x)\,dx$ is known, the probability of occurrence of x for any range of x is given by

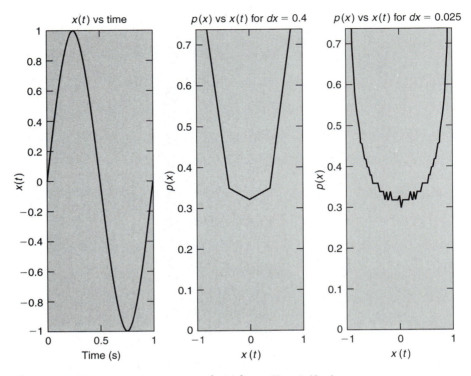

Figure 7.10   Approximations of $p(x)$ from $x(t) = \sin(2\pi t)$.

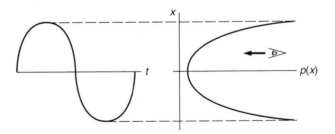

Figure 7.11   Projection of signal's amplitude densities.

$$p(x)\,dx = \lim_{\Delta x \to 0} p(x)\Delta x = \lim_{T \to \infty} \left[\frac{T_x}{T}\right]. \qquad \textbf{[7.34]}$$

Recognizing this, an alternative approach can be used to determine the quantity $p(x)\,dx$ by choosing a very small Δx, such as the thickness of a pencil line, as illustrated in Figure 7.12. If a horizontal line is moved along the amplitude axis at

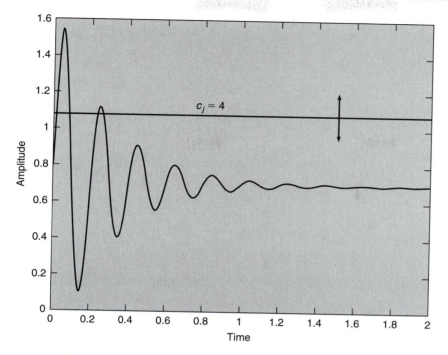

Figure 7.12 Graphical approach to determine $p(x)\, dx$.

constant-amplitude increments and number of times the line crosses the signal is determined for each amplitude increment, C_x, then

$$p(x)\, dx = \lim_{C \to \infty} \left[\frac{C_x}{C} \right],$$ **[7.35]**

where C is a very large number. Note that $p(x)\, dx$ was determined by this approach, *not* $p(x)$, as was done previously.

7.8 CENTRAL MOMENTS

Once the probability density function of a signal has been determined, this information can be used to determine the values of various parameters. These parameters can be found by computing the **central moments** of the probability density function. Computations of statistical moments are similar to those performed to determine mechanical moments, such as the moment of inertia of an object. The term *central* refers to the fact that the various statistical moments are computed with respect to the centroid or mean of the probability density function of the population.

The *mth central moment* is defined as

$$\langle (x - x')^m \rangle = E[(x - x')^m] = \mu_m \equiv \int_{-\infty}^{+\infty} (x - x')^m p(x) \, dx. \qquad \textbf{[7.36]}$$

The $\langle \; \rangle$ or $E[\;]$ denote the **expected value** or **expectation** of the quantity inside the brackets. This is the value that is expected (in the probabilistic sense) if the integral is performed.

When the centroid or mean, x', equals 0, the central moments are known as moments about the origin. Equation 7.36 becomes

$$\langle x^m \rangle \equiv \int_{-\infty}^{+\infty} x^m p(x) \, dx = \mu'_m. \qquad \textbf{[7.37]}$$

Further, the central moment can be related to the moment about the origin by the transformation

$$\mu_m = \sum_{i=0}^{m} (-1)^i \binom{m}{i} \mu_1^i \mu'_{m-i}, \quad \text{where} \quad \binom{m}{i} = \frac{m!}{i!(m-i)!}. \qquad \textbf{[7.38]}$$

The **zeroth central moment**, μ_0, is an identity

$$\mu_0 = \int_{-\infty}^{+\infty} p(x) \, dx = 1. \qquad \textbf{[7.39]}$$

Having $\mu_0 = 1$ assures that $p(x)$ is normalized correctly.

The **first central moment**, μ_1, leads to the definition of the mean value (the centroid of the distribution). For $m = 1$,

$$\left\langle (x - x')^1 \right\rangle = \int_{-\infty}^{+\infty} (x - x') p(x) \, dx. \qquad \textbf{[7.40]}$$

Expanding the left-hand side of Equation 7.40 yields

$$\langle (x - x') \rangle = \langle x \rangle - \langle x' \rangle = \langle x \rangle - x' = 0, \qquad \textbf{[7.41]}$$

because the expectation of x, $\langle x \rangle$, is the true mean value of the population, x'. Hence, $\mu_1 = 0$. Now expanding the right-hand side of Equation 7.40 reveals that

$$\int_{-\infty}^{+\infty} (x - x') p(x) \, dx = \int_{-\infty}^{+\infty} x p(x) \, dx - x' \int_{-\infty}^{+\infty} p(x) \, dx = \int_{-\infty}^{+\infty} x p(x) \, dx - x'. \qquad \textbf{[7.42]}$$

Because the right-hand side of the equation must equal zero, it follows that

$$x' = \int_{-\infty}^{+\infty} x p(x) \, dx. \qquad \textbf{[7.43]}$$

Equation 7.43 is used to compute the mean value of a distribution given its probability density function.

The **second central moment**, μ_2, defines the **variance**, σ^2, as

$$\mu_2 = \int_{-\infty}^{+\infty} (x - x')^2 p(x)\, dx = \sigma^2, \qquad \textbf{[7.44]}$$

which has units of x^2. The **standard deviation**, σ, is the square root of the variance. It describes the width of the probability density function.

The variance of x can be expressed in terms of the expectation of x^2, $E[x^2]$, and the square of the mean of x, x'^2. Equation 7.36 for this case becomes

$$\begin{aligned}
\sigma^2 &= E[(x - x')^2] \\
&= E[x^2 - 2xx' + x'^2] \\
&= E[x^2] - 2x' E[x] + x'^2 \\
&= E[x^2] - 2x'x' + x'^2 \\
&= E[x^2] - x'^2. \qquad \textbf{[7.45]}
\end{aligned}$$

So, when the mean of x equals 0, the variance of x equals the expectation of x^2. Further, if the mean and variance of x are known, then $E[x^2]$ can be computed directly from Equation 7.45.

STATEMENT

Example 7.8

Determine the *mean* power dissipated by a 2 Ω resistor in a circuit when the current flowing through the resistor has a mean value of $3\,a$ and a variance of $0.4\,a^2$.

SOLUTION

The power dissipated by the resistor is given by $P = I^2 R$, where I is the current and R is the resistance. The mean power dissipated is expressed as $E[P]$. Assuming that R is constant, $E[P] = E[I^2 R] = R E[I^2]$. Further, using Equation 7.45, $E[I^2] = \sigma_I^2 + I'^2$. So, $E[P] = R(\sigma_I^2 + I'^2) = 2(0.4 + 3^2) = 18.8$ W. Expressed with the correct number of significant figures (one), the answer is 20 W.

The **third central moment**, μ_3, is used in the definition of the **skewness**, Sk, where

$$\mathrm{Sk} = \frac{\mu_3}{\sigma^3} = \frac{1}{\sigma^3} \int_{-\infty}^{+\infty} (x - x')^3 p(x)\, dx. \qquad \textbf{[7.46]}$$

Defined in this manner, the skewness has no units. It describes the symmetry of the probability density function, where a positive skewness implies that the distribution is skewed or stretched to the right, as shown in Figure 7.13. For this case the probability density function's mean is greater than its mode, where the **mode** is the most frequently occurring value. A negative skewness implies the opposite (stretched to the left with mean < mode). The sign of the mean minus the mode is the sign of the skewness. For the normal distribution, Sk = 0 because the mean equals the mode.

The **fourth central moment**, μ_4, is used in the definition of the **kurtosis**, Ku, where

$$\mathrm{Ku} = \frac{\mu_4}{\sigma^4} = \frac{1}{\sigma^4} \int_{-\infty}^{+\infty} (x - x')^4 p(x)\, dx, \qquad \textbf{[7.47]}$$

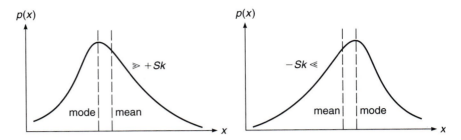

Figure 7.13 Distributions with positive and negative skewness.

which has no units. The kurtosis describes the peakedness of the probability density function. A *leptokurtic* probability density function has a slender peak, a *mesokurtic* one a middle peak, and a *platykurtic* one a flat peak.

For the normal distribution, Ku = 3. Sometimes, another expression is used for the kurtosis, where Ku* = Ku − 3 such that Ku* < 0 implies a probability density function that is flatter than the normal probability density function and Ku* > 0 implies one that is more peaked than the normal probability density function.

For the special case in which x is a normally distributed random variable, the mth central moments can be written in terms of the standard deviation, where $\mu_m = 0$

7.11 | MATLAB SIDEBAR

The MATLAB command moment(X,m) determines the mth central moment of the vector X. The second central moment is calculated using a divisor of N instead of $N - 1$, where N is the sample size. The variance is moment(X,2), the skewness moment(X,3)/moment(X,2)$^{1.5}$ and the kurtosis moment(X,4)/moment(X,2)^{2}. As an example, for $X = [1, 2, 3, 4, 5, 6, 7, 8, 9]$, moment$(X, 2) = 6.6667$, moment $(X, 3) = 0$, and moment$(X,4) = 78.6667$. This yields $\sigma^2 = 6.6667$, Sk = 0, and Ku = 78.6667/$6.6667^2 = 1.7700$.

MATLAB also has many functions that are used to statistically quantify a set of numbers. Information about any MATLAB function can be obtained by simply typing help xxx, where xxx denotes the specific MATLAB command. The MATLAB command mean(X) provides the mean of the vector X. If the matrix called data has more than one column, then each column can be vectorized using the statement X(i) = data(:,i), which assigns the vector $X(i)$ the elements in the ith column of the matrix data. Subsequently, one can find the minimum value of $X(i)$ by the command min(X(i)), the maximum by max(X(i)), the standard deviation by std(X(i)), the variance by [std(X(i))]^2 and the root-mean-square by norm(X(i))/sqrt(N). The command norm(X(i),p) is the same as the command sum(abs(X(i)).^p)^(1/p). The default value of p is 2 if a value for p is not entered into the argument of norm. Thus, norm(X(i)) is the same as norm(X(i),2).

7.12	**MATLAB SIDEBAR**

The user-defined function stats.m computes the values of the following statistics of the vector X: number of values, minimum, maximum, mean, standard deviation, variance, root-mean-square, skewness, and kurtosis. The MATLAB commands skewness(X) and kurtosis(X) are used to compute their values. For the vector X specified in the previous Sidebar, typing stats(X) results in the following values: number = 9, minimum = 1, maximum = 9, mean = 5, standard deviation = 2.7386, variance = 7.5000, rms = 5.6273, skewness = 0, and kurtosis = 1.7700.

---

when m is odd and > 1, and $\mu_m = 1 \cdot 3 \cdot 5 \cdots (m-1)\sigma^m$ when m is even and > 1. This formulation obviously is useful in determining higher-order central moments of a normally distributed variable when the standard deviation is known.

7.9 THE PROBABILITY DISTRIBUTION FUNCTION

The probability that a value x is less than or equal to some value of x^* is defined by the **probability distribution function**, $P(x)$. Sometimes this also is referred to as the cumulative probability distribution function. The probability distribution function is expressed in terms of the integral of the probability density function

$$P(x^*) = \Pr[x \le x^*] = \int_{-\infty}^{x^*} p(x)\, dx. \qquad \textbf{[7.48]}$$

From this, the probability that a value of x will be between the values of x_1^* and x_2^* becomes

$$\Pr[x_1^* \le x \le x_2^*] = P(x_2^*) - P(x_1^*) = \int_{x_1^*}^{x_2^*} p(x)\, dx. \qquad \textbf{[7.49]}$$

Note that $p(x)$ has units of $1/x$. The units of $P(x)$ are dimensionless.

An example $p(x)$ and $P(x)$ are shown in Figure 7.14. The probability density function is in the left figure and the probability distribution function in the right figure. The $P(x)$ values for each x-value are determined simply by finding the area under the $p(x)$ curve up to each x-value. For example, the area at the value of $x = 1$ is a triangular area equal to $0.5 \times 1.0 \times 1.0 = 0.5$. This is the corresponding value of the probability distribution function at $x = 1$. Note that the probability density in this example is not normalized. What does the maximum value of the probability distribution function have to be for the probability density function to be normalized correctly, as described in Section 7.8? The answer is unity. So, to normalize $p(x)$ correctly, all values of $p(x)$ should be divided by 1/3 in this example.

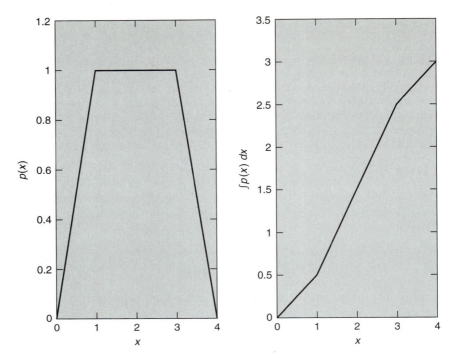

Figure 7.14   Example probability density (left) and probability distribution (right) functions.

REFERENCES

[1] I. Peterson. 1998. *The Jungles of Randomness, A Mathematical Safari*. New York: John Wiley and Sons.

[2] I. Stewart. 1998. What a Coincidence! *Scientific American* June: 95–96.

[3] W. A. Rosenkrantz. 1997. *Introduction to Probability and Statistics for Scientists and Engineers*. New York: McGraw-Hill.

[4] D. C. Montgomery and G. C. Runger. 1994. *Applied Statistics and Probability for Engineers*. New York: John Wiley and Sons.

[5] D. Salsburg. 2001. *The Lady Tasting Tea: How Statistics Revolutionized Science in the Twentieth Century*. New York: W.H. Freeman and Company.

[6] U. Hoffrage, S. Lindsey, R. Hertwig, and G. Gigerenzer. 2000. Communicating Statistical Information. *Science* 290: 2261–2262.

[7] Student. 1908. The Probable Error of a Mean. *Biometrika* 4: 1–25.

[8] A. Hald. 1990. *A History of Probability and Statistics and Their Application Before 1750*. New York: John Wiley and Sons.

[9] D. W. Scott. 1992. *Multivariate Density Estimation: Theory, Practice and Visualization*. New York: John Wiley and Sons.

[10] H. A. Sturgis. 1926. The Choice of a Class Interval. *Journal of the American Statistical Association* 21: 65–66.

[11] H. B. Mann and A. Wald. 1942. On the Choice of the Number of Class Intervals in the Application of the Chi Square Test. *Annals of Mathematical Statistics* September: 306–317.

[12] C. A. Williams, Jr. 1950. On the Choice of the Number and Width of Classes for the Chi-Square Test of Goodness of Fit. *American Statistical Association Journal* 45: 77–86.

[13] D. W. Scott. 1979. On Optimal and Data-Based Histograms. *Biometrika* 66: 605–610.

[14] A. Stuart and J. K. Ord. 1994. *Kendall's Advanced Theory of Statistics*, Vol. 1, *Distribution Theory*, 6th ed. New York: John Wiley and Sons.

[15] L. Kirkup. 1994. *Experimental Methods: An Introduction to the Analysis and Presentation of Data*. New York: John Wiley and Sons.

[16] A. Hald. 1952. *Statistical Theory with Engineering Applications*. New York: John Wiley and Sons.

[17] J. S. Bendat and A. G. Piersol. 1966. *Measurement and Analysis of Random Data*. New York: John Wiley and Sons.

REVIEW PROBLEMS

1. Assuming equal probability of being born any day of the year, match each of the following birthday possibilities (for one person) to its correct probability:

birthday occurring on the 31st of a month	0.0849
birthday occurring in August	0.329
birthday occurring Feb. 29, 1979 (for a person born in that year)	0
birthday occurring in a month with 30 days	0.0192

2. One of each U.S. coin currencies is placed into a container (a penny, a nickel, a dime, and a quarter). Given that the withdrawal of a coin from the container is random, find the correct value for each of the following described quantities:

(a) If two coins are drawn, find the probability of any one permutation occurring.

(b) If three coins are drawn without replacement, find the probability of the total being the maximum possible monetary value.

(c) If three coins are drawn with replacement, find the probability of the total being the maximum possible monetary value.

(d) If two coins are drawn with replacement, find the probability of the total number of cents being even.

3. A sports bar hosts a gaming night where students play casino games using play money. A business major has $1500 in play money and decides to test a strategy on the roulette wheel. The minimum bet is $100 with no maximum. He decides to bet that the ball will land on red each time the wheel is spun. On the first bet, he bets the minimum. For each consecutive spin of the wheel, he doubles his previous bet. He decides beforehand that he will play roulette the exact number of times that his cash stock would allow if he lost each time consecutively. What is the probability that he will run out of money before leaving the table?

4. An engineering student samples the wall pressure exerted by a steady-state flow through a pipe 1233 times using an analog-to-digital converter. Using the recommendations made in this chapter, how many equal-interval bins should the student use to create a histogram of the measurements? Respond to the nearest whole bin.

5. Given the probability density function pictured in Figure 7.15, compute the height, h, that conserves the zeroth central moment.

6. Compute the first central moment from the PDF pictured in Figure 7.16.

7. Compute the kurtosis of the PDF pictured in Figure 7.15.

8. Compute the skewness of the PDF pictured in Figure 7.15.

9. Compute the standard deviation of the PDF pictured in Figure 7.16.

10. A diagnostic test is designed to detect a cancer-precursor enzyme that exists in 1 out of every 1000 people. The test falsely identifies the presence of the enzyme in 50 out of 1000 people who actually do not have the enzyme. What is the percent chance that a person identified as having the enzyme actually does have the enzyme?

11. What is the chance that you will throw either a 3 or a 5 on the toss of a fair die?

    (a) 1/12, **(b)** 1/6, **(c)** 1/3, **(d)** 1/2, **(e)** 1/250

12. A pressure transducer's output is in units of volts. N samples of its signal are taken each second. The frequency density distribution of the sampled data has what units?

    (a) 1/volts, **(b)** volts times seconds, **(c)** volts/N, **(d)** none—it is nondimensional, **(e)** seconds

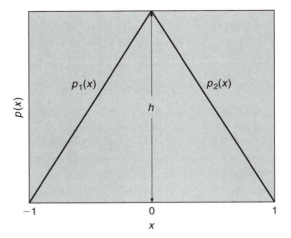

Figure 7.15   A triangular probability density function.

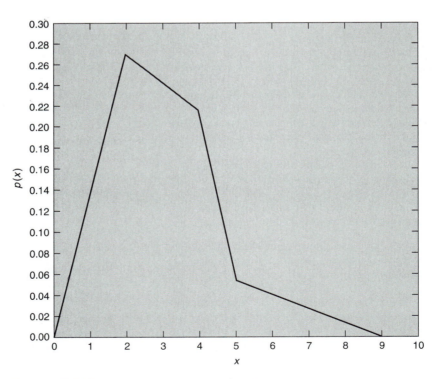

Figure 7.16   A probability density function.

13. What is the kurtosis?
(**a**) bad breath, (**b**) the fourth central moment,
(**c**) the mean minus the mode of a distribution,

(**d**) the name of a new, secret football play that
hopefully will make a difference next season,
(**e**) the square of the standard deviation

HOMEWORK PROBLEMS

1. Determine [a] the percent probability that at least two out of 19 students in a classroom will have a birthday on the *same* birth date of the year, [b] how many people would have to be in the room in order to have a greater-than-50 % chance to have a birthday on the *same* birth date of the year, and [c] the percent probability that at least 2 of the 19 students will have a birthday on a *specific* birth date of the year.

2. A cab was involved in a hit-and-run accident during the night near a famous midwestern university. Two cab companies, the Blue and the Gold, operate in the city near the campus. There are two facts: (1) 85 % of the cabs in the city are Gold and 15 % are Blue, and (2) a witness identified the cab as Blue. The court tested the reliability of the witness under the same circumstances that existed the night of the accident and concluded that the witness correctly identified each of the two colors 80 % of the time and failed to do so 20 % of the time. Determine the percent probability that the cab involved in the accident was Blue.

3. A diagnostic test is designed to detect a bad aircraft component whose prevalence is 1 in 1000. The test has a false positive rate of 5 %, where it identifies a *good* component as *bad* 5 % of the time. What is the

percent chance that a component identified as *bad* really is bad?

4. Use the data file `diam.dat`. This text file contains two columns (and approximately 2500 rows) of time (s) and diameter (μm) data. Copy it into your own directory. For the diameter data only (column 2), using MATLAB, plot [a] its histogram and [b] its frequency distribution. Use Sturgis's formula for the number of bins, K, as related to the number of data points, N: $K = 1 + 3.322 \log_{10} N$. (Hint: MATLAB's function `hist(x,k)` plots the histogram of x with k bins.) The statement `[a,b] = hist(x,k)` produces the column matrices a and b, where a contains the counts in each bin and b is the center location coordinate of each bin. MATLAB's function `bar(b,a/N)` will plot the frequency distribution, where N is the total number of x values.

5. Using the graph of the probability density function of an underlying population presented in Figure 7.16, determine [a] the percent probability that one randomly selected value from this population will lie between the values of 2 and 5. If a sample of 20 values are drawn randomly from this population, determine [b] how many will have values greater than 2 and [c] how many will have values greater than 5.

CHAPTER REVIEW

ACROSS

8. set of all members in both *A* and *B*

9. set of all members of *A* or *B* or both

11. the most frequently occurring value

12. continuous in both magnitude and time or space

13. with specific, fixed intervals both in magnitude and in time or space

DOWN

1. third central moment is used to define this

2. relates to the peakedness of the distribution

3. the integral of the probability density function is the probability (?) function

4. arrangement irrespective of order

5. continuous in magnitude and at fixed intervals of time or space

6. second central moment

7. probability that depends on a previous event

10. set comprised of "not the event"

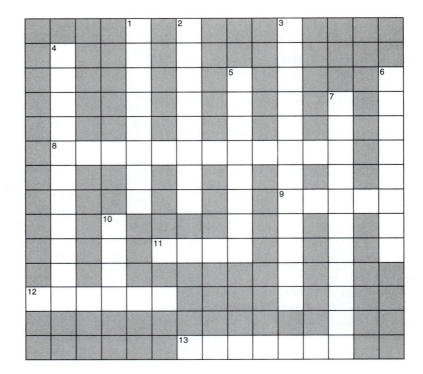

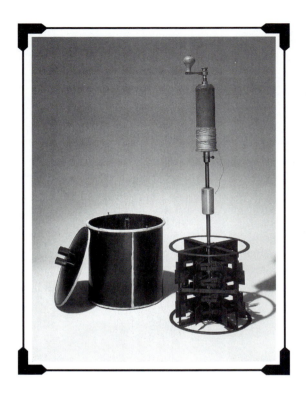

Joule's Water-Friction Apparatus

Credit: Science Museum / Science & Society Picture Library, London.

This particular apparatus laid the foundation for the principle of conservation of energy by experimentally determining the equivalency between mechanical work and heat. James Prescott Joule (1818–1889), an English scientist, was an extraordinary experimentalist. His apparatus was simple and elegant. A thermally well-insulated copper drum filled with water served as the calorimeter. Inside it was a brass paddle wheel that stirred the water but did not rotate it. The wheel's drive shaft was connected via a fine wire through a pulley to a weight. The change in potential energy of the weight as it fell over a certain distance was related to the increase in the temperature of the water to within 0.005 °F. Joule was a careful experimentalist who believed in repeating experiments until there was very little difference in the results. He concluded in 1850 that "the quantity of heat capable of increasing the temperature of 1 lb of water (weighed *in vacuo* and taken between 50 °F and 60 °F) by 1 °F requires for its evolution the expenditure of mechanical force represented by the fall of 772 lb through the space of 1 ft" (*Heat and Cold Catalog*, No. 63).

That is not exactly true, but it's probably more true than false.

Murray Winn,
St. Joseph County Republican Chairman, cited in *The South Bend Tribune*,
South Bend, IN, November 6, 2003.

… very many have striven to discover the cause of this direction … but they wasted oil and labor, because, not being practical in the research of objects in nature, being acquaint only with books, … , they constructed certain ratiocinations on a basis of mere opinions, and old-womanishly dreamt the things that were not.

William Gilbert, 1600,
cited in *De Magnete*. 1991. New York: Dover Press.

CHAPTER OUTLINE

8.1   CHAPTER OVERVIEW

Statistics are at the heart of many claims. How many times have you heard that one candidate is ahead of another by a certain percentage in the latest poll or that it is safer to fly than to drive? How confident can we be in such statements? Similar questions arise when interpreting the results of experiments.

In this chapter we will study statistics. We start by examining some classic distributions, including the binomial, Poisson, normal, Student's t, and χ^2. We will learn how to use them to determine the probabilities of events and various statistical quantities. We will examine statistical inference and learn how to estimate the characteristics of a population from finite information. Finally, we will investigate how experiments can be planned efficiently using methods of statistics. After finishing with this chapter, you will have most of the tools necessary to perform an experiment and to interpret its results correctly.

8.2   LEARNING OBJECTIVES

You should be able to do the following after completing this chapter:

- Describe the main attributes of the binomial, Poisson, and normal distributions
- Know the definition of the normalized z-variable and the % P coverage for the values of $z = 1, 2, 3,$ and 4
- Determine probability that a value will fall within a certain range of values using the z-variable table
- Determine probability that a value will fall within a certain range of values using the Student's t-variable table
- Know the definition of the standard deviation of the means, and be able to calculate it
- Know the definition of the degrees of freedom and the specific formulas for sample mean and variance calculations, and least-squares linear regression, and χ^2-analysis
- Know the definitions of the χ^2-variable and the level of significance
- Calculate the range that contains the true mean by using the Student's t-variable table
- Calculate the range that contains the true variance by using the χ^2-variable table
- Determine using χ^2-analysis the percent chance that a difference between expected and observed values is due to random effects
- Determine using χ^2-analysis the percent chance of agreement between a histogram of discrete data and an assumed probability density function
- Describe the methods of DOE and factorial analysis and how each can be used to plan experiments

8.3 VARIOUS PROBABILITY DENSITY FUNCTIONS

The concept of the probability density function was introduced in Chapter 7. There are many specific probability density functions. Each represents a different population that is characteristic of some physical process. In the following, a few of the more common ones will be examined.

Some probability density functions are for discrete processes (those having only discrete outcomes), such as the *binomial* probability density function. This describes the probability of the number of successful outcomes, n, in N repeated trials, given that only either success (with probability P) or failure (with probability $Q = 1 - P$) is possible. The binomial probability density function, for example, describes the probability of obtaining a certain sum of the numbers on a pair of dice when tossed or the probability of getting a particular number of heads and tails for a series of coin tosses. The *Poisson* probability density function models the probability of rarely occurring events. It can be derived from the binomial probability density function. Two examples of processes that can be modeled by the Poisson probability density function are the number of disintegrative emissions from an isotope and the number of micrometeroid impacts on a spacecraft. Although the outcomes of these processes are discrete whole numbers, the process is considered continuous because of the very large number of events considered. This essentially amounts to possible outcomes that span a large, continuous range of whole numbers.

Other probability density functions are for continuous processes. The most common one is the *normal (Gaussian)* probability density function. Many situations closely follow a normal distribution, such as the times of runners finishing a marathon, the scores on an exam for a very large class, and the IQs of everyone without a college degree (or with one). The *Weibull* probability density function is used to determine the probability of fatigue-induced failure times for components. The *lognormal* probability density function is similar to the normal probability density function but considers its variable to be related to the logarithm of another variable. The size distributions of raindrops are lognormally distributed as well as the populations of various biological systems. Most recently, scientists have suggested a new probability density function that can be used quite successfully to model the occurrence of clear-air turbulence and earthquakes. This probability density function is similar to the normal probability density function but skewed to the left to account for the observed higher frequency of more rarely occurring events.

8.3.1 BINOMIAL DISTRIBUTION

Consider first the **binomial distribution**. In a *repeated trials* experiment consisting of N *independent* trials with a probability of success, P, for an individual trial, the probability of getting exactly n successes (for $n \leq N$) is given by the binomial probability density function

---

8.1 | MATLAB SIDEBAR

The probability density and distribution functions for many common statistical distributions are available via MATLAB's statistics toolbox. These include the binomial (`bino`), Poisson (`poiss`), normal (`norm`), chi-square (`chi2`), lognormal (`logn`), Weibull (`weib`), uniform (`unif`), and exponential (`exp`) distributions. Three MATLAB commands used quite often are `pdf`, `cdf`, and `inv`, denoting the probability density, cumulative distribution, and inverse functions, respectively. MATLAB provides each of these functions for specific distributions, where different input arguments are required for each distribution. These functions carry a prefix signifying the abbreviated name of the distribution, as already indicated, and the suffix of either pdf, cdf, or inv. For example, `normpdf` is the normal distribution probability density function. MATLAB also provides generic `pdf` and `cdf` functions for which one of the function's arguments is the abbreviated name of the distribution. A word of caution: in MATLAB the symbol P denotes the integral of the probability density function from $-\infty$ to the argument of P. In this text, the meaning of P sometimes is different. $P(x)$ (see Equation 8.8) denotes the integral of the probability density function from $-$ the argument of P to $+$ the argument of P, whereas $P(x^*)$ (see Equation 7.48) has the same meaning as in MATLAB.

8.2 | MATLAB SIDEBAR

MATLAB's statistics toolbox contains a number of commands that simplify statistical analysis. Two tools, `disttool` and `randtool`, are easy to use. `disttool` plots the probability density and distribution functions for any 1 of 19 of the most common distributions. `randtool` generates a frequency density plot for a user-specified number of samples drawn randomly from any one of these distributions. In both tools, the mean and the standard deviation of the distribution are user specified.

8.3 | MATLAB SIDEBAR

The MATLAB command `binopdf(n,N,P)` computes the binomial probability density function value for getting n successes with success probability P for N repeated trials. For example, the command `binopdf(5,10,0.5)` yields the value 0.1762. That is, in a fair coin toss experiment, there is approximately an 18 % chance that there will be exactly 5 heads in 10 tosses.

$$p(n) = \left[\frac{N!}{(N-n)!n!} \right] P^n (1-P)^{N-n}. \qquad \textbf{[8.1]}$$

The mean and the variance are NP and NPQ, respectively, where Q is the probability of failure, which equals $1 - P$. The higher-order central moments of the skewness and kurtosis are $(Q - P)/(NPQ)^{0.5}$ and $3 + [(1 - 6PQ)/NPQ]$, respectively.

As shown in Figure 8.1, for a fixed N, as P becomes larger, the probability density function becomes skewed more to the right. For a fixed P, as N becomes

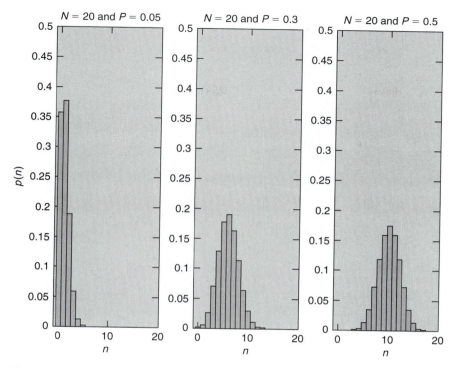

Figure 8.1 Binomial probability density functions for various N and P.

larger, the probability density function becomes more symmetric. Tending to the limit of large N and small but finite P, the probability density function approaches a normal probability density function. The MATLAB M-file `bipdfs.m` was used to generate this figure based on the `binopdf(n,N,P)` command.

STATEMENT

Suppose that there are five students taking an uncertainty analysis course. Typically, only 3/4 of the students who take such a course pass it. Determine the probabilities that out of these five, exactly 0, 1, 2, 3, 4, or 5 students will pass the course.

Example 8.1

SOLUTION

These probabilties are calculated using Equation 8.1, where $N = 5$, $P = 0.75$, and $n = 0, 1, 2, 3, 4,$ and 5. They are displayed immediately next and plotted in Figure 8.2.

$$p(0) = 1 \times 0.75^0 \times 0.25^5 = 0.0010$$
$$p(1) = 5 \times 0.75^1 \times 0.25^4 = 0.0146$$
$$p(2) = 10 \times 0.75^2 \times 0.25^3 = 0.0879$$
$$p(3) = 10 \times 0.75^3 \times 0.25^2 = 0.2637$$

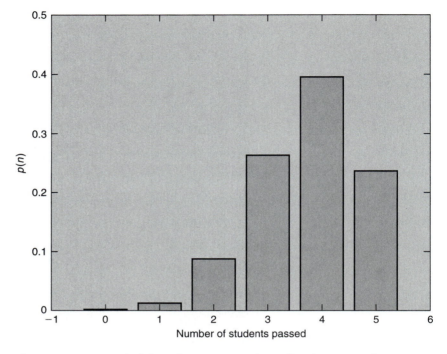

Figure 8.2   Probabilities for various numbers of students passed.

$$p(4) = 5 \times 0.75^4 \times 0.25^1 = 0.3955$$
$$p(5) = 1 \times 0.75^5 \times 0.25^0 = \underline{0.2373}$$
$$\text{sum} = 1.0000$$

8.3.2 POISSON DISTRIBUTION

Next consider the **Poisson distribution**. In the limit when N becomes very large and P becomes very small (close to zero, which implies a rare event) in such a way that the mean $(= NP)$ remains finite, the binomial probability density function very closely approximates the Poisson probability density function.

For these conditions, the Poisson probability density function enables us to determine the probability of n rare event successes (occurrences) out of a large number of N repeated trial experiments (during a series of N time intervals) with the probability P of an event success (during a time interval) as given by the probability density function

$$p(n) = \frac{(NP)^n}{n!} e^{-NP}. \qquad \textbf{[8.2]}$$

The MATLAB command `poisspdf(n,N*P)` can be used to calculate the probabilities given by Equation 8.2. The mean and variance both equal NP, noting $(1 - P) \approx 1$. The skewness and kurtosis are $(NP)^{-0.5}$ and $3 + 1/NP$, respectively. As NP is increased, the Poisson probability density function approaches a normal probability density function.

STATEMENT

Example 8.2

There are 2×10^{-20} α particles per second emitted from the nucleus of an isotope. This implies that the probability for an emission from a nucleus to occur is 1 in 2×10^{-20}. Assume that the total material to be observed is comprised of 10^{20} atoms. Emissions from the material are observed at 1-s intervals. Determine the resulting probabilities that a total of $0, 1, 2, \dots, 8$ emissions occur in the interval.

SOLUTION

The probabilities are calculated using Equation 8.2, where $N = 10^{20}$, $P = 2 \times 10^{-20}$, and $n = 0$ through 8. They are displayed immediately below and plotted in Figure 8.3.

$$p(0) = 0.135$$
$$p(1) = 0.271$$
$$p(2) = 0.271$$
$$p(3) = 0.180$$
$$p(4) = 0.090$$
$$p(5) = 0.036$$
$$p(6) = 0.012$$
$$p(7) = 0.003$$
$$p(8) = \underline{0.001}$$
$$\text{sum} = 0.999$$

8.3.3   NORMAL DISTRIBUTION

Finally, consider the **normal distribution** in more detail. In the limit when N becomes very large and P is finite, assuming that the variance remains constant, the binomial probability density function becomes the normal probability density function.

   Consider a random error to be comprised of a large number of N elementary errors of equal and infinitesimally small magnitude, e, with an equally likely chance of being either positive or negative, where $P = 1/2$. The normal distribution allows us to find the probability of occurrence of any error in the range from $-Ne$ to $+Ne$, where the probability density function is

$$p(x) = \frac{1}{\sqrt{2\pi NP(1 - P)}} \exp\left[\frac{-(x - NP)^2}{2NP(1 - P)}\right]. \qquad \textbf{[8.3]}$$

The mean and variance are the same as the binomial distribution, NP and NPQ, respectively. The higher-order central moments of the skewness and kurtosis are 0 and 3, respectively.

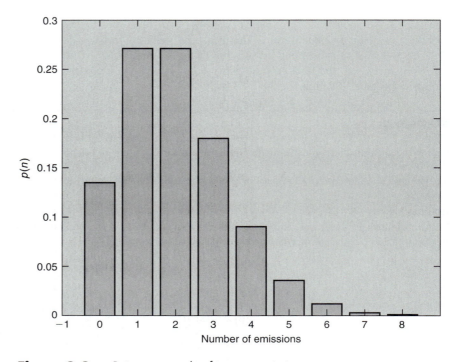

Figure 8.3   Poisson example of isotope emissions.

Utilizing expressions for the mean, x, and the variance, σ^2, in Equation 8.3, the probability density function assumes the more familiar form

$$p(x) = \frac{1}{\sigma\sqrt{2\pi}} \exp\left[-\frac{1}{2\sigma^2}(x - x')^2\right]. \qquad \textbf{[8.4]}$$

The normal probability density function is shown in the left-hand plot in Figure 8.4 in which $p(x)$ is plotted versus the nondimensional variable $z = (x - x')/\sigma$. Its maximum value equals 0.3989 at $z = 0$.

The normal probability density function is very significant. Many probability density functions tend to the normal probability density function when the sample size is large. This is supported by the central limit and related theorems. The central limit theorem can be stated loosely as follows [5]: Given a population of values with finite variance, if independent samples are taken from this population, all of size N, then the new population formed by the averages of these samples will tend to be governed by the normal probability density function, *regardless* of what distribution governed the original population. Alternatively, the central limit theorem states that whatever the distribution of the independent variables, subject to certain conditions, the probability density function of their sum approaches the normal probability density function (with a mean equal to the sum of their means and a variance equal to the sum of their variances) as N approaches infinity. The conditions are that (1) the variables are

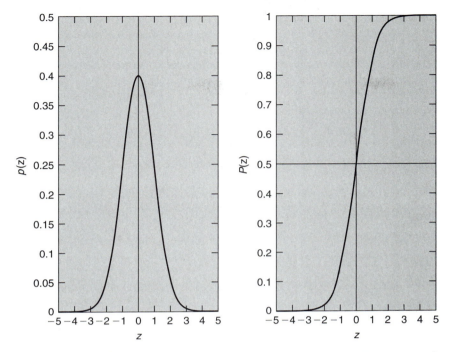

Figure 8.4   The probability density and distribution functions for the normal distribution.

expressed in a standardized, nondimensional format; (2) no single variate dominates; and (3) the sum of the variances tends to infinity as n tends to infinity. The central limit theorem also holds for certain classes of dependent random variables.

The normal probability density function describes well those situations in which the departure of a measurand from its central tendency is brought about by *a very large number* of *small* random effects. This is most appropriate for experiments in which all systematic errors have been removed and a large number of values of a measurand are acquired. This probability density function consequently has been found to be the appropriate probability density function for many types of physical measurements. In essence, a measurement subject to many small random errors will be distributed normally. Further, the mean values of finite samples drawn from a distribution other than normal will most likely be distributed normally, as assured by the central limit theorem.

Francis Galton (1822–1911) devised a mechanical system called a *quincunx* to demonstrate how the normal probability density function results from a very large number of small effects with each effect having the same probability of success or failure. This is illustrated in Figure 8.5. As a ball enters the quincunx and encounters the first effect, it falls a lateral distance e to either the right or the left. This event has caused it to depart slightly from its true center. After it encounters the second event, it can either return to the center or depart a distance of $2e$ from it. This process continues for a very large number, N, of events, resulting in a continuum of possible outcomes ranging from a value of $\bar{x} - Ne$ below the average value, $\bar{x}$, to a value of $\bar{x} + Ne$ above it. The key, of course, to arrive at such a continuum of normally distributed values is to have e small and N large. This illustrates why many phenomena are normally distributed. In many situations, there are a number of very small, uncontrollable effects always present that lead to this distribution.

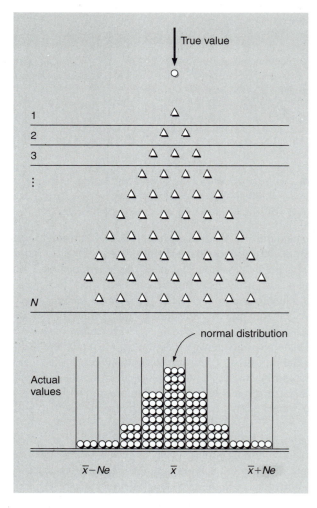

Figure 8.5   Galton's quincunx.

8.4 | MATLAB SIDEBAR

The normal-distribution-specific `pdf` function has three arguments: a particular x-value at which one seeks the pdf value, the mean value of the distribution, and the standard deviation of the distribution. For example, the MATLAB command `normpdf(x,xmean,sigma)` returns the value of the probability density function, $p(x)$, evaluated at the value of x of a normal probability distribution having a mean of xmean and a standard deviation of sigma. Specifically, typing the MATLAB command `normpdf(2, 0, 1)` yields $p(2) = 0.0540$. A similar result can be obtained using the generic MATLAB command `pdf(norm, 2, 0, 1)`.

The normal distribution `cdf` function also has similar arguments: a particular x-value at which you seek the cdf value, the mean value of the distribution

and the standard deviation of the distribution. This is none other than the integral of the `pdf` function from $-\infty$ to x^*, $P(x^*)$. So, typing the MATLAB command `normcdf(1, 0, 1)` gives $P(1) = 0.8413$. This says that 84.13 % of all normally distributed values are between $-\infty$ and one standard deviation *above* the mean. Typing `cdf(norm, 1, 0, 1)` yields the same value.

The normal distribution `inv` function gives the inverse of the `cdf` function It provides the value of x^* for the cdf value of P. Thus, typing the MATLAB command `norminv(0.9772, 0, 1)` results in the value of 2. This says that 97.72 % of all normally distributed values are *below* two standard deviations *above* the mean. There is *not* an equivalent generic MATLAB `inv` command.

8.5 | MATLAB SIDEBAR

The MATLAB command `normspec([lower, upper], mu, sigma)` generates a plot of the normal probability density function of mean mu and standard deviation sigma with the area shaded under the

function between the value limits `lower` and `upper`. It also determines the percent probability of having a normally distributed value between these limits.

8.4 NORMALIZED VARIABLES

For convenience in performing statistical calculations, the statistical variable often is nondimensionalized. For any statistical variable x, its **standardized normal variate**, β, is defined by

$$\beta = \frac{x - x'}{\sigma}, \qquad \textbf{[8.5]}$$

in which x' is the mean value of the population and σ is its standard deviation. In essence, the dimensionless variable β signifies how many standard deviations that x is from its mean value. When a *specific* value of x, say x_1, is considered, the standardized normal variate is called the **normalized z-variable**, z_1, as defined by

$$z_1 = \frac{x_1 - x'}{\sigma}. \qquad \textbf{[8.6]}$$

These definitions can be incorporated into the probability expression of a dimensional variable to yield the corresponding expression in terms of the nondimensional variable. The probability that the dimensional variable x will be in the interval $x' \pm \delta x$ can be written as

$$P(x' - \delta x \le x \le x' + \delta x) = \int_{x'-\delta x}^{x'+\delta x} p(x)\, dx. \qquad \text{[8.7]}$$

Note that the width of the interval is $2\delta x$, which in some previous expressions was written as Δx. Likewise,

$$P(-x_1 \le x \le +x_1) = \int_{-x_1}^{+x_1} p(x)\, dx. \qquad \text{[8.8]}$$

This general expression can be written specifically for a normally distributed variable as

$$P(-x_1 \le x \le +x_1) = \int_{-x_1}^{+x_1} \frac{1}{\sigma \sqrt{2\pi}} \exp\left[-\frac{1}{2\sigma^2}(x - x')^2\right] dx. \qquad \text{[8.9]}$$

Using Equations 8.5 and 8.6 and noting that $dx = \sigma\, d\beta$, Equation 8.9 becomes

$$P(-z_1 \le \beta \le +z_1) = \frac{1}{\sigma \sqrt{2\pi}} \int_{-z_1}^{+z_1} \exp\left(\frac{-\beta^2}{2}\right) \sigma\, d\beta,$$

$$= \frac{1}{\sqrt{2\pi}} \int_{-z_1}^{+z_1} \exp\left(\frac{-\beta^2}{2}\right) d\beta,$$

$$= 2\left[\frac{1}{\sqrt{2\pi}} \int_0^{+z_1} \exp\left(\frac{-\beta^2}{2}\right) d\beta\right]. \qquad \text{[8.10]}$$

The factor of 2 reflects the symmetry of the normal probability density function, which is shown in Figure 8.6, in which $p(x)$ is plotted as a function of the normalized z-variable. The term in the [ ] brackets is called the *normal error function*, denoted as $p(z_1)$. That is,

$$p(z_1) = \frac{1}{\sqrt{2\pi}} \int_0^{+z_1} \exp\left(\frac{-\beta^2}{2}\right) d\beta. \qquad \text{[8.11]}$$

The values of $p(z_1)$ will be presented shortly in Table 8.2 for various values of z_1. For example, there is a 34.13 % probability that a normally distributed variable z_1 will be within the range from $x_1 - x' = 0$ to $x_1 - x' = \sigma [p(z_1) = 0.3413]$. In other words, there is a 34.13 % probability that a normally distributed variable will be within one standard deviation *above* the mean. Note that the normal error function is one-sided because it represents to integral from 0 to $+z_1$. Some normal error function tables are two-sided and represent the integral from $-z_1$ to $+z_1$. Always check to see whether such tables are one-sided or two-sided.

Using the definition of z_1, the probability that a normally distributed variable, x_1, will have a value within the range $x' \pm z_1\sigma$ is

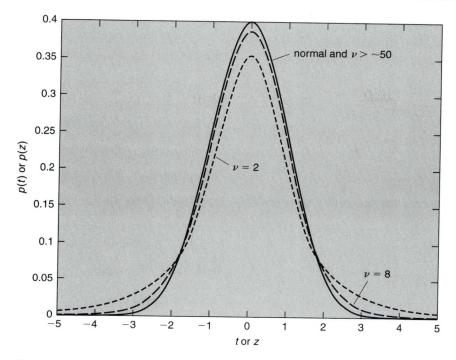

Figure 8.6 The normal and Student's t probability density functions.

$$2p(z_1) = \frac{2}{\sqrt{2\pi}} \int_0^{+z_1} \exp\left(-\frac{\beta^2}{2}\right) d\beta = \frac{P\%}{100}. \qquad \textbf{[8.12]}$$

In other words, there is P percent probability that the normally distributed variable, x_i, will be within $\pm z_P$ standard deviations of the mean. This can be expressed formally as

$$x_i = x' \pm z_P\sigma \quad (\% \ P). \qquad \textbf{[8.13]}$$

8.6 | MATLAB SIDEBAR

To obtain a value of the normal error function, $p(z_1)$, use MATLAB's cdf function. This is specified by

$$P(z_1) = \text{normcdf}\ (z_1, 0, 1) - 0.5.$$

Recall that normcdf is the integral of the normal probability density function from $-\infty$ to x^*, or to z_1 in this case. Because the normal error function is defined between the limits of a normal mean equal to zero and z_1, a value of 0.5 must be subtracted from normcdf to obtain the correct normal error function value. (Why 0.5?) For example, typing the MATLAB command normcdf(1.37,0,1)-0.5 returns the correct normal error function value of 0.4147.

The percent probabilities for various z_P values from 1 to 4 are presented in Table 8.1. As shown in the table, there is a 68.27 % chance that a normally distributed variable will be within ± one standard deviation of the mean and a 99.73 % chance that it will be within ± three standard deviations of the mean.

Example 8.3 | **STATEMENT**
Consider the situation in which a large number of voltage measurements are made. From this data, the mean value of the voltage is 8.5 V and that its variance is 2.25 V^2. Determine the probability that a single voltage measurement will fall in the interval between 10 and 11.5 V. That is, determine $P(10.0) \leq x \leq P(11.5)$.

SOLUTION
Using the definition of the probability distribution function, $P(10.0 \leq x \leq 11.5) = P(8.5 \leq x \leq 11.5) - P(8.5 \leq x \leq 10.0)$. The two probabilities on the right-hand side of this equation are found by determining their corresponding normalized z-variable values and then using Table 8.2.

First, $P(8.5 \leq x \leq 10.0)$:

$$z = \frac{x - x'}{\sigma} = \frac{10 - 8.5}{1.5} = 1 \Rightarrow P(8.5 \leq x \leq 10.0) = \frac{0.6827}{2} = 0.3413.$$

Then, $P(8.5 \leq x \leq 11.5)$:

$$z = \frac{11.5 - 8.5}{1.5} = 2 \Rightarrow P(8.5 \leq x \leq 11.5) = \frac{0.9545}{2} = 0.4772.$$

Thus, $P(10.0 \leq x \leq 11.5) = 0.4772 - 0.3413 = 0.1359$ or 13.59 %. Likewise, the probability that a single voltage measurement will fall in the interval between 10 and 13 V is 15.74 %.

Example 8.4 | **STATEMENT**
Based on a large database, the State Highway Patrol has determined that the average speed of Friday-afternoon drivers on an interstate is 67 mph with a standard deviation of 4 mph. How many drivers out of 1000 traveling on that interstate on Friday afternoon will be traveling in excess of 72 mph?

Table 8.1   Probabilities of some common z_P values.

z_P	% P
1	68.27
1.645	90.00
1.960	95.00
2	95.45
2.576	99.00
3	99.73
4	99.99

Table 8.2   Values of the normal error function.

$z_1 = \frac{x_1 - x'}{\sigma}$	0.00	0.01	0.02	0.03	0.04	0.05	0.06	0.07	0.08	0.09
0.0	.0000	.0040	.0080	.0120	.0160	.0199	.0239	.0279	.0319	.0359
0.1	.0398	.0438	.0478	.0517	.0557	.0596	.0636	.0675	.0714	.0753
0.2	.0793	.0832	.0871	.0910	.0948	.0987	.1026	.1064	.1103	.1141
0.3	.1179	.1217	.1255	.1293	.1331	.1368	.1406	.1443	.1480	.1517
0.4	.1554	.1591	.1628	.1664	.1700	.1736	.1772	.1808	.1844	.1879
0.5	.1915	.1950	.1985	.2019	.2054	.2088	.2123	.2157	.2190	.2224
0.6	.2257	.2291	.2324	.2357	.2389	.2422	.2454	.2486	.2517	.2549
0.7	.2580	.2611	.2642	.2673	.2704	.2734	.2764	.2794	.2823	.2852
0.8	.2881	.2910	.2939	.2967	.2995	.3023	.3051	.3078	.3106	.3133
0.9	.3159	.3186	.3212	.3238	.3264	.3289	.3315	.3340	.3365	.3389
1.0	.3413	.3438	.3461	.3485	.3508	.3531	.3554	.3577	.3599	.3621
1.1	.3643	.3665	.3686	.3708	.3729	.3749	.3770	.3790	.3810	.3830
1.2	.3849	.3869	.3888	.3907	.3925	.3944	.3962	.3980	.3997	.4015
1.3	.4032	.4049	.4066	.4082	.4099	.4115	.4131	.4147	.4162	.4177
1.4	.4192	.4207	.4222	.4236	.4251	.4265	.4279	.4292	.4306	.4319
1.5	.4332	.4345	.4357	.4370	.4382	.4394	.4406	.4418	.4429	.4441
1.6	.4452	.4463	.4474	.4484	.4495	.4505	.4515	.4525	.4535	.4545
1.7	.4554	.4564	.4573	.4582	.4591	.4599	.4608	.4616	.4625	.4633
1.8	.4641	.4649	.4656	.4664	.4671	.4678	.4686	.4693	.4699	.4706
1.9	.4713	.4719	.4726	.4732	.4738	.4744	.4750	.4758	.4761	.4767
2.0	.4772	.4778	.4783	.4788	.4793	.4799	.4803	.4808	.4812	.4817
2.1	.4821	.4826	.4830	.4834	.4838	.4842	.4846	.4850	.4854	.4857
2.2	.4861	.4864	.4868	.4871	.4875	.4878	.4881	.4884	.4887	.4890
2.3	.4893	.4896	.4898	.4901	.4904	.4906	.4909	.4911	.4913	.4916
2.4	.4918	.4920	.4922	.4925	.4927	.4929	.4931	.4932	.4934	.4936
2.5	.4938	.4940	.4941	.4943	.4945	.4946	.4948	.4949	.4951	.4952
2.6	.4953	.4955	.4956	.4957	.4959	.4960	.4961	.4962	.4963	.4964
2.7	.4965	.4966	.4967	.4968	.4969	.4970	.4971	.4972	.4973	.4974
2.8	.4974	.4975	.4976	.4977	.4977	.4978	.4979	.4979	.4980	.4981
2.9	.4981	.4982	.4982	.4983	.4984	.4984	.4985	.4985	.4986	.4986
3.0	.4987	.4987	.4987	.4988	.4988	.4988	.4989	.4989	.4989	.4990
3.5	.4998	.4998	.4998	.4998	.4998	.4998	.4998	.4998	.4998	.4998
4.0	.5000	.5000	.5000	.5000	.5000	.5000	.5000	.5000	.5000	.5000

SOLUTION

Assume that the speeds of the drivers follow a normal distribution. The 72 mph speed first converted into its corresponding z-variable value is

$$z = \frac{72 - 67}{4} = 1.2. \qquad \textbf{[8.14]}$$

Thus, we need to determine

$$\Pr[z > 1.2] = 1 - \Pr[z \le 1.2] = 1 - (\Pr[-\infty \le z \le 0] + \Pr[0 \le z \le 1.2]). \qquad \textbf{[8.15]}$$

From the one-sided z-variable probability table

$$\Pr[0 \le z \le 1.2] = 0.3849. \qquad \textbf{[8.16]}$$

Also, because the normal probability distribution is symmetric about its mean,

$$\Pr[-\infty \le z \le 0] = 0.5000. \qquad \textbf{[8.17]}$$

Thus,

$$\Pr[z > 1.2] = 1 - (0.5000 + 0.3949) = 0.1151. \qquad \textbf{[8.18]}$$

This means that approximately 115 of the 1000 drivers will be traveling in excess of 72 mph on that Friday afternoon.

8.7 | MATLAB SIDEBAR

The MATLAB command cdf can be used to determine the probability, $\tilde{P}(z)$, that a normally distributed value will be *between* $\pm z$. This probability is determined by

$$\tilde{P}(z) = \int_{-\infty}^{z} p(x)\,dx - \int_{-\infty}^{-z} p(x)\,dx$$
$$= \texttt{normcdf}\,(z, 0, 1) - \texttt{normcdf}\,(-z, 0, 1).$$

Using this formulation, $\tilde{P}(2) = 0.9545$. That is, 95.45 % of all normally distributed values will be *between* $z = \pm\,2$. Further, recalling that MATLAB's $P(z)$ denotes the inv value from $-\infty$ to z, a particular value of $-z$ can be obtained by typing the command $\texttt{normcdf}(\frac{1-P}{2}, 0, 1)$ and of $+z$ by $\texttt{norminv}(\frac{1+P}{2}, 0, 1)$. For example, typing the command $\texttt{norminv}(\frac{1-0.9545}{2}, 0, 1)$ gives the z-value of -2.00. A similar approach can be taken to find $\tilde{P}$ for other distributions.

8.5   STUDENT'S t DISTRIBUTION

It was about 100 years ago that William Gosset, a statistician working for the Guiness brewery, recognized a problem in using the normal distribution to describe the distribution of a small sample. As a consequence of his observations it was recognized that the normal probability density function *overestimated* the probabilities of the small-sample members near its mean and *underestimated* the probabilities far away from its mean. Using his data as a guide and working with the ratios of sample estimates, Gosset was able to develop a new distribution that better described how the members of a small sample drawn from a normal population were actually distributed. Because his employer would not allow him to publish his findings, he published them under the pseudonym "Student" [2]. His distribution was named the **Student's *t* distribution**. "Student" continued to publish significant works for over 30 years. Mr. Gosset did so well at Guiness that he eventually was put in charge of their entire Greater London operations [1].

The essence of what Gosset found is illustrated in Figure 8.6. The solid curve indicates the normal probability density function values for various z. It also represents the Student's *t* probability density function for various t and a large sample size ($N \cong 100$). The dashed curve shows the Student's *t* probability density function values for various t for a sample consisting of 9 members ($\nu = 8$), and the dotted curve for a sample of 3 members ($\nu = 2$). It is clear that as the sample size becomes smaller, the normal probability density function near its mean (where $z = 0$) overestimates the sample probabilities and, near its extremes (where $z > \approx 2$ and $z < \approx -2$), underestimates the sample probabilities. These differences can be quantified easily using the expressions for the probability density functions.

The probability density function of Student's *t* distribution is

$$p(t, \nu) = \frac{\Gamma[(\nu + 1)/2]}{\sqrt{\pi \nu}\,\Gamma(\nu/2)} \left(1 + \frac{t^2}{\nu}\right)^{-(\nu+1)/2}, \qquad \textbf{[8.19]}$$

where ν denotes the degrees of freedom and Γ is the gamma function, which has these properties:

$$\Gamma(n) = (n - 1)! \ \text{ for } \ n = \text{whole integer,}$$
$$\Gamma(m) = (m - 1)(m - 2) \cdots (3/2)(1/2)\sqrt{\pi} \ \text{ for } \ m = \text{half-integer,}$$
$$\Gamma(1/2) = \sqrt{\pi}.$$

Note in particular that $p = p(t, \nu)$ and, consequently, that there are an *infinite* number of Student's *t* probability density functions, one for each value of ν. This was suggested already in Figure 8.6 in which there were different curves for each value of N.

The statistical concept of **degrees of freedom** was introduced by R. A. Fisher in 1924 [1]. The number of degrees of freedom, ν, at any stage in a statistical calculation equals the number of recorded data, N, minus the number of different, independent restrictions (constraints), c, used for the required calculations. That is, $\nu = N - c$. For example, when computing the sample mean, there are no constraints ($c = 0$). This is because only the actual sample values are required (hence, no constraints) to determine the sample mean. So for this case, $\nu = N$. However, when either the sample

---

8.8 | **MATLAB Sidebar**

The MATLAB M-file `tz.m` was used to generate Figure 8.6. This M-file utilizes the MATLAB commands `normpdf` and `tpdf`. The M-file can be configured to plot Student's *t* probability density functions for any number of degrees of freedom.

standard deviation or the sample variance is computed, the value of the sample mean value is required (one constraint). Hence, for this case, $\nu = N - 1$. Because both the sample mean and sample variance are contained implicitly in t in Equation 8.19, $\nu = N - 1$. Usually, whenever a probability density function expression is used, values of the mean and the variance are required. Thus, $\nu = N - 1$ for these types of statistical calculations.

The expressions for the mean and standard deviation were developed in Chapter 7 for a continuous random variable. Analogous expressions can be developed for a discrete random variable. When N is very large,

$$x' = \lim_{N \to \infty} \frac{1}{N} \sum_{i=1}^{N} x_i \qquad \textbf{[8.20]}$$

and

$$\sigma^2 = \lim_{N \to \infty} \frac{1}{N} \sum_{i=1}^{N} (x_i - x')^2. \qquad \textbf{[8.21]}$$

When N is small,

$$\bar{x} = \frac{1}{N} \sum_{i=1}^{N} x_i \qquad \textbf{[8.22]}$$

and

$$S_x^2 = \frac{1}{N - 1} \sum_{i=1}^{N} (x_i - \bar{x})^2. \qquad \textbf{[8.23]}$$

Here $\bar{x}$ denotes the *sample mean*, whose value can (and usually does) vary from that of the *true mean*, x'. Likewise, S_x^2 denotes the *sample variance* in contrast to the *true variance* σ^2. The factor $N - 1$ occurs in Equation 8.23 as opposed to N to account for the one degree of freedom lost to calculate $\bar{x}$ beforehand.

Example 8.5

STATEMENT
Consider an experiment in which a finite sample of 19 values of a differential pressure are recorded. These are, in units of kPa: 4.97, 4.92, 4.93, 5.00, 4.98, 4.92, 4.91, 5.06, 5.01, 4.98, 4.97, 5.02, 4.92, 4.94, 4.98, 4.99, 4.92, 5.04, and 5.00. Estimate the range of pressure within which another pressure measurement would be at $P = 95\,\%$ given the recorded values.

SOLUTION
From this data, using the equations for the sample mean and the sample variance for small N,

$$\bar{p} = \frac{1}{19} \sum_{i=1}^{19} p_i = 4.97 \text{ and } S_p = \sqrt{\frac{1}{19 - 1} \sum_{i=1}^{19} (p_i - \bar{p})^2} = 0.046.$$

Now $\nu = N - 1 = 18$, which gives $t_{\nu, P} = t_{18, 95} = 2.101$ using Table 8.4 (presented shortly).

So,

$$p_i = \bar{p} \pm t_{\nu, P} S_p \ (\% \ P) \Rightarrow p_i = 4.97 \pm 0.10 \ (95 \ \%)$$

Thus, the next pressure measurement is estimated to be within the range of 4.87 kPa to 5.07 kPa at 95 % confidence.

Now, what if the sample had the same mean and standard deviation values but that they were determined from only five measurements? Then

$$t_{\nu, P} = t_{4, 95} = 2.770 \Rightarrow p_i = 4.97 \pm 0.13 \ (95 \ \%)$$

For this case, the next pressure measurement is estimated to be within the range of 4.84 kPa to 5.10 kPa at 95 % confidence. So, for the same confidence, a smaller sample size implies a broader range of uncertainty.

Further, what if the original sample size was used but only 50 % confidence was required in the estimate? Then

$$t_{\nu, P} = t_{18, 50} = 0.668 \Rightarrow p_i = 4.97 \pm 0.03 \ (50 \ \%)$$

For this case, the next pressure measurement is estimated to be within the range of 4.94 kPa to 5.00 kPa at 50 % confidence. Thus, for the same sample size but a lower required confidence, the uncertainty range is narrower. On the contrary, if 100 % confidence was required in the estimate, the range would have to extend over *all* possible values.

In a manner analogous to that done for the normalized z-variable in Equation 8.6, Student's t variable is defined as

$$t_1 = \frac{x_1 - \bar{x}}{S_x}. \qquad \textbf{[8.24]}$$

It follows there is % P probability that the normally distributed variable x_i in a small sample will be within $\pm t_{\nu, P}$ sample standard deviations from the sample mean. This can be expressed formally as

$$x_i = \bar{x} \pm t_{\nu, P} S_x \ (\% \ P). \qquad \textbf{[8.25]}$$

The interval $\pm t_{\nu, P} S_x$ is called the **precision interval**. The percentage probabilities for the $t_{\nu, P}$ values of 1, 2, 3, and 4 for three different values of ν are shown in Table 8.3. Thus, in a sample of nine ($\nu = N - 1 = 8$) there is a 65.34 % chance

Table 8.3   Probabilities for some typical $t_{\nu, P}$ values.

$t_{\nu, P}$	$\% P_{\nu=2}$	$\% P_{\nu=8}$	$\% P_{\nu=100}$
1	57.74	65.34	68.03
2	81.65	91.95	95.18
3	90.45	98.29	99.66
4	94.28	99.61	99.99

that a normally distributed variable will be within ± one sample standard deviation from the sample mean and a 99.61 % chance that it will be within ± four sample standard deviations from the sample mean. Also, as the sample size becomes smaller, the % P that a sample value will be within ± a certain number of sample standard deviations becomes less. This is because Student's t probability density function is slightly broader than the normal probability density function and extends out to larger values of t from the mean for smaller values of ν, as shown in Figure 8.6.

Another way to compare the Student's t distribution with the normal distribution is to examine the percent difference in the areas underneath their probability density functions for the same range of z and t values. This implicitly compares their probabilities, which can be done for various degrees of freedom. The results of such a comparison are shown in Figure 8.7. The probabilities are compared between t and z equal to 0 up to t and z equal to 5. It can be seen that the percent difference decreases as the number of degrees of freedom increases. At $\nu = 40$, the difference is less than 1 %. That is, the areas under their probability density functions over the specified range differ by less than 1 % when the number of measurements are approximately greater than 40.

The values for $t_{\nu, P}$ are given in Table 8.4. Using this table, for $\nu = 8$ there is a 95 % probability that a sample value will be within ±2.306 sample standard deviations

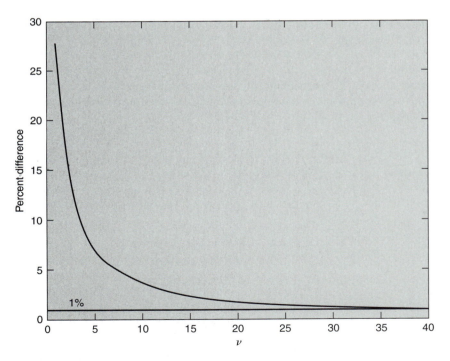

Figure 8.7   Comparison of Student's t and normal probabilities.

Table 8.4   Student's *t* variable values for different *P* and ν.

ν	$t_{\nu, P=50\%}$	$t_{\nu, P=90\%}$	$t_{\nu, P=95\%}$	$t_{\nu, P=99\%}$
1	1.000	6.341	12.706	63.657
2	0.816	2.920	4.303	9.925
3	0.765	2.353	3.192	5.841
4	0.741	2.132	2.770	4.604
5	0.727	2.015	2.571	4.032
6	0.718	1.943	2.447	3.707
7	0.711	1.895	2.365	3.499
8	0.706	1.860	2.306	3.355
9	0.703	1.833	2.262	3.250
10	0.700	1.812	2.228	3.169
11	0.697	1.796	2.201	3.106
12	0.695	1.782	2.179	3.055
13	0.694	1.771	2.160	3.012
14	0.692	1.761	2.145	2.977
15	0.691	1.753	2.131	2.947
16	0.690	1.746	2.120	2.921
17	0.689	1.740	2.110	2.898
18	0.688	1.734	2.101	2.878
19	0.688	1.729	2.093	2.861
20	0.687	1.725	2.086	2.845
21	0.686	1.721	2.080	2.831
30	0.683	1.697	2.042	2.750
40	0.681	1.684	2.021	2.704
50	0.680	1.679	2.010	2.679
60	0.679	1.671	2.000	2.660
120	0.677	1.658	1.980	2.617
∞	0.674	1.645	1.960	2.576

8.9 | **MATLAB SIDEBAR**

The MATLAB M-file `tzcompare.m` determines the percent difference in the probabilities between the normal and Student's *t* distributions that occurs with various degrees of freedom. This M-file utilizes MATLAB's `cdf` command. Figure 8.7 was made using this M-file.

of the sample mean. Likewise, for $\nu = 40$, there is a 95 % probability that a sample value will be within ± 2.021 sample standard deviations of the sample mean.

A relationship between Student's t variable and the normalized z-variable can be found directly by equating the x_i's of Equations 8.13 and 8.25. This is

$$t_{\nu, P} = \pm \left[\frac{(x' - \bar{x}) \pm z_P \sigma}{S_x} \right] \quad (\% \ P). \qquad [8.26]$$

Now, in the limit as $N \to \infty$, the sample mean, $\bar{x}$, tends to the true mean, x', and the sample standard deviation S_x tends to the true standard deviation σ. It follows from Equation 8.26 that $t_{\nu, P}$ tends to z_P. This is illustrated in Figure 8.8, in which the $t_{\nu, P}$ values for $P = 68.27$ %, 95.45 %, and 99.73 % are plotted versus ν. This figure was constructed using the MATLAB M-file tnuP.m. As shown in the figure, for increasing values of ν, the $t_{\nu, P}$ values for $P = 68.27$ %, 95.45 %, and 99.73 % approach the z_P values of 1, 2, and 3, respectively. In other words, Student's t distribution approaches the normal distribution as N tends to infinity.

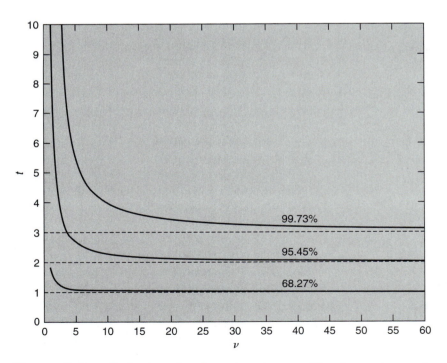

Figure 8.8   Student's t values for various degrees of freedom and percent probabilities.

8.10 | MATLAB SIDEBAR

The MATLAB functions `tpdf`, `tcdf`, and `tinv` are used for Student's t distribution. MATLAB's command `tpdf(t,ν)` returns the value of the probability density function, $p(t, \nu)$, evaluated at the value of t of Student's t distribution having ν degrees of freedom (recall that there are an infinite number of Student's t distribution, one for each degree of freedom value). Specifically, typing the MATLAB command `tpdf(2, 1000)` yields at value of 0.0541. Compare this to `normpdf(2, 0, 1)`, which gives the value of 0.0540. This shows that the probability density function value for Student's t distribution for a very large number (here 1000) of degrees of freedom is effectively the same as that of the normal distribution. However, for a low value of the number of degrees of freedom, say equal to 2, the probability

density function value of Student's t distribution differs significantly from its normal distribution counterpart (`tpdf(2, 2)` yields a value of 0.0680).

Exercises similar to those performed for the normal distribution can be done for Student's t distribution. Typing `tcdf(1, 1000)` gives the value of 0.8412 (compare this to the normal distribution value). Typing `tinv(0.9772, 1000)` results in the value of 2.0016 (again, compare this to the normal distribution value). Finally, typing `tinv(`$\frac{1+0.9545}{2}$`, 1000)` yields a value of 2.0025. This tells us that 95.45 % of all Student's t-distributed values be between ±2.0025 standard deviations from the mean of the distribution.

8.6   THE STANDARD DEVIATION OF THE MEANS

Consider a sample of N measurands. The region that contains the true mean, x', of the underlying population can be inferred from its same mean, $\bar{x}$, and standard deviation, S_x. This is done by statistically relating the sample to the population through the **standard deviation of the means (SDOM)**.

Assume that there are M sets (samples), each comprised of N measurands. A specific measurand value is denoted by x_{ij}, where $i = 1$ to N refers to the specific number within a set and $j = 1$ to M refers to the particular set. Each set will have a mean value, $\bar{x}_j$, where

$$\bar{x}_j = \frac{1}{N} \sum_{i=1}^{N} x_{ij}, \qquad \text{[8.27]}$$

and a sample standard deviation, S_{x_j}, where

$$S_{x_j} = \sqrt{\frac{1}{N-1} \sum_{i=1}^{N} (x_{ij} - \bar{x}_j)^2}. \qquad \text{[8.28]}$$

Now each $\bar{x}_j$ is a random variable. The central limit theorem assures us that the $\bar{x}_j$ values will be normally distributed about their mean value (the mean of the mean values), $\bar{\bar{x}}$, where

$$\bar{\bar{x}} = \frac{1}{M} \sum_{j=1}^{M} \bar{x}_j. \qquad \text{[8.29]}$$

This is illustrated in Figure 8.9.

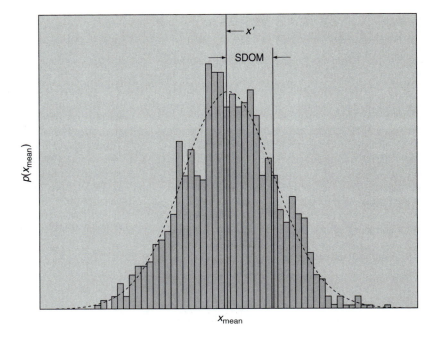

Figure 8.9   The probability density function of the mean values of x.

The standard deviation of the mean values (termed the *standard deviation of the means*), then will be

$$S_{\bar{x}} = \left[\frac{1}{M-1} \sum_{j=1}^{M} (\bar{x} - \bar{\bar{x}})^2 \right]^{1/2}.$$ **[8.30]**

It can be proven using Equations 8.28 and 8.30 [4] that

$$S_{\bar{x}} = \frac{S_x}{\sqrt{N}}.$$ **[8.31]**

This deceptively simple formula specifies the bounds within which the true mean value of the entire population is from the values of only *one* finite set. Formally,

$$x' = \bar{x} \pm t_{v,P} S_{\bar{x}} = \bar{x} \pm t_{v,P} \frac{S_x}{\sqrt{N}} \ (\% \ P).$$ **[8.32]**

This formula implies that the bounds within which x' is contained can be reduced, which means that the estimate of x' can be made more precise, by increasing N or by decreasing the value of S_x. There is a moral here. It is better to carefully plan an experiment to minimize the number of random effects beforehand and hence to reduce S_x rather than to spend the time taking more data to achieve the same bounds on x'.

The interval $\pm t_{\nu,P} S_{\bar{x}}$ is called the **precision interval of the true mean**. As N becomes large, from Equation 8.31 it follows that the SDOM becomes small and the sample mean value tends toward the true mean value. In this light, the precision interval of the true mean value can be viewed as a measure of the *uncertainty* in determining x'.

STATEMENT

Example 8.6

Consider the differential pressure transducer measurements in the previous example. What is the range within which the true mean value of the differential pressure, p', is contained?

SOLUTION

Equation 8.32 reveals that

$$p' = \bar{p} \pm t_{\nu,P} S_{\bar{p}} = \bar{p} \pm t_{\nu,P} \frac{S_p}{\sqrt{N}} (\% \ P)$$

$$= 4.97 \pm \frac{(2.101)(0.046)}{\sqrt{19}} = 4.97 \pm 0.02 \ (95 \ \%)$$

Thus, the true mean value is estimated at 95 % confidence to be within the range from 4.95 kPa to 4.99 kPa.

Finally, it is very important to note that although Equations 8.25 and 8.32 appear to be similar, they are uniquely different. Equation 8.25 is used to estimate the range within which another x_i value will be with a given confidence, whereas Equation 8.32 is used to estimate the range that contains the true mean value for a given confidence. Both equations use the values of the sample mean, standard deviation, and number of measurands in making these estimates.

8.7 POOLING SAMPLES

In some situations it may be necessary to combine the data gathered from M replicate experiments, each comprised of N measurands. The measurands can be **pooled** to form one set of MN measurands [3].

For the jth experiment

$$\bar{x}_j = \frac{1}{N} \sum_{i=1}^{N} x_{ij} \quad \text{and} \quad S_{x_j}^2 = \frac{1}{N-1} \sum_{i=1}^{N} (x_{ij} - \bar{x}_j)^2. \qquad \textbf{[8.33]}$$

From these expressions the following expressions can be developed. The mean of all $\bar{x}_j$'s, called the *pooled mean* of x, $\{\bar{x}\}$, the mean of the means, then becomes

$$\bar{\bar{x}} = \{\bar{x}\} = \frac{1}{M} \sum_{j=1}^{M} \bar{x}_j = \frac{1}{MN} \sum_{j=1}^{M} \sum_{i=1}^{N} x_{ij}. \qquad \textbf{[8.34]}$$

The *pooled variance* of x, $\{S_x^2\}$, is actually the average of the variances of the M experiments where $\{S_x^2\}$ is treated as a random variable and is given by

$$\{S_x^2\} = \frac{1}{M}\sum_{j=1}^{M} S_{x_j}^2 = \frac{1}{M(N-1)}\sum_{j=1}^{M}\sum_{i=1}^{N}(x_{ij} - \bar{x}_j)^2. \qquad \textbf{[8.35]}$$

The *pooled standard deviation*, $\{S_x\}$, is the positive square root of the pooled variance. The *pooled standard deviation of the means*, $\{S_{\bar{x}}\}$, is

$$\{S_{\bar{x}}\} = \frac{\{S_x\}}{\sqrt{MN}}. \qquad \textbf{[8.36]}$$

Now consider when the number of measurands varies in each experiment, when N is not constant. There are N_j measurands for the jth experiment. The resulting pooled statistical properties must be weighted by N_j. The *pooled weighted mean*, $\{\bar{x}\}_w$, is

$$\{\bar{x}\}_w = \frac{\sum_{j=1}^{M} N_j \bar{x}_j}{\sum_{j=1}^{M} N_j}. \qquad \textbf{[8.37]}$$

The *pooled weighted standard deviation*, $\{S_x\}_w$, is

$$\{S_x\}_w = \sqrt{\frac{\nu_1 S_{x_1}^2 + \nu_2 S_{x_2}^2 + \cdots + \nu_M S_{x_M}^2}{\nu}}, \qquad \textbf{[8.38]}$$

where

$$\nu = \sum_{j=1}^{M} \nu_j = \sum_{j=1}^{M}(N_j - 1). \qquad \textbf{[8.39]}$$

The *pooled weighted standard deviation of the means*, $\{S_{\bar{x}}\}_w$, is

$$\{S_{\bar{x}}\}_w = \frac{\{S_x\}_w}{\left[\sum_{j=1}^{M} N_j\right]^{1/2}}. \qquad \textbf{[8.40]}$$

8.8   HYPOTHESIS TESTING

Hypothesis testing [6] incorporates the tools of statistics into a decision-making process. In the terminology of statistics, a null hypothesis is indicated by H_0 and an alternative hypothesis by H_1. The alternative hypothesis is considered to be the complement of the null hypothesis. There is the possibility that H_0 could be rejected, that is, considered false, when it is actually true. This is called a *Type I error*. Conversely, H_0 could be accepted, that is, considered true, when it is actually false. This is termed a *Type II error*. Type II errors are of particular concern in engineering. Sound engineering decisions should be based on the assurance that Type II error is minimal. For example, if H_0 states that a structure will not fail when its load is less than a particular safety-limit load, then it is important to assess the probability that the structure can fail *below* the safety-limit load. This can be quantified by the power of the test, where the power is defined as $1 -$ probability of Type II error. For a fixed level of significance (see Section 8.9), the power increases as the sample size increases. Large values of power signify better precision. Null hypothesis decisions are summarized in Table 8.5.

Consider the rationale behind using statistical analyzes to determine whether or not the mean of a population, x', will have a particular value, x_o. In an experiment, each measurand value will be subject to small, random variations because of minor, uncontrolled variables. The null hypothesis would be $H_0 : x' = x_o$ and the alternative hypothesis $H_1 : x' \neq x_o$. Because the alternative hypothesis would be true if either $x' < x_o$ or $x' > x_o$, the appropriate hypothesis test would be a *two-sided t*-test. If the null hypothesis were either $H_0 : x' \leq x_o$ or $H_0 : x' \geq x_o$, then the appropriate hypothesis test would be a *one-sided t*-test. The modifier t implies that Student's t variable is used to assess the hypothesis. These tests implicitly require that all measurand values are provided such that their sample mean and sample standard deviation can be determined.

Decision of either hypothesis acceptance or rejection is made using Student's t distribution. For a one-sided t-test, if $H_0 : x' \leq x_o$, then its associated probability, $P[X \leq t]$, must be determined. X represents the value of a single sample that is drawn randomly from a t-distribution with $v = N - 1$ degrees of freedom. Likewise, if $H_0 : x' \geq x_o$, then its associated probability, $P[X \geq t]$ must be found. For a two-sided t-test, the sum of the probabilities $P[X \leq t]$ and $P[X \geq t]$ must be determined. This sum equals $2P[X \geq |t|]$ because of the symmetry of Student's t distribution. These probabilities are determined through Student's t value. For hypothesis testing, the particular t-value, termed the *t-statistic*, is based on the sample standard deviation of the means, where $t = (\bar{x} - x_o)/(S_x/\sqrt{N})$.

A p-value, sometimes referred to as the *observed* level of significance, is defined for the null hypothesis of a set of measurands as the probability of obtaining the measurand set or a set having less agreement with the hypothesis. The p-value is proportional to the plausibility of the null hypothesis. The criteria for accepting or rejecting the null hypothesis are listed on the next page.

Table 8.5 Null hypothesis decisions and their associated probabilities and errors.

	Accept H_0	Reject H_0
H_0 true	correct $(1 - \alpha)$	Type I error (α)
H_0 false	Type II error (β)	correct $(1 - \beta)$

- $p < 0.01$ indicates noncredible H_0, so reject H_0 and accept H_1
- $0.01 \le p \le 0.10$ is inconclusive, so acquire more data
- $p > 0.10$ indicates plausible H_0, so accept H_0 and reject H_1

Sometimes, $p = 0.05$ is used as a decision value to avoid an inconclusive result, where $p < 0.05$ implies plausibility and $p > 0.05$ signifies noncredibility. Keep in mind that only the plausibility, not the exact truth, of a null hypothesis can be ascertained. Rejecting the null hypothesis of a two-sided test means $x' \ne x_o$. Accepting the null hypothesis implies that x_o is a plausible value of x', but not necessarily that $x_o = x'$. So, rejecting a null hypothesis is more exact statistically than accepting a null hypothesis. Rejecting the null hypothesis $H_0 : x' \le x_o$ means $x' \ge x_o$. Accepting the null hypothesis indicates that, plausibly, $x' \le x_o$. Again, it is more exact statistically to reject the null hypothesis or, conversely, to accept the alternative hypothesis. Hence, it is better to pose the null hypothesis such that its alternative hypothesis most likely will be accepted.

Stated differently, if x' is on the side of x_o that favors the null hypothesis, then the hypothesis should be accepted. If it is not, then the plausibility of the hypothesis must be ascertained, based on the aforementioned p-value criteria. If the hypothesis test is one-sided, then $x' \le x_o$ means $t \le 0$ and $p > 0.50$, which indicates acceptance. Also, $x' \gg x_o$ means $t > 0$ and $p \sim 0$, which implies rejection. Further, x' slightly greater than x_o means t is slightly greater than 0 and p is nonzero and finite, signifying plausibility. If the hypothesis test is two-sided, the larger the value of $|t|$, the farther away x' is from x_0. The p-value is calculated with the probability that a measurand set with $x' = x_o$ has a t-statistic with an absolute value greater than $|t|$. Any measurand set with a t-statistic that is greater than $|t|$ or less than $-|t|$ has less agreement with the null hypothesis. The acceptance or rejection of the null hypothesis is based on the same aforementioned p-value criteria. The following serves to illustrate how the p-values specifically are determined for one-sided and two-sided hypothesis tests.

Consider the one-sided test where $H_0 : x' \le 10$. For this example, $\bar{x} = 12$, $S_x = 3$, and $N = 20$. Thus, the t-statistic value equals $(12-10)/(3\sqrt{20}) = 2.981$. For $\nu = 19$, the corresponding p-value equals $1 - P[t \le 2.981] = 1 - 0.9962 = 0.0038$. Thus, the null hypothesis is rejected and the alternative hypothesis, that x' is greater than 10, is accepted.

Next examine the two-sided test where $H_0 : x' = 10$. Using the same statistical parameters as in the previous example, where the t-statistic value is 2.891, the p-value equals $2P[t \ge |2.981|] = (2)(0.0038) = 0.0076$. Here the absolute value of the t-statistic, 2.981, is greater than the p-value, 0.0076. So, the measurand set with $x' = 10$ has less agreement with the null hypothesis. In fact, because $p < 0.01$, the null hypothesis is not credible.

More specificity about accepting or rejecting a null hypothesis can be obtained by associating this decision with a level of significance, α. Here, the null hypothesis is accepted if the p-value is larger than α and rejected if the p-value is less than α. The level of significance is the probability of a Type I error. The relationships between null hypothesis acceptance or rejection and their associated p- and α-values are presented in Table 8.6.

Table 8.6 Null hypothesis decisions and associated p and α values.

p-value	$\alpha = 0.01$	$\alpha = 0.05$	$\alpha = 0.10$
$p \geq 0.10$	accept	accept	accept
$0.05 < p < 0.10$	accept	accept	reject
$0.01 < p < 0.05$	accept	reject	reject
$p < 0.01$	reject	reject	reject

Example 8.7

STATEMENT

A test is conducted to assess the reliability of a transducer designed to indicate when the pressure in a vessel is 120 psi. The vessel pressure is recorded each time the sensor gives an indication. The test is repeated 20 times, resulting in a mean pressure at detection of 121 psi with a standard deviation of 3 psi. Determine the reliability of the transducer based on a 5 % level of significance.

SOLUTION

It is best statistically to test the null hypothesis that the transducer's detection level is 120 psi. The alternative hypothesis would be that the detection level is either less than or greater than 120 psi. This appropriate test is a two-sided t-test. For this case, the value of the t-statistic is

$$t = \frac{\bar{x} - x_o}{S_x / \sqrt{N}} = \frac{121 - 120}{3/\sqrt{20}} = 1.491.$$

For $v = 19$, the corresponding p-value equals $2P[|t| \geq 1.491] = (2)(0.0762) = 0.1524$. According to Table 8.6, the null hypothesis is acceptable. Thus, the transducer can be considered reliable. At this level of significance, the p-value would have to be less than 0.05 before the null hypothesis could be rejected and the transducer considered unreliable.

This type of analysis also can be employed to test the hypothesis that two measurand sets with *paired* samples (each having the same number of samples) come from the same population. This is illustrated by Example 8.8.

Example 8.8

STATEMENT

Ten thermal boundary-layer thickness measurements were made at a specific location along the length of a heat-exchanger plate. Ten other thickness measurements were made after the surface of the heat-exchanger plate was modified to improve its heat transfer. The results are shown in Table 8.7. Determine the percent confidence that the plate surface modification has no effect on the thermal boundary-layer thickness.

SOLUTION

Assume that the thicknesses follow a t-distribution. This implies that the differences of the thicknesses for each set, $\delta_{A-B} = \delta_A - \delta_B$, also follow a t-distribution. The mean and the standard deviation of the differences can be computed from the sample data. They are 0.043 mm and 0.171 mm, respectively. Now if both samples come from the same population (here,

Table 8.7 Boundary-layer thickness measurements.

Test No.	Set A δ_A (mm)	Set B δ_B (mm)	$\delta_A - \delta_B$ (mm)
1	3.0806	2.9820	0.0986
2	3.0232	2.9902	0.0330
3	2.9010	3.0728	−0.1718
4	3.1340	2.9107	0.2233
5	3.0290	2.9775	0.0514
6	3.1479	2.9348	0.2131
7	3.1138	2.9881	0.1257
8	2.9316	3.2303	−0.2987
9	2.8708	2.9090	−0.0382
10	2.9927	2.7979	0.1948

this would imply that the surface modification had no detectable effect on the boundary-layer thickness), then the difference of their true mean values must be zero. Thus, the problem can be rephrased as follows: What is the confidence that a parameter with a mean value of 0.043 mm, $\bar{\delta}$, and a standard deviation of 0.171 mm, $S_{\bar{\delta}}$, that are determined from 10 samples actually comes from a population whose mean value is zero?

This involves a two-sided hypothesis test. The null hypothesis is that the mean value of the differences is zero (that the surface modification has no detectable effect) and the alternative hypothesis is that it does (that the surface modification has a detectable effect). The t-statistic value for this case is

$$t = \frac{\sqrt{N}(\bar{\delta} - 0)}{S_{\bar{\delta}}} = \frac{\sqrt{10}(0.043)}{0.171} = 0.795.$$

For $v = 9$, the p-value equals $2P[t \geq |0.795|] = (2)(0.2235) = 0.4470$. Thus, there is approximately a 45 % chance that the surface modification has a detectable effect and a 55 % chance that it does not. So the present experiment gives an ambiguous result. If the mean of the thickness difference were smaller, say 0.020 mm, given everything else the same, then the p-value would be 0.7200, based on $t = 0.37$. Now there is more confidence in the hypothesis that the surface modification has no detectable effect. However, this still is not significant enough. In fact, to have 95 % confidence in the hypothesis, the mean of the thickness difference would have to be equal to 0.004 mm, given everything else the same. This type of analysis can be extended further to experiments involving unpaired samples with or without equal variances [6].

8.9 THE CHI-SQUARE DISTRIBUTION

The range that contains the true mean of a population can be estimated using the values from only a single sample of N measurands and Equation 8.32.

Likewise, there is an analogous way of estimating the range that contains the true variance of a population using the values from only a single sample of N measurands. The estimate involves using one more probability distribution, the chi-square distribution.

The chi-square distribution is used in many statistical calculations. For example, it can be used to determine the precision interval of the true variance, to quantify how well a sample matches an assumed parent distribution, and to compare two samples of same or different size with one another. The statistical variable, χ^2, represents the sum of the squares of the differences between the measured and expected values normalized by their variance. Thus, the value of χ^2 is dependent on the number of measurements, N, at which the comparison is made, and, hence, the number of degrees of freedom, $\nu = N - 1$. From this definition it follows that χ^2 is related to the standardized variable, $z_i = (x_i - x')/\sigma$, and the number of measurements by

$$\chi^2 = \sum_{i=1}^{N} z_i^2 = \sum_{i=1}^{N} \frac{(x_i - x')^2}{\sigma^2}. \qquad \text{[8.41]}$$

χ^2 can be viewed as a quantitative measure of the total deviation of all x_i values from their population's true mean value with respect to their population's standard deviation. This concept can be used, for example, to compare the χ^2 value of a sample with the value that would be expected for a sample of the same size drawn from a normally distributed population. Using the definition of the sample variance given in Equation 8.23, this expression becomes

$$\chi^2 = \frac{\nu S_x^2}{\sigma^2}. \qquad \text{[8.42]}$$

So, in the limit as $N \to \infty$, $\chi^2 \to \nu$.

The probability density function of χ^2 (for $\chi^2 \geq 0$) is

$$p(\chi^2, \nu) = \left[2^{\nu/2} \Gamma\left(\frac{\nu}{2}\right) \right]^{-1} (\chi^2)^{(\nu/2)-1} \exp\left(-\frac{\chi^2}{2}\right), \qquad \text{[8.43]}$$

where Γ denotes the gamma function given by

$$\Gamma\left(\frac{\nu}{2}\right) = \int_0^\infty x^{(\nu/2)-1} \exp(-x)\, dx = \left(\frac{\nu}{2} - 1\right)! \qquad \text{[8.44]}$$

and the mean and the variance of $p(\chi^2, \nu)$ are ν and 2ν, respectively. Sometimes values of χ^2 are normalized by the expected value ν. The appropriate parameter then becomes the **reduced chi-square variable**, which is defined as $\tilde{\chi}^2 \equiv \chi^2/\nu$. The mean value of the reduced chi-square variable then equals unity. Finally, note that there is a different probability density function of χ^2 for each value of ν.

The χ^2 probability density functions for three different values of ν are plotted versus χ^2 in Figure 8.10. The MATLAB M-file chipdf.m was used to construct this figure. The value of $p(\chi^2 = 10, \nu = 10)$ is 0.0877, whereas the value of

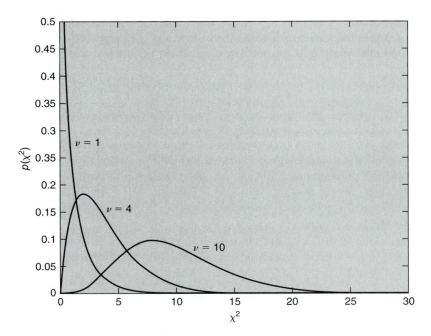

Figure 8.10   Three χ^2 probability density functions.

$p(\chi^2 = 1, \nu = 1)$ is 0.2420. For a sample of only $N = 2$ ($\nu = N - 1 = 1$), there is almost a 100 % chance that the value of χ^2 will be less than approximately 9. However, if $N = 11$ there is about a 50 % chance that the value of χ^2 will be less than approximately 9.

The corresponding probability distribution function, given by the integral of the probability density function from 0 to a specific value of χ^2, denoted by χ_α^2, is called the chi-square distribution with ν degrees of freedom. It denotes the probability $P(\chi_\alpha^2) = 1 - \alpha$ that $\chi^2 \leq \chi_\alpha^2$, where α denotes the **level of significance**. In other words, the area under a specific χ^2 probability *density* function curve from 0 to χ_α^2 equals $P(\chi_\alpha^2)$ and the area from χ_α^2 to ∞ equals α. The χ^2 probability density function for $\nu = 20$ is plotted in Figure 8.11. The MATLAB M-file chicdf.m was used to construct this figure. Three χ_α^2 values (for $\alpha = 0.05$, 0.50, and 0.95) are indicated by vertical lines. The lined area represents 5 % of the total area under the probability density curve, corresponding to $\alpha = 0.05$. The χ_α^2 values for various ν and α are presented in Table 8.8. Using this table, for $\nu = 20$, $\chi_{0.95}^2 = 10.9$, $\chi_{0.50}^2 = 19.3$, and $\chi_{0.05}^2 = 31.4$. That is, when $\nu = 20$, 50 % of all χ^2 values are less than or equal to 10.9.

The χ^2 probability distribution functions for the same values of ν used in Figure 8.10 are plotted versus χ^2 in Figure 8.12. For $N = 2$, there is a 99.73 % chance that the value of χ^2 will be less than 9. For $N = 11$, there is a 46.79 % chance

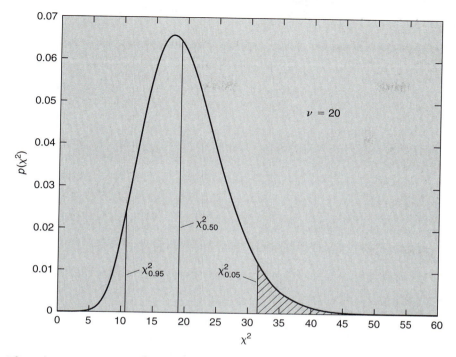

Figure 8.11   The χ^2 probability density function for $\nu = 20$.

that the value of χ^2 will be less than 9. Finally, for $\nu = 4$, as already determined using Table 8.8, a value of $\chi^2 = 3.36$ yields $P(\chi_\alpha^2) = 0.50$.

8.9.1   ESTIMATING THE TRUE VARIANCE

Consider a finite set of x_i values drawn randomly from a normal distribution having a true mean value x' and a true variance σ^2. It follows directly from Equation 8.42 and the definition of χ_α^2 that there is a probability of $1 - \alpha/2$ that $\nu S_x^2/\sigma^2 \leq \chi_{\alpha/2}^2$ or that $\nu S_x^2/\chi_{\alpha/2}^2 \leq \sigma^2$. Conversely, there is a probability of $1 - (1 - \alpha/2) = \alpha/2$ that $\sigma^2 \leq \nu S_x^2/\chi_{\alpha/2}^2$. Likewise, there is a probability of $\alpha/2$ that $\nu S_x^2/\sigma^2 \leq \chi_{1-\alpha/2}^2$ or that $\nu S_x^2/\chi_{1-\alpha/2}^2 \leq \sigma^2$. Thus, the true variance of the underlying population, σ^2, is *within* the **precision interval of the true variance**, from $\nu S_x^2/\chi_{\alpha/2}^2$ to $\nu S_x^2/\chi_{1-\alpha/2}^2$ with a probability $1 - (\alpha/2) - (\alpha/2) = P$. Also there is a probability $\alpha/2$ that it is *below* the lower bound of the precision interval, a probability $\alpha/2$ that it is *above* the upper bound of the precision interval, and a probability α that it is outside of the bounds of the precision interval. Formally, this is

$$\frac{\nu S_x^2}{\chi_{\alpha/2}^2} \leq \sigma^2 \leq \frac{\nu S_x^2}{\chi_{1-\alpha/2}^2} \quad (\% \ P). \qquad \textbf{[8.45]}$$

Table 8.8 χ_α^2 values for various ν and α.

ν	$\chi_{0.99}^2$	$\chi_{0.975}^2$	$\chi_{0.95}^2$	$\chi_{0.90}^2$	$\chi_{0.50}^2$	$\chi_{0.05}^2$	$\chi_{0.025}^2$	$\chi_{0.01}^2$
1	0.000	0.000	0.000	0.016	0.455	3.84	5.02	6.63
2	0.020	0.051	0.103	0.211	1.39	5.99	7.38	9.21
3	0.115	0.216	0.352	0.584	2.37	7.81	9.35	11.3
4	0.297	0.484	0.711	1.06	3.36	9.49	11.1	13.3
5	0.554	0.831	1.15	1.61	4.35	11.1	12.8	15.1
6	0.872	1.24	1.64	2.20	5.35	12.6	14.4	16.8
7	1.24	1.69	2.17	2.83	6.35	14.1	16.0	18.5
8	1.65	2.18	2.73	3.49	7.34	15.5	17.5	20.1
9	2.09	2.70	3.33	4.17	8.34	16.9	19.0	21.7
10	2.56	3.25	3.94	4.78	9.34	18.3	20.5	23.2
11	3.05	3.82	4.57	5.58	10.3	19.7	21.9	24.7
12	3.57	4.40	5.23	6.30	11.3	21.0	23.3	26.2
13	4.11	5.01	5.89	7.04	12.3	22.4	24.7	27.7
14	4.66	5.63	6.57	7.79	13.3	23.7	26.1	29.1
15	5.23	6.26	7.26	8.55	14.3	25.0	27.5	30.6
16	5.81	6.91	7.96	9.31	15.3	26.3	28.8	32.0
17	6.41	7.56	8.67	10.1	16.3	27.6	30.2	33.4
18	7.01	8.23	9.39	10.9	17.3	28.9	31.5	34.8
19	7.63	8.91	10.1	11.7	18.3	30.1	32.9	36.2
20	8.26	9.59	10.9	12.4	19.3	31.4	34.2	37.6
30	15.0	16.8	18.5	20.6	29.3	43.8	47.0	50.9
40	22.2	24.4	26.5	29.1	39.3	55.8	59.3	63.7
50	29.7	32.4	34.8	37.7	49.3	67.5	71.4	76.2
60	37.5	40.5	43.2	46.5	59.3	79.1	83.3	88.4
70	45.4	48.8	51.7	55.3	69.3	90.5	95.0	100.4
80	53.5	57.2	60.4	64.3	79.3	101.9	106.6	112.3
90	61.8	65.6	69.1	73.3	89.3	113.1	118.1	124.1
100	70.1	74.2	77.9	82.4	99.3	124.3	129.6	135.8

The width of the precision interval of the true variance in relation to the probability P can be examined further. First consider the two extreme cases. When $P = 1$ (100 %) then $\alpha = 0$, which implies that $\chi_0^2 = \infty$ and $\chi_1^2 = 0$. Thus, the sample variance precision interval is from 0 to ∞ according to Equation 8.45. That is, there is a 100 % chance that σ^2 will have a value between 0 and ∞. When $P = 0$ (0 %) then $\alpha = 1$, which implies that the sample variance precision interval are the same. That is, there is a 0 % chance the σ^2 will exactly equal one specific value out of an infinite number of possible values (when $\alpha = 1$ and $\nu \gg 1$, that unique value would be S_x^2). These two extreme-case examples illustrate the upper and lower limits of the sample

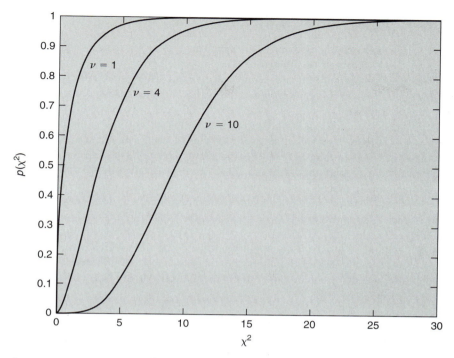

Figure 8.12   Three χ^2 probability distribution functions.

variance precision interval and its relation to P and α. As α varies from 0 to 1 (hence, P from 1 to 0), the precision interval width decreases from ∞ to 0. In other words, the probability, α, that the true variance is *outside* of the sample variance precision interval increases as the precision interval width decreases.

8.11 | MATLAB SIDEBAR

The MATLAB commands chi2pdf, chi2cdf, and chi2inv are used for chi-square distribution. MATLAB's command chi2pdf(chisq,nu) returns the value of the probability density function, $p(\chi^2, \nu)$, evaluated at the value of χ^2 of the chi-square distribution having ν degrees of freedom. Note that, like for Student's t distribution, there are an infinite number of chi-square distributions. Typing the MATLAB command chi2pdf(10,10) yields a value of 0.0877. Typing chi2cdf(10,10) gives a value of 0.5595, signifying that 55.95 % of all chi-square-distributed values for 10 degrees of freedom be between 0 and 10 (recall that there are no *negative* chi-square values). Or, in other words, there is a 44.15 % chance of finding a value greater than 10 for a 10-degree-of-freedom chi-square distribution. Similar to other specific-distribution inv functions, chi2inv(P,ν) returns the χ^2 value where between it and 0 are P % of all the ν-degree-of-freedom chi-square distribution values. For example, typing chi2inv(95.96,4) yields a value of 10.

8.9.2 ESTABLISHING A REJECTION CRITERION

The relation between the probability of occurrence of a χ^2 value being less than a specified χ^2 value can be utilized to ascertain whether or not effects other than random ones are present in an experiment or process. This is particularly relevant, for example, in establishing a rejection criterion for a manufacturing process or in an experiment. If the sample's χ^2 value exceeds the value of χ_α^2 based on the probability of occurrence $P = 1 - \alpha$, it is likely that systematic effects (biases) are present. In other words, the level of significance α also can be used as a chance indicator of random effects. A low value of α implies that there is very little chance that the noted difference is due to random effects and, thus, that a systematic effect is the cause for the discrepancy. In essence, a low value of α corresponds to a relatively high value of χ^2, which, of course, has little chance to occur randomly. It does, however, have *some* chance to occur randomly, which leads to the possibility of falsely identifying a random effect as being systematic, which is a Type II error. For example, a batch sample yielding a low value of α implies that the group from which it was drawn is suspect and probably (but not definitely) should be rejected. A high value of α implies the opposite, that the group probably (but not definitely) should be accepted.

Example 8.9

STATEMENT
This problem is adapted from [3]. A manufacturer of bearings has compiled statistical information that shows the true variance in the diameter of "good" bearings is 3.15 μm^2. The manufacturer wishes to establish a batch rejection criterion such that only small samples need to be taken and assessed to check whether or not there is a flaw in the manufacturing process that day. The criterion states that when a batch sample of 20 manufactured bearings has a sample variance > 5.00 μm^2 the batch is to be rejected, because, most likely, there is a flaw in the manufacturing process. What is the probability that a batch sample will be rejected even though the true variance of the population from which it was drawn was within the tolerance limits or, in other words, of making a Type II error?

SOLUTION
From Equation 8.42,

$$\chi_\alpha^2(v) = v\frac{S_x^2}{\sigma^2} = \frac{(20-1)(5.00)}{(3.15)} = 30.16.$$

For this value of χ^2 and $v = 19$, $\alpha \cong 0.05$ using Table 8.8. So, there is a 5 % chance that the discrepancy is due to random effects (that a new batch will be rejected even though its true variance is within the tolerance limits), or a 95 % chance that it is not. Thus, the standard for rejection is good. That is, the manufacturer should reject any sample that has $S_x^2 > 5$ μm^2. In doing so, he risks only a 5 % chance of falsely identifying a good batch as bad. If the χ^2 value equaled 11.7 instead, then there would be a 90 % chance that the discrepancy is due to random effects.

Now what if the size of the batch sample was reduced to $N = 10$? For this case, $\alpha = 0.0004$. So, there is a 0.04 % chance that the discrepancy is due to random effects. In other words, getting a χ^2 value of 30.16 with a batch sample of 10 instead of 20 gives us even more assurance that the criterion is a good one.

8.9.3   COMPARING OBSERVED AND EXPECTED DISTRIBUTIONS

In some situations, a sample distribution should be compared with an expected distribution to determine whether or not the expected distribution actually governs the underlying process. When comparing two distributions using a χ^2 analysis,

$$\chi^2 \approx \sum_{j=1}^{K} \frac{(O_j - E_j)^2}{E_j}, \qquad \textbf{[8.46]}$$

with O_j and E_j the number of observed and expected occurrences in the jth bin, respectively. The expected occurrence for the jth bin is the product of the total number of occurrences, N, and the probability of occurrence, P_j. The probability of occurrence is the difference of the probability *distribution* function's values at the jth bin's two endpoints. It also can be approximated by the product of the bin width and the probability *density* function value at the bin's midpoint value. Equation 8.46 follows from Equation 8.41 by noting that $\sigma^2 \sim \nu \sim E$. Strictly speaking, this expression is an *approximation* for χ^2 and is subject to the additional constraint that $E_j \geq 5$ [7]. The number of degrees of freedom, ν, are given by $\nu = K - (L + n)$, where K is the number of bins, preferably using Scott's formula for equal-width intervals that was described in Section 7.6. Here, $n = 2$ because two values are needed to compute the expected probabilities from the assumed distribution (one for the mean and one for the variance). There is an additional constraint ($L = 1$), because the number of expected values must be determined. Thus, whenever a χ^2 analysis of this type is performed, $\nu = K - 3$.

From this type of analysis, agreement between observed and expected distributions can be ascertained with a certain confidence. The percent probability that the expected distribution is the correct one is specified by α. By convention, when $\alpha < 0.05$, the disagreement between the sample and expected distributions is *significant* or the agreement is *unlikely*. When $\alpha < 0.01$, the disagreement between the sample and expected distributions is *highly significant* or the agreement is *highly unlikely*.

Example 8.10

STATEMENT

Consider a study conducted by a professor who wishes to determine whether or not the 300 undergraduate engineering students in his department are normal. He determines this by comparing the distribution of their heights to that expected for a normally distributed student population. His height data are presented in Table 8.9.

SOLUTION

For this case, $\nu = 8 - 3 = 5$, where $K = 8$ was determined using Scott's formula (actually, $K = 7.7$, which then is rounded up). The expected values are calculated for each bin by noting that $E_k = NP_k$, where $N = 300$. For example, for bin 2 where $-1.5\sigma \leq x \leq -\sigma$,

Table 8.9   Observed and expected heights.

Bin Number, k	Heights in Bin	Observed Number, O_k	Expected Number, E_k
1	less than $X - 1.5\sigma$	19	20.1
2	between $X - 1.5\sigma$ and $X - \sigma$	25	27.5
3	between $X - \sigma$ and $X - 0.5\sigma$	44	45.0
4	between $X - 0.5\sigma$ and X	59	57.5
5	between X and $X + 0.5\sigma$	60	57.5
6	between $X + 0.5\sigma$ and $X + \sigma$	45	45.0
7	between $X + \sigma$ and $X + 1.5\sigma$	30	27.5
8	above $X + 1.5\sigma$	18	20.1

$P_k = P(-1.5\sigma \leq x \leq -\sigma) = P(z_1 = -1.5) - P(z_1 = -1) = 0.4332 - 0.3413$ (using Table 8.2) $= 0.0919$. So, the expected number in bin 2 is $(0.0919)(300) = 27.5$. The results for every bin are shown in Table 8.9.

Substitution of these results into Equation 8.46 yields $\chi^2 = 0.904$. For the values of $\chi_\alpha^2 = 0.904$ and $\nu = 5$, from Table 8.8, $\alpha \simeq 0.97$. That is, the probability of obtaining this χ^2 value or less is ~97 %, under the assumption that the expected distribution is correct. Thus, agreement with the assumed normal distribution is *significant*. This is contrary to what the professor originally thought!

8.12   MATLAB SIDEBAR

A χ^2 analysis of data can be performed automatically. The MATLAB M-file chinormchk.m does this. This M-file reads a single-column data file and plots its frequency density distribution using Scott's formula for equal-width intervals. It also determines the corresponding values of the normal probability density function based on the mean and standard deviation of the data and then it plots them on the same graph. Finally, it performs a χ^2 analysis that yields the percent confidence that the expected distribution is the correct one.

An example of the analysis of 500 test scores using chinormchk.m is shown in Figure 8.13. The mean of the test scores was 6.0 and the standard deviation was 1.0. The analysis shows that there is a 19 % chance that the scores represent those drawn from a normal distribution. This is based on the values of $\chi^2 = 8.7$ and $\nu = 6$.

8.10   DESIGN OF EXPERIMENTS

Statistical tools can be used in experimental planning. The method of **design of experiments** (DOE) provides an assessment of an experiment's output sensitivity to its independent variables. In DOE terminology, this method assesses the sensitivity of the *result* (the measurand − the dependent variable) to various *factors* (independent

$\alpha = 19.0759\ \%;\ \chi^2 = 8.7067;\ N = 500;\ K = 9;\ \nu = 6$

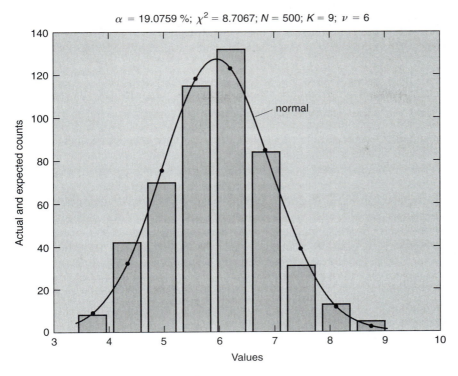

Figure 8.13   Analysis of 500 test scores using `chinormchk.m`.

variables) that comprise the *process* (experiment). The significance of DOE is that it can be carried out *before* an experiment is conducted. DOE, for example, can be used to identify the variables that most significantly affect the output. In essence, DOE provides an efficient way to help plan and conduct experiments.

Methods of DOE have been known for many years. According to Hald [9], fundamental work on DOE was carried out by R. A. Fisher and published in 1935 [8]. DOE, and the related topic, Taguchi methods, have become popular in recent years due to the quest through experimentation for improved quality in consumer and industrial products (for example, see [10], [11], and [12]).

The main objective of DOE is to determine how the factors influence the result of a process. The approach is to find this out by running trials (actual and/or computer trials), and measuring the process response for planned, controlled changes in the factors. A feature of DOE is that it provides a ranking of the sensitivity of the result to each factor; it ranks the factors in terms of their effects. It provides the direction of the sensitivity, whether a factor change increases or decreases the result. A major and important feature of DOE is that it provides this information with a minimum number of measurements or calculations. An additional advantage of DOE is that knowledge of the statistical nature of the input is unnecessary. If statistical information is available,

DOE can lead directly to an **ANalysis Of VAriance** (ANOVA) (for example, see [9]) and hypothesis tests for the factors and their interactions. If the statistical nature of some or all of the factors are unavailable, methods still exist to examine and rank the sensitivity.

The basics of DOE are as follows. Consider a hypothetical experiment with a result, y_j, depends on a number of factors, $A, B, C, \ldots$. Trials of the experiment are conducted for different, predetermined values of the factors, where the jth run produces the jth value of the result. The primary function of DOE is to determine a quantitative measure of the effect of each of the factor changes on the result. The output must be quantitative or measurable. The factors can be *quantitative*. They also can be *attribute* variables, such as hot or cold, fast or slow, and so forth. In the coverage here, factors will be allowed to take on only two values called a low level, indicated with a minus sign, and a high level, indicated with a plus sign. The high and low levels of the factors are selected by the experimenter to represent a practical range of values large enough to have an influence, yet small enough to determine the local behavior of the process. It is not uncommon to carry out one exploratory DOE to establish ranges of variables, and then perform another DOE, based on the results of the first, for a more refined analysis. In the case of attribute variables, such as fast and slow, the choice of high and low can be purely arbitrary. In the case of quantitative variables, the choice usually is intuitive, but it still remains arbitrary because the signs of the results reverse if the levels are reversed.

To illustrate the method of DOE, consider a hypothetical experiment in which an experimentalist wishes to assess the sensitivity of a light-level detector to two factors, the position of the detector from the surface of an object (factor A) and the time response of the detector (factor B). The percentage change in the amplitude of the detector's output from a reference amplitude is chosen as the result. Table 8.10 lists the factors and their levels.

Four trials are carried out. This provides a *complete* experiment, in which all four possible combinations of factor type and level are considered. In one trial, for example, the detector with the shortest time response, 20 ms (the low value of B), is placed near the surface, at 5 mm (the low value of A). The result for this case is an increase in the amplitude of 40.9 %, as displayed in Table 8.11. Examination of all results reveals that the greatest output is achieved by placing a 50-ms detector 10 mm from the surface of the object. DOE can be extended readily to consider more than two factors. To achieve a *complete* experiment, 2^k trials are required, where k is the number of factors, with two levels for each factor.

Table 8.10   Factors and their levels for the detector experiment.

	Low	High
A (mm)	5	10
B (ms)	20	50

Table 8.11 Percentage changes for four trials.

	A: low	A: high
B: low	40.9	47.8
B: high	42.4	50.2

8.11 FACTORIAL DESIGN

The method of DOE suggests a manner in which the contribution of each factor on an experimental result can be assessed. This is the method of **factorial design**. Because this method identifies the effect of each factor, it can be used to organize and minimize the number of experimental trials.

When dealing with the effects of changes in two factors, two levels and four runs, a measure of the effect, or *main effect*, ME, resulting from changing a factor from its low to high value can be estimated using the average of the two observed changes in the response. So for factor A,

$$\text{ME}_A = \frac{1}{2}[(y_2 - y_1) + (y_4 - y_3)]. \qquad [8.47]$$

Similarly, a measure of the effect of changing factor B from its low to high level can be estimated by averaging two corresponding changes as

$$\text{ME}_B = \frac{1}{2}[(y_3 - y_1) + (y_4 - y_2)]. \qquad [8.48]$$

These average effects from four runs have significantly greater reliability than changes computed from three runs. In addition, an interaction may exist between the factors. An effect of the interaction can be estimated from the four runs by taking the differences between the diagonal averages, where

$$\text{ME}_{AB} = \frac{1}{2}(y_1 + y_4) - \frac{1}{2}(y_2 + y_3) = \frac{1}{2}[(y_1 - y_2) + (y_4 - y_3)]. \qquad [8.49]$$

Finally, a measure of the overall level of the process can be based on the average as

$$\text{ME} = \frac{1}{4}[y_1 + y_2 + y_3 + y_4]. \qquad [8.50]$$

Equations 8.47 through 8.50 provide the basic structure of a factorial design with two levels per factor and four runs. Examine the form of the above main effect equations.

If the parentheses in Equations 8.47 through 8.49 are dropped and the responses are placed in the order of their subscripts, these equations become

$$\text{ME}_A = \frac{1}{2}[-y_1 + y_2 - y_3 + y_4], \qquad [8.51]$$

$$\mathrm{ME}_B = \frac{1}{2}[-y_1 - y_2 + y_3 + y_4],$$ [8.52]

and

$$\mathrm{ME}_{AB} = \frac{1}{2}[+y_1 - y_2 - y_3 + y_4].$$ [8.53]

A certain pattern of plus and minus signs from left to right appears in each equation. Table 8.12 lists the pattern of plus and minus signs in columns under each factor, A and B; the interaction, AB; and the overall gain, M. Note that the signs of the interaction are products of the signs of the factors.

A full set of trials often is repeated to permit estimation of the effects of the influence of uncontrolled variables. The preceding equations can be expressed more conveniently in terms of the sums or totals of the responses rather than the responses themselves. That is, for r runs or full sets of trials, the total, T_j, for the responses, y_{ji}, is given by adding the r responses as

$$T_j = \sum_{i=1}^{r} y_{ji}.$$ [8.54]

Experiments organized as here are referred to as 2^k designs, which yield four trials for two levels and two factors. A general form of Equations 8.51 through 8.53 that provides the estimates of the effects for k factors is [13]

$$\mathrm{ME}_j = \frac{1}{r2^{k-1}} \left[\sum_{i=1}^{2^k} \pm T_i \right],$$ [8.55]

where $j = A, B, AB, \dots$. For two factors, the proper signs for each term from left to right in Equation 8.55 are those signs in the column under the jth factor of Table 8.12. For example, for $r = 1$, $y_j = T_j$, if the main effect of factor B is to be estimated, then Equation 8.55 gives

$$\mathrm{ME}_B = \frac{1}{2}[-T_1 - T_2 + T_3 + T_4],$$ [8.56]

where the signs are those in the column under B in Table 8.12. For values of $k \geq 2$, a listing of the sequence of signs is given in statistics texts (for example, see [14]).

Table 8.12   Sign pattern for two factors, A and B.

Trial	A	B	AB	M	Response Total
1	−	−	+	+	T_1
2	+	−	−	+	T_2
3	−	+	−	+	T_3
4	+	+	+	+	T_4

Note that Equation 8.55 must be modified to calculate the overall mean, M, of the responses, where,

$$M = \frac{1}{r2^k} \left[\sum_{i=1}^{2^k} +T_i \right],$$ **[8.57]**

in which $j = A, B, AB, \ldots$, and all the signs are plus signs.

The sensitivity of a process to a factor level change generally differs from factor to factor. A small change in one factor may cause a large change in the response while another does not. Sensitivity is the slope of the response curve. In factorial design, the sensitivity, ζ, of the response to a certain factor is the main effect divided by the change in the factor. For example, in the case of factor B,

$$\zeta_B = \frac{ME_B}{B^+ - B^-}.$$ **[8.58]**

A problem can arise with the application of Equation 8.58 when the factors are attributes rather than numeric. As a result, sensitivity is usually viewed as being determined directly by the main effects themselves.

The example presented in Section 8.10 now can be analyzed using factorial analysis. Suppose that two runs were conducted, giving the eight results that are presented in Table 8.13. With this information, the main effects can be computed using Equation 8.55 and using the sign patterns in Table 8.12. For example, the main effect of factor A, detector response time, using Equation 8.55, is

$$ME_A = \frac{1}{4} [-82.5 + 87.7 - 84.4 + 96.7] = 4.4.$$ **[8.59]**

Similarly, $ME_B = 2.7$ and $ME_{AB} = 1.8$. These can be interpreted in the following manner. When going from the low to high level of factor A, that is, switching the detector position from 5 mm to 10 mm, there is a 4.4 % increase in the detector's amplitude. Similarly, when going from the low to high level of factor B, that is, changing from a 20-ms response detector to a 50-ms response detector, there is a 2.7 % increase in amplitude. The interaction main effect implies that changing from the

Table 8.13   Percentage changes for four trials, each conducted twice.

	A: low	A: high
B: low	$y_{11} = 40.9$	$y_{21} = 47.8$
	$y_{12} = 41.6$	$y_{22} = 39.9$
	$T_1 = 82.5$	$T_2 = 87.7$
B: high	$y_{31} = 42.4$	$y_{41} = 50.2$
	$y_{32} = 42.0$	$y_{42} = 46.5$
	$T_3 = 84.4$	$T_4 = 96.7$

combination of a 5-mm position of the 20-ms response detector to a 10-mm position of the 50-ms response detector increases the amplitude by 1.8 %. Finally, the average amplitude increase, M, is found from Equation 8.57, for $r = 2$ and $k = 2$, to be 43.9 %.

An inherent feature of any 2^k factorial design is that it is presented in a form that can be analyzed easily using ANOVA (for example, see [6], [14]), provided there are two or more full sets of runs, that is, $r > 1$. ANOVA yields an important piece of information. It determines if the effects of changing the levels of factors are statistically insignificant. If this is so, it means that uncontrolled variables were present in the experiment and caused changes greater than the controlled factor changes. If only one run is made, a 2^k design provides no measure of uncontrolled variations, known as the statistical error, and methods other than ANOVA must be used to measure significance.

REFERENCES

[1] D. Salsburg. 2001. *The Lady Tasting Tea: How Statistics Revolutionized Science in the Twentieth Century*. New York: W.H. Freeman and Company.

[2] Student. 1908. The Probable Error of a Mean. *Biometrika* 4: 1–25.

[3] R. Figliola and D. Beasley. 2002. *Theory and Design for Mechanical Measurements*, 3rd ed. New York: John Wiley and Sons.

[4] H. Young. 1962. *Statistical Treatment of Experimental Data*. New York: McGraw–Hill.

[5] J. Mandel. 1964. *The Statistical Analysis of Experimental Data*. New York: Dover.

[6] A. J. Hayter. 2002. *Probability and Statistics for Engineers and Scientists*, 2nd ed. Pacific Grove, Calif.: Duxbury/Thomson Learning.

[7] J. L. Devore. 2000. *Probability and Statistics for Engineering and the Sciences*. Pacific Grove, Calif.: Duxbury/Thomson Learning.

[8] R. A. Fisher. 1935. *The Design of Experiments*. Edinburgh: Oliver and Boyd.

[9] A. Hald. 1952. *Statistical Theory with Engineering Applications*. New York: John Wiley and Sons.

[10] L. B. Barrentine. 1999. *Introduction to Design of Experiments: A Simplified Approach*. Milwaukee, Wisc.: ASQ Quality Press.

[11] R. F. Gunst and R. L. Mason. 1991. *How to Construct Fractional Factorial Experiments*. Milwaukee, Wisc.: ASQ Quality Press.

[12] D. C. Montgomery. 2000. *Design and Analysis of Experiments*, 5th ed. New York: John Wiley and Sons.

[13] I. Guttman, S. Wilks, and J. Hunter. 1982. *Introductory Engineering Statistics*, 3rd ed. New York: John Wiley and Sons.

[14] D. C. Montgomery and G. C. Runger. 1994. *Applied Statistics and Probability for Engineers*. New York: John Wiley and Sons.

REVIEW PROBLEMS

1. Given 1233 instantaneous pressure measurements that are distributed normally about a mean of 20 psi with a standard deviation of 0.5 psi, what is the probability that a measured value will be between 19 psi and 21 psi?

2. What is the probability, in decimal form, that a normally distributed variable will be within 1.500 standard deviations of the mean?

3. A laser pinpoints the target for an advanced aircraft weapons system. In a system test, the fighter aircraft simulates targeting a flight test aircraft equipped with an optical receiver. Data recorders show that the standard deviation of the angle of the beam trajectory is $0.1400°$ with a mean of $0°$. The uncertainty in the angle of the beam trajectory is caused by precision errors, and the angle is distributed normally. What is the probability, in decimal form, that the aircraft laser system will hit a target 10.00 cm wide at a range of 100.0 m? (Consider only one dimension).

4. The average age of the dining clientele at a restaurant is normally distributed about a mean of 52 with a standard deviation of 20. What is the probability, in decimal form, of finding someone between the ages of 18 and 21 in the restaurant?

5. Each of 10 engineering students measures the diameter of a spherical ball bearing using dial calipers. The values recorded are [0.2503, 0.2502, 0.2501, 0.2497, 0.2504, 0.2496, 0.2500, 0.2501, 0.2494, 0.2502]. With knowledge that the diameters of the ball bearings are distributed normally, find the probability that the diameter of a bearing is within $+0.0005$ and -0.0005 of the mean based on the sample statistics. Respond in decimal form.

6. A series of acceleration measurements is normally distributed about a mean of 5.000 m/s^2 with a standard deviation of 0.2000 m/s^2. Find the value such that the probability of any value occurring below that value is 95 %.

7. The following data set, a sample of a normally distributed variable, is recorded from an accelerometer [9.81, 9.87, 9.85, 9.78, 9.76, 9.80, 9.89, 9.77, 9.78, 9.85]. Estimate the range of acceleration within which the next measured acceleration would be at $P = 99$ %.

8. From the data given in the previous problem, what range contains the true mean value of the acceleration? Respond with the absolute value of the range limits in accordance with the correct number of significant figures.

9. If normal distribution (population) statistics are used (incorrectly!) to compute the sample statistics, find the percent probability that the next measured acceleration will be within the sample precision interval computed in the previous accelerometer problem. Compute statistics as if they were population statistics.

10. A pressure pipeline manufacturer has performed wall thickness measurements for many years and knows that the true variance of the wall thickness of a 120 psi pipe is 0.500 mm^2. If the rejection criterion for sample variance is 1.02 mm^2 for a single wall, find the probability that the rejection criterion is good for a sample size of 16. Respond as a percent to the nearest whole percentage point.

11. A student determines, with 95 % confidence, that the true mean value based on a set of 61 values equals 6. The sample mean equals 4. Determine the value of the sample standard deviation to the nearest hundredth. Remember that the standard deviation has a positive value.

12. The following values of x were measured in an experiment: $x = [5, 1, 3, 6]$. Determine with 95 % confidence the upper value of the range within which the next data point will be to the nearest hundredth.

13. A student determines, with 95 % confidence, that the true mean value based on a set of N values equals 8. The sample mean equals 7. Assuming that N is very large (say > 100), determine the value of the standard deviation of the means to the nearest hundredth. Remember that the standard deviation of the means has a positive value.

14. The mean and standard deviation of a normally distributed population are 105 and 2, respectively. Determine the percent probability that a member of the population will have a value between 101 and 104. Give your answer in % to the nearest tenth.

15. The scores of the students who took the last year's SAT math exam were normally distributed with a mean of 580 and a standard deviation of 60. Determine the percentage of students who scored greater than 750 to the nearest hundredth of a percent.

HOMEWORK PROBLEMS

1. A February 14, 1997, *Observer* article cited a NCAA report on a famous midwest university's admission gap between all 1992–95 entering freshmen and the subset of entering freshman football team members. The article reported that the mean SAT scores were 1220 for all entering freshmen and 894 for the football team members. Assume that the standard deviations of the SAT scores were 80 and 135 for all freshman and all football team members, respectively. Determine [a] the percentage of all freshmen who scored greater than 1300 on their SATs, [b] the percentage of football players who scored greater than 1300 on their SATs, and [c] the number of football players who scored greater than half of all of the freshmen class, assuming that there were 385 football players. State all assumptions.

2. Assume that students who took the SAT math exam were normally distributed about a mean value of 580 with a standard deviation of 60. Determine what percentage of the students scored higher than 750 on the exam.

3. Using MATLAB, determine for a class of 69 the percent probability to four significant figures of getting a test score within ±1.5 standard deviations of the mean, assuming that the test scores are distributed according to [a] Student's *t* distribution and [b] the normal distribution.

4. During an experiment, an aerospace engineering student measures a wind tunnel's velocity N times. The student reports the following information, based on 90 % confidence, about the finite data set: mean velocity = 25.00 m/s; velocity standard deviation = 1.50 m/s; uncertainty in velocity = ±2.61 m/s. Determine [a] N, [b] the standard deviation of the means based on this data set (in m/s), [c] the uncertainty, at 95 % confidence, in the estimate of the true mean value of the velocity (in m/s), and [d] the interval about the sample mean over which 50 % of the data in this set will be (in m/s).

5. An aerospace engineering student performs an experiment in a wind tunnel to determine the lift coefficient of an airfoil. The student takes 61 measurements of the vertical force using a force balance, which yield a sample mean value of 44.20 N and a sample variance of 4.00 N². Determine [a] the percent probability that an additional measurement will be between 45.56 N and 48.20 N, [b] the range (in N) over which the true mean value will be assuming 90 % confidence, and [c] the range (in N²) over which the true variance will be assuming 90 % confidence.

6. The data presented in the table below was obtained during a strain gage force balance calibration, where $F(N)$ denotes the applied force in N and $E(V)$ the measured output voltage in V. Determine a suitable least-squares fit of the data using the appropriate functions in MATLAB. Quantify the best fit through the standard error of the fit, S_{yx}. Which polynomial fit has the lowest value of S_{yx}? Which polynomial fit is the most suitable (realistic) based on your knowledge of strain gage system calibrations? Plot each polynomial fit on a separate graph and include error bars for the y-variable, with the magnitude of error bar estimated at 95 % confidence and based on S_{yx}. Use the M-file plotfit.m. This M-file requires three columns of data, where the third column is the measurement uncertainty in y. Table 8.14 does not give those, so they must be added. Assume that each y measurement uncertainty is 5 % of the measured $E(V)$ value. S_{yx} is actually the curve-fit uncertainty. plotfit.m shows this uncertainty as dotted lines (not as error bars as in the problem statement). plotfit.m also prints out the S_{yx} value and calculates the precision interval $(t \cdot S_{yx})$ to help plot the dotted lines. Calculate the y measurement uncertainties and use them as a third column input to plotfit.m, then run plotfit.m for each order of the curve fit desired. Also be sure to label the axes appropriately.

Table 8.14   Strain gage calibration data.

$F(N)$	$E(V)$
0.4	2.7
1.1	3.6
1.9	4.4
3.0	5.2
5.0	9.2

7. An airplane manufacturer intends to establish a component acceptance criterion that is based on sound statistical methods. Preliminary tests on 61 acceptable components have determined that the mean load to produce component failure is 500 psi with a standard deviation of 25 psi. Based on this information, provide [a] an estimate, with 99 % confidence, of the value of the next (the 62nd) measured load to produce failure; [b] an estimate, with 99 % confidence, of the true mean load to produce failure; and [c] an estimate, with 98 % confidence, of the true variance. Finally, the manufacturer wants to be 99 % confident that if the batch sample meets the test criterion that the group from which the batch sample came is acceptable. [d] Determine the *range* of sample standard deviation values (in psi) that the batch sample can have and still meet the test criterion.

8. The sample mean of 21 golf ball weights equals 0.42 N and the sample variance equals 0.04 N^2. Determine the range (in N^2) that contains the true variance, with 90 % confidence.

9. The following values of x were measured in an experiment, $x = \{5, 3, 1, 6\}$. Find the range within which the next data point will fall, with 95 % confidence.

10. The mean and standard deviation of a normally distributed population of values x are $x' = 105$ and $\sigma = 2$. Find the percent probability that a value of x will fall in the range between 101 and 104.

11. The expected number of occurrences, E_k, (assuming a normal distribution) and the observed number of occurrences, O_k, for 40 measurements are given in Table 8.15. Use the χ^2 test to determine the probability that the discrepancies between the observed and expected data are due to chance alone. Choose your answer from these choices: [a] between 95 % and 90 %, [b] between 90 % and 50 %, [c] between 50 % and 5 %, and [d] between 5 % and 0 %.

Table 8.15   Expected and observed occurrences.

E_k	6.4	13.6	13.6	6.4
O_k	8	10	16	6

12. For the set of data [1, 3, 5, 7, 9] determine, with 95 % confidence, the range that contains [a] the true mean and [b] the true variance.

CHAPTER REVIEW

ACROSS

2. interval that characterizes the range within which a next value will occur

3. distribution characterizing successes and failures

6. type of χ^2 variable

9. device to demonstrate the normal distribution

10. a subset of the population

12. the most common distribution

13. probability that a specific number will occur out of all numbers

DOWN

1. level equal to one minus the probability

2. grouped into one set of measurands

4. statistical method used to estimate population parameters

5. the square root of the variance of the means

7. type of distribution that characterizes deviation from a population

8. distribution characterizing rarely occurring events

11. approximate Student's t-variable for large N at 95 % confidence

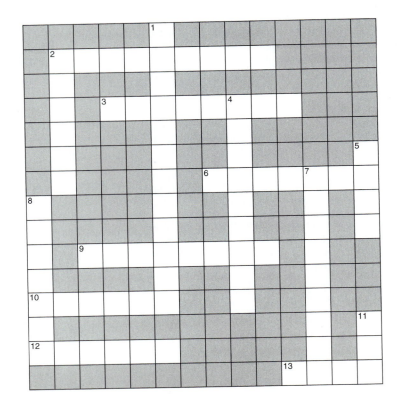

Babbage's Differencing Machine No. 2

Credit: Science Museum / Science & Society Picture Library, London.

The impetus for inventing this device, which is the forefather of the computer, was to reduce the uncertainty of numerical values in tables. The device was conceived by Charles Babbage (1791–1871) in England to calculate a series of numerical values and automatically print results. This was to overcome the errors present in tables that led to losses in large amounts of money and life at that time. There were three sources of error in producing tables: (1) calculation, (2) transcription, and (3) typesetting and printing. Babbage's machine was designed to mitigate all of these errors. The descriptor *differencing* referred to the method of finite differences, which reduced calculations to a series of additions. Shown in the photograph is the front, three-quarter view of Difference Engine No. 2. This engine was constructed by Reg Crick and Barrie Holloway of the Science Museum in London from drawings designed by Babbage between 1847 and 1849, and was completed in 1991 for the bicentennial year of Babbage's birth. It is the first of Babbage's calculating engines to be completed. It consists of 4000 parts and weighs over 3 tons. It has seven orders of difference and is designed to calculate to thirty figures. The machine is operated by the handle seen on the right (*Charles Babbage and His Calculating Engines*, Doron Swade, Science Museum, London, 1991).

UNCERTAINTY ANALYSIS

Whenever we choose to describe some device or process with mathematical equations based on physical principles, we always leave the real world behind, to a greater or lesser degree. ... These approximations may, in individual cases, be good, fair, or poor, but some *discrepancy between modeled and real behavior* always *exists.*

E. O. Doebelin. 1995.
Engineering Experimentation. New York: McGraw-Hill.

But in truth with more complicated instances, there is no more common error than to assume that, because prolonged and accurate mathematical calculations have been made, the application of the result to some fact of nature is absolutely certain.

Alfred N. Whitehead. 1948.
An Introduction to Mathematics. Oxford: Oxford University Press.

... it is a very certain truth that, whenever we are unable to identify the most true options, we should follow the most probable ...

René Descartes. 1637.
Discourse on Method.

CHAPTER OUTLINE

9.1   CHAPTER OVERVIEW

Uncertainty is one part of life that cannot be avoided. Its presence is a constant reminder of our limited knowledge and inability to control each and every factor that influences us. This especially holds true in the physical sciences. Whenever a process is quantified, by either modeling or experiments, uncertainty is present. In the beginning of this chapter the uncertainties present in modeling and experiments are identified. Agreement between the two is described in the context of uncertainty. Measurement uncertainties are studied in detail. Conventional methods on how to characterize, quantify, and propagate them are presented. The generic cases of single or multiple measurements of a measurand or a result are considered. Finally, numerical uncertainties associated with measurements are discussed.

9.2   LEARNING OBJECTIVES

You should be able to do the following after completing this chapter:

- Know the sources of modeling uncertainty
- Know the definitions of modeling uncertainty and error, measurement uncertainty, and error
- Differentiate between systematic and random errors, accuracy, and precision
- Differentiate between repetition and replication
- Estimate the relative systematic uncertainty for an instrument
- Know the definition of design-stage uncertainty and its common instrument errors
- Estimate the design-stage uncertainty of an entire measurement system
- Differentiate between general and detailed uncertainty analysis
- Estimate the uncertainty in a (1) single-measurement measurand, (2) single-measurement result, (3) multiple-measurement measurand, and (4) multiple-measurement result
- Estimate the discretization error of a derivative and of an integral

9.3   UNCERTAINTY

Aristotle first addressed uncertainty over 2300 years ago when he pondered the certainty of an outcome. However, it was not until the late 18th century that scientists considered the quantifiable effects of errors in measurements [1]. Continual progress on characterizing uncertainty has been made since then. Within the last 50 years, various methodologies on quantifying measurement uncertainty have been proposed [2]. In 1993, an international experimental uncertainty standard was developed by the International Organization for Standards (ISO) [3]. Its methodology now has been adopted by most of the international scientific community. In 1997, the National Conference of Standards Laboratories (NCSL) produced a U.S. Guide almost identical to the ISO Guide [4], "to promote consistent international methods in the expression of measurement uncertainty within U.S. standardization, calibration, laboratory accreditation, and metrology services." The American National Standards Institute with the American Society of Mechanical Engineers (ANSI/ASME) [5] and the American Institute of Aeronautics and Astronautics (AIAA) [6] also have new standards that follow the ISO Guide. These new standards differ from the ISO Guide only in that they use different names for the two categories of errors, random and systematic instead of Type A and Type B, respectively.

How is uncertainty categorized in the physical sciences? Whenever a physical process is quantified, uncertainties associated with modeling and computer simulation and/or with measurements can arise. Modeling and simulation uncertainties occur during the phases of "conceptual modeling of the physical system, mathematical modeling of the conceptual model, discretization and algorithm selection for the mathematical model, computer programming of the discrete model, numerical solution of the computer program model, and representation of the numerical solution" [7]. Such predictive uncertainties can be subdivided into modeling and numerical uncertainties [10]. Modeling uncertainties result from the assumptions and approximations made in mathematically describing the physical process. For example, modeling uncertainties occur when empirically based or simplified submodels are used as part of the overall model. Modeling uncertainties perhaps are the most difficult to quantify, particularly those that arise during the conceptual modeling phase. Numerical uncertainties occur as a result of numerical solutions to mathematical equations. These include discretization, round-off, nonconvergence, artificial dissipation, and related uncertainties. No standard for modeling and simulation uncertainty has been established internationally. Experimental or measurement uncertainties are inherent in the measurement stages of calibration and data acquisition. Numerical uncertainties also can occur in the analysis stage of the acquired data.

The terms **uncertainty** and error each have different meanings in modeling and experimental uncertainty analysis. *Modeling uncertainty* is defined as a *potential* deficiency due to a lack of knowledge and *modeling error* as a *recognizable* deficiency *not* due to a lack of knowledge [7]. According to Kline [8], **measurement error** is the difference between the true value and the measured value. It is a specific value. **Measurement uncertainty** is an estimate of the error in a measurement. It represents a range of possible values that the error might assume for a specific measurement.

By convention, the reported value of x is expressed with the same precision as its uncertainty U_x, such as 1.25 ± 0.05. The magnitude of U_x depends on the assumed confidence, the uncertainties that contribute to U_x, and how the contributing uncertainties are combined. The approach taken to determine U_x involves adopting an uncertainty standard, such as that presented in the ISO Guide, identifying and categorizing all of the contributory uncertainties, assuming a confidence for the estimate, and then, finally, combining the contributory uncertainties to determine U_x. The types of error that contribute to measurement uncertainty must be identified first. The remainder of this chapter focuses almost primarily on measurement uncertainty analysis. Its associated numerical uncertainties are considered at the end of this chapter.

9.4 COMPARING THEORY AND MEASUREMENT

Given that both modeling and experimental uncertainties exist, what does agreement between the two mean? How should such a comparison be illustrated? Conventionally, the experimental uncertainty is denoted in a graphical presentation by error bars centered on measurement values and the modeling uncertainty by dashed or dotted curves on both sides of the theoretical curve. Both uncertainties should be estimated with the same statistical confidence. An example is shown in Figure 9.1. When all data points and their error bars are within the model's uncertainty curves, then the experiment and theory are said to agree *completely* within the assumed confidence. When some data points and either part or all of their error bars are within the predictive uncertainty curves, then the experiment and theory are said to agree *partially* within the assumed confidence. There is *no* agreement when all of the data points and their error bars are outside of the predictive uncertainty curves.

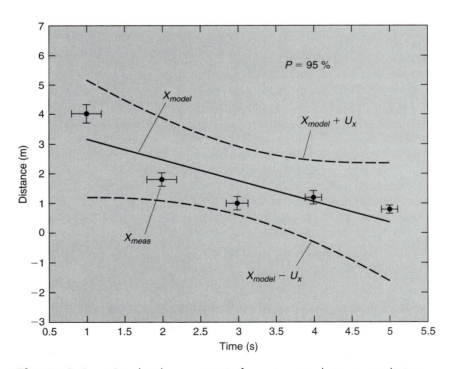

Figure 9.1 Graphical presentation of a comparison between predictive and experimental results.

| **9.1** | **MATLAB SIDEBAR** |

Experimental results can be plotted with x and y error bars using the MATLAB M-file `ploterror.m`. This m-file uses a MATLAB M-file `eplot.m` written by Andri M. Gretarsson to draw error bars in both the x- and y-directions. The MATLAB command `errorbar` only draws an error bar in the x-direction.

`ploterror.m` reads a four-column text file of x, x-error, y, and y-error values. It plots x,y error bars centered on each data point, like that shown in Figure 9.1.

Does agreement between experiment and theory imply that both correctly represent the process under investigation? Not all of the time. Caution must be exercised whenever such comparisons are made. Agreement between theory and experiment necessarily does not imply correctness. This issue has been addressed by Aharoni [11], who discusses the interplay between good and bad theory and experiments. There are several possibilities. A bad theory can agree with bad data. The wrong theory can agree with a good experiment by mere coincidence. A correct theory may disagree with the experiment simply because an unforeseen variable was not considered or controlled during the experiment. Therefore, agreement does not necessarily ensure correctness.

Caution also must be exercised when arguments are made to support agreement between theory and experiment. Scientific data can be misused [12]. The data may have been presented selectively to fit a particular hypothesis while ignoring other hypotheses. Indicators may have been chosen with units to support an argument, such as the number of automobile deaths instead of the number per mile traveled. Inappropriate scales may have been used to exaggerate an effect. Taking the logarithm of a variable appears graphically as having less scatter. This can be deceptive. Other illogical mistakes may have been made, such as confusing cause and effect, and implicitly using unproven assumptions. Important factors and details may have been neglected that could lead to a different conclusion. *Caveat emptor!*

STATEMENT

Lemkowitz et al. [12] illustrate how different conclusions can be drawn from the same data. Consider the number of fatalities per 100 million passengers for two modes of transport given in the 1992 British fatality rates. For the automobile, there were 4.5 fatalities per journey and 0.4 fatalities per km. For the airplane, there were 55 fatalities per journey and 0.03 fatalities per km. Therefore, on a per km basis airplane travel had approximately 10 times fewer fatalities than the automobile. Yet, on a per journey basis, the auto had approximately 10 times fewer fatalities. Which of the two modes of travel is safer and why?

Example 9.1

SOLUTION

There is no unique answer to this question. In fact, on a per hour basis, the automobile and airplane have the *same* fatality rate, which is 15 fatalities per 100 million passengers. Perhaps driving for shorter distances and flying for longer distances is safer than the converse.

9.5   UNCERTAINTY AS AN ESTIMATED VARIANCE

When measurements are made under fixed conditions, the recorded values of a variable still will vary to an extent. This implicitly is caused by small variations in uncontrolled variables. The extent of these variations can be characterized by the variance of the values. Further, as the number of acquired values becomes large ($N \geq 30$), the sample variance approaches its true variance, σ^2, and the distribution evolves to a normal distribution. Thus, if uncertainty is considered to represent the range within which an acquired value will occur, then uncertainty can be viewed as an estimate of the true variance of a normal distribution.

Consider the most general situation of a result, r, where $r = r(x_1, x_2, \ldots)$ and $x_1, x_2, \ldots$ represent measured variables whose distributions are normal. In uncertainty analysis, a **result** is defined as a variable that is not measured but is related functionally to the **measurands**. The result's variance is defined as

$$\sigma_r^2 = \lim_{N \to \infty} \left[\frac{1}{N} \sum_{i=1}^{N} (r_i - r')^2 \right],$$ **[9.1]**

where N is the number of determinations of r based on the set of $x_1, x_2, \ldots$ measurands. The difference between a particular r_i value and its mean value, r', can be expressed in terms of a Taylor series expansion and the measurands' differences by

$$(r_i - r') \simeq (x_{1_i} - x_1') \left(\frac{\partial r}{\partial x_1} \right) + (x_{2_i} - x_2') \left(\frac{\partial r}{\partial x_2} \right) + \cdots.$$ **[9.2]**

In this equation the higher-order terms involving second derivatives and beyond are assumed to be negligible. Equation 9.2 can be substituted into Equation 9.1 to yield

$$\begin{aligned}
\sigma_r^2 &\simeq \lim_{N \to \infty} \frac{1}{N} \sum_{i=1}^{N} \left[(x_{1_i} - x_1') \left(\frac{\partial r}{\partial x_1} \right) + (x_{2_i} - x_2') \left(\frac{\partial r}{\partial x_2} \right) + \cdots \right]^2 \\
&\simeq \lim_{N \to \infty} \frac{1}{N} \sum_{i=1}^{N} \left[(x_{1_i} - x_1')^2 \left(\frac{\partial r}{\partial x_1} \right)^2 + (x_{2_i} - x_2')^2 \left(\frac{\partial r}{\partial x_2} \right)^2 \right. \\
&\quad \left. + 2(x_{1_i} - x_1')(x_{2_i} - x_2') \left(\frac{\partial r}{\partial x_1} \right) \left(\frac{\partial r}{\partial x_2} \right) + \cdots \right].
\end{aligned}$$ **[9.3]**

The first two terms on the right-hand side of Equation 9.3 are related to the variances of x_1 and x_2, where

$$\sigma_{x_1}^2 = \lim_{N \to \infty} \left[\frac{1}{N} \sum_{i=1}^{N} (x_{1_i} - x_1')^2 \right]$$ **[9.4]**

and

$$\sigma_{x_2}^2 = \lim_{N \to \infty} \left[\frac{1}{N} \sum_{i=1}^{N} (x_{2_i} - x_2')^2 \right].$$ **[9.5]**

The third term is related to the **covariance** of x_1 and x_2, σ_{x_1,x_2}, where

$$\sigma_{x_1,x_2} = \lim_{N \to \infty} \left[\frac{1}{N} \sum_{i=1}^{N} (x_{1_i} - x_1')(x_{2_i} - x_2') \right]. \qquad \textbf{[9.6]}$$

When x_1 and x_2 are statistically independent, $\sigma_{x_1 x_2} = 0$. Substituting Equations 9.4, 9.5, and 9.6 into Equation 9.3 gives

$$\sigma_r^2 \simeq \sigma_{x_1}^2 \left(\frac{\partial r}{\partial x_1} \right)^2 + \sigma_{x_2}^2 \left(\frac{\partial r}{\partial x_2} \right)^2 + 2\sigma_{x_1 x_2} \left(\frac{\partial r}{\partial x_1} \right)\left(\frac{\partial r}{\partial x_2} \right) + \cdots. \qquad \textbf{[9.7]}$$

Equation 9.7 can be extended to relate the uncertainty in a result as a function of measurand uncertainties. Defining the squared uncertainty u_i^2 as an estimate of the variance σ_i^2, Equation 9.7 becomes

$$u_c^2 \simeq u_{x_1}^2 \left(\frac{\partial r}{\partial x_1} \right)^2 + u_{x_2}^2 \left(\frac{\partial r}{\partial x_2} \right)^2 + 2u_{x_1 x_2} \left(\frac{\partial r}{\partial x_1} \right)\left(\frac{\partial r}{\partial x_2} \right) + \cdots. \qquad \textbf{[9.8]}$$

This equation shows that the uncertainty in the result is a function of the estimated variances (uncertainties) u_{x_1} and u_{x_2} and their estimated covariance $u_{x_1 x_2}$. It forms the basis for more detailed uncertainty expressions that are developed in the remainder of this chapter and used to estimate the overall uncertainty in a variable. u_c^2 is called the **combined estimated variance**. The **combined standard uncertainty** is u_c. This is denoted by u_r for a result and by u_m for a measurand. In order to determine the combined standard uncertainty, the types of errors that contribute to the uncertainty must be examined first.

9.6   SYSTEMATIC AND RANDOM ERRORS

When a single measurement is performed, a number is assigned that represents the magnitude of the sensed physical variable. Because the measurement system used is not perfect, an error is associated with that number. If the system's components have been calibrated against more accurate standards, these standards have their own inaccuracies. The act of calibration itself introduces further uncertainty. All these factors contribute to the measurement uncertainty of a single measurement. Further, when the measurement is repeated, its value most likely will not be the same as it was the first time. This is because small, imperceptible changes in variables that affect the measurement have occurred in the interim, despite any attempts to perform a controlled experiment. Fortunately, almost all experimental uncertainties can be estimated, provided there is a consistent framework that identifies the types of uncertainties and establishes how to quantify them. The first step in this process is to identify the types of errors that give rise to measurement uncertainty.

Following the convention of the 1998 ANSI/ASME guidelines [5], the errors that arise in the measurement process can be categorized into either **systematic**

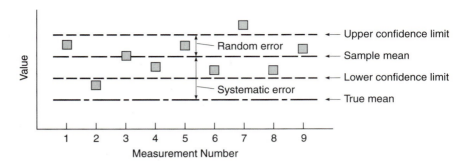

Figure 9.2   Nine recorded values of the same measurand.

(bias) or **random** (precision) **errors**. Systematic errors sometimes can be difficult to detect and can be found and minimized through *calibration*, which involves the comparison with a true, known value. They determine the **accuracy** of the measurement. Further, they lack any statistical information. The systematic error of the experiment whose results are illustrated in Figure 9.2 is the difference between the true mean value and the sample mean value. In other words, if the estimate of a quantity does not equal the actual value of the quantity, then the quantity is biased. Random errors are related to the scatter in the data obtained under fixed conditions. They determine the **precision**, or repeatability, of the measurement. The random error of the experiment whose results are shown in Figure 9.2 is the difference between a confidence limit (either upper or lower) and the sample mean value. This confidence limit is determined from the standard deviation of the measured values, the number of measurements, and the assumed percent confidence. Random errors are statistically quantifiable. Therefore, an ideal experiment would be highly accurate *and* highly repeatable. High repeatability alone does not imply minimal error. An experiment could have hidden systematic errors and yet highly repeatable measurements, thereby always yielding approximately the same, yet inaccurate, values. Experiments having no bias but poor precision also are undesirable. In essence, systematic errors can be minimized through careful calibration. Random errors can be reduced by repeated measurements and the careful control of conditions.

| Example 9.2 | **STATEMENT** |

Some car rental agencies use an onboard global positioning system (GPS) to track an auto. Assume that a typical GPS's precision is 2 % and its accuracy is 5 %. Determine the combined standard uncertainty in position indication that the agency would have if (1) it uses the GPS system as is and (2) it recalibrates the GPS to within an accuracy of 1 %.

SOLUTION

Denote the precision uncertainty by u_p and the accuracy as u_a. The combined uncertainty, u_c, obtained by applying Equation 9.8, is

$$u_c = \sqrt{u_a^2 + u_p^2}.$$ **[9.9]**

For case [1], $u_c = \sqrt{29} = 5.39$. For case [2], $u_c = \sqrt{5} = 2.24$. So the recalibration decreases the combined uncertainty by 3.15 %.

Examine two circumstances that help to clarify the difference between systematic and random errors. First, consider a stop watch being used to measure the time that a runner crosses the finish line. If the physical reaction times of a timer who is using the watch are equally likely to be over or under by some average reaction time in starting and stopping the watch, the measurement of a runner's time will have a random error associated with it. This error could be assessed statistically by determining the variation in the recorded times of a same event whose time is known exactly by another method. If the watch does not keep exact time, there is a systematic error in the time record itself. This error can be ascertained by comparing the mean value of recorded times of a same event with the exact time. Further, if the watch's inaccuracy is comparable to the timer's reaction time, the systematic error may be hidden within (or negligible with respect to) the random error. In this situation, the two types of errors are very hard to distinguish.

Next consider an analog panel meter that is used to record a signal. If the meter is read consistently from one side by an experimenter who records the reading, this method introduces a systematic error into the measurements. If the meter is read head-on, but not exactly so in every instance, then the experimenter introduces a random error into the recorded value.

For both examples, the systematic and random errors combine to determine the overall uncertainty. This is illustrated in Figure 9.2, which shows the results of an experiment. A sample of $N = 9$ recordings of the measurand, x, were made under fixed operating conditions. The *true* mean value of x, x_{true}, would be the sample mean value of the nine readings if no uncertainties were present. However, because of systematic and random errors, the two mean values differ. Fortunately, an estimate of the true mean value can be made to within certain confidence limits by using the sample mean value and methods of uncertainty analysis.

9.7 MEASUREMENT PROCESS ERRORS

The experimental measurement process itself introduces systematic and random errors. Most experiments can be categorized either as timewise or as sample-to-sample experiments. In a timewise experiment, measurand values are recorded sequentially in time. In a sample-to-sample experiment, measurand values are recorded for multiple samples. The random error determined from a series of repeated measurements in a timewise experiment performed under steady conditions results from small, uncontrollable factors that vary during the experiment and influence the measurand. Some errors may not vary over short time periods but will over longer periods. So, the effect of the measurement time interval must be considered [5]. In the analogous sample-to-sample experiment, the random error arises from both sample-to-sample measurement system variability and variations due to small, uncontrollable factors during the measurement process.

Errors that are not related directly to measurement system errors can be identified through repeating and replicating an experiment. In measurement uncertainty analysis, **repetition** implies that measurements are repeated in a particular experiment under the *same* operating conditions. **Replication** refers to the duplication of an experiment having *similar* experimental conditions, equipment, and facilities. The specific manner in which an experiment is replicated helps to identify various kinds of error. For example, replicating an experiment using a similar measurement system and the same conditions will identify the error resulting from using similar equipment. The definitions of repetition and replication differ from those commonly found in statistics texts (for example, see [14]), which consider an experiment repeated n times to be replicated $n + 1$ times, with no changes in the fixed experimental conditions, equipment, or facility.

The various kinds of errors can be identified by viewing the experiment in the context of different *orders* of replication levels. At the **zeroth-order replication level**, only the errors inherent in the measurement system are present. This corresponds to either absolutely steady conditions in a timewise experiment or a single, fixed sample in a sample-to-sample experiment. This level identifies the smallest error that a given measurement system can have. At the **first-order replication level**, the additional random error introduced by small, uncontrolled factors that occur either timewise or from sample to sample are assessed. At the **Nth-order replication level**, further systematic errors beyond the first-order level are considered. These, for example, could come from using different but similar equipment.

Measurement process errors originate during the calibration, measurement technique, data acquisition, and data reduction phases of an experiment [5]. Calibration errors can be systematic or random. Large systematic errors are reduced through calibration usually to the point where they are indistinguishable with inherent random errors. Uncertainty propagated from calibration against a more accurate standard still reflects the uncertainty of that standard. The order of standards in terms of increasing calibration errors proceeds from the primary standard through interlaboratory, transfer, and working standards. Typically, the uncertainty of a standard used in a calibration is **fossilized**, that is, it is treated as a fixed systematic error in that calibration and in any further uncertainty calculations [13]. Data acquisition errors originate from the measurement system's components, including the sensor, transducer, signal conditioner, and signal processor. These errors are determined mainly from the elemental uncertainties of the instruments. Data reduction errors arise from computational methods, such as a regression analysis of the data, and from using finite differences to approximate derivatives and integrals. Other errors come from the techniques and methods used in the experiment. These can include uncertainties from uncontrolled environmental effects, inexact sensor placement, instrument disturbance of the process under investigation, operational variability, and so on. All these errors can be classified as either systematic or random errors.

9.8 QUANTIFYING UNCERTAINTIES

Before an overall uncertainty can be determined, analytical expressions for systematic and random errors must be developed. These expressions come from probabilistic considerations. Random error results from a large number of very small, uncontrollable effects that are independent of one another and with each influencing the measurand. These effects yield measurand values that differ from one another, even under fixed operating conditions. It is reasonable to assume that the resulting random errors will follow a Gaussian distribution. This distribution is characterized through the mean and standard deviation of the random error. Because uncertainty is an estimate of the error in a measurement, it can be characterized by the standard deviation of the random error. Formally, the **random uncertainty** (precision limit) in the value of the measurand x is

$$P_x = t_{\nu_{P_x},C} \cdot S_{P_x},$$ [9.10]

where S_{P_x} is the standard deviation of the random error, $t_{\nu_{P_x},C}$ is the Student's t variable based on ν_{P_x} degrees of freedom, and C is the percent confidence. Likewise, the random uncertainty in the *average* value of the measurand x determined from N measurements is

$$P_{\bar{x}} = t_{\nu_{P_x},C} \cdot S_{P_{\bar{x}}} = t_{\nu_{P_x},C} \cdot \frac{S_{P_x}}{\sqrt{N}}.$$ [9.11]

Usually Equation 9.10 is used for experiments involving the single measurement of a measurand and Equation 9.11 for multiple measurements of a measurand. The degrees of freedom for the random uncertainty in x are

$$\nu_{P_x} = N - 1.$$ [9.12]

Systematic uncertainty can be treated in a similar manner. Although systematic errors are assumed to remain constant, their estimation involves the use of statistics. Following the ISO guidelines, systematic errors are assumed to follow a Gaussian distribution. The **systematic uncertainty** (bias limit) in the value of the measurand x is denoted by B_x. The value of B_x has a **reliability** of ΔB_x. Typically, a manufacturer provides a value for an instrument's accuracy. This number is assumed to be B_x. The value of the reliability is an estimate of the accuracy and is expressed in units of B_x. Hence, the lower its value, the more the confidence that is placed in the reported value of the accuracy. Formally, the systematic uncertainty in the value of x is

$$B_x = t_{\nu_{B_x},C} \cdot S_{B_x},$$ [9.13]

where S_{B_x} is the standard deviation of the systematic error. Equation 9.13 can be rearranged to determine S_{B_x} for a given confidence and value of B_x. For example, $S_{B_x} \cong B_x/2$ for 95 % confidence. Finally, according to the ISO guidelines, the degrees of freedom for the systematic uncertainty in x are

$$\nu_{B_x} \cong \frac{1}{2}\left(\frac{\Delta B_x}{B_x}\right)^{-2} = \frac{1}{2}\left(\frac{\Delta S_{B_x}}{S_{B_x}}\right)^{-2}.$$ [9.14]

The quantity $\Delta B_x/B_x$ is termed the **relative systematic uncertainty** of B_x. More certainty in the estimate of B_x implies a smaller ΔB_x and, hence, a larger ν_{B_x}. 100 % certainty corresponds to $\nu_{B_x} = \infty$. This effectively means that an infinite number of measurements are needed to assume 100 % certainty in the stated value of B_x.

Example 9.3

STATEMENT

A manufacturer states that the accuracy of a pressure transducer is 1 psi. Assuming that the reliability of this value is 0.5 psi, determine the relative systematic uncertainty in a pressure reading using this transducer and the degrees of freedom in the systematic uncertainty of the pressure.

SOLUTION

According to the definition of the relative systematic uncertainty, $\Delta B_x/B_x = 0.5/1 = 0.5$ or 50 %. From Equation 9.14, the degrees of freedom for the systematic uncertainty in the pressure are $\frac{1}{2}(0.5)^{-2} = 2$. This is a relatively lower number, which reflects the high relative systematic uncertainty.

9.9   MEASUREMENT UNCERTAINTY ANALYSIS

Experimental uncertainty analysis involves the identification of errors that arise during all stages of an experiment and the propagation of these errors into the overall uncertainty of a desired result. The specific goal of uncertainty analysis is to obtain a value of the **overall uncertainty** (expanded uncertainty), U_x, of the variable, x, such that either the next-measured value, x_{next}, or the true value, x_{true}, can be estimated. x_{next} represents the value of the next member of the sample of acquired data. x_{true} is the value of the mean of the population from which the sample was drawn. For the case of a single measurement or result based on x, this is expressed as

$$x_{\text{next}} = x \pm U_x \quad (\%C), \qquad [9.15]$$

in which the estimate is obtained with $\%C$ confidence. For the case of a multiple measurement or result based on x, this becomes

$$x_{\text{true}} = \bar{x} \pm U_x \quad (\%C), \qquad [9.16]$$

in which $\bar{x}$ denotes the sample average of x. For either case, the overall uncertainty, U_x, can be expressed as

$$U_x = k \cdot u_c, \qquad [9.17]$$

where k is the **coverage factor** and u_c the combined standard uncertainty. According to the ISO guidelines, the coverage factor is represented by the Student's t variable that is based on the number of effective degrees of freedom. That is,

$$U_x = t_{\nu_{\text{eff}},C} \cdot u_c. \qquad [9.18]$$

This assumes a normally distributed measurement error with zero mean and σ^2 variance. A zero mean implies that all significant systematic errors have been identified and removed from the measurement system prior to acquiring data. How these uncertainties contribute to the combined standard uncertainty and how they determine ν_{eff} depends on the type of experimental situation encountered. The value of ν_{eff} is determined knowing the values of ν_{P_x} and ν_{B_x} given by Equations 9.12 and 9.14, respectively.

There are four situations that typify most experiments:

1. The single-measurement measurand experiment, in which the value of the measurand is based on a single measurement (example: a measured temperature);

2. The single-measurement result experiment, in which the value of a result depends on single values of one or more measurands (example: the volume of a cylinder based on its length and diameter measurements);

3. The multiple-measurement measurand experiment, in which the mean value of a measurand is determined from a number of repeated measurements (example: the average temperature determined from a series of temperature measurements); or

4. The multiple-measurement result experiment, in which the mean value of a result depends on the values of one or more measurands, each determined from the same or a different number of measurements (example: the mean density of a perfect gas determined from N_1 temperature and N_2 pressure measurements).

Two standard types of uncertainty analysis can be used for experiments, general and detailed. Which type of uncertainty analysis applies to a particular experiment is not fixed. General uncertainty analysis usually applies to the first two situations and detailed uncertainty analysis to the latter two. Coleman and Steele [10] give a thorough presentation of both types.

General uncertainty analysis is a simplified approach that considers each measurand's overall uncertainty and its propagation into the final result. It does not consider the specific systematic and random errors that contribute to the overall uncertainty. This type of analysis typically is done during the planning stage of an experiment. It helps to identify sources of error and their contribution to the overall uncertainty. It also aids in determining whether or not a particular measurement system is appropriate for a planned experiment.

Detailed uncertainty analysis is a more thorough approach that identifies the systematic and random errors contributing to each measurand's overall uncertainty. The propagation of systematic and random errors into the final result is computed in parallel. The framework of detailed uncertainty analysis in this text is consistent with that presented in the ISO Guide [3]. This type of uncertainty analysis usually is done for more-involved experimental designs and follows the calibration, data acquisition, and data reduction phases of an experiment.

In the following, each of the four common situations will be examined in more detail using the uncertainty analysis approach that is most appropriate.

9.10   GENERAL UNCERTAINTY ANALYSIS

General uncertainty analysis is most applicable to experimental situations involving either a single-measurement measurand or a single-measurement result. The uncertainty of a single-measurement measurand is related to its instrument uncertainty, which is determined from calibration, and to the resolution of instrument used to read the measurand value. The uncertainty of a single-measurement result comes directly from the uncertainties of its associated measurands. The expressions for these uncertainties follow directly from Equation 9.8.

For the case of J measurands, the combined standard uncertainty in a result becomes

$$u_r^2 \simeq \sum_{i=1}^{J} (\theta_i)^2 \, u_{x_i}^2 + 2 \sum_{i=1}^{J-1} \sum_{j=i+1}^{J} (\theta_i)(\theta_j) \, u_{x_i, x_j}, \qquad \textbf{[9.19]}$$

where

$$u_{x_i, x_j} = \sum_{k=1}^{L} (u_i)_k (u_j)_k, \qquad \textbf{[9.20]}$$

with L being the number of elemental error sources that are common to measurands x_i and x_j, and $\theta_i = \partial r / \partial x_i$. θ_i is known as the **absolute sensitivity coefficient**. This coefficient should be evaluated at the expected value of x_i. It is important to note that the covariances in Equation 9.19 should not be ignored simply for convenience when performing an uncertainty analysis. Variable interdependence should be assessed. This occurs through common factors, such as ambient temperature and pressure or an instrument used for different measurands.

When the covariances are negligible, Equation 9.19 for J *independent* variables simplifies to

$$u_r^2 \simeq \sum_{i=1}^{J} \left(\theta_i u_{x_i} \right)^2, \qquad \textbf{[9.21]}$$

where u_{x_i} is the **absolute uncertainty**. The values of the result's uncertainty will follow Student's t distribution [3], based on the number of *effective degrees of freedom*, ν_{eff}, with

$$\nu_{\text{eff}} = \nu_r = \frac{u_r^4}{\sum_{i=1}^{J} (\theta_i^4 u_{x_i}^4)/\nu_i} = \frac{\left(\sum_{i=1}^{J} \theta_i^2 u_{x_i}^2 \right)^2}{\sum_{i=1}^{J} (\theta_i^4 u_{x_i}^4)/\nu_i}, \qquad \textbf{[9.22]}$$

where $\nu_i = N_i - 1$ is the number of degrees of freedom for u_{x_i} and $\nu_{\text{eff}} \leq \sum_{i=1}^{J} \nu_i$. This equation, known as the Welch-Satterthwaite formula, was presented originally by Welch [9]. The value of ν_{eff} obtained from the formula is rounded to the nearest whole number.

Equation 9.19 can be applied to estimate the uncertainty in a measurand. In this case, all the absolute sensitivity coefficients equal unity, and Equation 9.19 reduces to

$$u_m^2 \simeq \sum_{i=1}^{J} u_{x_i}^2 + 2 \sum_{i=1}^{J-1} \sum_{j=i+1}^{J} u_{x_i,x_j}, \qquad \textbf{[9.23]}$$

where u_m is the measurand uncertainty. Further, when the covariances are negligible, Equation 9.23 simplifies to

$$u_m^2 \simeq \sum_{i=1}^{J} u_{x_i}^2. \qquad \textbf{[9.24]}$$

The corresponding number of effective degrees of freedom becomes

$$\nu_{\text{eff}} = \nu_m = \frac{u_m^4}{\sum_{i=1}^{J}(u_{x_i}^4/\nu_i)} = \frac{\left(\sum_{i=1}^{J} u_{x_i}^2\right)^2}{\sum_{i=1}^{J}(u_{x_i}^4/\nu_i)}. \qquad \textbf{[9.25]}$$

Example 9.4

STATEMENT

Two pressure transducers are used to determine the pressure difference, $\Delta P = P_2 - P_1$, between a wind tunnel's reference pressure tap and a static pressure tap on the surface of an airfoil placed inside the wind tunnel. Both transducers are identical and each has a reported accuracy of 1 %, as determined by the manufacturer's calibration. An experimenter decides to recalibrate these transducers against a laboratory standard that has 0.3 % accuracy. Further, the recalibration method itself introduces an additional 0.5 % uncertainty. Determine the combined standard uncertainty in ΔP for two situations: one when the experimenter does not recalibrate the transducers (case A) and the other when she does (case B).

SOLUTION

Both situations involve the determination of the uncertainty in a measurand. When the transducers are not recalibrated, each transducer has an uncertainty of 1 % and the uncertainties are independent. According to Equation 9.21, $u_{\Delta P_A}^2 = u_{P_1}^2 + u_{P_2}^2$. Thus, $u_{\Delta P_A} = \sqrt{0.01^2 + 0.01^2} \simeq 0.014$ or 1.4 %. When the transducers are recalibrated against the *same* standard, their uncertainties are correlated. Hence, the covariant term in Equation 9.19 must be considered. Thus, $u_{\Delta P_B}^2 = u_{P_1}^2 + u_{P_2}^2 + 2\theta_1\theta_2 u_{P_1,P_2}$. Here $\theta_1 = \partial u_{P_1}/\partial u_{\Delta P_B} = -1$ and $\theta_2 = \partial u_{P_2}/\partial u_{\Delta P_B} = 1$.

The uncertainty resulting from the recalibration according to Equation 9.20 is

$$u_{P_1,P_2} = \sum_{k=1}^{2}(u_i)_k(u_j)_k = (u_{P_1})_1(u_{P_2})_1 + (u_{P_1})_2(u_{P_2})_2$$
$$= (0.3)(0.3) + (0.5)(0.5) \simeq 0.3 \%.$$

Thus, applying Equation 9.19, $u_{\Delta P_B} = \sqrt{0.01^2 + 0.01^2 - (2)(0.3)} \simeq 0.012$ or 1.2 %. So, the recalibration of the transducers reduces the uncertainty in the pressure difference from 1.4 % to 1.2 %.

Because this situation involves the uncertainty of a measured *difference*, one of the θ_i's is negative. This reduces the uncertainty in the difference *if* the instruments are calibrated against the *same* standard under the *same* conditions (time, place, and so forth). This reduction seems counterintuitive at first, that one can reduce the measurement uncertainty of a difference through recalibration, which itself introduces uncertainty. However, the uncertainty of a measured difference can be reduced because recalibration using the same standard and conditions introduces a *dependent* systematic error. This error effectively obviates the independent systematic errors of each instrument.

Each single-measurement situation is considered in the following.

9.10.1 SINGLE-MEASUREMENT MEASURAND EXPERIMENT

Many times it is desirable to estimate the uncertainty of a single measurand taken using a certain instrument. Typically this is done *before* conducting an experiment. The contributory errors are considered fossilized, hence systematic. The expression for the combined standard uncertainty is Equation 9.23.

This particular type of uncertainty is known as the design-stage uncertainty, u_d, which is analogous to the combined standard uncertainty. Often it is used to choose an instrument that meets the accuracy required for a measurement. It is expressed as a function of the zero-order uncertainty of the instrument, u_0, and the instrument uncertainty, u_I, as

$$u_d = \sqrt{u_0^2 + u_I^2},$$
[9.26]

which usually is computed at the 95 % confidence level.

Instruments have resolution, readability, and errors. The **resolution** of an instrument is the smallest *physically indicated* division that the instrument displays or is marked. The zero-order uncertainty of the instrument, u_0, is set arbitrarily to be equal to one-half the resolution, based on 95 % confidence. Equation 9.26 shows that the design-stage uncertainty can never be less than u_0, which would occur when $u_0 \gg u_I$. In other words, even if the instrument is perfect and has no instrument errors, its output must be read with some finite resolution and, therefore, some uncertainty.

The **readability** of an instrument is the closeness with which the scale of the instrument is read by an experimenter. This is a subjective value. Readability does not enter into assessing the uncertainty of the instrument.

The instrument uncertainty usually is stated by the manufacturer and results from a number of possible elemental instrument uncertainties, e_i. Examples of e_i are hysteresis, linearity, sensitivity, zero-shift, repeatability, stability, and thermal-drift errors. Thus,

$$u_I = \sqrt{\sum_{i=1}^{N} e_i^2}.$$
[9.27]

Instrument errors (elemental errors) are identified through calibration. An elemental error is an error that can be associated with a *single* uncertainty source. Usually, it is related to the full-scale output (FSO) of the instrument, which is its maximum output value. The most common instrument errors are the following:

1. Hysteresis:

$$\tilde{e}_H = \frac{e_{H,\max}}{\text{FSO}} = \frac{|y_{\text{up}} - y_{\text{down}}|_{\max}}{\text{FSO}}. \qquad \textbf{[9.28]}$$

The **hysteresis** error is related to $e_{H,\max}$, which is the greatest deviation between two output values for a given input value that occurs when performing an up-scale, down-scale calibration. This is a single calibration proceeding from the minimum to the maximum input values, then back to the minimum. Hysteresis error usually arises from having a physical change in part of the measurement system on reversing the system's input. Examples include the mechanical sticking of a moving part of the system and the physical alteration of the environment local to the system, such as a region of recirculating flow called a separation bubble. This region remains attached to an airfoil upon decreasing its angle of attack from the region of stall.

2. Linearity:

$$\tilde{e}_L = \frac{e_{L,\max}}{\text{FSO}} = \frac{|y - y_L|_{\max}}{\text{FSO}}. \qquad \textbf{[9.29]}$$

Linearity error is a measure of how linear is the best-fit of the instrument's calibration data. It is defined in terms of its maximum deviation distance, $|y - y_L|_{\max}$.

3. Sensitivity:

$$\tilde{e}_K = \frac{e_{K,\max}}{\text{FSO}} = \frac{|y - y_{\text{nom}}|_{\max}}{\text{FSO}}. \qquad \textbf{[9.30]}$$

Sensitivity error is characterized by the greatest change in the slope (static sensitivity) of the calibration fit.

4. Zero-shift:

$$\tilde{e}_Z = \frac{e_{Z,\max}}{\text{FSO}} = \frac{|y_{\text{shift}} - y_{\text{nom}}|_{\max}}{\text{FSO}}. \qquad \textbf{[9.31]}$$

Zero-shift error refers to the greatest possible shift that can occur in the intercept of the calibration fit.

5. Repeatability:

$$\tilde{e}_R = \frac{2S_x}{\text{FSO}}. \qquad \textbf{[9.32]}$$

Repeatability error is related to the precision of the calibration. This is determined by repeating the calibration many times for the same input values. S_x represents the precision interval of the data for a particular value of x.

6. Stability:

$$\tilde{e}_S = \frac{e_{S,\max} \cdot \Delta t}{\text{FSO}}. \qquad \textbf{[9.33]}$$

Stability error is related to $e_{S,\max}$, which is the greatest deviation in the output value for a fixed input value that could occur during operation. This deviation is expressed in units of FSO/Δt, with Δt denoting the time since instrument purchase or calibration. Stability error is a measure of how much the output can drift over a period of time for the same input.

7. Thermal-drift:

$$\tilde{e}_T = \frac{e_{T,\max}}{\text{FSO}}. \qquad \textbf{[9.34]}$$

Thermal-drift error is characterized by the greatest deviation in the output value for a fixed input value, $e_{T,\max}$, that could occur during operation because of variations in the environmental temperature. Stability and thermal-drift errors are similar in behavior to the zero-shift error.

The instrument uncertainty, u_I, combines all the known instrument errors:

$$u_I = \sqrt{\sum e_i^2} = \text{FSO} \cdot \sqrt{\tilde{e}_H^2 + \tilde{e}_L^2 + \tilde{e}_K^2 + \tilde{e}_Z^2 + \tilde{e}_R^2 + \tilde{e}_S^2 + \tilde{e}_T^2 + \tilde{e}_{\text{other}}^2}, \qquad \textbf{[9.35]}$$

where $\tilde{e}_{\text{other}}$ denotes any other instrument errors. All $\tilde{e}_i$'s expressed in Equation 9.35 are dimensionless.

How are these elemental errors actually assessed? Typically, hysteresis and linearity errors are determined by performing a *single* up-scale, down-scale calibration. The results of this type of calibration are displayed in the left graph of Figure 9.3.

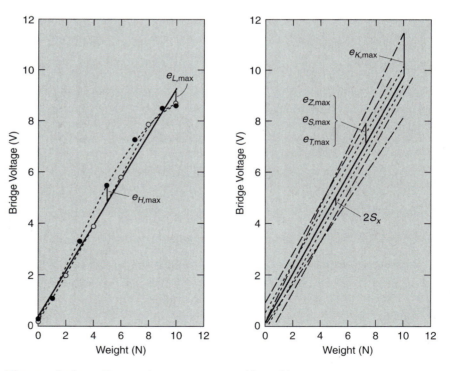

Figure 9.3   Elemental errors ascertained by calibration.

In that graph, the up-scale results are plotted as open circles and the down-scale results as stars. The dotted lines are linear interpolations between the data. Hysteresis is evident in this example by down-scale output values that are higher than their up-scale counterparts. The best-fit curve of the data is indicated by a solid line. Both the hysteresis and linearity errors are assessed with respect to the best-fit curve.

Sensitivity, repeatability, zero-shift, stability, and thermal-drift errors are ascertained by performing a *series* of calibrations and then determining each particular error by comparisons between the calibrations. The results of a series of calibrations are shown in the right graph of Figure 9.3. The solid curve represents the best-fit of the data from *all* the calibrations. The dotted curves indicate the limits within which a calibration is repeatable with 95 % confidence. The repeatability error is determined from the difference between either dotted curve and the best-fit curve. The dash-dotted curves identify the calibration curves that have the maximum and minimum slopes. The sensitivity error is assessed in terms of the greatest difference between the minimum or maximum sensitivity curve and the best-fit curve. The dashed curves denote shifts that can occur in the calibration because of zero-shift, stability, and thermal-drift. Each error can have a different value and is determined from the greatest-difference calibration curve that occurs with each effect, with respect to the best-fit curve.

Examples 9.5 and 9.6 illustrate the effects of instrument errors on measurement uncertainty.

Example 9.5

STATEMENT

A pressure transducer is connected to a digital panel meter. The panel meter converts the pressure transducer's output in volts back to pressure in psi. The manufacturer provides the following information about the panel meter:

Resolution:	0.1 psi
Repeatability:	0.1 psi
Linearity:	within 0.1 % of reading
Drift:	less than 0.1 psi/6 months within the 32 °F to 90 °F range

The only information given about the pressure transducer is that it has "an accuracy of within 0.5 % of its reading."

Estimate the combined standard uncertainty in a measured pressure at a nominal value of 100 psi at 70 °F. Assume that the transducer's response is linear with an output of 1 V for every psi of input.

SOLUTION

The uncertainty in the measured pressure, $(u_d)_{mp}$, is the combination of the uncertainties of the transducer, $(u_d)_t$, and the panel meter, $(u_d)_{pm}$. This can be expressed as

$$(u_d)_{mp} = \sqrt{[(u_d)_t]^2 + [(u_d)_{pm}]^2}.$$

For the transducer,

$$(u_d)_t = \sqrt{u_{I_t}^2 + u_{o_t}^2} = u_{I_t} = 0.005 \times 100 \text{ psi} = 0.50 \text{ psi}.$$

For the panel meter,

$$(u_d)_{pm} = \sqrt{u_{I_{pm}}^2 + u_{o_{pm}}^2}.$$

Now,

$$u_{o_{pm}} = 0.5 \text{ resolution} = 0.05 \text{ psi},$$

$$u_{I_{pm}} = \sqrt{e_1^2 + e_2^2 + e_3^2},$$ **[9.36]**

where

$$e_1 \text{ (repeatability)} = 0.1 \text{ psi},$$
$$e_2 \text{ (linearity)} = 0.1 \text{ \% reading} = 0.001 \times 100 \text{ V}/(1 \text{ V/psi}) = 0.1 \text{ psi},$$
$$e_3 \text{ (drift)} = 0.1 \text{ psi}/6 \text{ months} \times 6 \text{ months} = 0.1 \text{ psi},$$

which implies that

$$u_{I_{pm}} = 0.17 \text{ psi},$$
$$(u_d)_{pm} = 0.18 \text{ psi},$$
$$(u_d)_{mp} = \sqrt{0.50^2 + 0.18^2} = 0.53 \text{ psi}.$$

Note that most of the combined standard uncertainty comes from the transducer. So, to improve the accuracy of the measurement system, a more accurate transducer is required.

Example 9.6

STATEMENT

An analog-to-digital (A/D) converter with the specifications listed below (see Chapter 6 for terminology) is to be used in an environment in which the A/D converter's temperature may change by ±10 °C. Estimate the contributions of conversion and quantization errors to the combined standard uncertainty in the digital representation of an analog voltage by the converter.

E_{FSR}	0 V to 10 V
M	12 bits
Linearity	±3 bits/E_{FSR}
Temperature drift	1 bit/5 °C

SOLUTION

The instrument uncertainty is the combination of uncertainty due to quantization errors, e_Q, and to conversion errors, e_c,

$$(u_I)_E = \sqrt{e_Q^2 + e_I^2}.$$

The resolution of a 12-bit A/D converter with a full scale range of 0 V to 10 V is given by (see Chapter 6)

$$Q = \frac{E_{FSR}}{2^{12}} = \frac{10}{4096} = 2.4 \text{ mV/bit}.$$

The quantization error per bit is found to be

$$e_Q = 0.5Q = 1.2 \text{ mV}.$$

The conversion error is affected by two elements:

$$\text{linearity error} = e_1 = 3 \text{ bits} \times 2.4 \text{ mV/bit}$$
$$= 7.2 \text{ mV}$$
$$\text{temperature error} = e_2 = \frac{1 \text{ bit}}{5 \text{ °C}} \times 10 \text{ °C} \times 2.4 \text{ mV/bit}$$
$$= 4.8 \text{ mV}.$$

Thus, an estimate of the conversion error is

$$e_I = \sqrt{e_1^2 + e_2^2}$$
$$= \sqrt{(7.2 \text{ mV})^2 + (4.8 \text{ mV})^2} = 8.6 \text{ mV}.$$

The combined standard uncertainty in the digital representation of the analog value due to the quantization and conversion errors becomes

$$(u_I)_E = \sqrt{(1.2 \text{ mV})^2 + (8.6 \text{ mV})^2}$$
$$= 8.7 \text{ mV}.$$

Here the conversion errors dominate the uncertainty. So, a higher-resolution converter is not necessary to reduce the uncertainty. A converter having smaller instrument errors is required.

Examples 9.5 and 9.6 illustrate the process of design-stage uncertainty estimation. Once the components of the measurement system have been chosen, uncertainty analysis can be extended to consider other types of errors that can affect the measurement, such as temporal variations in the system's output under fixed conditions. This involves multiple measurements, which is the topic of Section 9.11.

9.10.2 SINGLE-MEASUREMENT RESULT EXPERIMENT

Section 9.10.1 considered estimating the uncertainties of a measurand. But what about the uncertainty in a result? This uncertainty was introduced beforehand and is given by Equation 9.21. From this equation, uncertainty expressions can be developed for specific analytical relations [16]:

1. If $r = Bx$, where B is a constant, then

$$u_r = |B| u_x. \qquad \textbf{[9.37]}$$

If r is directly proportional to a measurand through a constant of proportionality B, then the uncertainty in r is the product of the absolute value of the proportionality constant and the measurand uncertainty.

2. If $r = x + \cdots + z - (u + \cdots + w)$, then

$$u_r = \sqrt{(u_x)^2 + \cdots + (u_z)^2 + (u_u)^2 + \cdots + (u_w)^2}. \qquad \textbf{[9.38]}$$

If r is related directly to all of the measurands, then the uncertainty in r is the combination in quadrature of the measurands' uncertainties.

3.　If $r = (x \ldots z)/(u \ldots w)$, then

$$\frac{u_r}{|r|} = \sqrt{\left(\frac{u_x}{x}\right)^2 + \cdots + \left(\frac{u_z}{z}\right)^2 + \left(\frac{u_u}{u}\right)^2 + \cdots + \left(\frac{u_w}{w}\right)^2}. \qquad \textbf{[9.39]}$$

The quantity $u_r/|r|$ is the **fractional uncertainty** of a result. If r is related directly and/or inversely to all of the measurands, then the fractional uncertainty in r is the combination in quadrature of the measurands' fractional uncertainties.

4.　If $r = x^n$, then

$$\frac{u_r}{|r|} = |n|\frac{u_x}{|x|}. \qquad \textbf{[9.40]}$$

This equation follows directly from Equation 9.39.

Estimation of the uncertainty in a result is shown in Example 9.7.

Example 9.7

STATEMENT

The coefficient of restitution, e, of a ball can be determined by dropping the ball from a known height, h_a, onto a surface and then measuring its return height, h_b (as described in Example 1.1). For this experiment $e = \sqrt{h_b/h_a}$. If the uncertainty in the height measurement, u_h, is 1 mm, $h_a = 1.000$ m, and $h_b = 0.800$ m, determine the combined standard uncertainty in e.

SOLUTION

Direct application of Equation 9.21 yields

$$u_e = \sqrt{\left(\frac{\partial e}{\partial h_b} u_{h_b}\right)^2 + \left(\frac{\partial e}{\partial h_a} u_{h_a}\right)^2}.$$

Now,

$$(\partial e/\partial h_b) = \frac{1/h_a}{2\sqrt{h_b/h_a}} \quad \text{and} \quad \left(\frac{\partial e}{\partial h_a}\right) = \frac{-h_b/h_a^2}{2\sqrt{h_b/h_a}}.$$

Substitution of the known values into the preceding equation for u_e gives $u_e = \sqrt{(5.59 \times 10^{-4})^2 + (4.47 \times 10^{-4})^2} = 7.16 \times 10^{-4} = 0.0007$.

Often there are experiments involving results that have angular dependencies. The values of these results can vary significantly with angle because of the presence of trigonometric functions in the denominators of their uncertainty expressions. Examples 9.8 and 9.9 illustrate this point.

9.2 | MATLAB SIDEBAR

The previous problem can be solved by the MATLAB M-file uncerte.m. This M-file uses MATLAB's function diff(x,y) that symbolically determines the partial derivative of x with respect to y. A more general M-file that determines the uncertainty in a result can be constructed using this format. First, the symbols for the measured variables are declared and the expression for the result is provided by the commands

```
syms ha hb
e=sqrt(hb/ha);
```

Next, typical values for the variables and the elemental uncertainties are given. Then the uncertainty expression is stated in terms of the diff function:

```
u_e=sqrt((diff(e,ha)*u_ha).^2
    +diff(e,hb)*u_hb).^2);
```

Finally, the uncertainty is computed using successive substitutions

```
u_e_1=subs(u_e,ha,s_ha);
uncertainty_in_e=subs(u_e,hb,s_hb);
```

in which the typical values s_{ha} and s_{hb} are substituted for ha and hb, respectively. The result obtained is $u_e = 7.1589 \times 10^{-4}$, which agrees with that calculated in Example 9.7.

9.3 | MATLAB SIDEBAR

Determining the uncertainty in a result that depends on a number of elemental uncertainties often is tedious and subject to calculational error. One alternative approach is to expand an M-file, such as uncerte.m, to include more than a couple of elemental errors. The MATLAB M-file uncertvhB.m was written to determine the velocity of a pendulum at the bottom of a swing (this is part of the laboratory exercise on Measurement, Modeling, and Uncertainty). The velocity is a function of seven variables. Such an M-file can be generalized to handle many uncertainties.

STATEMENT

Example 9.8

A radar gun determines the speed of a directly oncoming auto to within 4 %. However, if the gun is used off-angle, an additional uncertainty arises. Determine the gun's off-angle uncertainty, u_{oa}, as a function of the angle at which the car is viewed. What would be the combined uncertainty in the speed if the off-angle, θ_{oa}, equals 70°? Finally, what is the overall uncertainty in the speed assuming 95 % confidence?

SOLUTION

A schematic of this problem is shown in Figure 9.4. Assume that the gun acquires a reading within a very short time period, Δt. The actual speed, s_{ac}, is the ratio of the actual highway distance traveled, L_{ac}, during the time period to Δt. Similarly, the apparent speed, s_{ap}, equals $L_{ap}/\Delta t$. From trigonometry,

$$L_{ac} = L_{ap} \cdot \sin(\theta). \qquad \textbf{[9.41]}$$

Substitution of the speed definitions into this equation yields

$$s_{ac} = s_{ap} \cdot \sin(\theta). \qquad \textbf{[9.42]}$$

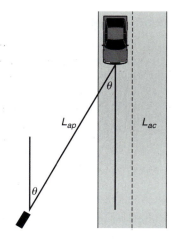

Figure 9.4   Radar detection of a car's speed.

The off-angle uncertainty can be defined as

$$u_{oa} = \frac{|s_{ac} - s_{ap}|}{s_{ac}} = \frac{|\sin(\theta) - 1|}{\sin(\theta)}.$$ **[9.43]**

Note that when $\theta = 90°$, $\sin(90) = 1$, which yields $u_{oa} = 0$. This is when the radar gun is pointed directly along the highway at the car. When $\theta = 70°$, $\sin(70) = 0.940$, which yields $u_{oa} = (|0.940 - 1|)/0.940 = 0.064$ or 6.4 %. This uncertainty must be combined in quadrature with the radar gun's instrument uncertainty, $u_I = 0.04$, to yield the combined uncertainty, u_c,

$$u_c = \sqrt{u_I^2 + u_{oa}^2} = \sqrt{0.04^2 + 0.64^2} = 0.075.$$ **[9.44]**

So, the combined uncertainty is 7.5 % or almost twice the instrument uncertainty. This uncertainty increases as the off-angle increases. Assuming 95 % confidence, the overall uncertainty is approximately twice the combined uncertainty or 15 %. Thus, assuming that the indicated speed is 70 mph, the actual speed could be as low as approximately 60 mph $(70 - 0.15 \times 70)$ or as high as approximately 80 mph $(70 + 0.15 \times 70)$.

Example 9.9   **STATEMENT**

This problem is adapted from one in [16]. An experiment is constructed (see Figure 9.5) to determine the index of refraction of an unknown transparent glass. Find the fractional uncertainty, $\Delta n/n$, in the index of refraction, n, as determined using Snell's law, where $n = \sin \theta_i / \sin \theta_r$. Assume that the measurements of the angles are uncertain by $\pm 1°$ or 0.02 rad.

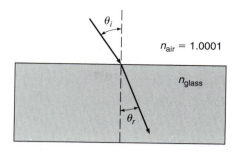

Figure 9.5   A light refraction experiment based on Snell's law.

SOLUTION

It follows that

$$\frac{\Delta n}{n} = \sqrt{\left(\frac{\partial \sin \theta_i}{\sin \theta_i}\right)^2 + \left(\frac{\partial \sin \theta_r}{\sin \theta_r}\right)^2}.$$

Now,

$$\partial \sin \theta = \left|\frac{d \sin \theta}{d\theta}\right| \partial\theta = |\cos \theta| \partial\theta \quad \text{(in rad)}.$$

So,

$$\frac{\partial \sin \theta}{|\sin \theta|} = |\cot \theta| \partial\theta \quad \text{(in rad)}.$$

These considerations yield the following uncertainties:

$\theta_i \pm 1°$	$\theta_r \pm 1°$	$\sin \theta_i$	$\sin \theta_r$	n	$\frac{\partial \sin \theta_i}{|\sin \theta_i|}$	$\frac{\partial \sin \theta_r}{|\sin \theta_r|}$	$\frac{\Delta n}{n}$
20	13.0	0.342	0.225	1.52	5 %	8 %	9 %
40	23.5	0.643	0.399	1.61	2 %	4 %	5 %

Note that the percentage uncertainty in n decreases with increasing the angle of incidence. In fact, as the angle of incidence approaches that of normal incidence ($\theta_i = 0°$), the uncertainty tends to infinity.

Many times, experimental uncertainty analysis involves a series of uncertainty calculations that lead to the uncertainty of a desired result. In that situation, usually it is best to perform the analysis in steps, identifying the uncertainties in intermediate results. This not only helps to avoid mistakes made in calculations but also aids in identifying the variables that contribute significantly to the desired result's uncertainty. Examples 9.10 and 9.11 illustrate this point.

Example 9.10

STATEMENT

Determine the combined standard uncertainty in the density of air, assuming $\rho = P/RT$. Assume negligible uncertainty in R ($R_{air} = 287.04$ J/kg · K). Let $T = 24$ °C $= 297$ K and $P = 760$ mm Hg.

SOLUTION

The uncertainty in the density (a result) becomes

$$u_\rho = \sqrt{\left(\frac{\partial\rho}{\partial T}u_T\right)^2 + \left(\frac{\partial\rho}{\partial P}u_P\right)^2}$$

$$= \sqrt{\left(\frac{-P}{RT^2}u_T\right)^2 + \left(\frac{1}{RT}u_P\right)^2},$$

where

$$u_P = \frac{1}{2}(1 \text{ mm Hg}) = \frac{1}{2}\left(\frac{1.01 \times 10^5 \text{ Pa}}{760 \text{ mm Hg}} \times 1 \text{ mm Hg}\right)$$

$$= \frac{1}{2}(133 \text{ Pa}) = 67 \text{ Pa},$$

and

$$u_T = 0.5(1\,°C) = 0.5(1 \text{ K}) = 0.5 \text{ K}.$$

Thus,

$$u_\rho = \left\{\left[\frac{101\,325}{(287.04)(297)^2}(0.5)\right]^2 + \left[\frac{1}{(287.04)(297)}(67)\right]^2\right\}^{1/2}$$

$$= \left(4.00 \times 10^{-6} + 0.62 \times 10^{-6}\right)^{1/2},$$

$$= 2.15 \times 10^{-3} \text{ kg/m}^3.$$

Finally,

$$\rho = \frac{P}{RT} = \frac{101\,325}{(287.04)(297)} = 1.19 \text{ kg/m}^3$$

$$\Rightarrow \frac{u_\rho}{\rho} = \frac{2.15 \times 10^{-3}}{1.19} = 0.19 \text{ \%}.$$

This is a typical value for the combined standard uncertainty in the density of air as determined from pressure and temperature measurements in a contemporary laboratory.

Example 9.11

STATEMENT

Consider an experiment in which the static pressure distribution around the circumference of a cylinder in a cross flow in a wind tunnel is measured. Determine the combined standard uncertainty in the pressure coefficient, C_p, as defined by the equation

$$C_p \equiv \frac{P - P_\infty}{\frac{1}{2}\rho V_\infty^2}. \qquad\qquad \textbf{[9.45]}$$

Assume that the pressure difference $P - P_\infty$ is measured as Δp using an inclined manometer with

$$u_{\Delta p} = 0.06 \text{ in. H}_2\text{O} = 15 \text{ N/m}^2 \quad (\Delta P = 996 \text{ N/m}^2),$$

$$u_\rho = 2.15 \times 10^{-3} \text{ kg/m}^3 \quad (\rho = 1.19 \text{ kg/m}^3), \quad \text{and}$$

$$u_{V_\infty} = 0.31 \text{ m/s} \quad (V_\infty = 40.9 \text{ m/s}).$$

SOLUTION

Now, as is clear to see from Equation 9.45, the pressure coefficient is a *result* and is a function of the density, the change in pressure and the freestream velocity. In short,

$$C_p = f(\Delta p, \rho, V_\infty).$$

Therefore, applying Equation 9.21 yields

$$\Rightarrow u_{C_P} = \left[\left(\frac{\delta C_p}{\delta \Delta p} u_{\Delta p} \right)^2 + \left(\frac{\delta C_p}{\delta \rho} u_\rho \right)^2 + \left(\frac{\delta C_p}{\delta V_\infty} u_{V_\infty} \right)^2 \right]^{1/2}$$

$$= \left[\left(\frac{2}{\rho V_\infty^2} u_{\Delta p} \right)^2 + \left(-\frac{2\Delta p}{\rho^2 V_\infty^2} u_\rho \right)^2 + \left(-\frac{4\Delta p}{\rho V_\infty^3} u_{V_\infty} \right)^2 \right]^{1/2}$$

$$= \left\{ \left[\frac{(2)(15)}{(1.19)(40.9)^2} \right]^2 + \left[\frac{(2)(996)(2.15 \times 10^{-3})}{(1.19)^2(40.9)^2} \right]^2 + \left[\frac{(4)(996)(0.31)}{(1.19)(40.9)^3} \right]^2 \right\}^{1/2}$$

$$= \left(2.27 \times 10^{-4} + 3.27 \times 10^{-6} + 2.30 \times 10^{-4} \right)^{1/2}$$

$$= 0.021.$$

Alternatively, C_p as the ratio of two transducer differential pressures,

$$C_p \equiv \frac{P - P_\infty}{\frac{1}{2}\rho V_\infty^2} = \frac{\Delta P}{\Delta P_{p-s}} \quad (\Delta p = 996 \text{ N/m}^2). \qquad \textbf{[9.46]}$$

Assume that $u_{\Delta p} = u_{\Delta p_{p-s}} = 15 \text{ N/m}^2$. The equation for the uncertainty in C_p when Equation 9.46 is used becomes

$$u_{C_p} = \left[\left(\frac{\partial C_p}{\partial \Delta P} u_{\Delta P} \right)^2 + \left(\frac{\partial C_p}{\partial \Delta P_{p-s}} u_{\Delta P_{p-s}} \right)^2 \right]^{1/2}$$

$$= \left[\left(\frac{1}{\Delta P_{p-s}} u_{\Delta P} \right)^2 + \left(-\frac{\Delta P}{\Delta P_{p-s}^2} u_{\Delta P_{p-s}} \right)^2 \right]^{1/2}$$

$$= \left[2 \left(\frac{15}{996} \right)^2 \right]^{1/2}$$

$$= 0.021.$$

The latter measurement approach is easier to determine C_p than the former. When designing an experimental procedure in which the pressure coefficient needs to be determined, it is preferable to ratio the two transducer differential pressures.

Finally, one may be interested in estimating the uncertainty of a result that can be found by different measurement approaches. This process is quite useful during the planning stage of an experiment. For example, take the simple case of an experiment designed to determine the volume of a cylinder, V. One approach would be to measure the cylinder's height, h, and diameter, d, and then compute its volume based on the expression $V = \pi \cdot d^2 \cdot h/4$. An alternative approach would be to obtain its weight, w, and then compute its volume according to the expression $V = w/(\rho \cdot g)$, where ρ is the density of the cylinder and g the local gravitational acceleration. Which approach is chosen depends on the uncertainties in d, h, w, ρ, and g.

Example 9.12

STATEMENT

An experiment is being designed to determine the mass of a cube of Teflon. Two approaches are under consideration. Approach A involves determining the mass from a measurement of the cube's weight; approach B from measurements of the cube's length, width, and height (l, w, and h, respectively). For approach A, $m = W/g$; for approach B, $m = \rho V$, where m is the mass, W is the weight, g is the gravitational acceleration ($= 9.81$ m/s^2), ρ is the density ($= 2200$ kg/m^3), and V is the volume ($= l \cdot w \cdot h$). The fractional uncertainties in the measurements are: W (2 %); g (0.1 %); ρ (0.1 %); and l, w, and h (1 %). Determine which approach has the least uncertainty in the mass.

SOLUTION

Because both equations for the mass involve products or quotients of the measurands, the fractional uncertainty of the mass can be computed using Equation 9.39. For approach A,

$$\left(\frac{u_m}{|m|}\right)_A = \sqrt{\left(\frac{u_W}{W}\right)^2 + \left(\frac{u_g}{g}\right)^2} = \sqrt{(0.02)^2 + (0.001)^2} = 0.020 = 2.0 \text{ %}.$$

For approach B, the fractional uncertainty must be determined in the volume. This is

$$\frac{u_V}{|V|} = \sqrt{\left(\frac{u_l}{l}\right)^2 + \left(\frac{u_w}{w}\right)^2 + \left(\frac{u_h}{h}\right)^2} = \sqrt{3 \cdot 0.01^2} = 0.017 = 1.7 \text{ %}.$$

This result can be incorporated into the fractional uncertainty calculation for the mass

$$\left(\frac{u_m}{|m|}\right)_B = \sqrt{\left(\frac{u_\rho}{\rho}\right)^2 + \left(\frac{u_V}{V}\right)^2} = \sqrt{(0.001)^2 + (0.017)^2} = 0.017 = 1.7 \text{ %}.$$

Thus, approach B has the least uncertainty in the mass. Note that the uncertainties in g and ρ both are negligible in these calculations.

9.11   DETAILED UNCERTAINTY ANALYSIS

Detailed uncertainty analysis is appropriate for measurement situations that involve multiple measurements, either for a measurand or a result. Systematic and random errors are identified in this approach. Their contributions to the overall uncertainty are treated separately in the analysis until they are combined at the end. Detailed uncertainty analysis is performed *after* a statistically viable set of measurand values has been obtained under fixed operating conditions. Multiple measurements usually are made for one or both of two reasons. One is to assess the uncertainty present in an experiment due to uncontrollable variations in the measurands. The other is to obtain sufficient data such that the average value of a measurand can be estimated. Thus, detailed uncertainty calculations are more extensive than for the cases of a single-measurement measurand or result.

In general, for the situation in which both systematic and random errors are considered, application of Equation 9.19 for a result based on J measurands leads directly to

$$u_r^2 = \sum_{i=1}^{J} \theta_i^2 S_{B_i}^2 + 2 \sum_{i=1}^{J-1} \sum_{j=i+1}^{J} \theta_i \theta_j S_{B_i, B_j} + \sum_{i=1}^{J} \theta_i^2 S_{P_i}^2 + 2 \sum_{i=1}^{J-1} \sum_{j=i+1}^{J} \theta_i \theta_j S_{P_i, P_j},$$

[9.47]

where

$$S_{B_i}^2 = \sum_{k=1}^{M_B} (S_{B_i})_k^2,$$

[9.48]

$$S_{P_i}^2 = \sum_{k=1}^{M_P} (S_{P_i})_k^2,$$

[9.49]

$$S_{B_i, B_j} = \sum_{k=1}^{L_B} (S_{B_i})_k (S_{B_j})_k,$$

[9.50]

and

$$S_{P_i, P_j} = \sum_{k=1}^{L_P} (S_{P_i})_k (S_{P_j})_k.$$

[9.51]

M_B is the number of elemental systematic uncertainties, M_P is the number of elemental random uncertainties. $(S_{B_i})_k$ represents the kth elemental error out of M_B elemental errors contributing to S_{B_i}, the estimate of the systematic error of the ith measurand. Equation 9.48 in analogous to Equation 9.27 for the case of a single-measurement measurand. Further, L_B is the number of systematic errors that are common to B_i and B_j, and L_P is the number of random errors that are common to P_i and P_j. $S_{B_i}^2$ and $S_{P_i}^2$ are estimates of systematic and random errors of the ith measurand, respectively. S_{B_i, B_j} and S_{P_i, P_j} are estimates of the covariances of the systematic and random errors of the ith and jth measurands, respectively.

The number of effective degrees of freedom of a multiple-measurement result, based upon the Welch-Satterthwaite formula, is

$$\nu_r = \frac{\left[\sum_{i=1}^{J} \left(\theta_i^2 S_{B_i}^2 + \theta_i^2 S_{P_i}^2 \right) \right]^2}{\sum_{i=1}^{J} \left\{ (\theta_i^4 S_{P_i}^4 / \nu_{P_i}) + \left[\sum_{k=1}^{M_B} \theta_i^4 (S_{B_i})_k^4 / \nu_{(S_{B_i})_k} \right] \right\}},$$

[9.52]

where contributions are made from each ith measurand. The random and systematic numbers of degrees of freedom are given by

$$\nu_{P_i} = N_i - 1, \qquad\qquad \textbf{[9.53]}$$

where N_i denotes N measurements of the ith measurand and

$$\nu_{(S_{B_i})_k} \cong \frac{1}{2}\left[\frac{\Delta(S_{B_i})_k}{(S_{B_i})_k}\right]^{-2}. \qquad\qquad \textbf{[9.54]}$$

When S_{B_i} and S_{P_i} have similar values, the value of ν_r, as given by Equation 9.52, is approximately that of ν_{P_i} if $\nu_{B_i} \gg \nu_{P_i}$. Conversely, the value of ν_r is approximately that of ν_{B_i} if $\nu_{P_i} \gg \nu_{B_i}$. Further, if $S_{B_i} \ll S_{P_i}$, then the value of ν_r is approximately that of ν_{P_i}. The converse also is true.

Once ν_r is determined from Equation 9.52, the overall uncertainty in the result, U_r, can be found. This can be expressed using the definition of the overall uncertainty (Equation 9.18) and Equation 9.47 as

$$U_r^2 = t_{\nu_r,C}^2 u_r^2 = B_r^2 + (t_{\nu_r,C} S_r)^2 = B_r^2 + P_r^2, \qquad\qquad \textbf{[9.55]}$$

where

$$B_r = t_{\nu_r,C}\left(\sum_{i=1}^{J}\theta_i^2 S_{B_i}^2 + 2\sum_{i=1}^{J-1}\sum_{j=i+1}^{J}\theta_i\theta_j S_{B_i,B_j}\right)^{1/2} \qquad\qquad \textbf{[9.56]}$$

and

$$P_r = t_{\nu_r,C}\left(\sum_{i=1}^{J}\theta_i^2 S_{P_i}^2 + 2\sum_{i=1}^{J-1}\sum_{j=i+1}^{J}\theta_i\theta_j S_{P_i,P_j}\right)^{1/2}. \qquad\qquad \textbf{[9.57]}$$

The first term on the right-hand side of Equation 9.56 is the sum of the estimated variances of the systematic errors. The second is the sum of the estimated covariances of the systematic errors. Likewise, for Equation 9.57, the first term is the estimated variance of the random errors and the second is the estimated covariance of the random errors. The covariance of two systematic errors is zero if the errors are not correlated, or are independent of each other. Nonzero covariances arise when the error sources are common, such as when two pressure transducers are calibrated against the same standard. Correlated random errors can be identified by examining the amplitude variations in time of two measurands. When they follow the same trends, they may be correlated. Usually, however, the covariances of the systematic errors are assumed negligible in most uncertainty analyses.

When there are no correlated uncertainties,

$$u_r^2 = \sum_{i=1}^{J}\theta_i^2 S_{B_i}^2 + \sum_{i=1}^{J}\theta_i^2 S_{P_i}^2, \qquad\qquad \textbf{[9.58]}$$

where

$$B_r = t_{\nu_r,C}\left(\sum_{i=1}^{J}\theta_i^2 S_{B_i}^2\right)^{1/2} \qquad\qquad \textbf{[9.59]}$$

and

$$P_r = t_{v_r,C} \left(\sum_{i=1}^{J} \theta_i^2 S_{P_i}^2 \right)^{1/2}. \qquad \textbf{[9.60]}$$

For the situation when an experiment is replicated M times and a mean result is determined from the M individual results, the expression for S_r in Equation 9.55 is replaced by either

$$S_r = \sqrt{\frac{1}{M-1} \sum_{k=1}^{M} (r_k - \bar{r})^2} \qquad \textbf{[9.61]}$$

or

$$S_{\bar{r}} = \frac{S_r}{\sqrt{M}}, \qquad \textbf{[9.62]}$$

depending on which outcome is desired (r or $\bar{r}$, respectively) [10]. The Welch-Satterthwaite formula is then

$$v_r = \frac{\left[S_r^2 + \sum_{i=1}^{J} \left(\theta_i^2 S_{B_i}^2 \right) \right]^2}{(S_r^4/v_{S_r}) + \sum_{i=1}^{J} \sum_{k=1}^{M_B} \left\{ \left[\theta_i^4 (S_{B_i})_k^4 / v_{(S_{B_i})_k} \right] \right\}}, \qquad \textbf{[9.63]}$$

with $v_{S_r} = M - 1$.

Equations 9.47 and 9.52 can be applied to estimate the uncertainty in a multiple-measurement measurand. For this case, these equations reduce to

$$u_m^2 = \sum_{i=1}^{J} S_{B_i}^2 + 2 \sum_{i=1}^{J-1} \sum_{j=i+1}^{J} S_{B_i,B_j} + \sum_{i=1}^{J} S_{P_i}^2 + 2 \sum_{i=1}^{J-1} \sum_{j=i+1}^{J} S_{P_i,P_j} \qquad \textbf{[9.64]}$$

and

$$v_m = \frac{\left[\sum_{i=1}^{J} \left(S_{B_i}^2 + S_{P_i}^2 \right) \right]^2}{\sum_{i=1}^{J} \left\{ (S_{P_i}^4/v_{P_i}) + \left[\sum_{k=1}^{M_B} (S_{B_i})_k^4 / v_{(S_{B_i})_k} \right] \right\}}. \qquad \textbf{[9.65]}$$

Further simplification occurs where there are no correlated uncertainties. Equation 9.64 becomes

$$u_m^2 = \sum_{i=1}^{J} S_{B_i}^2 + \sum_{i=1}^{J} S_{P_i}^2. \qquad \textbf{[9.66]}$$

When estimating the uncertainty in a *mean* value, S_{P_i} is replaced in Equation 9.66 by $S_{\bar{P}_i}$, where

$$S_{\bar{P}_i} = \sqrt{\sum_{i=1}^{M_P} \frac{S_{P_i}^2}{N_{P_i}}}. \qquad \textbf{[9.67]}$$

For most engineering and scientific experiments $v \geq 9$ [10]. When this is the case, it is reasonable to assume for 95 % confidence that $t_{v,95} \cong 2$ (when $v \geq 9$, $t_{v,95}$

is within 10 % of 2). This implies that

$$U_{x,95} \cong 2u_c = 2\sqrt{S_{B_i}^2 + S_{P_i}^2} = \sqrt{B_i^2 + P_i^2}, \qquad \textbf{[9.68]}$$

using Equations 9.17, 9.10, and 9.13 where $S_{B_i}^2$ is shorthand notation for $\sum_{i=1}^{N} S_{B_i}^2$ and $S_{P_i}^2$ for $\sum_{i=1}^{N} S_{P_i}^2$. Equation 9.68 is known as the *large-scale approximation*. Its simplicity is that the overall uncertainty can be estimated directly from the systematic and random uncertainties. This is illustrated in Example 9.13.

Example 9.13

STATEMENT

A load cell is used to measure the central load applied to a structure. The accuracy of the load cell as provided by the manufacturer is stated to be 2.3 N. The experimenter, based on previous experience in using that manufacturer's load cells, estimates the relative systematic uncertainty of the load cell to be 10 %. A series of 11 measurements are made under fixed conditions, resulting in a standard deviation of the random error equal to 11.3 N. Determine the overall uncertainty in the load cell measurement at the 95 % confidence level.

SOLUTION

The overall uncertainty can be determined using Equations 9.65 and 9.68. The standard deviations and the degrees of freedom for both the systematic and random errors need to be determined first. Here $\nu_{P_x} = N - 1 = 10$ and, from Equation 9.14, $\nu_{B_x} = \frac{1}{2}(0.10)^{-2} = 50$. Thus, at 95 % confidence, $t_{10,95} = 2.228$ and $t_{50,95} = 2.010$ for the random and systematic errors, respectively. Now $S_{P_x} = P_x/t_{10,95} = 11.3/2.228 = 5.072$ from Equation 9.10, and, from Equation 9.13, $S_{B_x} = B_x/t_{50,95} = 2.3/2.010 = 1.144$. Substitution of these values into Equation 9.65 yields

$$\nu_{\text{eff}} = \frac{(5.072^2 + 1.144^2)^2}{(5.072^4/10) + (1.144^4/50)} = 11.02 \cong 11.$$

Because $\nu_m \geq 9$, the large-scale approximation at 95 % confidence can be used, where

$$U_{x,95} \cong \sqrt{B_i^2 + P_i^2} = \sqrt{2.3^2 + 11.3^2} = 11.5 \text{ N}.$$

In Sections 9.11.1 and 9.11.2, the application of the preceding equations to multiple-measurement measurand and result experiments is presented.

9.11.1   MULTIPLE-MEASUREMENT MEASURAND EXPERIMENT

Consider the uncertainty estimation of a measurand involving multiple measurements done to assess the contribution of temporal variability (temporal precision error) of a measurand under fixed conditions. For this situation the temporal variations of a measurand are treated as a random error and all other errors are considered to be systematic. Example 9.14 illustrates this process.

Example 9.14

STATEMENT

The supply pressure of a blow-down facility's plenum is to be maintained at set pressure for a series of tests. A compressor supplies air to the plenum through a regulating valve that controls the set pressure. A dial gauge (resolution: 1 psi; accuracy: 0.5 psi) is used to monitor pressure in the vessel. Thirty trials of pressurizing the vessel to a set pressure of 50 psi are attempted to estimate pressure controllability. The results show that the standard deviation in the set pressure is 2 psi. Estimate the combined standard uncertainty at 95 % confidence in the set pressure that would be expected during normal operation.

SOLUTION

The uncertainty in the set pressure reading results from the instrument uncertainty, u_d, and the additional uncertainty arising from the temporal variability in the pressure under controlled conditions, u_1,

$$u_p = \sqrt{u_d^2 + u_1^2},$$

where $u_1 = P_x = t_{v,P} S_x$ according to Equation 9.10. The instrument uncertainty is

$$u_d = \sqrt{u_o^2 + u_I^2},$$

where

$$u_o = \frac{1}{2}(1.0 \text{ psi}) = 0.5 \text{ psi},$$

$$u_I = 0.5 \text{ psi},$$

$$\Rightarrow u_d = \sqrt{0.5^2 + 0.5^2} = 0.7 \text{ psi}.$$

Also,

$$u_1 = t_{29,95} S_x = (2.047)(2) = 4.1 \text{ psi},$$

where $v = N - 1 = 29$ for this case. So,

$$u_p = \sqrt{0.7^2 + 4.1^2} = 4 \text{ psi}.$$

Note that the uncertainty primarily is the result of repeating the set pressure and not the resolution or accuracy of the gauge. Why does u_p have only one significant figure?

Next, consider the multiple-measurement situation of estimating the uncertainty in the range that contains the true value of a measurand. This is shown in Example 9.15, in which the contributory systematic and random errors are specified, having their elemental uncertainties already summed using Equations 9.48 and 9.49.

Example 9.15

STATEMENT

The stress on a loaded electric-powered remotely-piloted-vehicle (RPV) wing is measured using a system consisting of a strain gage, Wheatstone bridge, amplifier, and data acquisition system. The following systematic and random uncertainties arising from calibration, data acquisition, and data reduction are

Calibration:	$S_{B_1} = 1.0$ N/cm^2	$S_{P_1} = 4.6$ N/cm^2	$N_{P_1} = 15 \Rightarrow \nu_{P_1} = 14$
Data acquisition:	$S_{B_2} = 2.1$ N/cm^2	$S_{P_2} = 10.3$ N/cm^2	$N_{P_2} = 38 \Rightarrow \nu_{P_2} = 37$
Data reduction:	$S_{B_3} = 0.0$ N/cm^2	$S_{P_3} = 1.2$ N/cm^2	$N_{P_3} = 9 \Rightarrow \nu_{P_3} = 8$

Assume 100 % reliability in the values of all systematic errors and that there are no correlated uncertainties. Determine for 95 % confidence the range that contains the true mean value of the stress, σ', given that the average value is $\bar{\sigma} = 223.4$ N/cm^2.

SOLUTION
The systematic variances are

$$S_B^2 = S_{B_1}^2 + S_{B_2}^2 + S_{B_3}^2 = 1^2 + 2.1^2 + 0^2 = 5.4 \text{ N}^2/\text{cm}^4.$$

The random variances are

$$S_{\bar{P}}^2 = \frac{S_{P_1}^2}{N_{P_1}} + \frac{S_{P_2}^2}{N_{P_2}} + \frac{S_{P_3}^2}{N_{P_3}} = \frac{4.6^2}{15} + \frac{10.3^2}{38} + \frac{1.2^2}{9} = 4.4 \text{ N}^2/\text{cm}^4.$$

It follows directly from Equation 9.66 that $u_\sigma^2 = 5.4 + 4.4 = 9.8$ N^2/cm^4. So, $u_\sigma = 3.1$ N/cm^2.

The number of effective degrees of freedom are determined using Equation 9.65, which becomes

$$\nu_\sigma = \frac{(1^2 + 2.1^2 + 0^2 + 4.6^2 + 10.3^2 + 1.2^2)^2}{[(4.6^4/14) + (10.3^4/37) + (1.2^4/8)]} = \frac{17\,982.8}{336.4} = 53.5 = 54,$$

noting that each $\nu_{B_i} = \infty$ because of its 100 % reliability. This yields $t_{54,95} \cong 2$.

Thus, the true mean value of the stress is

$$\sigma' = \bar{\sigma} \pm U_\sigma \quad \text{where } U_\sigma = t_{54,95} \cdot u_\sigma = 6.2 \text{ N/cm}^2 \quad (95\ \%).$$

Thus,

$$\sigma' = 223.4 \pm 6.2 \text{ N/cm}^2 \quad (95\ \%).$$

9.11.2   MULTIPLE-MEASUREMENT RESULT EXPERIMENT

The uncertainty estimation of a result based on multiple measurements of measurands can be made using Equations 9.47 through 9.54. This estimation is slightly more complicated than for the multiple-measurement measurand because it involves determinations of the absolute sensitivity coefficients. Example 9.16 illustrates the process in which the true mean value of a result is estimated.

Example 9.16

STATEMENT
This problem is adapted from [15]. An experiment is performed in which the density of a gas, assumed to be ideal, is determined from different numbers of separate pressure and temperature measurements. The gas is contained within a vessel of fixed volume. Pressure

is measured within 1 % accuracy; temperature to within 0.6 °R. Twenty measurements of pressure ($N_p = 20$) and ten of temperature ($N_T = 10$) were performed, yielding the following data:

$$\bar{p} = 2253.91 \text{ psfa}, \quad S_{P_p} = 167.21 \text{ psfa},$$
$$\bar{T} = 560.4 \text{ °R}, \quad S_{P_T} = 3.0 \text{ °R},$$

where psfa denotes pound-force per square foot absolute. Estimate the true mean value of the density at 95 % confidence.

SOLUTION

Assuming ideal gas behavior, the sample mean density, $\bar{\rho}$, becomes

$$\bar{\rho} = \frac{\bar{p}}{R \cdot \bar{T}} = 0.074 \text{ lbm/ft}^3.$$

There are two sources of error in this experiment. One is from the variability in the pressure and temperature readings, as signified by the values of S_{P_p} and S_{P_T} given above. These are random errors. The other is from the specified instrument inaccuracies. These are systematic errors. They are

$$S_{B_p} = 1 \% = 22.5 \text{ psfa}, \quad S_{B_T} = 0.6 \text{ °R},$$
$$S_{P_{\bar{p}}} = \frac{S_{P_p}}{\sqrt{20}} = 37.4 \text{ psfa}, \quad S_{P_{\bar{T}}} = \frac{S_{P_T}}{\sqrt{10}} = 0.9 \text{ °R},$$

where the degrees of freedom are $\nu_P = 19$ and $\nu_T = 9$. Because the true mean value estimate is made from multiple measurements, the random uncertainties are based on the standard deviations of their means. Thus, the combined random error becomes

$$S_{\bar{\rho}} = \sqrt{\left(\frac{\partial \rho}{\partial T} S_{P_{\bar{T}}}\right)^2 + \left(\frac{\partial \rho}{\partial P} S_{P_{\bar{p}}}\right)^2} = \sqrt{\left(\frac{-\bar{p}}{R \cdot \bar{T}^2} S_{P_{\bar{T}}}\right)^2 + \left(\frac{1}{R \cdot \bar{T}} S_{P_{\bar{p}}}\right)^2}$$

$$= \sqrt{(1.2 \cdot 10^{-4})^2 + (1.2 \cdot 10^{-3})^2} = 0.0012 \text{ lbm/ft}^3.$$

Likewise, the combined systematic error becomes 0.0007 lbm/ft^3. Using Equation 9.47, assuming no correlated uncertainties, this yields

$$u_\rho = \sqrt{0.0012^2 + 0.0007^2} = 0.0014 \text{ lbm/ft}^3.$$

Further, assuming 100 % certainty in the stated accuracies of the instruments, the effective number of degrees of freedom as determined from Equation 9.52 is

$$\nu = \frac{\left\{\left[(\partial\rho/\partial T)S_{P_{\bar{T}}}\right]^2 + \left[(\partial\rho/\partial P)S_{P_{\bar{p}}}\right]^2\right\}^2}{\dfrac{\left[(\partial\rho/\partial T)S_{P_{\bar{T}}}\right]^4}{\nu_T} + \dfrac{\left[(\partial\rho/\partial P)S_{P_{\bar{p}}}\right]^4}{\nu_P}} = 23.$$

This yields $t_{23,95} = 2.06$.

Thus, the true mean value of the pressure is

$$\rho' = \bar{\rho} \pm U_\rho \quad \text{where } U_\rho = t_{23,95} \cdot u_\rho = 0.0029 \text{ lbm/ft}^3 \ (95 \%).$$

Thus,

$$\rho' = 0.074 \pm 0.003 \text{ lbm/ft}^3 \quad (95\,\%),$$

which is an uncertainty of $\pm 3.4\,\%$.

9.12 UNCERTAINTY ANALYSIS SUMMARY

The most common uncertainty estimation situations involve a single-measurement measurand, a single-measurement result, a multiple-measurement measurand, or a multiple-measurement result. Expressions for the uncertainty in either a single- or multiple-measurement result contain absolute sensitivity coefficients. These coefficients are evaluated at typical measurand values. Those expressions for the uncertainty in either a single- or multiple-measurement measurand differ only in that the values of all absolute sensitivity coefficients become unity. When multiple measurements are considered, random uncertainties are expressed in terms of the standard deviations of the means of their random errors (Equation 9.11). This is the main difference between the single- and multiple-measurement cases.

The objective of any uncertainty analysis is to obtain an *estimate* of the overall uncertainty, U_x. The summary expression containing U_x involves either x_{next} or x_{true} (Equations 9.15 and 9.16). The overall uncertainty is expressed as the product of Student's t variable based on the number of effective degrees of freedom (evaluated with $\%C$ confidence), $t_{\nu_{\text{eff}},C}$, and the combined standard uncertainty, u_c (Equation 9.18). The values of ν_{eff} and u_c depend on the particular uncertainty estimation situation. When the effective number of degrees of freedom is ≥ 9, the overall uncertainty can be estimated using the large-scale approximation (Equation 9.68) for 95 % confidence where $U_x = 2u_c$. This greatly simplifies the steps required to estimate U_x.

For single-measurement situations, generalized uncertainty analysis is the most appropriate (Section 9.10). No differentiation is made between systematic and random uncertainties. Expressions are available for the combined standard uncertainty of either a measurand or a result with (Equations 9.23 and 9.19) and without (Equations 9.24 and 9.21) correlated uncertainties. These expressions involve standard uncertainties for the measurands that originate primarily from instrument uncertainties determined from previous calibrations and assessments. There are associated expressions for the effective number of degrees of freedom (Equations 9.25 and 9.22 for a measurand and result, respectively).

The uncertainties for multiple-measurement situations are assessed best using detailed uncertainty analysis (Section 9.11). Errors are categorized as either systematic, S_{B_i}, or random, S_{P_i}. Expressions are developed for the combined standard uncertainty of either a measurand or a result with (Equations 9.64 and 9.47) and without (Equations 9.66 and 9.58) correlated uncertainties. There are associated expressions for the effective number of degrees of freedom (Equations 9.65 and 9.52 for a measurand and a result, respectively).

In summary, the overall uncertainty of a measurand or of a result can be arrived at via the following steps:

---

1. *Determine which experimental situation applies.* Is the uncertainty estimate for a measurand or a result and is it based on single or multiple measurements?
2. *Identify all measurands and, if applicable, all results.*
3. *Identify all factors affecting the measurands and, if applicable, the results.* What instruments are used? What information is available on their calibrations? Are there any circumstances that lead to correlated uncertainties, such as the same instrument used for two different measurands?
4. *Define all functional relationships between the measurands and, if applicable, the results.* What are the nominal values of each measurand and result? Be sure to use the same system of units throughout all calculations.
5. *If the uncertainty in a result is estimated, determine the values of the absolute sensitivity coefficients from the functional relationships and nominal values.*
6. *Identify all uncertainties.* What are the instrument errors, systematic and random errors? Are there temporal or spatial variations in the measurands that contribute to uncertainty?
7. *Calculate and propagate all uncertainties for the measurands and, if applicable, the results.* Proceed from estimates of the elemental uncertainties and calculate the systematic and random uncertainties.
8. *Propagate all uncertainties to obtain the combined standard uncertainty.*
9. *Determine the number of effective degrees of freedom.*
10. *Determine the value of the coverage factor based on the assumed confidence and the number of effective degrees of freedom.* Use a value of two for the coverage factor if 95 % confidence is assumed and $\nu_{\text{eff}} \geq 9$.
11. *Determine the overall uncertainty.*
12. *Present your findings in the proper form.* $x \pm U_x$ (% C).

STATEMENT | **Example 9.17**

Very accurate weight standards were used to calibrate a force-measurement system. Nine calibration measurements were made, including repetition of the measurement five times at 5 N of applied weight. The results are presented in Table 9.1. Determine [a] the static sensitivity of the calibration curve at 3.5 N, [b] the random uncertainty with 95 % confidence in the value of the output voltage based on the data obtained for the 5 N application cases, [c] the range within which the true mean of the voltage is for the 5 N application cases at *90 %* confidence, [d] the range within which the true variance of the voltage is for the 5 N application cases at *90 %* confidence, [e] the standard error of the fit based on all of the data (see Equation 10.29), and [f] the design-stage uncertainty of the instrument at 95 % confidence assuming that 0.04 V instrument uncertainty was obtained through calibration.

SOLUTION

Examination of the data reveals that the average value of the output voltage for the five values of the 5 N case is 9.0 V. Thus, the data can be fitted the best by the line $E = 2W + 1$. [a] The sensitivity of a linear fit is the slope, 2 V/N, which is the same for all applied-weight values. [b] The random uncertainty for the 5 N cases with 95 % uncertainty is $P_{5N} = t_{v, P=95\%} S_{P, 5N}$. Here, $S_{P, 5N} = 0.1$ V, $v = 4$, and $t_{4, 95} = 2.770$. This implies $P_{5N} = 0.277$ V. [c] The range within which the true mean value is contained extends $\pm t_{v, P=90\%} S_{P, 5N} / \sqrt{N} = 5$ from the sample mean value of 9 V. Here, $t_{4, 90} = 2.132$. So the range is from $9 - (2.132)(0.1)/\sqrt{5}$ V to $9 + (2.132)(0.1)/\sqrt{5}$ V, or from 8.905 V to 9.095 V. [d] The range of the true variance is

$$\frac{v S_{P, 5N}^2}{\chi_{\alpha/2}^2} \le \sigma_{5N}^2 \le \frac{v S_{P, 5N}^2}{\chi_{1-\alpha/2}^2}.$$

$P = 0.90$, which implies $\alpha = 0.10$. So, $\chi_{\alpha/2}^2$ for $v = 4$ equals 9.49 and $\chi_{1-\alpha/2}^2$ equals 0.711. Substitution of these values yields

$$\frac{(4)(0.01)}{9.49} \le \sigma_{5N}^2 \le \frac{(4)(0.01)}{0.711},$$

or

$$4.22 \times 10^{-3} \text{ V}^2 \le \sigma_{5N}^2 \le 56.26 \times 10^{-3} \text{ V}^2.$$

This also gives $0.065 \text{ V} \le \sigma_{5N}^2 \le 0.237$ V. [e] The first four and one of the five 5-N applied-weight case values are on the best-fit line. Therefore, only four of the five 5-N-case values contribute to the standard error of the fit. Thus

$$S_{yx} = \sqrt{\frac{(0.1)^2 + (0.1)^2 + (-0.1)^2 + (0.1)^2}{9 - 2}} = 0.0756 = 0.1 \text{ V}.$$

Table 9.1   Force-measurement system calibration data.

Applied Weight, W (N)	Output Voltage, E (V)
1	1.0
2	3.0
3	5.0
4	7.0
5	9.1
5	8.9
5	8.9
5	9.1
5	9.0

[f] The resolution of the voltage equals 0.1 V from inspection of the data. This implies that $u_o = 0.05$ V. This uncertainty is combined in quadrature with the instrument uncertainty of 0.04 V to yield a design-stage uncertainty equal to $0.064 = 0.06$ V.

9.13   FINITE-DIFFERENCE UNCERTAINTIES

There are additional uncertainties that need to be considered. These occur when experiments are conducted to determine a result that depends on the integral or derivative of measurand values obtained at discrete locations or times. The actual derivative or integral only can be estimated from this discrete information. A discretization (truncation) error results. For example, consider an experiment in which the velocity of a moving object is determined from measurements of time as the object passes known spatial locations. The actual velocity may vary nonlinearly *between* the two locations but it can only be approximated using the **finite** information available. Similarly, an actual velocity gradient in a flow only can be estimated from measured velocities at two adjacent spatial locations. Examples involving integral approximations are the lift and drag of an object, determined from a finite number of pressure measurements along the surface of the object, and the flow rate of a fluid through a duct, determined from a finite number of velocity measurements over a cross section of the duct.

The discretization errors of integrals and derivatives can be estimated, as described in Section 9.13.1. Numerical round-off errors also can occur in such determinations. For most experimental situations, however, discretization errors far exceed numerical round-off errors. When measurements are relatively few, the discretization error can be comparable to the measurement uncertainty. There are many excellent references that cover finite-difference approximation methods and their errors ([17], [18], [19], and [20]).

9.13.1   DERIVATIVE APPROXIMATION

If the values of a measurand are known at several locations or times, its actual derivative can be approximated. This is accomplished by representing the actual derivative in terms of a finite-difference approximation based on, most commonly, a Taylor series expansion. The amount of error is determined by the order of the expansion method used.

Suppose $f(x)$ is a continuous function with all of its derivatives defined at x. The next, or forward, value of $f(x + \Delta x)$ can be estimated using a Taylor series expansion of $f(x + \Delta x)$ about the point x:

$$f(x + \Delta x) = f(x) + \Delta x f'(x) + \frac{(\Delta x)^2}{2} f''(x) + \frac{(\Delta x)^3}{6} f'''(x) + \cdots . \qquad \textbf{[9.69]}$$

Equation 9.69 can be rearranged to solve for the derivative

$$f'(x) = \frac{f(x + \Delta x) - f(x)}{\Delta x} - \frac{(\Delta x)}{2} f''(x) - \frac{(\Delta x)^2}{6} f'''(x) + \cdots . \qquad \textbf{[9.70]}$$

The first term on the right-hand side is the finite-difference representation of $f'(x)$ and the subsequent terms define the discretization error. A finite-difference representation is termed nth order when the leading term in the discretization error is proportional to $(\Delta x)^n$. Thus, Equation 9.70 is known as the first-order, forward-difference expression for $f'(x)$. If the actual $f(x)$ can be expressed as a polynomial of the first degree, then $f'' = 0$, the finite-difference representation of $f'(x)$ exactly equals the actual derivative, and there is no discretization error. In general, an nth-order method is exact for polynomials of degree n.

Example 9.18	**STATEMENT**

STATEMENT

The velocity profile of a fluid flowing between two parallel plates spaced a distance $2h$ apart is given by the expression $u(y) = u_o[1 - (y/h)^2]$, where y is the coordinate perpendicular to the plates. Determine the exact value of $u(0.2h)/u_o$ and compare it with the finite-difference values obtained from the Taylor series expansion that result as each term is included additionally in the series.

SOLUTION

The exact value, found from direct substitution of $y = 0.2h$ into the velocity profile, is $u(0.2h)/u_o|_{exact} = 0.96$. For the Taylor series given by Equation 9.69, the derivatives must be computed. The result is $u'(y) = -2u_o y/h^2$, $u''(y) = -2u_o/h^2$, and $u'''(y) = 0$. Noting that $0.2h = \Delta x$ for this case, substitutions into Equation 9.69 yield $u(0.2h)/u_o|_{series} = 1 - 0.08 + 0.04 + 0 + \cdots$. So, three terms are required in the series in this case to give the exact result; fewer terms result in a difference between the exact and series values.

Similarly, the first-order, backward-difference expression for $f'(x)$ is

$$f'(x) = \frac{f(x) - f(x - \Delta x)}{\Delta x} + \frac{(\Delta x)}{2} f''(x) - \frac{(\Delta x)^2}{6} f'''(x) + \cdots. \qquad \textbf{[9.71]}$$

Equation 9.70 can be added to Equation 9.71 to yield

$$f'(x) = \frac{f(x + \Delta x) - f(x - \Delta x)}{2\Delta x} - \frac{(\Delta x)^2}{6} f'''(x) + \cdots, \qquad \textbf{[9.72]}$$

resulting in a second-order, central-difference expression for $f'(x)$. Other expressions for second-order, central-difference, and central-mixed-difference second and higher derivatives can be obtained following a similar approach [18].

Usually second-order accuracy is sufficient for experimental analysis. Assuming this, the discretization error, e_d, of the first derivative approximated by a second-order central-difference estimate using values at two locations $(x - \Delta x$ and $x + \Delta x)$ is

$$e_d \simeq f''' \frac{(\Delta x)^2}{6}, \qquad \textbf{[9.73]}$$

where f''' is evaluated somewhere in the interval, usually at its maximum value. A problem arises, however, because the value of f''' is not known. So, only the order of magnitude of e_d can be estimated. Formally, the uncertainty in a first derivative approximated by a second-order central-difference method is

$$u_{f'(x)} \simeq C_{f'(x)}(\Delta x)^2, \qquad \textbf{[9.74]}$$

where $C_{f'(x)}$ is a constant with same units as f'''. $C_{f'(x)}$ can be assumed to be of order one as a first approximation. The important point to note is that the discretization error is proportional to $(\Delta x)^2$. So, if Δx is reduced by 1/2, the discretization error is reduced by 1/4.

9.13.2 INTEGRAL APPROXIMATION

Many different numerical methods can be used to determine the integral of a function. The method chosen depends on the accuracy required, if the values of the function are known at its endpoints, if the numerical integration is done using equal-spaced intervals, and so on. The trapezoidal rule is used most commonly for situations in which the intervals are equally spaced and the function's values are known at its endpoints.

The trapezoidal rule approximates the area under the curve $f(x)$ over the interval between a and b by the area of a trapezoid,

$$\int_a^b f(x) = \frac{b-a}{2}[f(b) + f(a)] + E, \qquad \textbf{[9.75]}$$

where $E = (\Delta x)^3 f''/24$, with f'' evaluated somewhere in the interval from a to b. This rule can be extended to N points,

$$\int_{a=x_1}^{b=x_N} f(x)\, dx = \Delta x \left[\frac{1}{2}f(x_1) + f(x_2) + \cdots + f(x_{N-1}) + \frac{1}{2}f(x_N) \right] + \sum_{i=1}^{N} E_i$$

$$= \Delta x \left\{ \sum_{i=1}^{N} f(x_i) - \left[\frac{1}{2}f(x_1) + \frac{1}{2}f(x_N) \right] \right\} + \sum_{i=1}^{N} E_i, \qquad \textbf{[9.76]}$$

9.4 | MATLAB SIDEBAR

The MATLAB M-file `differ.m` numerically differentiates one user-specified variable, y, with respect to another variable, x, using the MATLAB function `diff`. The `diff` function computes the first-order, forward-difference of a variable. The derivative of $y = f(x)$ is then determined as `diff(y)/diff(x)`. The M-file is constructed to receive a user-specified data file from which two columns are specified as the variables of interest. It plots the $\Delta y / \Delta x$ and y values versus the x values. An example output of `integ.m` is shown in Figure 9.6, in which nine (x, y) pairs comprised the data file.

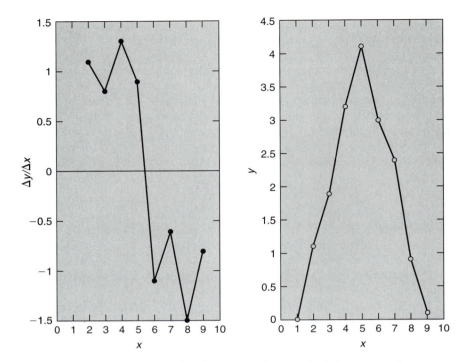

Figure 9.6   The output plot of `differ.m` for nine (x,y) data pairs of a triangular waveform.

where $\Delta x = (b - a)/N$. The total discretization error, e_d, then becomes

$$e_d = \sum_{i=1}^{N} E_i \simeq \frac{1}{24} \sum_{i=1}^{N} (\Delta x)^3 f'' = \frac{N}{24}(\Delta x)^3 f'' = \frac{(b-a)^3}{24N^2}f''. \qquad \textbf{[9.77]}$$

Thus, the uncertainty in applying the extended trapezoidal rule to approximate an integral is

$$u_{\int f(x)} \simeq C_{\int f(x)} N(\Delta x)^3, \qquad \textbf{[9.78]}$$

where $C_{\int f(x)}$ is a constant with the same units as f''. $C_{\int f(x)}$ can be assumed to be of order one as a first approximation.

A numerical estimate of $C_{\int f(x)}$ can be made if some expression for f'' can be found. A second-order central second-difference approximation can be used, where

$$f''(x_i) = \frac{f(x_{i+1}) - 2f(x_i) + f(x_{i-1})}{(\Delta x)^2}. \qquad \textbf{[9.79]}$$

This introduces an additional error of $O[f''''(x_i)(\Delta x)^2]$. So, using Equation 9.79 does not improve the accuracy of the estimate; it simply provides a convenient way

9.5 | **MATLAB SIDEBAR**

The MATLAB M-file `integ.m` numerically integrates one user-specified variable, y, with respect to another variable, x, using the MATLAB function `trapz(x,y)`. The `trapz` function sums the areas of the trapezoids that are formed from the x and y values. The M-file is constructed to receive a user-specified data file from which two columns are specified as the variables of interest. It plots the data and lists the value of calculated area. More accurate MATLAB functions, such as quad and quad8, perform integration using quadrature methods in which the intervals of integration are varied. An example output of `integ.m` is shown in Figure 9.7, in which nine (x, y) pairs comprised the data file.

to estimate f''. Now the discretization error also can be written alternatively as

$$e_d = \frac{b-a}{24N} \sum_{i=1}^{N} f''(x_i)(\Delta x)^2.$$

[9.80]

Substitution of Equation 9.79 into Equation 9.80 gives

$$u_{\int f(x)} \simeq \frac{b-a}{24N} \left\{ \sum_{i=1}^{N} [f(x_{i+1}) - 2f(x_i) + f(x_{i-1})]^2 \right\}^{1/2}.$$

[9.81]

Note that the terms with the brackets represent the discretization errors in the individual f'' estimates, which are combined in quadrature.

Examples 9.19, 9.20, and 9.21 illustrate how uncertainties arising from the finite-difference approximation of an integral factor into the uncertainty of a result.

STATEMENT

| **Example 9.19**

Continuing with the experiment presented in Example 9.11, determine the uncertainties in the lift coefficient, C_L, and the drag coefficient, C_D, of the cylinder. The lift and drag coefficients are determined from 36 static pressure measurements around the cylinder's circumference done in $10°$ increments.

SOLUTION

The lift coefficient is given by the equation

$$C_L = -\frac{1}{2} \int C_p(\theta) \sin(\theta)\, d\theta.$$

[9.82]

Because there are only 36 discrete measurements of the static pressure, the integral in Equation 9.82 must be approximated by a finite sum using the trapezoidal rule. The general equation for the trapezoidal rule is

$$\int_a^b f(x)\, dx \cong \frac{b-a}{N} \left[\frac{f(a)}{2} + f(x_2) + \cdots + f(x_{N-1}) + \frac{f(b)}{2} \right].$$

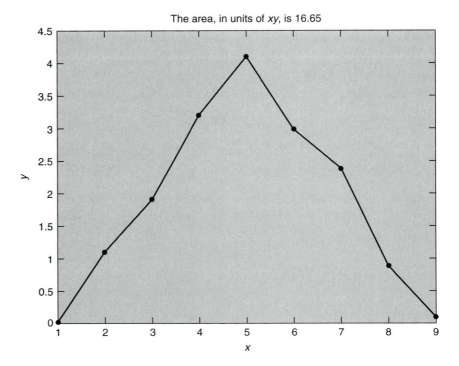

Figure 9.7   The output plot of `integ.m` for nine (x,y) data pairs of a triangular waveform.

Applying this formula to Equation 9.82 yields

$$C_L \cong -\frac{\pi}{72}\left[C_p(\theta=0)+2C_p\left(\theta=\frac{10\pi}{180}\right)\sin\left(\frac{10\pi}{180}\right)\right.$$

$$+\cdots+C_p\left(\theta=\frac{350\pi}{180}\right)\sin\left(\frac{350\pi}{180}\right)$$

$$\left.+C_p\left(\theta=\frac{360\pi}{180}\right)\sin\left(\frac{360\pi}{180}\right)\right]\ (N=36).$$

Because the uncertainty in calculating C_L is an uncertainty in a result,

$$u_{C_L}=\sqrt{\left(\frac{\partial C_L}{\partial C_p}u_{C_p}\right)^2+\left(\frac{\partial C_L}{\partial \sin\theta}u_{\sin\theta}\right)^2}.$$

Now,

$$u_{\sin\theta}=\frac{\partial \sin\theta}{\partial\theta}u_\theta=\cos\theta u_\theta.$$

Therefore,

$$u_{C_L} = \sqrt{(u_{C_p} \sin \theta)^2 + (C_p(\theta) u_\theta \cos \theta)^2}.$$

This formulation must be applied to the finite-series approximation. Doing so leads to

$$u_{C_L} = \frac{\pi}{72} \left\{ (\sin(\theta = 0)u_{C_p})^2 + (C_p(\theta = 0)\cos(\theta = 0)u_\theta)^2 \right.$$

$$+ \left[2\sin\left(\theta = \frac{10\pi}{180}\right) u_{C_p} \right]^2 + \left[2C_p\left(\theta = \frac{10\pi}{180}\right) \cos\left(\frac{10\pi}{180}\right) u_\theta \right]^2 + \cdots$$

$$+ \left[2\sin\left(\theta = \frac{350\pi}{180}\right) u_{C_p} \right]^2 + \left[2C_p\left(\theta = \frac{350\pi}{180}\right) \cos\left(\frac{350\pi}{180}\right) u_\theta \right]^2$$

$$+ \left[\sin\left(\theta = \frac{360\pi}{180}\right) u_{C_p} \right]^2 + \left[C_p\left(\theta = \frac{360\pi}{180}\right) \cos\left(\frac{360\pi}{180}\right) u_\theta \right]^2 \right\}^{1/2}.$$

This can be evaluated using a spreadsheet or MATLAB. For the case when $u_{C_p} = 0.021$ and $u_\theta = \pi/360$ ($\pm 0.5°$), $u_{C_L} = 0.0082$. Likewise, u_{C_D} can be evaluated. The expression is similar to u_{C_L} as just given but with cos and sin reversed. For $u_{C_p} = 0.021$ and $u_\theta = \pi/180$ ($\pm 1.0°$), $u_{C_D} = 0.0081$. Now assume that the experiment was performed within a Reynolds number range that yields $C_D \sim 1$. Thus, % $u_{C_D} \cong 0.8$ %.

It is important to note that these u_{C_L} and u_{C_D} uncertainties do not include their finite-series approximation uncertainties. These are determined in Example 9.21.

Example 9.20

STATEMENT

Determine the uncertainty in the drag of the cylinder that was studied in Examples 9.11 and 9.19, where the drag, D, is defined as

$$D \equiv C_D \frac{1}{2} \rho V_\infty^2 A_{\text{frontal}}, \qquad \qquad \textbf{[9.83]}$$

with $A_{\text{frontal}} = \text{diameter} \cdot \text{length}$.

SOLUTION

In the experiment $\frac{1}{2}\rho V_\infty^2$ was measured as ΔP. Thus,

$$u_D = \left[\left(\frac{\partial D}{\partial C_D} u_{C_D} \right)^2 + \left(\frac{\partial D}{\partial \Delta P} u_{\Delta P} \right)^2 + \left(\frac{\partial D}{\partial d} u_d \right)^2 + \left(\frac{\partial D}{\partial L} u_L \right)^2 \right]^{1/2}$$

$$= [(\Delta P\, dL u_{C_D})^2 + (C_D\, dL u_{\Delta P})^2 + (C_D \Delta P L u_d)^2 + (C_D \Delta P\, d u_L)^2]^{1/2}.$$

Now, given that

$$\Delta P = 4 \text{ in. } H_2O = 996 \text{ N/m}^2, \quad u_{\Delta P} = 15 \text{ N/m}^2,$$

$$d = 1.675 \text{ in.} = 0.0425 \text{ m}, \quad u_d = 0.005 \text{ in.} = 1.27 \times 10^{-4} \text{ m},$$
$$L = 16.750 \text{ in.} = 0.425\,45 \text{ m}, \quad u_L = 0.005 \text{ in.} = 1.27 \times 10^{-4} \text{ m},$$
$$C_D \cong 1, \quad u_{C_D} = 0.0092,$$

then

$$u_D = [(996)(0.0425)(0.425\,45)(0.0092)^2$$
$$(1)(0.0425)(0.425\,45)(15)^2$$
$$(1)(996)(0.425\,45)(1.27 \times 10^{-4})^2$$
$$(1)(996)(0.0425)(1.27 \times 10^{-4})^2]^{1/2}$$
$$= (0.166^2 + 0.271^2 + 0.054^2 + 0.005^2)^{1/2}$$
$$= 0.32.$$

In order to get a percentage error, the nominal value of the drag is computed, where

$$D \cong (1)(996)(0.0425)(0.42545) = 18.0.$$

Thus, the percentage error in the drag is 1.7 %, and $D = 18.0 \pm 0.3$ N.

Example 9.21

STATEMENT

Recall the experiment presented in Example 9.19 in which 36 static pressure measurements were made around a cylinder's circumference. Determine the uncertainties in the lift and drag coefficients that arise when the extended trapezoidal rule is used to approximate the integrals involving the pressure coefficient, where

$$C_L = -\frac{1}{2} \int_0^{2\pi} C_P(\theta) \, \sin(\theta) \, d\theta$$

[9.84]

and

$$C_D = -\frac{1}{2} \int_0^{2\pi} C_P(\theta) \, \cos(\theta) \, d\theta.$$

Compare these numerical uncertainties to their respective measurement uncertainties, which were obtained previously. Finally, determine the overall uncertainties in C_L and C_D.

SOLUTION

Equation 9.81 implies

$$u_{\int f(x), C_L} \simeq \frac{2\pi}{24N} \left\{ \sum_{i=1}^{N} [g(\theta_{i+1}) - 2g(\theta_i) + g(\theta_{i-1})]^2 \right\}^{1/2}$$

[9.85]

and

$$u_{\int f(x),C_D} \simeq \frac{2\pi}{24N} \left\{ \sum_{i=1}^{N} [h(\theta_{i+1}) - 2h(\theta_i) + h(\theta_{i-1})]^2 \right\}^{1/2},$$

where

$$g(\theta) = C_P(\theta) \sin(\theta),$$
$$h(\theta) = C_P(\theta) \cos(\theta) \text{ and}$$
$$N = 36.$$

These uncertainties can be evaluated using a spreadsheet or MATLAB, yielding $u_{\int f(x),C_L} = 0.0054$ and $u_{\int f(x),C_D} = 0.0042$. These uncertainties are approximately one-half of the C_L and C_D measurement uncertainties.

Combining the C_L and C_D measurement and numerical approximation uncertainties gives

$$u_{C_L} = (0.0082^2 + 0.0054^2)^{1/2} = 0.0098,$$
$$u_{C_D} = (0.0081^2 + 0.0042^2)^{1/2} = 0.0092.$$

9.13.3 UNCERTAINTY ESTIMATE APPROXIMATION

In some situations, the direct approach to estimating the uncertainty in a result can be complicated if the mathematical expression relating the result to the measurands is algebraically complex. An alternative approach is to approximate numerically the partial derivatives in the uncertainty expression, thereby obtaining a more tractable expression that is amenable to spreadsheet or program analysis.

The partial derivative term, $\partial q / \partial x_i$, can be approximated numerically by the finite-difference expression

$$\frac{\partial q}{\partial x_i} \approx \frac{\Delta q}{\Delta x_i} = \frac{q|_{x_i + \Delta x_i} - q|_{x_i}}{\Delta x_i}. \tag{9.86}$$

This approximation is first-order accurate, as seen by examining Equation 9.70. Thus, its discretization error is of order $\Delta x f''(x)$. Use of Equation 9.86 yields the forward finite-difference approximation to Equation 9.21:

$$u_q \approx \sqrt{\left(\frac{\Delta q}{\Delta x_1} u_{x_1} \right)^2 + \left(\frac{\Delta q}{\Delta x_2} u_{x_2} \right)^2 + \cdots + \left(\frac{\Delta q}{\Delta x_k} u_{x_k} \right)^2}. \tag{9.87}$$

The value of Δx_i is chosen to be small enough such that the finite-difference expression closely approximates the actual derivative. Typically, $\Delta x_i = 0.01 x_i$ is a good starting value. The value of Δx_i then should be decreased until appropriate convergence in the value of u_q is obtained.

Example 9.22

STATEMENT

In Example 9.10, the uncertainty in the density for air was determined directly by the expression

$$u_\rho = \sqrt{\left(\frac{\partial \rho}{\partial T} u_T\right)^2 + \left(\frac{\partial \rho}{\partial P} u_P\right)^2} = 2.15 \times 10^{-3} \text{ kg/m}^3,$$

where $u_P = 67$ Pa, $u_T = 0.5$ K, $\rho = 1.19$ kg/m^2, $T = 297$ K, $P = 101\,325$ Pa, and air was assumed to be an ideal gas ($\rho = P/RT$). Determine the uncertainty in ρ by application of Equation 9.87.

SOLUTION

The finite-difference expression for this case is

$$u_\rho \approx \sqrt{\left(\frac{\Delta \rho}{\Delta P} u_P\right)^2 + \left(\frac{\Delta \rho}{\Delta T} u_T\right)^2},$$

where $\Delta\rho/\Delta P = [\rho|_{P+\Delta P} - \rho|_P]/\Delta P$ and $\Delta\rho/\Delta T = [\rho|_{T+\Delta T} - \rho|_T]/\Delta T$. Letting $\Delta P = 0.01 P$ and $\Delta T = 0.01 T$ yields $\Delta\rho/\Delta P = [(P+\Delta P)/RT - P/RT]/\Delta P = 1/RT = \rho/P = 1.17 \times 10^{-5}$ and $\Delta\rho/\Delta T = [P/R(T - \Delta T) - P/RT]/\Delta T = -\rho/(T + \Delta T) = 3.97 \times 10^{-3}$.

Substitution of these values into the equation gives $u_\rho = \sqrt{0.62 \times 10^{-6} + 3.93 \times 10^{-6}} = 2.13 \times 10^{-3}$ kg/m^3. This agrees to within 1 % of the value of 2.15×10^{-3} found using the direct method.

REFERENCES

[1] S. M. Stigler. 1986. *The History of Statistics*. Cambridge: Harvard University Press.

[2] S. J. Kline and F. A. McClintock. 1953. Describing Uncertainties in Single-Sample Experiments. *Mechanical Engineering*. January: 3–9.

[3] 1993. *Guide to the Expression of Uncertainty in Measurement*. ISBN 92-67-10188-9. Geneva: International Organization for Standardization (ISO) [corrected and reprinted in 1995].

[4] 2000. *U.S. Guide to the Expression of Uncertainty in Measurement*. ANSI/NCSL Z540-2-1997. Boulder, Colo.: NCSL International.

[5] 1998. *Test Uncertainty*. ANSI/ASME PTC 19.1-1998. New York: ASME.

[6] 1995. *Assessment of Wind Tunnel Data Uncertainty*. AIAA Standard S-071-1995. New York: AIAA.

[7] W. L. Oberkampf, S. M. DeLand, B. M. Rutherford, K. V. Diegert, and K. F. Alvin. 2000. Estimation of Total Uncertainty in Modeling and Simulation. *Sandia Report SAND2000-0824* Albuquerque, N.M.: Sandia National Laboratories.

[8] S. J. Kline. 1985. The Purposes of Uncertainty Analysis. *Journal of Fluids Engineering* 107: 153–164.

[9] B. L. Welch. 1947. The Generalization of "Student's" Problem When Several Different Population Variances Are Involved. *Biometrika* 34: 28–35.

[10] H. W. Coleman and W. G. Steele. 1999. *Experimentation and Uncertainty Analysis for Engineers*, 2nd ed. New York: Wiley Interscience.

[11] A. Aharoni. 1995. Agreement Between Theory and Experiment. *Physics Today* June: 33–37.

[12] S. Lemkowitz, H. Bonnet, G. Lameris, G. Korevaar, and G. J. Harmsen. 2001. How "Subversive" is Good University Education? A Toolkit of Theory and Practice for Teaching Engineering Topics with Strong Normative Content, Like Sustainability. *ENTREE 2001 Proceedings* 65–82. Florence, Italy. November.

[13] R. J. Moffat. 1998. Describing the Uncertainties in Experimental Results. *Experimental Thermal and Fluid Science* 1: 3–17.

[14] A. Hald. 1952. *Statistical Theory with Engineering Applications*. New York: John Wiley and Sons.

[15] R. Figliola and D. Beasley. 2000. *Theory and Design for Mechanical Measurements*, 3rd ed. New York: John Wiley and Sons.

[16] J. R. Taylor. 1982. *An Introduction to Error Analysis*. Mill Valley, Calif.: University Science Books.

[17] W. H. Press, S. A. Teukolsy, W. T. Vetterling, and B. P. Flannery. 1992. *Numerical Recipes*, 2nd ed. New York: Cambridge University Press.

[18] J. D. Anderson, Jr. 1995. *Computational Fluid Dynamics*. New York: McGraw-Hill.

[19] S. Nakamura. 1996. *Numerical Analysis and Graphic Visualization with MATLAB*. Upper Saddle River, N.J.: Prentice Hall.

[20] W. E. Boyce and R. C. Di Prima. 1997. *Elementary Differential Equations and Boundary Value Problems*, 6th ed. New York: John Wiley and Sons.

REVIEW PROBLEMS

1. A researcher is measuring the length of a microscopic scratch in a microphone diaphragm using a stereoscope. A ruler incremented into ten-thousandths of an inch is placed next to the microphone as a distance reference. If the stereoscope magnification is increased 10 times, what property of the distance measurement system has been improved?

(a) sensitivity, (b) precision, (c) readability,
(d) least count

2. A multimeter, with a full-scale output of 5 V, retains two decimal digits of resolution. For instance, placing the multimeter probes across a slightly used AA battery results in a readout of 1.35 V. Through calibration, the following maximum instrument uncertainties are established: (1) sensitivity: 0.5 % of FSO; (2) offset: 0.1 % of FSO. What is the total design stage uncertainty in volts based on this information to 95 % confidence? (*Note:* The readout of the instrument dictates that the uncertainty should be expressed to three decimal places.)

3. Three students are playing darts. The results of the first round are shown in Figure 9.8, where the circle in the center is the bullseye. Circles = Player 1; squares = Player 2; triangles = Player 3. In terms of hitting the bullseye, which player best demonstrates precision, but not accuracy?

(a) Player 1 (circles), (b) Player 2 (squares),
(c) Player 3 (triangles)

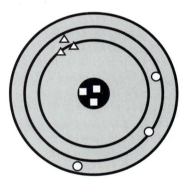

Figure 9.8 Dartboard.

4. Compare the precision of a metric ruler, incremented in millimeters, with a standard customary measure ruler, incremented into 16ths of an inch. How much more precise is the more precise instrument? Express your answer in millimeters and consider the increments on the rulers to be exact.

5. For a circular rod with density, ρ, diameter, D, and length, L, derive an expression for the uncertainty in computing its moment of inertia about the rod's end from the measurement of those three quantities. If $\rho = 2008$ kg/m^3 ± 1 kg/m^3, $D = 3.60$ mm ± 0.05 mm, and $L = 2.83$ m ± 0.01 m, then compute the uncertainty in the resulting moment of inertia to the correct number of significant figures in the SI unit kg-m^2. Significant figures are based on the measured quantities. The formula for the moment of inertia of a circular rod about its end is $I = \rho\pi D^2 L^3 / 12$.

6. The velocity of the outer circumference of a spinning disk may be measured in two ways. Using an optical sensing device, the absolute uncertainty in the velocity is 0.1 %. Using a strobe and a ruler, the uncertainty in the angular velocity is 0.1 rad/s and the uncertainty in the diameter of the disk is 1 mm. Select the measurement method with the *least* uncertainty for the two methods if the disk is 0.25 m in diameter and is spinning at 10 rpm. Express the numerical result (the least uncertainty) in m/s to three significant digits.

7. Using a pair of calipers with 0.001-in. resolution, a machinist measures the diameter of a pinball seven times with the following results (in units of inches): 1.026, 1.053, 1.064, 1.012, 1.104, 1.098, and 1.079. He uses the average of the measurements as the correct value. Compute the uncertainty in the average diameter in inches to one-thousandth of an inch.

8. Match the following examples of experimental error (numbered 1 through 4) and uncertainty to the best categorization of that example (lettered a through d):

    (1) Temperature fluctuations in the laboratory
    (2) Physically pushing the pendulum through its swing during one experimental trial
    (3) The numerical bounds of the scatter in the distance traveled by the racquet ball about the mean value
    (4) Releasing the pendulum from an initial position so that the shaft has a small initial angle to the vertical for each trial

[a] uncertainty, [b] systematic error, [c] experimental mistake, [d] random error.

9. A geologist finds a rock of unknown composition and desires to measure its density. To measure the volume, she places the rock in a graduated cylinder, which is graduated into milliliter increments, half-filled with water, so that the rock is submerged in the water. She removes the rock from the cylinder and directly measures the rock's mass using a scale with a digital readout resolution of 0.1 g. No information is provided from the manufacturer about the scale uncertainty. She records the volume, V, and mass, m, as follows: $V = 40.5$ mL and $m = 143.1$ g. What is the uncertainty in the density value computed from these measurements? Express your answer as a percentage with one significant figure.

10. In addition to the digital display, a manometer has an analog voltage output that is proportional to the sensed differential pressure in in. H_2O. A researcher calibrates the analog output by applying a series of known pressures across the manometer ports and observing both the analog and digital LED output for each input pressure. A linear fit to the data yields $P = 1.118V + 0.003$, where P is pressure in in. H_2O and V is the analog voltage in volts. If for zero input pressure, the manufacturer specifies that no output voltage should result, what magnitude of systematic error has been discovered? Express your answer as a percentage of the full-scale output to one significant figure.

11. The mean dynamic pressure in a wind tunnel is measured using the manometer described in Problem 10 and a multimeter to measure the output voltage. The recorded voltages are converted to pressures using the calibration fit presented in Problem 10. The resolution of the multimeter is 1 mV. The manufacturer specifies the following instrument errors: sensitivity = 0.5 mV and linearity = 0.5 mV. If 150 multimeter readings with a standard deviation of 0.069 V are acquired with a mean voltage of 0.312 V, what is the uncertainty in the resulting computed mean pressure in in. H_2O? Express your answer to the precision of the digital readout of the manometer.

12. A technician uses a graduated container of water to measure the volume of odd-shaped objects. The changes in the density of the water caused by ambient temperature and pressure fluctuations directly contribute to [a] systematic error, [b] redundant error, [c] random error, and [d] multiple measurement error.

13. To calibrate a digital scale, a graduate student orders a set of very accurate weights. The formula for the moment of inertia of a circular rod about its end is $I = \rho\pi D^2 L^3 / 12$. The desired accuracy of the scale is 0.1 g. The manufacturer of the weights states that the mass of each weight is accurate to 0.04 g. What is the maximum number of weights that can be used in combination to calibrate the scale?

14. A test engineer performs a first-run experiment to measure the time required for a 2005 prototype car to travel 1/4-mile beginning from rest. When the car begins motion, a green light flashes in the engineer's field of vision, signaling him to start the time count with a hand-held stopwatch. Similarly, a red light flashes when the car reaches the finish line. The resulting times from four trials are 13.42 s,

13.05 s, 12.96 s, and 12.92 s. Outside of the test environment, another engineer measures the first test engineer's reaction time to the light signals. The results of the test show that the test engineer overanticipates the green light and displays slowed reaction to the red light. Both reaction times were measured to be 0.13 s. Compute the average travel time in seconds, correcting for the systematic error in the experimental procedure.

15. A digital manometer measures the differential pressure across two inputs. The range of the manometer is 0 in. H_2O to 0.5 in. H_2O. The LED readout resolves pressure into 0.001 in. H_2O. Based on calibration, the manufacturer specifies the following instrument errors: hysteresis error = 0.1% of FSO (full-scale output); linearity = 0.25% of FSO; sensitivity = 0.1% of FSO. Determine the design-stage uncertainty of the digital manometer in in. H_2O to the least significant digit resolved by the manometer.

16. The smallest division marked on the dial of a pressure gage is 2 psi. The accuracy of the pressure gage as stated by the manufacturer is ±1 psi. Determine the design-stage uncertainty in psi and express it with the correct number of significant figures.

17. A student conducts an experiment in which the panel meter displaying the measurement system's output in volts fluctuates up and down in time. Being a conscientious experimenter, the student decides to estimate the temporal random error of the measurement. She takes 100 repeated measurements and finds that the standard deviation equals a whopping 1.0 V! Estimate the temporal random error in volts at 95 % confidence and express the answer with the correct number of significant figures.

HOMEWORK PROBLEMS

1. The supply reservoir to a water clock is constructed from a tube of circular section. The tube has a nominal length of 52 cm ± 0.5 cm, an outside diameter of 20 cm ± 0.04 cm, and an inside diameter of 15 cm ± 0.08 cm. Determine the *percentage* uncertainty in the calculated volume.

2. A mechanical engineer is asked to design a cantilever beam to support a concentrated load at its end. The beam is of circular section and has a length, L, of 6 ft and a diameter, d, of 2.5 in. The concentrated load, F, of 350 lbf is applied at the beam end, perpendicular to the length of the beam. If the uncertainty in the length is ±1.5 in., in the diameter is ±0.08 in., and in the force is ± 5 lbf, what is the uncertainty in the calculated bending stress, σ? [*Hint*: $\sigma = 32FL/(\pi d^3)$]. Further, if the uncertainty in the bending stress may be no greater than 6 %, what maximum uncertainty may be tolerated in the diameter measurement if the other uncertainties remain unchanged?

3. An electrical engineer must decide on a power usage measurement method that yields the least uncertainty. There are two alternatives to measuring the power usage of a dc heater: (1) heater resistance and voltage drop can be measured simultaneously and then the power computed, or (2) heater voltage drop and current can be measured simultaneously and then the power computed. The manu-

facturers' specifications of the available instruments are as follows: ohmmeter (resolution 1 Ω and % reading uncertainty = 0.5 %); ammeter (resolution 0.5 A and % reading uncertainty = 1 %); voltmeter (resolution 1 V and % reading uncertainty = 0.5 %). For loads of 10 W, 1 kW, and 10 kW each, determine the best method based on an appropriate uncertainty analysis. Assume nominal values as necessary for resistance and current based on a fixed voltage of 100 V.

4. A new composite material is being developed for an advanced aerospace structure. The material's density is to be determined from the mass of a cylindrical specimen. The volume of the specimen is determined from diameter and length measurements. It is estimated that the mass, m, can be determined to be within 0.1 lbm using an available balance scale, the length, L, to within 0.05 in. and the diameter, D, to within 0.0005 in. Estimate the zero-order design-stage uncertainty in the determination of the density. Which measurement would contribute most to the uncertainty in the density? Which measurement method should be improved first if the estimate in the uncertainty in the density is unacceptable? Use nominal values of $m = 4.5$ lbm, $L = 6$ in., and $D = 4$ in. Next, multiple measurements are performed yielding

$\bar{D} = 3.9924$ in. $\bar{m} = 4.4$ lbm $\bar{L} = 5.85$ in.
$S_D = 0.0028$ in. $S_m = 0.1$ lbm $S_L = 0.10$ in.
$N = 3$ $N = 21$ $N = 11$

Using this information and what was given initially, provide an estimate of the true density at 95 % confidence. Compare the uncertainty in this result to that determined in the design stage.

5. High-pressure air is to be supplied from a large storage tank to a plenum located immediately before a supersonic convergent-divergent nozzle. The engineer designing this system must estimate the uncertainty in the plenum's pressure measurement system. This system outputs a voltage that is proportional to pressure. It is calibrated against a transducer standard (certified accuracy: within ±0.5 psi) over its 0 psi to 100 psi range with the results given below. The voltage is measured with a voltmeter (instrument error: within ±10 μV; resolution: 1 μV). The engineer estimates that installation effects can cause the indicated pressure to be off by another ±0.5 psi. Estimate the uncertainty at 95 % confidence in using this system based on the following given information:

E(mV)	0.004	0.399	0.771	1.624	2.147	4.121
p(psi)	0.1	10.2	19.5	40.5	51.2	99.6

6. One approach to determining the volume of a cylinder is to measure its diameter and length and then calculate the volume. If the length and diameter of the cylinder are measured at four different locations using a micrometer with an uncertainty of 0.05 in. with 95 % confidence, determine the percentage uncertainty in the volume. The four diameters in units of inches are 3.9920, 3.9892, 3.9961, and 3.9995; those of the length are 4.4940, 4.4991, 4.5110, and 4.5221.

7. Given $y = ax^2$ and that the uncertainty in a is 3 % and that in x is 2 %, determine the % uncertainty in y for the nominal values of $a = 2$ and $x = 0.5$.

8. The lift force on a Wortmann airfoil is measured five times under the same experimental conditions. The acquired values are 10.5 N, 9.4 N, 9.1 N, 11.3 N, and 9.7 N. Assuming that the only uncertainty in the experiment is a temporal random error as manifested by the spread of the data, determine the uncertainty (in ±N) at the 95 % confidence level of the true mean value of the lift force.

9. A pressure transducer specification sheet lists the following instrument errors, all in units of % span, where the span for the particular pressure transducer is 10 in. H_2O: combined null and sensitivity shift = ±1.00, linearity = ±2.00, and repeatability and hysteresis = ±0.25. Estimate [a] the transducer's zero-order uncertainty in the pressure in units of in. H_2O and [b] the % zero-order uncertainty in a pressure reading of 1 in. H_2O. [c] Would this be a suitable transducer to use in an experiment in which the pressure ranged from 0 to 2 in. H_2O and the pressure reading must be accurate to within ±10 %?

10. The mass of a golf ball is measured using an electronic balance that has a resolution of 1 mg and an instrument uncertainty of 0.5 %. Thirty one measurements of the mass are made, yielding an average mass of 45.3 g and a standard deviation of 0.1 g. Estimate the [a] zero-order, [b] design-stage, and [c] first-order uncertainties in the mass measurement. What uncertainty contributes the most to the first-order uncertainty?

11. A group of students wish to determine the density of a cylinder to be used in a design project. They plan to determine the density from measurements of the cylinder's mass, length, and diameter, which have instrument resolutions of 0.1 lbm, 0.05 in., and 0.0005 in., respectively. The balance used to measure the weight has an instrument uncertainty (accuracy) of 1 %. The rulers used to measure the length and diameter are negligible instrument uncertainties. Nominal values of the mass, length, and diameter are 4.5 lbm, 6.00 in., and 4.0000 in., respectively. [a] Estimate the zero-order uncertainty in the determination of the density. [b] Which measurement contributes the most to this uncertainty? [c] Estimate the design-stage uncertainty in the determination of the density.

12. The group of students in the previous problem now perform a series of measurements to determine the actual density of the cylinder. They perform 20 measurements of the mass, length, and diameter that yield average values for the mass, length, and diameter equal to 4.5 lbm, 5.85 in., and 3.9924 in., respectively, and standard deviations equal to 0.1 lbm, 0.10 in., and 0.0028 in., respectively. Using this information and that presented in the previous problem, estimate [a] the average density of the cylinder in $lbm/in.^3$; [b] the systematic errors of the mass, length, and diameter measurements; [c] the random errors of the mass, length, and diameter measurements; [d] the combined systematic errors of the density; [e] the combined random errors of the density; [f] the uncertainty in the density estimate at 95 %

confidence (compare this to the design-stage uncertainty estimate, which should be smaller); and [g] an estimate of the true density at 95 % confidence.

13. An aerospace engineering student takes only four electric manometer readings [2 V, 4 V, 6 V, and 8 V] during a wind tunnel experiment. Write the correct formulas for and determine at 95 % confidence the range within which the following would lie: [a] the next voltage measurement and [b] the true mean voltage. Express your answer with the correct number of significant figures. State all approximations and their justifications.

14. Given King's law, $E^2 = A + B\sqrt{U}$, and the uncertainties in A, B, and U of 5, 4, and 6 %, respectively, determine the % uncertainty in E. Assume nominal values of $A = 3$, $B = 1/3$, and $U = 9$.

15. The resistivity ρ of a piece of wire must be determined. To do this, the relationship

$$R = \frac{\rho L}{A}$$

can be used and the appropriate measurements made. Nominal values of R, L, and the diameter of the wire, d, are 50 Ω, 10 ft, and 0.050 in., respectively. The error in L must be held to no more than 0.125 in. and R will be measured with a voltmeter accuracy of ± 0.2 % of the reading. How accurately will d need to be measured if the uncertainty in ρ is not to exceed 0.5 %?

16. The tip deflection of a cantilever beam with rectangular cross section subjected to a point load at the tip is given by the formula

$$\delta = \frac{PL^3}{3EI}, \quad \text{where } I = \frac{bh^3}{12}.$$

Here, P is the load, L is the length of the beam, E is the Young's modulus of the material, b is the width of the cross section, and h is the height of the cross section. If the uncertainties in P, L, E, b, and h are all 2 %, estimate the percent uncertainty in δ.

CHAPTER REVIEW

ACROSS

2. estimate of accuracy
4. precision
7. variance between two variables
8. treated as a fixed systematic error
10. (?) difference
11. variable related functionally to measurands
12. bias

DOWN

1. error when up-scale and down-scale calibrations differ
3. smallest, physically indicated division of an instrument
5. not measurement uncertainty
6. finite representation of a population
8. degrees of (?)
9. change in an output value over time

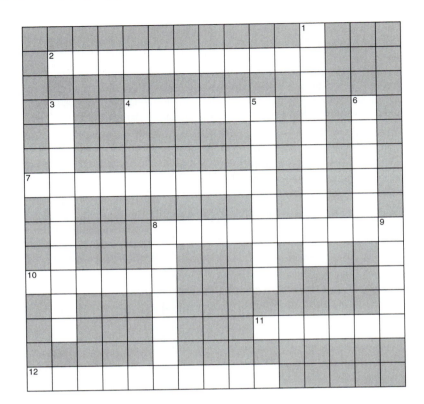

The Theodolite

The theodolite was used in the 19th century to help map the world. Distances of miles could be measured to within a fraction of an inch using this device and triangulation methods. This particular two-foot geodetic theodolite was made by Troughton and Simms and first used in 1828 for checking the Lough Foyle base in Northern Ireland. It also was used in conjunction with the Ramsden 3-ft geodetic theodolite for many years in preparation of the Primary Triangulation of Great Britian, the network of reference points that underpin the Ordinance Survey maps. It was the measurement of a quadrant of the arc length of the meridian passing from the North Pole to the Equator through Paris that lead Legendre to develop the method of least squares.

REGRESSION AND CORRELATION

Of all the principles that can be proposed for this purpose, I think there is none more general, more exact, or easier to apply, than that which we have used in this work; it consists of making the sum of the squares of the errors a minimum. By this method, a kind of equilibrium is established among the errors which, since it prevents the extremes from dominating, is appropriate for revealing the state of the system which most nearly approaches the truth.

Adrien-Marie Legendre. 1805.
Nouvelles méthodes pour la détermination des orbites des comètes. Paris.

Two variable organs are said to be co-related when the variation of the one is accompanied on the average by more or less variation of the other, and in the same direction.

Sir Francis Galton. 1888.
Proceedings of the Royal Society of London. 45: 135–145.

CHAPTER OUTLINE

10.1   CHAPTER OVERVIEW

This chapter introduces two important areas of data analysis: *regression* and *correlation*. Regression analysis establishes a mathematical relation between two or more variables. Typically, it is used to obtain the best fit of data with an analytical expression. Correlation analysis quantifies the extent to which one variable is related to another, but it does not establish a mathematical relation between them. Statistical methods can be used to determine the confidence levels associated with regression and correlation estimates.

We begin this chapter by considering the least-squares approach to regression analysis. This approach enables us to obtain a best-fit relation between variables. We focus on linear regression analysis first, then curvilinear and higher-order regression analysis. The statistical parameters that are used to characterize regression are introduced next. Then we consider regression analysis as applied to experiments along with their associated uncertainties and confidence limits. Finally we examine correlation analysis by considering how two discrete random variables are correlated with one another. We study how a continuous random variable is correlated with itself and with another continuous random variable.

10.2   LEARNING OBJECTIVES

You should be able to do the following after completing this chapter:

- Describe the concept and mathematical approach of least-squares regression analysis
- Perform a linear least-squares regression analysis on discrete data
- Differentiate the various regression parameters
- Determine the standard error of the fit of a linear fit of discrete data
- Know the common cases that occur when fitting data having uncertainty
- Determine the confidence intervals for the sample mean value, a new y-value, the precision interval, and an x-value for a given y-value
- Describe the concept of the correlation coefficient and the coefficient of determination
- Determine the percent chance of no correlation given the linear correlation coefficient and the number of discrete data
- Calculate the auto- and cross-correlation functions and coefficients for a specified periodic signal

10.3 LEAST-SQUARES APPROACH

Toward the end of the 18th century scientists faced an interesting problem. This was how to find the best agreement between measurements and an analytical model that contained the measured variables, given that repeated measurements were made, but with each containing error. Jean-Baptiste-Joseph Delambre (1749–1822) and Pierre-François-André Méchain (1744–1804) of France, for example [1][2], were in the process of measuring a $10°$ arc length of the meridian quadrant passing from the North Pole to the Equator through Paris. The measure of length for their newly proposed *Le Système International d'Unités*, the meter, would be defined as 1/10 000 000 the length of the meridian quadrant. So, the measured length of this quadrant had to be as accurate as possible.

Because it was not possible to measure the entire length of the $10°$ arc, measurements were made in arc lengths of approximately 65 000 modules (1 module $\cong$ 12.78 ft). From these measurements, an analytical expression involving the arc length and the astronomically determined latitudes of each of the arc's endpoints, the length of the meridian quadrant was determined. The solution essentially involved solving four equations containing four measured arc lengths with their associated errors for two unknowns, the ellipticity of the earth and a factor related to the diameter of the earth. Although many scientists proposed different solution methods, it was Adrien-Marie Legendre (1752–1833), a French mathematician, who arrived at the most accurate determination of the meter using the method of least squares, equal to 0.256 480 modules ($\sim$ 3.280 ft). Ironically, it was the more politically astute Pierre-Simon Laplace's (1749–1827) value of 0.256 537 modules ($\sim$ 3.281 ft) based on a less-accurate method that was adopted as the basis for the meter. Current geodetic measurements show that the quadrant from the North Pole to the Equator through Paris is 10 002 286 m long, rendering the meter as originally defined to be in error by 0.2 mm or 0.02 %.

Legendre's method of least squares, which originally appeared as a four-page appendix in a technical paper on comet orbits, was more far-reaching than simply determining the length of the meridian quadrant. It prescribed the methodology that would be used by countless scientists and engineers to this day. His method was elegant and straightforward; simply to express the errors as the squares of the differences between all measured and predicted values and then determine the values of the coefficients in the governing equation that minimize these errors. To quote Legendre [1], "we are led to a system of equations of the form

$$E = a + bx + cy + fz + \cdots, \qquad \textbf{[10.1]}$$

in which $a, b, c, f, \ldots$ are known coefficients, varying from one equation to the other, and $x, y, z, \ldots$ are unknown quantities, to be determined by the condition that each value of E is reduced either to zero, or to a very small quantity."

In the present notation, for a linear system

$$e_i = a + bx_i + cy_i = y_{c_i} - y_i, \qquad \textbf{[10.2]}$$

where e_i is the ith error for each of i equations based on the measurement pair $[x_i, y_i]$ and the general analytical expression $y_{c_i} = a + bx_i$ with $c = -1$. Using Legendre's method, the minimum of the sum of the squares of the e_i's would be found by varying the values of coefficients a and b. Formally, these coefficients are known as **regression coefficients** and the process of obtaining their values is called **regression analysis**.

10.4   LEAST-SQUARES REGRESSION ANALYSIS

Least-squares regression analysis follows a very logical approach in which the coefficients of an analytical expression that best fits the data are found through the process of error minimization. The best fit occurs when the sum of the squares of the differences (the errors or residuals) between each y_{c_i} value calculated from the analytical expression and its corresponding measured y_i value is a minimum (the differences are squared to avoid adding compensating negative and positive differences). The best fit would be obtained by continually changing the coefficients (a_0 through a_m) in the analytical expression until the differences are minimized. This, however, can be quite tedious unless a formal approach is taken and some simplifying assumptions are made.

Consider the data presented in Figure 10.1. The goal is to find the values of the a coefficients in the analytical expression $y_c = a_0 + a_1 x + a_2 x^2 + \cdots + a_m x^m$ that

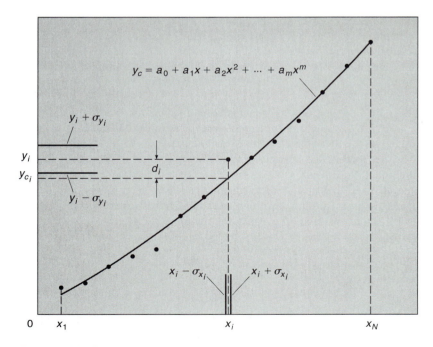

Figure 10.1   Least-squares regression analysis.

best fit the data. To proceed formally, D is defined as the sum of the squares of all the vertical distances (the d_i's) between the measured and calculated values of y (between y_i and y_{c_i}), as

$$D = \sum_{i=1}^{N} d_i^2 = \sum_{i=1}^{N}(y_i - y_{c_i})^2 = \sum_{i=1}^{N}[y_i - (a_0 + a_1 x_i + \cdots + a_m x_i^m)]^2. \quad \textbf{[10.3]}$$

Implicitly, it is assumed in this process that y_i is normally distributed with a true mean value of y_i' and a true variance of $\sigma_{y_i}^2$. The independent variable x_i is assumed to have no or negligible variance. Thus, $x_i = x'$, where x' denotes the true mean value of x. Essentially, the value of x is fixed, known, and with no variance, and the value of y is sampled from a normally distributed population. Thus, all of the uncertainty results from the y-value. If this were not the case, then the y_{c_i} value corresponding to a particular y_i value would not be vertically above or below it. This is because the x_i value would fall within a range of values. Consequently, the distances would not be vertical but rather at some angle with respect to the ordinate axis. Hence, the regression analysis approach being developed would be invalid.

Now D is to be minimized. That is, the value of the sum of the *squares* of the distances is to be the *least* of all possible values. This minimum is found by setting the total derivative of D equal to zero. This actually is a minimization of χ^2 (see [3]). Thus,

$$dD = 0 = \frac{\partial D}{\partial a_0} \, da_0 + \frac{\partial D}{\partial a_1} \, da_1 + \cdots + \frac{\partial D}{\partial a_m} \, da_m. \quad \textbf{[10.4]}$$

For this equation to be satisfied, the following set of $m + 1$ equations must be solved for $m + 1$ unknowns:

$$\frac{\partial D}{\partial a_0} = 0 = \frac{\partial}{\partial a_0} \sum_{i=1}^{N} d_i^2,$$

$$\frac{\partial D}{\partial a_1} = 0 = \frac{\partial}{\partial a_1} \sum_{i=1}^{N} d_i^2,$$

$$\vdots \qquad\qquad\qquad \textbf{[10.5]}$$

$$\frac{\partial D}{\partial a_m} = 0 = \frac{\partial}{\partial a_m} \sum_{i=1}^{N} d_i^2.$$

This set of equations leads to what are called the **normal equations** (named by Carl Friedrich Gauss).

10.5 LINEAR ANALYSIS

The most simple type of least-squares regression analysis that can be performed is for the linear case. Assume that y is linearly related to x by the expression $y_c = a_0 + a_1 x$.

10.1 | **MATLAB SIDEBAR**

The MATLAB command p = polyfit(x,y,m) uses least-squares regression analysis on equal-length vectors of [x,y] data to find the coefficients, a_0 through a_m, of the polynomial $p(x) = a_m x^m + a_{m-1} x^{m-1} + \cdots + a_1 x + a_0$ and places them in a row vector p of descending order. The polynomial p is evaluated at $x = x*$ using the MATLAB command polyval(p,x*). These commands, when used in conjunction with MATLAB's plot command, allow one to plot the data along with the regression fit, as was done to construct Figure 10.2.

Proceeding along the same lines, for this case only two equations (here $m + 1 = 1 + 1 = 2$) must be solved for two unknowns, a_0 and a_1, subject to the constraint that D is minimized.

When $dD = 0$,

$$\frac{\partial D}{\partial a_0} = 0 = \frac{\partial}{\partial a_0}\left\{\sum_{i=1}^{N}[y_i - (a_0 + a_1 x_i)]^2\right\} = -2\sum_{i=1}^{N}(y_i - a_0 - a_1 x_i). \quad \textbf{[10.6]}$$

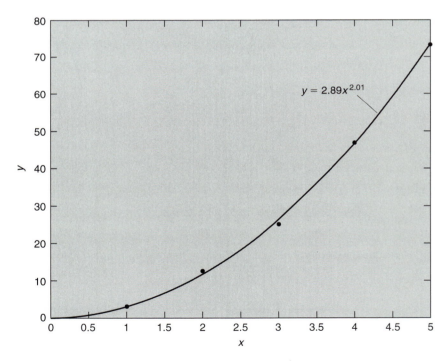

Figure 10.2   Regression fit of the model $y = ax^b$ with data.

Carrying through the summations on the right-hand side of Equation 10.6 yields

$$\sum_{i=1}^{N} y_i = a_0 N + a_1 \sum_{i=1}^{N} x_i. \qquad \textbf{[10.7]}$$

Also,

$$\frac{\partial D}{\partial a_1} = 0 = \frac{\partial}{\partial a_1} \left\{ \sum_{i=1}^{N} [y_i - (a_0 + a_1 x_i)]^2 \right\} = -2 \sum_{i=1}^{N} x_i (y_i - a_0 - a_1 x_i).$$

$$\textbf{[10.8]}$$

This gives

$$\sum_{i=1}^{N} x_i y_i = a_0 \sum_{i=1}^{N} x_i + a_1 \sum_{i=1}^{N} x_i^2. \qquad \textbf{[10.9]}$$

Thus, the two normal equations become Equations 10.7 and 10.9. These can be rewritten as

$$\overline{y} = a_0 + a_1 \overline{x} \qquad \textbf{[10.10]}$$

and

$$\overline{xy} = a_0 \overline{x} + a_1 \overline{x^2}. \qquad \textbf{[10.11]}$$

From the first normal equation it can be deduced that a linear least-squares regression analysis fit will *always* pass through the point $(\overline{x}, \overline{y})$. Equations 10.7 and 10.9 can be solved for a_0 and a_1 to yield

$$a_0 = \frac{\sum_{i=1}^{N} x_i^2 \sum_{i=1}^{N} y_i - \sum_{i=1}^{N} x_i \sum_{i=1}^{N} x_i y_i}{\Delta}, \qquad \textbf{[10.12]}$$

$$a_1 = \frac{N \sum_{i=1}^{N} x_i y_i - \sum_{i=1}^{N} x_i \sum_{i=1}^{N} y_i}{\Delta}, \qquad \textbf{[10.13]}$$

$$\Delta = N \sum_{i=1}^{N} x_i^2 - \left(\sum_{i=1}^{N} x_i \right)^2. \qquad \textbf{[10.14]}$$

Linear regression analysis also can be used for a higher-order expression if the variables in expression can be transformed to yield a linear expression. This sometimes is referred to as curvilinear regression analysis. Such variables are known as **intrinsically linear variables**. For this case, a least-squares linear regression analysis is performed on the transformed variables. Then the resulting regression coefficients are transformed back to yield the desired higher-order fit expression. For example, if $y = ax^b$, then $\log_{10}(y) = \log_{10}(a) + b \cdot \log_{10}(x)$. So, the least-squares linear regression fit of the data pairs $[\log_{10}(x), \log_{10}(y)]$ will yield a line of intercept $\log_{10}(a)$ and slope b. The resulting best-fit values of a and b can be determined and then used in the original expression.

Example 10.1

STATEMENT

An experiment is conducted to validate a physical model of the form $y = ax^b$. Five $[x, y]$ pairs of data are acquired: [1.00, 2.80; 2.00, 12.5; 3.00, 25.2; 4.00, 47.0; 5.00 73.0]. Find the regression coefficients a and b using a *linear* least-squares regression analysis.

SOLUTION

First express the data in the form of $[\log_{10}(x), \log_{10}(y)]$ pairs. This yields the transformed data pairs [0.000, 0.447; 0.301, 1.10; 0.477, 1.40; 0.602, 1.67; 0.699, 1.86]. A linear regression analysis of the transformed data yields the best-fit expression: $\log_{10}(y) = 0.461 + 2.01 \cdot \log_{10}(x)$. This implies that $a = 2.89$ and $b = 2.01$. Thus, the best-fit expression for the data in its original form is $y = 2.89x^{2.01}$. This best-fit expression is compared with the original data in Figure 10.2.

A similar approach can be taken when using a linear least-squares regression analysis to fit the equation $E^2 = A + B\sqrt{U}$, which is King's law. This law relates the voltage, E, of a constant-temperature anemometer to a fluid's velocity, U. A regression analysis performed on the data pairs $[E^2, \sqrt{U}]$ will yield the best-fit values for A and B. This is considered in homework problem 7.

10.6 HIGHER-ORDER ANALYSIS

Higher-order ($m > 2$) regression analysis can be performed in a manner similar to that developed for linear least-squares regression analysis. This will result in $m + 1$ algebraic normal equations. These can be solved most easily using methods of linear algebra to obtain the expressions for the $m + 1$ regression coefficients.

For higher-order regression analysis, the coefficients a_0 through a_m in the expression

$$a_0 + a_1 x_i + a_2 x_i^2 + \cdots + a_m x_i^m = y_{c_i} \qquad \textbf{[10.15]}$$

are found using the method of minimizing D as described in the previous section. The resulting equations are

$$a_0 N + a_1 \sum_{i=1}^{N} x_i + a_2 \sum_{i=1}^{N} x_i^2 + \cdots + a_m \sum_{i=1}^{N} x_i^m = \sum_{i=1}^{N} y_i$$

$$a_0 \sum_{i=1}^{N} x_i + a_1 \sum_{i=1}^{N} x_i^2 + a_2 \sum_{i=1}^{N} x_i^3 + \cdots + a_m \sum_{i=1}^{N} x_i^{m+1} = \sum_{i=1}^{N} x_i y_i$$

$$a_0 \sum_{i=1}^{N} x_i^2 + a_1 \sum_{i=1}^{N} x_i^3 + a_2 \sum_{i=1}^{N} x_i^4 + \cdots + a_m \sum_{i=1}^{N} x_i^{m+2} = \sum_{i=1}^{N} x_i^2 y_i \quad \textbf{[10.16]}$$

$$\vdots$$

$$a_0 \sum_{i=1}^{N} x_i^m + a_1 \sum_{i=1}^{N} x_i^{m+1} + a_2 \sum_{i=1}^{N} x_i^{m+2} + \cdots + a_m \sum_{i=1}^{N} x_i^{2m} = \sum_{i=1}^{N} x_i^m y_i.$$

In expanded matrix notation, the set of Equations 10.16 becomes

$$
\begin{bmatrix}
N & \sum x_i & \sum x_i^2 & \cdots & \sum x_i^m \\
\sum x_i & \sum x_i^2 & \sum x_i^3 & \cdots & \sum x_i^{m+1} \\
\sum x_i^2 & \sum x_i^3 & \sum x_i^4 & \cdots & \sum x_i^{m+2} \\
\vdots & \vdots & \vdots & \ddots & \vdots \\
\sum x_i^m & \sum x_i^{m+1} & \sum x_i^{m+2} & \cdots & \sum x_i^{2m}
\end{bmatrix}
\begin{bmatrix}
a_0 \\ a_1 \\ a_2 \\ \vdots \\ a_m
\end{bmatrix}
=
\begin{bmatrix}
\sum y_i \\ \sum x_i y_i \\ \sum x_i^2 y_i \\ \vdots \\ \sum x_i^m y_i
\end{bmatrix},
$$

where the summations are from $i = 1$ to N. Or, in matrix notation, this becomes

$$[E][a] = [F]. \qquad \textbf{[10.17]}$$

$[E]$, $[a]$, and $[F]$ represent the matrices shown in the expanded form. The solution to Equation 10.17 for the regression coefficients is

$$[a] = [E]^{-1}[F]. \qquad \textbf{[10.18]}$$

$[E]^{-1}$ is the inverse of the coefficient matrix.

STATEMENT | **Example 10.2**

An experiment is conducted in which a ball is dropped with an initial downward velocity, v_0, from a tall tower. The distance fallen, $y(m)$, is recorded at time, t, intervals of one second for a period of six seconds. The resulting time, distance data are [0, 0; 1, 7; 2, 21; 3, 48; 4, 81; 5, 131; 6, 185]. The distance equation is $y = y_0 + v_0 t + 0.5gt^2$, where g is the local gravitational acceleration. This equation is of the form $y = a_0 + a_1 t + a_2 t^2$. Using higher-order regression

---

10.2 | MATLAB SIDEBAR

Usually, when applying regression analysis to data, the number of data points exceeds the number of regression coefficients, as $N > m + 1$. This results in more independent equations than there are unknowns. The system of equations is *overdetermined*. Because the resulting matrix is not square, standard matrix inversion methods and Cramer's method will not work. An exact solution may or may not exist. Fortunately, MATLAB can solve this set of equations implicitly, yielding either the exact solution or a least-squares solution. The solution is accomplished by simply using MATLAB's matrix division operator (\). This operator uses Gaussian elimination as opposed to forming the inverse of the matrix. This is more efficient and has greater numerically accuracy. For N $[x_i, y_i]$ data pairs there will be N equations to evaluate, which in expanded matrix notation are

$$
\begin{bmatrix}
1 & x_1 & x_1^2 & \cdots & x_1^m \\
1 & x_2 & x_2^2 & \cdots & x_2^m \\
\vdots & \vdots & \vdots & \ddots & \vdots \\
1 & x_N & x_N^2 & \cdots & x_N^m
\end{bmatrix}
\begin{bmatrix}
a_0 \\ a_1 \\ a_2 \\ \vdots \\ a_m
\end{bmatrix}
=
\begin{bmatrix}
y_1 \\ y_2 \\ y_3 \\ \vdots \\ y_N
\end{bmatrix}.
$$

In matrix notation this becomes $[X][a] = [Y]$, where $[X]$ is an $(N \times m + 1)$ matrix, $[a]$ is an $(m + 1 \times 1)$ matrix, and $[Y]$ is an $(N \times 1)$ matrix. The solution to this matrix equation is achieved using the command $X \backslash Y$ after the matrices $[X]$ and $[Y]$ have been entered.

analysis, determine the values of the regression coefficients a_0, a_1, and a_2. From their values, determine the values of v_0 and g. Finally, plot the data with the regression fit.

SOLUTION

The solution is obtained using MATLAB's left-division method by typing $t \backslash y$, where $[t]$ is

$$\begin{bmatrix} 1 & 0 & 0 \\ 1 & 1 & 1 \\ 1 & 2 & 4 \\ 1 & 3 & 9 \\ 1 & 4 & 16 \\ 1 & 5 & 25 \\ 1 & 6 & 36 \end{bmatrix}$$

and $[y]$ is

$$\begin{bmatrix} 0 \\ 7 \\ 21 \\ 48 \\ 81 \\ 131 \\ 185 \end{bmatrix}.$$

The resulting regression coefficient matrix is

$$\begin{bmatrix} 0.5238 \\ 0.3214 \\ 5.0833 \end{bmatrix}.$$

Thus, the best-fit expression is

$$0.5238 + 0.3214t + 5.0833t^2 = y$$

The data and the above equation are shown in Figure 10.3. Also, $v_0 = 0.3214$ and $g = 10.1666$.

10.7   REGRESSION PARAMETERS

There are several statistical parameters that can be calculated from a set of data and its best-fit relation. Each of these parameters quantifies a different relationship between the quantities found from the data (the individual values x_i and y_i and the mean values $\overline{x}$ and $\overline{y}$) and from its best-fit relation (the calculated values).

Those quantities that are calculated directly from the data include the sum of the squares of x, S_{xx}, the sum of the squares of y, S_{yy}, and the sum of the product of x and y, S_{xy}. Their expressions are

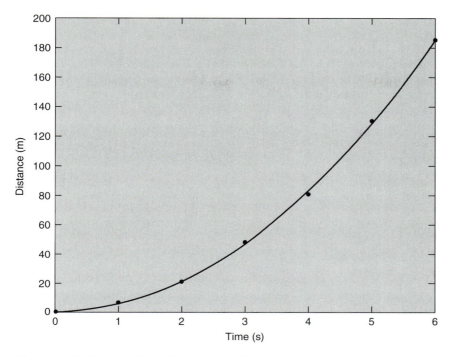

Figure 10.3   Higher-order regression fit example.

$$S_{xx} \equiv \sum_{i=1}^{N}(x_i - \overline{x})^2 = \sum_{i=1}^{N}x_i^2 - N\overline{x}^2, \qquad [\mathbf{10.19}]$$

$$S_{yy} \equiv \sum_{i=1}^{N}(y_i - \overline{y})^2 = \sum_{i=1}^{N}y_i^2 - N\overline{y}^2, \qquad [\mathbf{10.20}]$$

and

$$S_{xy} \equiv \sum_{i=1}^{N}(x_i - \overline{x})(y_i - \overline{y}) = \sum_{i=1}^{N}x_i y_i - N\overline{x} \cdot \overline{y}. \qquad [\mathbf{10.21}]$$

All three of these quantities can be viewed as measures of the square of the differences or product of the differences between the x_i and y_i values and their corresponding mean values. Equations 10.19 and 10.21 can be used with the normal equations of a linear least-squares regression analysis to simplify the expressions for the linear case's best-fit slope and intercept, where

$$b = \frac{S_{xy}}{S_{xx}} \qquad [\mathbf{10.22}]$$

and

$$a = \bar{y} - b\bar{x}. \tag{10.23}$$

Those quantities calculated from the data and the regression fit include the sum of the squares of the regression, SSR, the sum of the squares of the error, SSE, and the sum of the squares of the total error, SST. Their expressions are

$$\text{SSR} \equiv \sum_{i=1}^{N} (y_{c_i} - \bar{y})^2, \tag{10.24}$$

$$\text{SSE} \equiv \sum_{i=1}^{N} (y_i - y_{c_i})^2, \tag{10.25}$$

and

$$\text{SST} \equiv \text{SSE} + \text{SSR} = \sum_{i=1}^{N} (y_i - y_{c_i})^2 + \sum_{i=1}^{N} (y_{c_i} - \bar{y})^2. \tag{10.26}$$

All three of these can be viewed as quantitative measures of the square of the differences between the $\bar{y}$ and y_i values and their corresponding y_{c_i} values. SSR is also known as the *explained* variation and SSE as the *unexplained* variation. Their sum, SST, is called the *total* variation. SSR is a measure of the amount of variability in y_i accounted for by the regression line and SSE of the remaining amount of variation not explained by the regression line.

It can be shown further (see [5]) that

$$\text{SST} = \sum_{i=1}^{N} (y_i - \bar{y})^2 = S_{yy}. \tag{10.27}$$

The combination of Equations 10.26 and 10.27 yields what is known as the sum of squares partition [5] or the analysis of variance identity [6]:

$$\sum_{i=1}^{N} (y_i - \bar{y})^2 = \sum_{i=1}^{N} (y_i - y_{c_i})^2 + \sum_{i=1}^{N} (y_{c_i} - \bar{y})^2. \tag{10.28}$$

This expresses the three quantities of interest (y_i, y_{c_i}, and $\bar{y}$) in one equation.

An additional parameter that characterizes the quality of the best-fit is the standard error of the fit, S_{yx},

$$S_{yx} \equiv \sqrt{\frac{\text{SSE}}{v}} = \sqrt{\frac{\text{SSE}}{N-2}} = \sqrt{\frac{\sum_{i=1}^{N} (y_i - y_{c_i})^2}{N-2}}. \tag{10.29}$$

This is equivalent to the standard deviation of the measured y_i values with respect to their respective calculated y_{c_i} values, where $v = N - (m + 1) = N - 2$ for $m = 1$.

Example 10.3

STATEMENT

For the set of $[x, y]$ data pairs [0.5, 0.6; 1.5, 1.6; 2.5, 2.3; 3.5, 3.7; 4.5, 4.2; 5.5, 5.4], determine $\bar{x}, \bar{y}, S_{xx}, S_{yy}$, and S_{xy}. Then determine the intercept and the slope of the regression line using Equations 10.22 and 10.23 and compare the values to those found by performing a linear least-squares regression analysis. Next, using the regression fit equation determine the values of y_{c_i}. Finally calculate SSE, SSR, and SST. Show, using the results of these calculations, that SST = SSR + SSE.

SOLUTION

Direct calculations yield $\bar{x} = 3.00$, $\bar{y} = 2.97$, $S_{xx} = 17.50$, $S_{yy} = 15.89$, and $S_{xy} = 16.60$. The intercept and the slope values are $a = 0.1210$ and $b = 0.9486$ from Equations 10.23 and 10.22, respectively. The same values are found from regression analysis. Thus, from the equation $y_{c_i} = 0.1210 + 0.9486x_i$, the y_{c_i} values are 0.5952, 1.5438, 2.4924, 3.4410, 4.3895, and 5.3381. Direct calculations then give SSE = 0.1470, SSR = 15.7463, and SST = 15.8933. This shows that SSR + SSE = 15.7463 + 0.1470 = 15.8933 = SST, which follows from Equation 10.28.

Historically, regression originally was called *reversion*. Reversion referred to the tendency of a variable to revert to the average of the population from which it came. It was Francis Galton who first elucidated the property of reversion [14] by demonstrating how certain characteristics of a progeny revert to the population average more than to the parents. So, in general terms, regression analysis relates variables to their mean quantities.

10.8 UNCERTAINTY FROM MEASUREMENT ERROR

One of the major contributors to the differences between the measured and calculated y-values in a regression analysis is measurement error. This can be understood best by examining the linear case.

For an error-free experiment in which the data pairs $[x_i, y_i]$ are linearly related, the best-fit relation would be

$$y_i' = \alpha + \beta x_i', \qquad \text{[10.30]}$$

in which α and β are the true intercept and slope, respectively, and y_i' is the true mean value of y_i associated with the true mean value of x_i, x_i'. For an experiment in which measurement errors are present, one can write

$$x_i = x_i' + \epsilon_x \qquad \text{[10.31]}$$

and

$$y_i = y_i' + \epsilon_y, \qquad \text{[10.32]}$$

where x_i and y_i denote the actual, measured values and ϵ_x and ϵ_y are their *measurement* errors. Here, it is assumed that the value of all of the x_i errors is the same and equal to ϵ_x, and the value of all of the y_i errors is the same and equal to ϵ_y. That is, the x_i

and y_i errors are independent of the particular data pair. This is true if each of the y_i measurements result from an independent measurement situation. Using Equations 10.31 and 10.32, Equation 10.30 becomes

$$y_i = \alpha + \beta x_i + (\epsilon_y - \beta \epsilon_x) = y_{c_i} + E_y. \qquad \textbf{[10.33]}$$

The terms in parentheses represent the error term for y_i, which is denoted by E_y. Thus, the value of y_{c_i} will have an error of E_y with respect to its measured value, y_i. This error results from possible measurement errors in x and y or both.

This error is characterized best through its variance, σ_E^2. A subtle yet important point is that the variance of x_i is the same as that of ϵ_x and that the variance of y_i is the same as that of ϵ_y. This is because both x_i' and y_i' have no error. Thus, the variance in x_i is characterized by the variance in its error. This also is true for y_i. These variances are denoted by σ_x^2 and σ_y^2. If ϵ_y and $\beta \epsilon_x$ are statistically independent, then the variance of the combined errors, σ_{E_y}, is given by [4]

$$\sigma_{E_y}^2 = \sigma_y^2 + \beta^2 \sigma_x^2. \qquad \textbf{[10.34]}$$

This equation is valid only when either $\epsilon_x = 0$ or x is controlled such that its randomness is constrained. If either of these conditions are not met, then $\sigma_{E_y}^2$ cannot be subdivided into these two components. Then, the individual contributions of the ϵ_x and ϵ_y due to the difference between the measured and calculated value of y cannot be ascertained.

So, measurement errors lead to variances in x and y. These variances contribute to the combined variance, σ_{E_y}. It is σ_{E_y} that contributes to the differences between the y_i and y_{c_i} values.

10.9   DETERMINING THE APPROPRIATE FIT

Even determining the linear best-fit for a set of data and its associated precision can be more involved than it first appears. How to determine a linear best-fit of data already has been discussed. Here, implicitly it was assumed that the measurement uncertainties in x were negligible with respect to those in y and that the assumed mathematical expression was the most appropriate one to model the data. However, many common situations involving regression usually are more complicated. Examine the various cases that can occur when fitting data having uncertainty with a least-squares regression analysis.

There are six cases to consider, as listed in Table 10.1. Each assumes a level of measurement uncertainty in x, u_x, and in y, u_y, and whether or not the order of the regression is correct. The term *correct* implies that the underlying physical model that governs the relationship between x and y has the same order as the fit. The last two cases (5 and 6), in which both x and y have comparable uncertainties ($u_x \sim u_y$), are more difficult to analyze. Often, only special situations of these two cases are considered [10]. Each of the six cases is now discussed in more detail.

Table 10.1 Cases involving uncertainties and the type of fit.

Case	u_x	u_y	Fit
1	0	0	correct
2	0	0	incorrect
3	0	$\neq 0$	correct
4	0	$\neq 0$	incorrect
5	$\neq 0$	$\neq 0$	correct
6	$\neq 0$	$\neq 0$	incorrect

- Case 1: This corresponds to the ideal case in which there are no uncertainties in x and y ($u_x = u_y = 0$) and the order of the fit is the same as that of the underlying physical model (a correct fit). For example, consider a vertically oriented, linear spring with a weight, W, attached to its end. The spring will extend downward from its unloaded equilibrium position a distance x proportional to W, as given by Hooke's law, $W = -kx$, where k is the spring constant and negative x corresponds to positive displacement (extension). Assuming that the experiment is performed without error, a first-order (linear) regression analysis would yield a perfect fit of the data with an intercept equal to zero and a slope equal to $-k$. Because there are no measurement errors in either x or y, the values of the intercept and slope will be true values, even if the data set is finite.

- Case 2: This case involves an error-free experiment in which the data is fit with an incorrect order. For example, continuing with the spring-weight example, the work done by the weight to extend the spring, $W \cdot x$, could be plotted versus its displacement. This work equals the stored energy of the spring, E, which equals $0.5kx^2$. A linear regression fit of $W \cdot x$ versus x would result in a fit that does not correspond to the correct underlying physical model, as shown in Figure 10.4. A second-order fit would be appropriate because $E \sim x^2$. The resulting differences between the data and the linear fit come solely from the incorrect choice of the fit. These differences, however, easily could be misinterpreted as the result of errors in the experiment, as is the case for the data shown in Figure 10.4. Obviously, it is important to have a good understanding of most appropriate order of the fit *before* the regression analysis is performed.

10.3 | MATLAB SIDEBAR

The MATLAB M-file `plotfit.m` performs an mth order least-squares regression analysis on a set of $[x, y, ey]$ data pairs (where ey is the measurement error) and plots the regression fit and the P % confidence intervals for the y_i estimate. It also calculates S_{yx} and plots the data with its error bars.

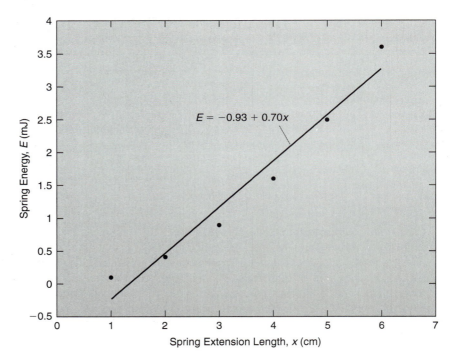

Figure 10.4   Example of case 2.

- Case 3: For this case there is uncertainty in y but not in x and the correct order of the fit is used. This is the type of situation encountered when regression analysis first was considered. The resulting differences between the measured and calculated y-values result from the measurement uncertainties in y. Consequently, a correct regression fit will agree with the data to within its measurement uncertainty.

    When the correct physical model is not known *a priori*, the standard approach is to increase the order of the fit within reason until an acceptable fit is obtained. What is acceptable is somewhat arbitrary. Ideally, all data points inclusive of their uncertainties should agree with the fit to within the confidence intervals specified by Equation 10.44. Although an nth order polynomial will fit $n - 1$ data points exactly, this usually does not correspond to a physically realizable model. Very seldom does a physical law involve more than a fourth power of a variable. In fact, high-degree polynomial fits characteristically exhibit large excursions *between* data points and have coefficients that require many significant figures for repeatable accuracy [13]. So, caution should be exercised when using higher-order fits. Whenever possible, the order of the fit should correspond to the order of the physical model.

- Case 4: This case considers the situation in which there is uncertainty in y but not in x and an incorrect correct order of the fit is used. Two uncertainties in the calculated y-values result in relation to the true fit. One is from the measurement uncertainty in y and the other is from the use of an incorrect model. Here it is difficult to determine directly the contribution of each uncertainty to the overall uncertainty. A systematic study involving either more accurate measurements of y or the use of a different model would be necessary to determine this.

Finally, there are two other cases that arise in which there is uncertainty in both x and y. The presence of both of these uncertainties leads to a best fit that is different from that when there is only uncertainty in y. This is illustrated in Figure 10.5 in which two regression fits are plotted for the same data. The dashed line represents the fit that considers only the uncertainty in y that was established using a linear least-squares regression analysis. The solid line is the fit that considers uncertainty in both x and y that was established using Deming's method (see [9]), which is considered in the following case. It is easy to see that when uncertainty is present in both x and y, a fit established using the linear least-squares regression analysis that does not consider the uncertainty in x will *not* yield the best fit.

Whenever $u_x \sim u_y$ and no further constraints are placed on them, more-involved regression techniques must be used to determine the best fit of the data (for example, see [12]). This topic is beyond the scope of this text. However, Mandel [10] has examined two special and practical situations in which uncertainty is present in x and linear regression analysis can be applied. These will be examined now.

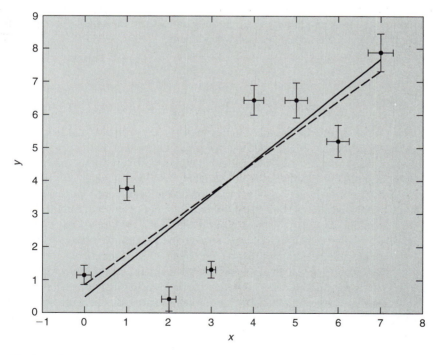

Figure 10.5   Two regression fits of the same data.

- Case 5: The general situation for this case involves uncertainties in both x and y and a correct order of the fit.

  For the first special situation in which the ratio of the variances of the x and y errors, $\lambda = \sigma_x^2/\sigma_y^2$, is known *a priori*, a linear best-fit equation can be determined using Deming's method of minimizing the *weighted* sum of squares of x and y. Further, estimates of the variances of the x and y can be obtained.

  The slope of the regression line calculated by this method is

$$b = \frac{\lambda S_{yy} - S_{xx} + \sqrt{(S_{xx} - \lambda S_{yy})^2 + 4\lambda S_{xy}^2}}{2\lambda S_{xy}} \qquad \textbf{[10.35]}$$

and the intercept is given by the normal Equation 10.23.

The estimates for the variances of the x and y errors are, respectively,

$$\widetilde{S}_x^2 = \left(\frac{\lambda}{1 + \lambda b^2}\right) \frac{S_{yy} - 2b S_{xy} + b^2 S_{xx}}{N - 2} \qquad \textbf{[10.36]}$$

and

$$\widetilde{S}_y^2 = \left(\frac{1}{1 + \lambda b^2}\right) \frac{S_{yy} - 2b S_{xy} + b^2 S_{xx}}{N - 2}. \qquad \textbf{[10.37]}$$

Note that Equations 10.36 and 10.37 differ only by the factor λ. These equations can be used to estimate the final uncertainties in x and y for P % confidence. These are the uncertainties in estimating x and y from the fit (as opposed to the measurement uncertainties in x and y). They are

$$u_{x_{\text{final}}} = t_{N-2,P} \widetilde{S}_x \qquad \textbf{[10.38]}$$

and

$$u_{y_{\text{final}}} = t_{N-2,P} \widetilde{S}_y. \qquad \textbf{[10.39]}$$

Using these equations, a regression fit can be plotted with data and its error bars, as shown in Figure 10.6, in addition to determining values of λ, $u_{x_{\text{final}}}$, and $u_{y_{\text{final}}}$. These values are 0.25, ± 4.0985, and ± 2.0492, respectively, for the data presented in the figure. The estimates for the variances of x and y are $\widetilde{S}_x^2 = 0.7014$ and $\widetilde{S}_y^2 = 2.8055$. The estimates of the final uncertainties in x and y appear relatively large at first sight. This is the result of the relatively large scatter in the data. So, for a specified value of x in this case, the value of y will be within ± 4.0985 of its best-fit value 95 % of the time. Likewise, for a specified value of y, the value of x will be within ± 2.0492 of its best-fit value 95 % of the time.

The second special situation considers when x is a controlled variable. This is known as the Berkson case, in which the value of x is set as close as possible to its desired value, thereby constraining its randomness. This corresponds, for example, to a static calibration in which there is some uncertainty in x but the value of x is specified for each calibration point. For this situation a standard linear least-squares regression fit of the data is valid. Further, estimates can be made for all of the uncertainties presented beforehand for case 3. The interpretation of

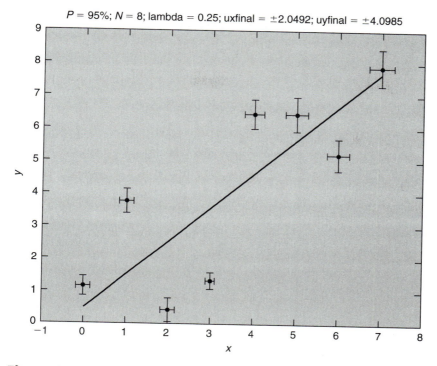

P = 95%; N = 8; lambda = 0.25; uxfinal = ±2.0492; uyfinal = ±4.0985

Figure 10.6 Example of case 5 when λ is known.

the uncertainties, however, is somewhat different [10]. The uncertainty in y with respect to the regression fit must be interpreted according to Equation 10.34.

- Case 6: This is the most complicated case in which there are uncertainties in both x and y and an incorrect order of the fit is used. The same analytical approaches can be taken here as were done for the special situations in case 5. However, the interpretation of the uncertainties is confounded further as a result of the additional uncertainty introduced by the incorrect order of the fit.

10.4 | MATLAB SIDEBAR

The MATLAB M-file calexey.m performs a linear regression analysis based on Deming's method on a set of [x, y, ex, ey] data pairs (where ex and ey are the measurement errors in the x and y, respectively) and plots the regression fit. It also plots the data with their measurement error bars and calculates the value of λ and the final estimates of the uncertainties in x and y, as given by Equations 10.38 and 10.39. This was used to generate Figure 10.6.

10.10   CONFIDENCE INTERVALS

There are additional uncertainties to consider that can contribute to differences between the measured and calculated y values. These arise from the finite acquisition of data in an experiment. The presence of these additional uncertainties affects the confidence associated with various estimates related to the fit. For example, in some situations, the inverse of the best-fit relation established through calibration is used to determine unknown values of the independent variable and its associated uncertainty. A typical example would be to determine the value and uncertainty of an unknown force from a voltage measurement using an established voltage-versus-force calibration curve. To arrive at such estimates, the sources of these additional uncertainties must be examined first.

For simplicity, focus on the situation where the correct order of the fit is assumed and there is no measurement error in x. Here, $\sigma_{E_y} = \sigma_y$. That is, the uncertainty in determining a value of y from the regression fit is solely due to the measurement error in y.

Consider the following situation, as illustrated in Figure 10.7, in which best-fits for two sets of data obtained under the same experimental conditions are plotted along with the data. Observe that different values of y_i are obtained for the same value of

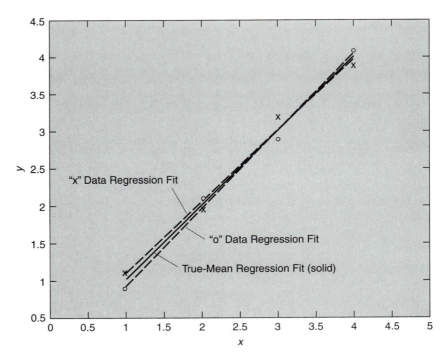

Figure 10.7   Linear regression fits for finite samples.

x_i each time the measurement is repeated (in this case there are two values of y_i for each x_i). This is because y is a random variable drawn from a normally distributed population. Because x is not a random variable, it is assumed to have no uncertainty. So, in all likelihood, the best-fit expression of the first set of data, $y = a_1 + b_1 x$, will be different from the second best-fit expression, $y = a_2 + b_2 x$, having different values for the intercepts ($a_1 \neq a_2$) and for the slopes ($b_1 \neq b_2$).

The true-mean regression line is given by Equation 10.30 in which $x = x'$. The true intercept and true slope values are those of the underlying population from which the finite samples are drawn. From another perspective, the true-mean regression line would be that found from the least-squares linear regression analysis of a very large set of data ($N \gg 1$).

Recognizing that such finite sampling uncertainties arise, how do they affect the estimates of the true intercept and true slope? The estimates for the true intercept and true slope values can be written in terms of the earlier expressions for S_{xx} and S_{yx} [5, 6]. The estimate of the true intercept of the true-mean regression line is

$$\alpha = a \pm t_{N-2,P} S_{yx} \sqrt{\frac{1}{N} + \frac{\bar{x}^2}{S_{xx}}}. \qquad \textbf{[10.40]}$$

The estimate of the true slope of the true-mean regression line is

$$\beta = b \pm t_{N-2,P} S_{yx} \sqrt{\frac{1}{S_{xx}}}. \qquad \textbf{[10.41]}$$

As N becomes larger, the sizes of the confidence intervals for the true intercept and true slope estimates become smaller. The value of a approaches that of α, and the value of b approaches that of β. This simply reflects the former statement, that any regression line based on several N will approach the true-mean regression line as N becomes large.

STATEMENT   **Example 10.4**

For the set of $[x, y]$ data pairs [1.0, 2.1; 2.0, 2.9; 3.0, 3.9; 4.0, 5.1; 5.0, 6.1] determine the linear best-fit relation using the method of least-squares regression analysis. Then estimate at 95 % confidence the values of the true intercept and the true slope.

SOLUTION

The best-fit relation is $y = 0.96 + 1.02x$ for $N = 5$ with $S_{yx} = 0.12$, $S_{xx} = 10$, $\bar{x} = 3$, and $t_{3,95} = 3.1824$. This yields $\alpha = 0.96 \pm 0.40$ (95 %) and $\beta = 1.02 \pm 0.12$ (95 %).

The values of some other useful quantities also can be estimated [5, 6, 7]. The estimate of the sample mean value of a large number of y_i values for a given value of x_i, denoted by $\bar{y}_i$ and also known as the mean response, is

$$\bar{y}_i = y_{c_i} \pm t_{N-2,P} S_{yx} \sqrt{\frac{1}{N} + \frac{(x_i - \bar{x})^2}{S_{xx}}}. \qquad \textbf{[10.42]}$$

Note that the greater the difference between x_i and $\bar{x}$, the greater the uncertainty in estimating $\bar{y}_i$. This leads to confidence intervals that are hyperbolic, as shown in Figure 10.8 by the curves labeled b (based on 95 % confidence) that are positioned above and below the regression line labeled a. The confidence interval is the smallest at $x = \bar{x}$. Also, because of the factor $t_{v,P}$, the confidence interval width will decrease with decreasing percent confidence.

The range within which a new y-value, y_n, added to the data set will be for a new value x_n is

$$y_n = y_{n_{c_i}} \pm t_{N-2,P} S_{yx} \sqrt{1 + \frac{1}{N} + \frac{(x_n - \bar{x})^2}{S_{xx}}}. \qquad \textbf{[10.43]}$$

This interval is marked by the hyperbolic curves labeled d (based on 95 % confidence). Note that the hyperbolic curves are farther from the regression line for this case than for the mean response case. This is because Equation 10.43 estimates a single new value of y, whereas Equation 10.42 estimates the mean of a large number of y values.

Finally, the range within which a y_i value probably will be, with respect to its corresponding y_{c_i} value, is

$$y_i = y_{c_i} \pm t_{v,P} S_{yx}, \qquad \textbf{[10.44]}$$

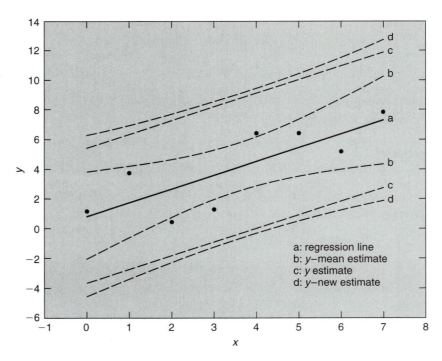

Figure 10.8   Various confidence intervals for linear regression estimates.

where $t_{v,P} S_{yx}$ denotes the precision interval. This expression establishes the confidence intervals that always should be plotted whenever a regression line is present. Basically, Equation 10.44 defines the limits within which P % of a large number of measured y_i values will be with respect to the y_{c_i} value for a given value of x_i. Its confidence intervals are denoted by the lines labeled c (for $P = 95$ %), which are parallel to the regression line. Equation 10.44 also can be used for a higher mth-order regression fit to establish the confidence intervals provided that v is determined by $v = N - (m + 1)$ and that the general expression for S_{yx} given in Equation 10.29 is used.

Several useful inferences can be drawn from Equation 10.44. For a fixed number of measurements, N, the extent of the precision interval increases as the percent confidence is increased. The extent of the precision interval must be greater if more confidence is required in the estimate of y_i. For a given confidence, as N is increased, the extent of the precision interval decreases. A smaller precision interval is required to estimate y_i if a greater number of measurements is acquired.

STATEMENT | **Example 10.5**

For the set of $[x, y]$ data pairs $[0.00, 1.15; 1.00, 3.76; 2.00, 0.41; 3.00, 1.30; 4.00, 6.42; 5.00, 6.42; 6.00, 5.20; 7.00, 7.87]$ determine the linear best-fit relation using least-squares regression analysis. Then estimate at 95 % confidence the intervals of $\bar{y}_i$, y_n, and y_i according to Equations 10.42 through 10.44 for $x = 2.00$.

SOLUTION

From Equation 10.42 it follows directly that $\bar{y}_i = 2.68 \pm 1.93$. That is, there is a 95 % chance that the mean value of a large number of measured y_i values for $x = 2.00$ will fall within ± 1.93 of the y_{c_i} value of 2.68. Further, from Equation 10.43, $y_n = 2.68 \pm 4.95$, which implies that there is a 95 % chance that a new measurement of y for $x = 2.00$ will fall within ± 4.95 of 2.68. Finally, from Equation 10.44, $y_i = 2.68 \pm 4.56$. The confidence intervals for this data set for the range $0 \le x \le 7$ are shown in Figure 10.8.

Another confidence interval related to the regression fit can be established for the estimate of a value of x for given a value of y. This situation is encountered when a calibration curve is used to determine unknown x values. Figures 10.9 and 10.10 each display a linear regression fit of the data (labeled by a) along with two different confidence intervals for $P = 95$ %. In addition to the usual estimate of the range within which a y_i value will be with respect to its calculated value (labeled by b), there is another estimate, the x-from-y estimate with its confidence interval (labeled

---

10.5 | MATLAB SIDEBAR

The MATLAB M-file `confint.m` performs a linear least-squares regression analysis on a set of $[x, y, ey]$ data pairs (where ey is the y-measurement error) and plots the regression fit and the associated confidence intervals as given by Equations 10.42 through 10.44. This was used to generate Figure 10.8 for the data set given in Example 10.5.

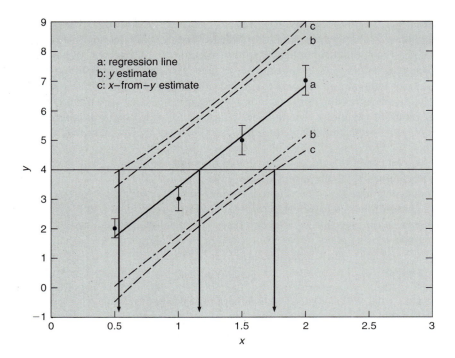

Figure 10.9   Regression fit with relatively large sensitivity.

by c). This new estimate's confidence interval should be greater in extent than that for the y estimate. This is because additional uncertainties arise when the best-fit is used to project from a chosen y-value back to an unknown x-value.

The confidence interval for the estimate of x from y is represented by hyperbolic curves. The uncertainty forming the basis of this confidence interval results from three different uncertainties associated with y and the best-fit expression: from the measurement uncertainty in y, from the uncertainty in the true value of the intercept, and from the uncertainty in the true value of the slope. The latter two result from determining the regression fit based on a finite amount of data. In essence, the hyperbolic curves can be viewed as bounds for the area within which all possible finite regression fits with their standard y-estimate confidence intervals are contained. When one projects from a chosen y-value back to the x-axis, one does not know on which regression fit the projected x-value is based. The chosen y-value could have resulted from an x-value different than the one used to establish the fit. This new confidence interval accounts for this. The three contributory uncertainties *cannot* be combined in quadrature to yield the final uncertainty because the intercept and slope uncertainties are not statistically independent from one another. So, a more rigorous approach must be taken to determine this confidence interval. This was done by Finney [8], who established this confidence interval to be

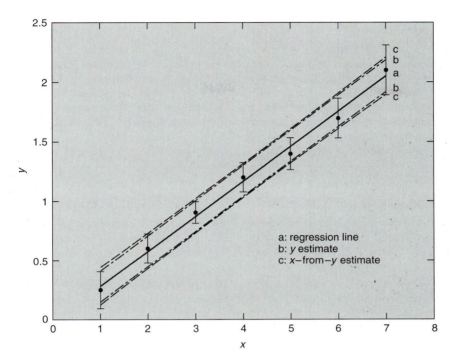

a: regression line
b: y estimate
c: x–from–y estimate

Figure 10.10   Regression fit with relatively small sensitivity.

$$y = y_c \pm t_{v,P} S_{yx} \sqrt{\frac{1}{n} + \frac{1}{N} + \frac{(x - \bar{x})^2}{S_{xx}}}, \qquad \text{[10.45]}$$

where n denotes the number of replications of y measurements for a particular value of x ($n = 1$ for the examples shown in Figures 10.10 and 10.9).

A comparison of Figures 10.9 and 10.10 reveals several important facts. When the magnitude of the uncertainty in y is relatively small, the confidence limits are closer to the regression line. When there is more scatter in the data, both intervals are wider. Fewer $[x, y]$ data pairs result in a relatively larger difference between the confidence limits. The sensitivity of y with respect to x (the slope of the regression line) plays an important role in determining the level of uncertainty in x in relation to the x-from-y

10.6 | **MATALAB SIDEBAR**

The MATLAB M-file `caley.m` performs a linear least-squares regression analysis on a set of $[x, y, ey]$ data pairs (where ey is the y-measurement error) and plots the regression fit and the associated confidence intervals as given by Equations 10.44 and 10.45. This was used to generate Figures 10.9 and 10.10.

10.7 | MATLAB SIDEBAR

The MATLAB M-file `caleyII.m` determines the range in the x-from-y estimate for a user-specified value of y in addition to those tasks done by the MATLAB M-file `caley.m`. This is accomplished by using Newtonian iteration [11] to solve for the x_{lower} and x_{upper} estimates that are shown in Figure 10.11. The MATLAB M-file `caleyIII.m` extends this type of analysis one step farther by determining the percent uncertainty in the x-from-y estimate for the entire range of y-values. It plots the standard regression fit with the data and also the x-from-y estimate uncertainty versus x. These two plots are shown in Figure 10.12.

confidence interval. Lower sensitivities result in relatively large uncertainties in x. For example, the uncertainty range in x for a value of $y = 4.0$ in Figure 10.9 is from approximately 0.53 to 1.76, as noted by the arrows in the figure, for a calculated value of $x = 1.18$. Note that the range of this uncertainty is *not* symmetric with respect to the calculated value of x.

Caution should be exercised when claims are made about trends in the data. Any claim must be made within the context of measurement uncertainty that is assessed at a

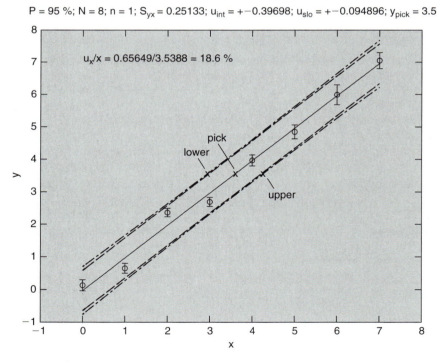

$P = 95\ \%;\ N = 8;\ n = 1;\ S_{yx} = 0.25133;\ u_{int} = +-0.39698;\ u_{slo} = +-0.094896;\ y_{pick} = 3.5$

$u_x/x = 0.65649/3.5388 = 18.6\ \%$

Figure 10.11 Regression fit with indicated x-from-y estimate uncertainty.

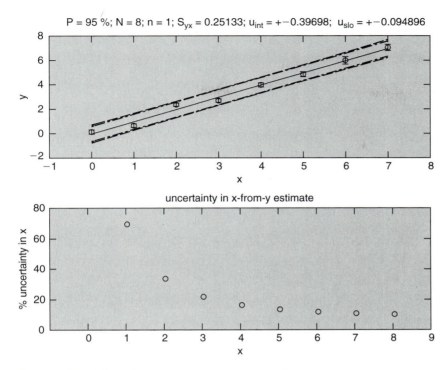

P = 95 %; N = 8; n = 1; S_{yx} = 0.25133; u_{int} = +−0.39698; u_{slo} = +−0.094896

Figure 10.12   Regression fit and *x*-from-*y* estimate uncertainty.

particular confidence level. An example is illustrated in Figure 10.13. The same values
of five trials are plotted in each of the two figures. The trend in the values appears
to increase with increasing trial number. In the top figure, the error bars represent
the measurement uncertainty assessed at a 95 % level of confidence. The solid line
suggests an increasing trend, whereas the dotted line implies a decreasing trend. Both
claims are valid to within the measurement uncertainty at 95 % confidence. In the
bottom figure, the error bars represent the measurement uncertainty assessed at a 68 %
level of confidence. It is now possible to exclude the claim of a decreasing trend and
support only that of an increasing trend. This, however, has been done at the cost of
reducing the confidence level of the claim. In fact, if the level of confidence is reduced
even further, the claim of an increasing trend cannot be supported. Thus, a specific
trend in comparison with others can be supported only through accurate experimentation in which the error bars are small and there is a high level of confidence.

10.11   MULTIVARIABLE LINEAR ANALYSIS

Linear least-squares regression analysis can be extended to situations involving more
than one independent variable. This is known as multivariable linear regression

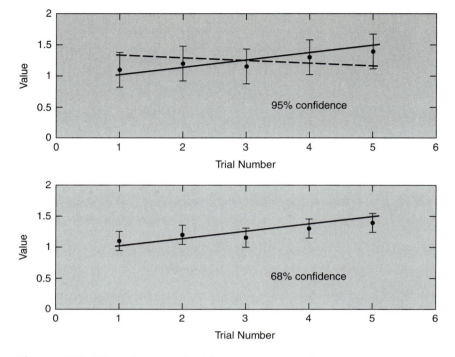

Figure 10.13   Data trends with respect to uncertainty.

analysis and results in $m + 1$ algebraic equations with $m + 1$ regression coefficient unknowns. This system of equations can be solved using methods of linear algebra.

Multivariable linear regression analysis for a system of three independent variables [15] with the regression coefficients a_0 through a_3 in the expression

$$a_0 + a_1 x_i + a_2 y_i + a_3 z_i = R_{c_i} \qquad \textbf{[10.46]}$$

yields a system of four equations:

$$a_0 N + a_1 \sum_{i=1}^{N} x_i + a_2 \sum_{i=1}^{N} y_i + a_3 \sum_{i=1}^{N} z_i = \sum_{i=1}^{N} R_i,$$

$$a_0 \sum_{i=1}^{N} x_i + a_1 \sum_{i=1}^{N} x_i^2 + a_2 \sum_{i=1}^{N} x_i y_i + a_3 \sum_{i=1}^{N} x_i z_i = \sum_{i=1}^{N} R_i x_i,$$

$$a_0 \sum_{i=1}^{N} y_i + a_1 \sum_{i=1}^{N} x_i y_i + a_2 \sum_{i=1}^{N} y_i^2 + a_3 \sum_{i=1}^{N} y_i z_i = \sum_{i=1}^{N} R_i y_i, \quad \textbf{[10.47]}$$

$$a_0 \sum_{i=1}^{N} z_i + a_1 \sum_{i=1}^{N} x_i z_i + a_2 \sum_{i=1}^{N} y_i z_i + a_3 \sum_{i=1}^{N} z_i^2 = \sum_{i=1}^{N} R_i z_i.$$

10.8 | MATLAB SIDEBAR

In many experimental situations, the number of data points exceeds the number of independent variables, leading to an overdetermined system of equations. Here, as was the case for higher-order regression analysis solutions, MATLAB can solve this set of equations, using MATLAB's left-division method. For N $[x_i, y_i, z_i, R_i]$ data points there will be N equations to evaluate, which in expanded matrix notation are

In matrix notation this becomes $[G][a] = [R]$, where $[G]$ is an $(N \times 4)$ matrix, $[a]$ is a (4×1) matrix, and $[R]$ is an $(N \times 1)$ matrix. The solution to this matrix equation is achieved using the command $G \backslash R$ after the matrices $[G]$ and $[R]$ have been entered.

$$\begin{bmatrix} 1 & x_1 & y_1 & z_1 \\ 1 & x_1 & y_2 & z_2 \\ \vdots & \vdots & \vdots & \vdots \\ 1 & x_N & y_N & z_N \end{bmatrix} \begin{bmatrix} a_0 \\ a_1 \\ a_2 \\ a_3 \end{bmatrix} = \begin{bmatrix} R_1 \\ R_2 \\ R_3 \\ \vdots \\ R_N \end{bmatrix}.$$

In expanded matrix notation, the set of Equations 10.47 becomes

$$\begin{bmatrix} N & \sum x_i & \sum y_i & \sum z_i \\ \sum x_i & \sum x_i^2 & \sum x_i y_i & \sum x_i z_i \\ \sum y_i & \sum x_i y_i & \sum y_i^2 & \sum y_i z_i \\ \sum z_i & \sum x_i z_i & \sum y_i z_i & \sum z_i^2 \end{bmatrix} \begin{bmatrix} a_0 \\ a_1 \\ a_2 \\ a_3 \end{bmatrix} = \begin{bmatrix} \sum R_i \\ \sum R_i x_i \\ \sum R_i y_i \\ \sum R_i z_i \end{bmatrix},$$

where the summations are from $i = 1$ to N. Or, in matrix notation, this becomes

$$[G][a] = [R]. \tag{10.48}$$

$[G]$, $[a]$, and $[R]$ represent the matrices shown in the expanded form. The solution to Equation 10.48 for the regression coefficients is

$$[a] = [G]^{-1}[R]. \tag{10.49}$$

$[G]^{-1}$ is the inverse of the coefficient matrix.

STATEMENT

An experiment is conducted in which the values of three independent variables, x, y, and z, are selected and then the resulting value of the dependent variable R is measured. This procedure is repeated six times for different combinations of x-, y-, and z-values. The $[x, y, z, R]$ data values are [1, 3, 1, 17; 2, 4, 2, 24; 3, 5, 1, 25; 4, 4, 2, 30; 5, 3, 1, 24; 6, 3, 2, 31]. Determine the regression coefficients for the multivariable regression fit of the data. Then, using the resulting best-fit expression, determine the calculated values of R in comparison to their respective measured values.

Example 10.6

SOLUTION

The solution is obtained using MATLAB's left-division method by typing $G \backslash R$, where $[G]$ for this example is

$$\begin{bmatrix} 1 & 1 & 3 & 1 \\ 1 & 2 & 4 & 2 \\ 1 & 3 & 5 & 1 \\ 1 & 4 & 4 & 2 \\ 1 & 5 & 3 & 1 \\ 1 & 6 & 3 & 2 \end{bmatrix}$$

and $[R]$ is

$$\begin{bmatrix} 17 \\ 24 \\ 25 \\ 30 \\ 24 \\ 31 \end{bmatrix}.$$

The resulting regression coefficient matrix is

$$\begin{bmatrix} 3.4865 \\ 2.0270 \\ 2.2162 \\ 4.3063 \end{bmatrix}.$$

Thus, the best-fit expression is

$$3.4865 + 2.0270x + 2.2162y + 4.3063z = R. \qquad \textbf{[10.50]}$$

The calculated values of R are obtained by typing G*ans after the regression coefficient solution is obtained. The values are [16.4685; 25.0180; 24.9550; 29.0721; 24.5766; 30.9099]. All calculated values agree with their respective values to within a difference of less than 1.0.

10.12 LINEAR CORRELATION ANALYSIS

It was not until late in the 19th century that scientists considered how to quantify the extent of the relation between two random variables. It was Francis Galton in his landmark paper published in 1888 [16] who quantitatively defined the word *co-relation*, now known as correlation. In that paper, he presented for the first time the method for calculating the **correlation coefficient** and its confidence limits. He was able to correlate the height (stature) of 348 adult males to their forearm (cubit) lengths. This data is presented in Table 10.2. Galton designated the coefficient by the symbol r, which "measures the closeness of co-relation." This symbol still is used today for the correlation coefficient.

Galton purposely presented his data in a particular tabular form, as shown in Table 10.2. In this manner, a possible co-relation between statue and cubit became immediately obvious to the reader, as indicated by the larger numbers along the table's diagonal. All that was left after realizing a co-relation was to quantify it. Galton approached this in an *ad hoc* manner by computing the mean value for each row (the mean cubit length for each stature) as well as the overall mean (the mean cubit length

Table 10.2   Galton's data of stature (S) versus cubit (C) length [16] (units of inches).

	C < 16.5	16.5 < C < 17.0	17.0 < C < 17.5	17.5 < C < 18.0	18.0 < C < 18.5	18.5 < C < 19.0	19.0 < C < 19.5	19.5 < C
S > 71	—	—	—	1	3	4	15	7
71 > S > 70	—	—	—	1	5	13	11	—
70 > S > 69	—	1	1	2	25	15	6	—
69 > S > 68	—	1	3	7	14	7	4	2
68 > S > 67	—	1	7	15	28	8	2	—
67 > S > 66	—	1	7	18	15	6	—	—
66 > S > 65	—	4	10	12	8	2	—	—
65 > S > 64	—	5	11	2	3	—	—	—
64 > S	9	12	10	3	1	—	—	—

for all statures). He then expressed this data in terms of standard units (the number of probable measurement error units from the overall mean). He plotted the standardized unit values for each row (the standardized cubit lengths) versus the standardized unit value of the row (the standardized stature). This yielded the regression of cubit length on stature. He followed a similar approach to determine the regression of stature on cubit length by interchanging the rows and columns of data. He then established the composite best linear fit by eye and approximated the slope's value to be equal to 0.8, which was the regression coefficient. This approach was formalized later by the statisticians Francis Edgeworth and Karl Pearson.

But what exactly is the correlation coefficient and how can it be calculated? This relates to the general process of **correlation analysis**. In this section, only *linear* correlation analysis is considered. In general, two random variables, x and y, are correlated if x's values can be related to the y's values to some extent. In the left graph of Figure 10.14, the variables show no correlation, whereas in the right graph, they are correlated moderately.

The extent of linear dependence between x and y is quantified through the correlation coefficient. This coefficient is related to the population variances of x and y, σ_x and σ_y, and the **population covariance**, σ_{xy}. The population correlation coefficient is defined as

$$\rho \equiv \frac{\sigma_{xy}}{\sqrt{\sigma_x \sigma_y}},\qquad\text{[10.51]}$$

where

$$\sigma_{xy} \equiv E[(x - x')(y - y')] = E(xy) - x'y',\qquad\text{[10.52]}$$

$$\sigma_x \equiv \sqrt{E[(x - x')^2]}\qquad\text{[10.53]}$$

and

$$\sigma_y \equiv \sqrt{E[(y - y')^2]}.\qquad\text{[10.54]}$$

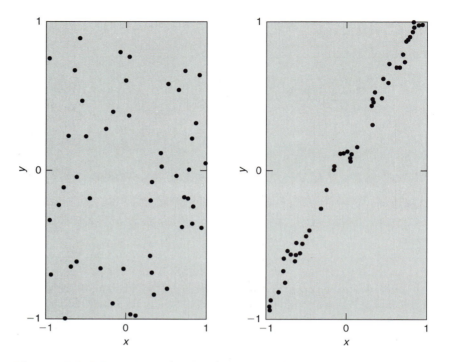

Figure 10.14   Uncorrelated and correlated data.

$E[\ ]$ denotes the expectation or mean value of a quantity, which for any statistical parameter q raised to a power m involving N discrete values is

$$E[q^m] \equiv \lim_{N \to \infty} \frac{1}{N} \sum_{i=1}^{N} q_i^m. \qquad \textbf{[10.55]}$$

Two parameters, q and r, are statistically independent when

$$E[q^m \cdot r^n] = E[q^m] \cdot E[r^n], \qquad \textbf{[10.56]}$$

where m and n are powers.

The covariance is the mean value of the product of the deviations of x and y from their true mean values. The population correlation coefficient is simply the ratio of the population covariance to the product of the x and y population variances. It measures the strength of the linear relationship between x and y. When $\rho = 0$, x and y are *uncorrelated*, which implies that y is *independent* of x. When $\rho = \pm 1$, there is *perfect* correlation, where $y = a \pm bx$ for *all* $[x, y]$ pairs.

The **sample correlation coefficient**, r, is an estimate of the **population correlation coefficient**, ρ. That is, the population correlation coefficient can be estimated but not determined exactly because a sample is finite and a population is infinite. The sample correlation coefficient is defined in a manner analogous to Equation 10.51 as

$$r = \frac{S_{xy}}{\sqrt{S_{xx} S_{yy}}}. \qquad \textbf{[10.57]}$$

Squaring both sides of this equation, substituting Equation 10.22 for the slope of the regression line, and then taking the square root of both sides, Equation 10.57 becomes

$$b = r\sqrt{\frac{S_{yy}}{S_{xx}}}. \qquad \textbf{[10.58]}$$

Thus, the slope of the regression fit equals the linear correlation coefficient times a scale factor. The scale factor is simply the square root of the ratio of the spread of the y values to the spread of the x values. So, b and r are related closely, but they are *not* the same.

Using Equations 10.19, 10.20, and 10.21, Equation 10.57 can be rewritten as

$$r = \frac{\sum_{i=1}^{N} (x_i - \bar{x})(y_i - \bar{y})}{\sqrt{\sum_{i=1}^{N}(x_i - \bar{x})^2 \sum_{i=1}^{N}(y_i - \bar{y})^2}}. \qquad \textbf{[10.59]}$$

Equation 10.59 is known as the *product-moment formula*, which automatically keeps the proper sign of r. From this, r is calculated directly from the data without performing any regression analysis. It is evident from both Equations 10.58 and 10.59 that r is a function, not only of the specific x_i and y_i values, but also of N. This point will be addressed shortly.

STATEMENT   **Example 10.7**

The Center on Addiction and Substance Abuse at Columbia University recently conducted a study on college-age drinking. They reported the following average drinks per week (DW) of alcohol consumption in relation to the average GPA (grade point average) for a large population of college students: [3.6, A; 5.5, B; 7.6, C; 10.6, D or F]. Using an index of $A = 4$, $B = 3$, $C = 2$, and D or $F = 0.5$, determine the linear best-fit relation and the value of the linear correlation coefficient.

SOLUTION

Using the MATLAB M-file plotfit.m, a linear relation with $r = 0.99987$ and GPA $= 5.77 - 0.50$DW can be determined for the range of $0 \le$ GPA ≤ 4. These cautionary results are presented in Figure 10.15.

A more physical interpretation of r can be made by examining the quantity r^2, which is known as the **coefficient of determination**. Note that r given by Equation 10.57 says nothing about the relation that best-fits x and y. It can be shown [5] that

$$\text{SSE} = S_{yy}(1 - r^2) = \text{SST}(1 - r^2). \qquad \textbf{[10.60]}$$

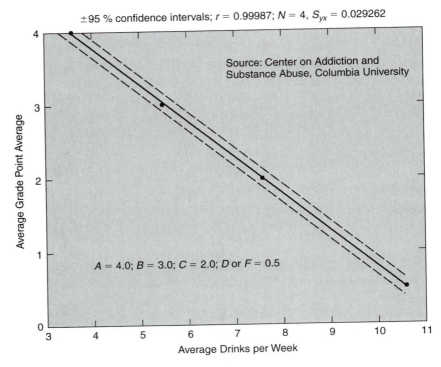

$\pm 95\%$ confidence intervals; $r = 0.99987$; $N = 4$, $S_{yx} = 0.029262$

Source: Center on Addiction and
Substance Abuse, Columbia University

$A = 4.0$; $B = 3.0$; $C = 2.0$; D or $F = 0.5$

Average Grade Point Average

Average Drinks per Week

Figure 10.15   College-student alcohol consumption.

From Equations 10.26 and 10.60, it follows that

$$r^2 = 1 - \frac{\text{SSE}}{S_{yy}} = \frac{\text{SSR}}{S_{yy}}. \qquad \textbf{[10.61]}$$

Equation 10.61 shows that the coefficient of determination is the ratio of the explained squared variation to the total squared variation. Because SSE and S_{yy} are always nonnegative, $1 - r^2 \geq 0$. So, the coefficient of determination is bounded as $0 \leq r^2 \leq 1$. It follows directly that $-1 \leq r \leq 1$. When the correlation is *perfect*, there is no unexplained squared variation (SSE $= 0$) and $r = \pm 1$. Further, when there is no fit, all the y_i values are the same because they are completely independent of x. That is, all $y_{c_i} = \bar{y}$ and, by Equation 10.25, SSE $= 0$. Thus, $r = 0$. Values of $|r| > 0.99$ imply a *very significant* correlation; values of $|r| > 0.95$ imply a *significant* correlation. On the other extreme, values of $|r| < 0.05$ imply an *insignificant* correlation; values of $|r| < 0.01$ imply a *very insignificant* correlation.

Another expression for r that relates it to the results of a regression analysis fit can be obtained. Substituting Equations 10.20 and 10.24 into Equation 10.61 yields

$$r = \sqrt{\frac{\sum_{i=1}^{N}(y_{c_i} - \overline{y})^2}{\sum_{i=1}^{N}(y_i - \overline{y})^2}}. \qquad \textbf{[10.62]}$$

This equation relates r to the y_{c_i} values obtained from regression analysis. This is in contrast to Equation 10.59, which yields r directly from data. These two equations help to underscore an important point. Correlation analysis and regression analysis are separate and distinct statistical approaches. Each are performed independently from the other. The results of a linear regression analysis, however, can be used for correlation analysis.

Caution should be exercised in interpreting various values of the linear correlation coefficient. For example, a value of $r \sim 0$ simply means that the two variables are not *linearly* correlated. They could be highly correlated *nonlinearly*. Further, a value of $r \sim \pm 1$ implies that there is a strong linear correlation. But the correlation could be casual, such as a correlation between the number of cars sold and pints of Guiness consumed in Ireland. Both are related to Ireland's population, but not to each other directly. Also, even if the linear correlation coefficient value is close to unity, that does not imply necessarily that the fit is the most appropriate. Return for a moment to the example of the linear spring (case 2). Although the spring's energy is related fundamentally to the square of its extension, a linear correlation coefficient value of 0.979 results when correlating the spring's energy with its extension. This high value implies a strong linear correlation between energy and extension, but it does *not* imply that a linear relation is the most appropriate one.

Finally, when attempting to establish a correlation between two variables it is important to recognize the possibility that two uncorrelated variables can appear to be correlated simply by chance. This circumstance makes it imperative to go one step more than simply calculating the value of r. One must also determine the probability that N measurements of two *uncorrelated* variables will give a value of r equal to or larger than any particular r_o. This probability is determined by

$$P_N(|r| \geq |r_o|) = \frac{2\Gamma[(N-1)/2]}{\sqrt{\pi}\,\Gamma[(N-2)/2]} \int_{|r_o|}^{1} (1-r^2)^{(N-4)/2}\,dr = f(N, r), \qquad \textbf{[10.63]}$$

where Γ denotes the gamma function. If $P_N(|r| \geq |r_o|)$ is small, then it is unlikely that the variables are uncorrelated. That is, it is likely that they are correlated. Thus, $1 - P_N(|r| \geq |r_o|)$ is the probability that two variables are correlated given $|r| \geq |r_o|$. If $1 - P_N(|r| \geq |r_o|) > 0.95$, then there is a *significant* correlation and if $1 - P_N(|r| \geq |r_o|) > 0.99$, then there is a *very significant* correlation. Values of $1 - P_N(|r| \geq |r_o|)$ versus the number of measurements, N, are shown in Figure 10.16. For example, a value of $r_o = 0.6$ gives a 60 % chance of correlation for $N = 4$ and a 99.8 % chance of correlation for $N = 25$. Thus, whenever citing a value of r it is imperative to present the % confidence of the correlation and the number of data points on which it is based. Reporting a value of r alone is ambiguous.

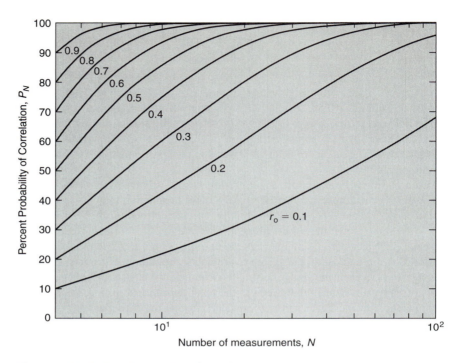

Figure 10.16   Probability of correlation.

10.9 │ **MATLAB SIDEBAR**

The M-file `corrprob.m` and its associated function-file `f1.m` calculates the probability $P_N(|r| \geq |r_o|)$. The definite integral is found using the `quad8` function, which is an adaptive, recursive, Newton-Côtes eight-panel method. As an example, for $N = 6$ and $r_o = 0.979$ `corrprob.m` gives $P_N = 0.1$ %. Thus, the correlation is *very significant*.

10.10 │ **MATLAB SIDEBAR**

The MATLAB M-file `PNplot.m` constructs a plot of $P_N(|r| \geq |r_o|)$ versus the number of measurements, N, for a user-specified value of r_o. By using MATLAB's `holdon` command, a figure such as Figure 10.16 can be generated for various values of r_o.

10.13 SIGNAL CORRELATIONS IN TIME

Thus far, the application of correlation analysis to discrete information has been considered. Correlation analysis also can be applied to information that is continuous in time.

Consider two signals, $x(t)$ and $y(t)$, of two experimental variables. Assume that these signals are stationary and ergodic. These terms are defined in Chapter 11. For a stationary signal, the statistical properties determined by ensemble averaging values for an arbitrary time from the beginning of a number of the signal's time history records are independent of the time chosen. Further, if these average values are the same as those found from the time-average over a single time history record, then the signal is also ergodic. So, an ergodic signal is also a stationary signal. By examining how the amplitude of either signal's time history record at some time compares to its amplitude at another time, important information, such as on the repeatability of the signal, can be gathered. This can be quantified through the **autocorrelation** function of the signal, which literally correlates the signal with its self (thus the prefix *auto*). The amplitudes of the signals also can be compared to one another to examine the extent of their *co-relation*. This is quantified through the cross-correlation function, in which the cross product of the signals is examined.

10.13.1 AUTOCORRELATION

For an ergodic signal $x(t)$, the autocorrelation function is the average value of the product $x(t) \cdot x(t + \tau)$, where τ is some time delay. Formally, the **autocorrelation function**, $R_x(\tau)$, is defined as

$$R_x(\tau) \equiv E[x(t) \cdot x(t + \tau)] = \lim_{T \to \infty} \int_0^T x(t) x(t + \tau) \, dt. \qquad \textbf{[10.64]}$$

Because the signal is stationary, $R_x(\tau)$, its mean and its variance are independent of time. So,

$$E[x(t)] = E[x(t + \tau)] = x' \qquad \textbf{[10.65]}$$

and

$$\sigma_{x(t)}^2 = \sigma_{x(t+\tau)}^2 = \sigma_x^2 = E[x^2(t)] - x'^2. \qquad \textbf{[10.66]}$$

Analogous to Equation 10.52, the **autocorrelation coefficient** can be defined as

$$\rho_{xx}(\tau) \equiv \frac{E\{[x(t) - x'][x(t + \tau) - x']\}}{\sigma_x^2}. \qquad \textbf{[10.67]}$$

The numerator in Equation 10.67 can be expanded to yield

$$\rho_{xx}(\tau) = \frac{E[x(t) \cdot x(t + \tau)] - x'E[x(t + \tau)] - x'E[x(t)] + x'^2}{\sigma_x^2}. \qquad \textbf{[10.68]}$$

Substitution of Equations 10.64 and 10.65 into Equation 10.68 results in an expression that relates the autocorrelation function to its coefficient

$$\rho_{xx}(\tau) = \frac{R_x(\tau) - x'^2}{\sigma_x^2} \qquad \textbf{[10.69]}$$

or

$$R_x(\tau) = \rho_{xx}(\tau)\sigma_x^2 + x'^2. \qquad \textbf{[10.70]}$$

Some limits can be placed on the value of $R_x(\tau)$. Because $-1 \leq \rho_{xx}(\tau) \leq 1$, $R_x(\tau)$ is bounded as

$$-\sigma_x^2 + x'^2 \leq R_x(\tau) \leq \sigma_x^2 + x'^2. \qquad \textbf{[10.71]}$$

Now it can be shown (see Chapter 7) that

$$E[x^2] = \sigma_x^2 + x'^2 \qquad \textbf{[10.72]}$$

by expanding $E[(x - x')^2]$. So, the maximum value that $R_x(\tau)$ can have is $E[x^2]$. It follows from Equation 10.64 that

$$R_x(0) = E[x^2]. \qquad \textbf{[10.73]}$$

That is, the maximum value of $R_x(\tau)$ occurs at $\tau = 0$. Using Equation 10.72 and the definition of the autocorrelation coefficient (Equation 10.69),

$$\rho_{xx}(0) = 1. \qquad \textbf{[10.74]}$$

Further, as $\tau \to \infty$ there is no correlation between $x(t)$ and $x(t + \tau)$ because $x(t)$ is the signal of a random variable. That is,

$$\rho_{xx}(\tau \to \infty) = 0, \qquad \textbf{[10.75]}$$

which implies that

$$R_x(\tau \to \infty) = x'^2. \qquad \textbf{[10.76]}$$

Finally, $R_x(\tau)$ is an even function because

$$R_x(-\tau) = E[x(t)x(t - \tau)] = E[x(t - \tau)x(t)] = E[x(t^*)x(t^* + \tau)] = R_x(\tau), \qquad \textbf{[10.77]}$$

where $t^* = t - \tau$, noting $x(t)$ is stationary. So, $R_x(\tau)$ is symmetric about the $\tau = 0$ axis.

A generic autocorrelation function and its corresponding autocorrelation coefficient having these properties is displayed in Figure 10.17. Values of $R_x(\tau)$ and $\rho_{xx}(\tau)$ that are greater than their respective limiting values as $\tau \to \infty$ indicate a positive correlation at that particular value of τ. Conversely, a negative correlation is indicated for values less than that limiting value. Values equal to the limiting value signify no correlation. Note that $\rho_{xx}(\tau)$ experiences decreasing oscillations about a value of 0 as $\tau \to \infty$. This always will be the case for a stationary signal provided there are no deterministic components in the signal other than a nonzero mean.

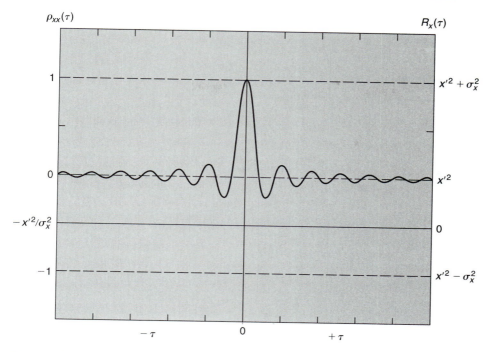

Figure 10.17   Typical autocorrelation function and coefficient.

Example 10.8

STATEMENT

Determine the autocorrelation coefficient for the signal $x(t) = A \cdot \sin(\omega t)$. Then plot the coefficient for values of $\omega = 1$ rad/s and $A = 1$.

SOLUTION

The autocorrelation function needs to be examined only for the range $0 \geq \tau \geq T/2\pi$ because $x(t)$ in this example is a periodic function of period $T = 2\pi/\omega$. Equation 10.64 for this periodic function becomes

$$R_x(\tau) = \frac{A^2}{T} \int_0^{2\pi/\omega} \sin(\omega t) \sin(\omega t + \phi) \, dt,$$

where $\phi = \omega\tau$. Performing and evaluating the integral,

$$R_x(\tau) = \frac{A^2}{T} [\sin^2(\omega t) \cos(\phi) + \sin(\omega t) \cos(\omega t) \sin(\phi)]_0^{2\pi/\omega} = \frac{1}{2} A^2 \cos(\omega \tau).$$

Now $\sigma_x = \sqrt{A/2}$, so, according to Equation 10.67,

$$\rho_{xx} = \cos(\omega \tau).$$

The plot of $\rho_{xx}(\tau)$ for values of $\omega = 1$ rad/s and $A = 1$ is presented in Figure 10.18. It shows that the sine function has a positive, perfect autocorrelation at values of $\tau = T = 2\pi$ when $\omega = 1$ and a negative, perfect autocorrelation at values of $\tau = T/2 = \pi$. Further, when $\tau = T/4 = \pi/2$ or $\tau = 3T/4 = 3\pi/2$ there is no correlation of the sine function with itself.

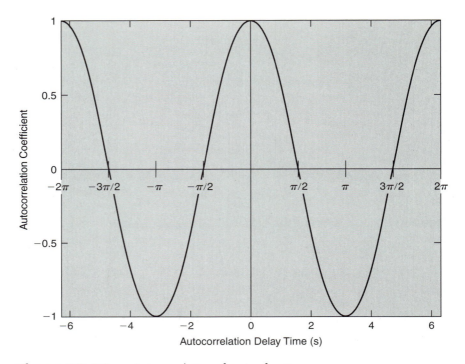

Figure 10.18   Autocorrelation of a sine function.

10.13.2 CROSS-CORRELATION

Expressions for the cross-correlation function and coefficient can be developed in the same manner as that done for the case of autocorrelation.

For the stationary signals $x(t)$ and $y(t)$ there are two **cross-correlation functions** defined as

$$R_{xy}(\tau) \equiv E[x(t) \cdot y(t + \tau)] \qquad \textbf{[10.78]}$$

and

$$R_{yx}(\tau) \equiv E[y(t) \cdot x(t + \tau)]. \qquad \textbf{[10.79]}$$

$R_{xy}(\tau)$ denotes the cross-correlation of x with y and $R_{yx}(\tau)$ that of y with x. Further, because the signals are stationary,

$$R_{xy}(\tau) = E[x(t - \tau)y(t)] = R_{yx}(-\tau) \qquad \textbf{[10.80]}$$

and

$$R_{yx}(\tau) = E[y(t - \tau)x(t)] = R_{xy}(-\tau). \qquad \textbf{[10.81]}$$

So, in general, $R_{xy}(\tau) \neq R_{yx}(\tau)$ and both are not even with respect to τ.

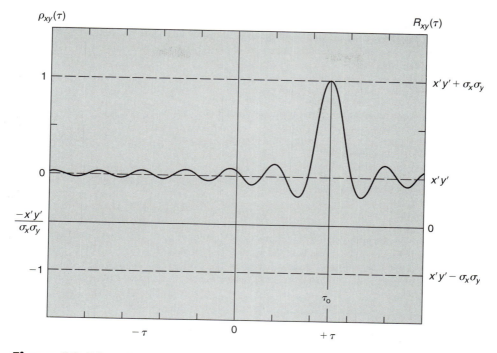

Figure 10.19   Typical cross-correlation of two signals.

The corresponding two **cross-correlation coefficients** are defined as

$$\rho_{xy}(\tau) \equiv \frac{R_{xy}(\tau) - x'y'}{\sigma_x \sigma_y} \qquad\qquad \textbf{[10.82]}$$

and

$$\rho_{yx}(\tau) \equiv \frac{R_{yx}(\tau) - x'y'}{\sigma_x \sigma_y}, \qquad\qquad \textbf{[10.83]}$$

where both coefficients are bounded between values of -1 and 1. Thus, both functions are bounded between values of $-\sigma_x \sigma_y + x'y'$ and $\sigma_x \sigma_y + x'y'$. Finally, as $\tau \to \infty$ both functions tend to the value of $x'y'$ because no correlation between the random signals $x(t)$ and $y(t)$ would be expected at that limit.

Typically, two signals will experience a maximum cross-correlation at some value of $\tau = \tau_o$, which corresponds to a phase lag between the two signals, where $\phi = \omega \tau_o$. This is shown for a typical cross-correlation in Figure 10.19.

STATEMENT

Determine the cross-correlation coefficient $\rho_{xy}(\tau)$ for the signals $x(t) = A \cdot \sin(\omega t)$ and $y(t) = B \cdot \cos(\omega t)$. Then plot the coefficient for the value of $\omega = 1$ rad/s.

Example 10.9

SOLUTION

The cross-correlation function needs to be examined only for the range $0 \leq \tau \leq T/2\pi$ because $x(t)$ and $y(t)$ are periodic functions of period $T = 2\pi/\omega$. Equation 10.78 for these periodic functions becomes

$$R_{xy}(\tau) = \frac{AB}{T} \int_0^{2\pi/\omega} \sin(\omega t) \cos(\omega t + \phi) \, dt,$$

where $\phi = \omega\tau$. Performing and evaluating the integral,

$$R_{xy}(\tau) = \frac{AB}{\tau} \left[-\frac{1}{4}\omega \cos(2\omega t + \phi) - \frac{1}{2}t \sin(\phi) \right]_0^{2\pi/\omega} = \frac{-AB}{2} \sin(\phi).$$

Now $\sigma_x = A/\sqrt{2}$ and $\sigma_y = B/\sqrt{2}$, so, according to Equation 10.82,

$$\rho_{xy}(\tau) = -\sin(\omega\tau).$$

Note the minus sign in this expression. The plot of $\rho_{xy}(\tau)$ for the value of $\omega = 1$ rad/s is given in Figure 10.20. For a delay time value of $\tau = \pi/2$, $\rho_{xy}(\tau) = 1$. This is because the value of $\cos(t + \pi/2)$ exactly equals that of $\sin(t)$. Similar reasoning can be used to explain the value of $\rho_{xy}(\tau) = -1$ when $\tau = 3\pi/2$, where the cosine and sine values are equal but opposite in sign.

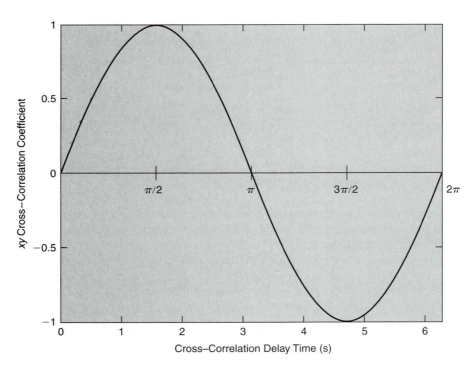

Figure 10.20   Cross-correlation of sine and cosine functions.

10.11 | MATLAB SIDEBAR

The MATLAB M-file `sigcor.m` determines and plots the autocorrelations and cross-correlation of discrete data that is user-specified. The file contains an arbitrary length of time, x- and y-values in three columns. This M-file uses MATLAB's `xcorr` command. The command `xcorr(x,'flag')` performs the autocorrelation of x and the command `xcorr(x, y,'flag')` calculates the cross-correlation of x with y. The argument 'flag' of `xcorr` permits the user

to specify how the correlations are normalized. The M-file `sigcor.m` uses the argument 'coeff' to normalize the correlations such that the autocorrelations at zero time lag are identically 1.0. An example plot generated by `sigcor.m` for a file containing eight sequential measurements of x and y data is shown in Figure 10.21. Note that both autocorrelations have a value of 1.0 at zero time lag.

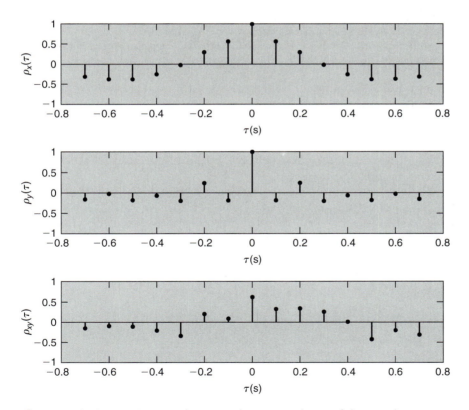

Figure 10.21   Autocorrelations and cross-correlation of discrete data.

REFERENCES

[1] S. M. Stigler. 1986. *The History of Statistics*. Cambridge: Harvard University Press.

[2] K. Alder. 2002. *The Measure of All Things*. London: Little, Brown.

[3] P. R. Bevington and D. K. Robinson. 1992. *Data Reduction and Error Analysis for the Physical Sciences*. New York: McGraw-Hill.

[4] J. R. Taylor. 1982. *An Introduction to Error Analysis*. Mill Valley: University Science Books.

[5] W. A. Rosenkrantz. 1997. *Introduction to Probability and Statistics for Scientists and Engineers*. New York: McGraw-Hill.

[6] D. C. Montgomery and G. C. Runger. 1994. *Applied Statistics and Probability for Engineers*. New York: John Wiley and Sons.

[7] I. Miller and J. E. Freund. 1985. *Probability and Statistics for Engineers*, 3rd ed. Englewood Cliffs, N.J.: Prentice Hall.

[8] D. J. Finney. 1952. *Probability Analysis*. Cambridge: Cambridge University Press.

[9] W. E. Deming. 1943. *Statistical Adjustment of Data*. New York: John Wiley and Sons.

[10] J. Mandel. 1984. *The Statistical Analysis of Experimental Data*. New York: Dover.

[11] S. Nakamura. 1996. *Numerical Analysis and Graphic Visualization with MATLAB*. Upper Saddle River, N.J.: Prentice Hall.

[12] W. H. Press, S. A. Teukolsky, W. T. Vetterling, and B. P. Flannery. 1992. *Numerical Recipes*, 2nd ed. Cambridge: Cambridge University Press.

[13] W. J. Palm, III. 1999. *MATLAB for Engineering Applications*. New York: McGraw-Hill.

[14] F. Galton. 1892. *Hereditary Genius: An Inquiry into its Laws and Consequences*, 2nd ed. London: Macmillian.

[15] H. Coleman and W. G. Steele. 1999. *Experimentation and Uncertainty Analysis for Engineers*, 2nd ed. New York: Wiley Interscience.

[16] F. Galton. 1888. Co-relations and Their Measurement, Chiefly from Anthropometric Data. *Proceedings of the Royal Society of London* 45: 135–145.

REVIEW PROBLEMS

1. Consider the following set of three (x, y) data pairs: $(0, 0)$, $(3, 2)$, and $(6, 7)$. Determine the intercept of the best-fit line for the data to two decimal places.

2. Consider the following set of three (x, y) data pairs: $(0, 0)$, $(3, 0)$ and $(9, 5)$. Determine the slope of the best-fit line for the data to two decimal places.

3. What is the last name of the famous mathematician who developed the method of least squares?

4. Consider the following set of three (x, y) data pairs: $(1.0, 1.7)$, $(2.0, 4.3)$, and $(3.0, 5.7)$. A linear least-squares regression analysis yields the best-fit equation $y = -0.10 + 2.00x$. Determine the standard error of the fit rounded off to two decimal places.

5. Consider the following set of three (x, y) data pairs: $(1.0, 1.7)$, $(2.0, 4.3)$, and $(3.0, 5.7)$. A linear least-squares regression analysis yields the best-fit equation $y = -0.10 + 2.00x$. Determine the precision interval based on 95 % confidence rounded off to two decimal places. Assume a value of 2 for Student's t factor.

6. An experimenter determines the precision interval, PI_1, for a set of data by performing a linear least-squares regression analysis. This interval is based on three measurements and 50 % confidence. Then the same experiment is repeated under identical conditions and a new precision interval, PI_2, is determined based on 15 measurements and 95 % confidence. The ratio of PI_2 to PI_1 is [a] less than one, [b] greater than one, [c] equal to one, [d] could be any of the above.

HOMEWORK PROBLEMS

1. Show that a least-squares linear regression analysis fit always goes through the point $(\bar{x}, \bar{y})$.

2. Starting with the equation $y_i - \bar{y} = (y_i - y_{c_i}) + (y_{c_i} - \bar{y})$ and using the normal equations, prove that $\sum_{i=1}^{N}(y_i - \bar{y})^2 = \sum_{i=1}^{N}(y_i - y_{c_i})^2 + \sum_{i=1}^{N}(y_{c_i} - \bar{y})^2$.

3. Find the best linear equation that fits the data shown in Table 10.3.

Table 10.3   Calibration data.

x	10	20	30	40
y	5.1	10.5	14.7	20.3

4. Determine the best-fit values of the coefficients a and b in the expression $y = 1/(a + bx)$ for the $[x, y]$ data pairs [1.00, 1.11; 2.00, −0.91; 3.00, −0.34; 4.00, −0.20; 5.00, −0.14].

5. For an ideal gas, $pV^\gamma = C$. Using regression analysis, determine the best-fit value for γ given the data shown in Table 10.4.

Table 10.4   Ideal gas data.

p (psi)	V (in.^{3})
16.6	50
39.7	30
78.5	20
115.5	15
195.3	10
546.1	5

6. The data presented in Table 10.5 was obtained during the calibration of a cantilever-beam force-measurement system, like that shown in Figure 4.11. The beam is instrumented with four strain gages that serve as the legs of a Wheatstone bridge. In the table $F(N)$ denotes the applied force, $E(V)$ is the measured output voltage, and $u_E(V)$ is

the measurement uncertainty in E. Based on your knowledge of how such a system operates, what order of the fit would model the physics of the system most appropriately? Perform a regression analysis of the data for various orders of the fit. What is the order of the fit that has the lowest value of S_{yx}? What is the order of the fit that has the smallest precision interval, $\pm t_{v,P} \cdot S_{yx}$? That is required to have the actual fit curve agree with *all* of the data to within the uncertainty of E?

Table 10.5   Strain gage force balance calibration data.

$F(N)$	$E(V)$	$u_E(V)$
0.4	2.7	0.1
1.1	3.6	0.2
1.9	4.4	0.2
3.0	5.2	0.3
5.0	9.2	0.5

7. A hot-wire anemometry system probe, inserted into a wind tunnel, is used to measure the tunnel's centerline velocity, U. The output of the system is a voltage, E. During a calibration of this probe, the data listed in Table 10.6 was acquired. Assume that the uncertainty in the voltage measurement is 2 % of the indicated value. Using a linear least-squares regression analysis, determine the best-fit values of A and B in the relation $E^2 = A + B\sqrt{U}$. Finally, plot the fit with 95 % confidence intervals and the data with error bars as voltage versus velocity. Is the assumed relation appropriate?

Table 10.6   Hot-wire probe calibration data.

Velocity (m/s)	Voltage (V)
0.00	3.19
3.05	3.99
6.10	4.30
9.14	4.48
12.20	4.65

8. The April 3, 2000, issue of *Time Magazine* published the body mass index (BMI) of each Miss America from 1922 to 1999. The BMI is defined as "the weight divided by the square of the height." The author argues, based on the data, that Miss America may dwindle away to nothing if the BMI-versus-year progression continues. Perform a linear least-squares regression analysis on the data and de-termine the linear regression coefficient. How statistically justified is the author's claim? Also determine how many Miss Americas have BMIs that are *below* the World Health Organization's cutoff for undernutrition, which is a BMI equal to 18.6. Use the data file `missamer.dat` that con-tains two columns, the year and the BMI.

CHAPTER REVIEW

ACROSS

1. (?) error of the fit
6. former name for regression
8. type of variable that can be transformed into a linear one
10. father of the least-squares method
11. father of the word *correlation*
12. type of least-square equations named by Gauss

DOWN

2. related to the product of itself at one time to that at another time
3. type of analysis to fit data with curves
4. an interval related to probability
5. type of correlation when the correlation coefficient is greater than 0.95
7. 1/10 000 000 of a meridian quadrant in SI units
9. regression analysis order greater than one

The Differential Analyzer

Credit: Science Museum / Science & Society Picture Library, London.

The machine shown is approximately half of an original differential analyzer, known as the Manchester machine. It was built by Douglas Hartree (1897–1958) in 1935, who based it on the differential analyzer constructed by Vannevar Bush at the Massachusetts Institute of Technology. A differential analyzer consists of a number of interconnecting integrating mechanisms. Its purpose is to provide an analog display of a numerical solution to a differential equation. Each integrating mechanism consists of an arm with a pointer that follows an input graph, a wheel connected to the other end of the arm, a rotating disk upon which the wheel rests, and an arm connected to the shaft of the disk on one end and to an output pen on the other, which plots the integral of the input curve.

11

SIGNAL CHARACTERISTICS

. . . there is a tendency in all observations, scientific and otherwise, to see what one is looking for . . .

D. J. Bennett. 1998.
Randomness. Cambridge: Harvard University Press.

But you perceive, my boy, that it is not so, and that facts, as usual, are very stubborn things, overruling all theories.

Professor VonHardwigg in
Voyage au centre de la terra by Jules Gabriel Verne, 1864.

CHAPTER OUTLINE

11.1 CHAPTER OVERVIEW

One of the key requirements in performing a successful experiment is a knowledge of signal characteristics. Signals contain vital information about the process under investigation. Much information can be extracted from them, provided the experimenter is aware of the methods that can be used and their limitations. In this chapter, the types of signals and their characteristics are identified. Formulations of the statistical parameters of signals are presented. Fourier analysis and synthesis are introduced and used to find the amplitude, frequency, and power content of signals. These tools are applied to continuous signals, first to some classic periodic signals and then to aperiodic signals. In Chapter 12, these methods are extended to digital signal analysis.

11.2 LEARNING OBJECTIVES

You should be able to do the following after completing this chapter:

- Know the meaning of signal, amplitude, frequency, time history record, ensemble, and waveform
- Know the classifications of signals
- Know the definition of an ensemble-averaged quantity
- Know the definitions of and be able to calculate the mean, mean square, root-mean-square, variance, and standard deviation of a deterministic signal from discrete data
- Know the definition of the trigonometric Fourier series and the Fourier coefficients
- Determine the Fourier series of a deterministic function
- Plot the amplitude spectrum and the power spectrum of a Fourier series
- Know the abscissa and ordinate units of the amplitude, power, and power density spectra plots

11.3 SIGNAL CHARACTERIZATION

In the context of measurements, a signal is a measurement system's representation of a physical variable that is sensed by the system. More broadly, it is defined as a detectable, physical quantity or impulse (as a voltage, current, or magnetic field strength) by which messages and information can be transmitted [1]. The information contained in a signal is related to its size and extent. The size is characterized by the **amplitude** (magnitude) and the extent (timewise or samplewise variation) by the **frequency**. The actual shape of a signal is called its **waveform**. A plot of a signal's

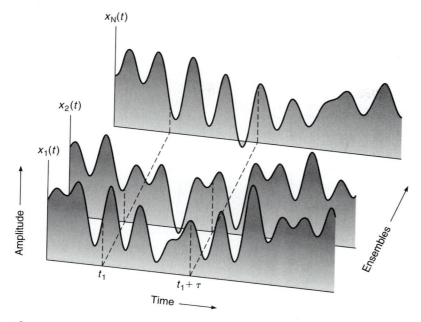

Figure 11.1 An ensemble of N time history records.

amplitude versus time is called a **time history record**. A collection of N time history records is called an **ensemble**, as illustrated in Figure 11.1. An ensemble also can refer to a set of many measurements made of a single entity, such as the weight of an object determined by each student in a science class, and of many entities of the same kind made at the same time, such as everyone's weight on New Year's morning.

Signals can be classified as either **deterministic** or **nondeterministic** (*random*). A deterministic signal can be described by an explicit mathematical relation. Its future behavior, therefore, is predictable. Each time history record of a random signal is unique. Its future behavior cannot be determined exactly but to within some limits with a certain confidence.

Deterministic signals can be classified into static and dynamic signals, which are subdivided further, as shown in Figure 11.2. Static signals are steady *in time*. Their amplitude remains constant. Dynamic signals are either periodic or aperiodic. A **periodic** signal, $y(t)$, repeats itself at regular intervals, nT, where $n = 1, 2, 3, \ldots$. Analytically, this is expressed as

$$y(t + T) = y(t) \qquad \textbf{[11.1]}$$

for all t. The smallest value of T for which Equation 11.1 holds true is called the **fundamental period**. If signals $y(t)$ and $z(t)$ are periodic, then their product $y(t)z(t)$ and the sum of any linear combination of them, $c_1 y(t) + c_2 z(t)$, are periodic.

A **simple** periodic signal has one period. A **complex** periodic signal has more than one period. An **almost-periodic** signal is comprised of two or more sinusoids

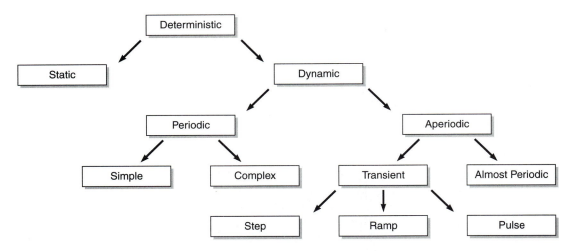

Figure 11.2   Deterministic signal subdivisions (adapted from [2]).

of arbitrary frequencies. However, if the ratios of all possible pairs of frequencies are rational numbers, then an almost-periodic signal is periodic.

Nondeterministic signals are classified as shown in Figure 11.3. Properties of the ensemble of the nondeterministic signals shown in Figure 11.1 can be computed by taking the average of the instantaneous property values acquired from each of the time histories at an arbitrary time, t_1. The ensemble mean value, $\mu_x(t_1)$, and the ensemble autocorrelation function (see Chapter 10 for more on the autocorrelation), $R_x(t_1, t_1 + \tau)$, are

$$\mu_x(t_1) = \lim_{N \to \infty} \frac{1}{N} \sum_{i=1}^{N} x_i(t_1) \qquad \textbf{[11.2]}$$

and

$$R_x(t_1, t_1 + \tau) = \lim_{N \to \infty} \frac{1}{N} \sum_{i=1}^{N} x_i(t_1) x_i(t_1 + \tau), \qquad \textbf{[11.3]}$$

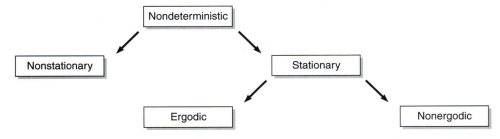

Figure 11.3   Nondeterministic signal subdivisions (adapted from [2]).

in which τ denotes an arbitrary time measured from time t_1. Both equations represent **ensemble averages**. This is because $\mu_x(t_1)$ and $R_x(t_1, t_1 + \tau)$ are determined by performing averages over the ensemble at time t_1.

If the values of $\mu_x(t_1)$ and $R_x(t_1, t_1 + \tau)$ change with t_1, then the signal is **nonstationary**. Otherwise, it is **stationary** (stationary in the *wide* sense). A nondeterministic signal is considered to be **weakly stationary** when only $\mu_x(t_1) = \mu_x$ and $R_x(t_1, t_1 + \tau) = R_x(\tau)$, that is, when only the signal's ensemble mean and autocorrelation function are time invariant. In a more restrictive sense, if all other ensemble higher-order moments and joint moments (see Chapter 7 for more about moments) also are time invariant, the signal is **strongly stationary** (stationary in the *strict* sense). So, the term stationary means that each of a signal's ensemble-averaged statistical properties are constant with respect to t_1. It does *not* mean that the amplitude of the signal is constant over time. In fact, a random signal is never stationary in time!

For a single time history, the temporal mean value, μ_x, and the temporal autocorrelation coefficient, $R_x(\tau)$, are

$$\mu_x = \lim_{T \to \infty} \frac{1}{T} \int_0^T x(t)\, dt \qquad \textbf{[11.4]}$$

and

$$R_x(\tau) = \lim_{T \to \infty} \frac{1}{T} \int_0^T x(t)x(t + \tau)\, dt. \qquad \textbf{[11.5]}$$

For most stationary data, the ensemble averages at an arbitrary time, t_1, will equal their corresponding temporal averages computed for an arbitrary single time history in the ensemble. When this is true, the signal is **ergodic**. If the signal is periodic, then the limits in Equations 11.4 and 11.5 do not exist because averaging over one time period is sufficient. Ergodic signals are important because all of their properties can be determined by performing time averages over a *single* time history record. This greatly simplifies data acquisition and reduction. Most random signals representing stationary physical phenomena are ergodic.

A finite record of data of an ergodic random process can be used in conjunction with probabilistic methods to quantify the statistical properties of an underlying process. For example, it can be used to determine a random variable's true mean value within a certain confidence limit. These methods also can be applied to deterministic signals, which are considered next.

11.4 SIGNAL VARIABLES

Most waveforms can be written in terms of sums of sines and cosines, as will be shown later in Section 11.7. Before examining more complex waveform expressions, the variables involved in simple waveform expressions must be defined. This can be done by examining the following expression for a simple, periodic sine function:

$$y(t) = C \sin(n\omega t + \phi) = C \sin(2\pi n f t + \phi), \qquad \textbf{[11.6]}$$

in which the argument of the sine is in units of radians. The variables and their units are as follows:

- C: amplitude [units of $y(t)$]
- n: number of cycles [dimensionless]
- ω: *circular* frequency [rad/s]
- f: *cyclic* frequency [cycles/s = Hz]
- t: time [s]
- T: period $(= 2\pi/\omega = 1/f)$ [s/cycle]
- ϕ: phase [rad] where $\phi = 2\pi(t/T) = 2\pi(\theta°/360°)$

Also note that 2π rad $= 1$ cycle $= 360°$ and $\sin(\omega t + \pi/2) = \cos(\omega t)$. The top plot in Figure 11.4 displays the signal $y(t) = \sin(\pi t)$. Its period equals $2\pi/\pi = 2$ s, as seen in the plot. The preceding definitions can be applied readily to determine the frequencies of a periodic signal, as in Example 11.1.

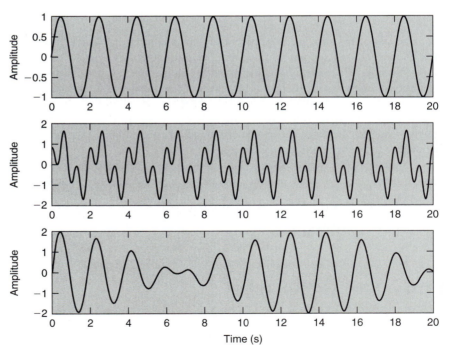

Figure 11.4   Various signals comprised of sines and cosines.

| **11.1** | **MATLAB SIDEBAR** |

MATLAB has a number of built-in functions that generate standard signals, such as square, triangular, and sawtooth waves, and the trigonometric functions. A square wave with a frequency of f Hz and an amplitude that varies from $-A$ to $+A$ over the time period from 0 to 7 s is generated and plotted using the MATLAB command sequence

```
t = 0:0.001:7;
sq = A*square(2*pi*f*t);
plot(t,sq)
```

The MATLAB sawtooth(t,width) function produces a sawtooth wave with period 2*pi. The fraction of the period at which sawtooth's peak occurs is specified its width argument, whose value varies between 0 and 1. A triangular wave is produced when width equals 0.5. A sawtooth wave with a period of 2*pi, its peak occurring at 0.25*2*pi and an amplitude that varies from $-A$ to $+A$ over the time period from 0 to 7 s is generated and plotted using the MATLAB command sequence

```
t = 0:0.001:7;
sw = A*sawtooth(t,0.25);
plot(t,sw)
```

STATEMENT | **Example 11.1**

Determine the circular and cyclic frequencies for the signal $y(t) = 10 \sin(628t)$.

SOLUTION

Using the above definitions,

circular frequency, $\omega = 628$ rad/s (assuming $n = 1$ cycle)

cyclic frequency, $f = \frac{\omega}{2\pi} = \frac{628}{2\pi} = 100$ cycles/s $= 100$ Hz.

When various sine and cosine waveforms are combined by addition, more complex waveforms result. Such waveforms occur in many practical situations. For example, the differential equations describing the behavior of many systems have sine and cosine solutions of the form

$$y(t) = A \cos(\omega t) + B \sin(\omega t). \qquad \textbf{[11.7]}$$

By introducing the phase angle, ϕ, $y(t)$ can be expressed as either a cosine function,

$$y(t) = C \cos(\omega t - \phi), \qquad \textbf{[11.8]}$$

or a sine function,

$$y(t) = C \sin\left(\omega t - \phi + \frac{\pi}{2}\right) = C \sin(\omega t + \phi^*), \qquad \textbf{[11.9]}$$

where C, ϕ, and ϕ^* are given by

$$C = \sqrt{A^2 + B^2}, \qquad \textbf{[11.10]}$$

$$\phi = \tan^{-1}\left(\frac{B}{A}\right)$$

[11.11]

and

$$\phi^* = \tan^{-1}\left(\frac{A}{B}\right),$$

[11.12]

noting that $\phi^* = (\pi/2) - \phi$. Reducing the waveform in Equation 11.7 to either Equation 11.8 or Equation 11.9 often is useful in interpreting results. The middle plot in Figure 11.4 shows the signal $y(t) = \sin(\pi t) + 0.8\cos(3\pi t)$. This signal is complex and has two frequencies, $\omega_1 = \pi$ and $\omega_2 = 3\pi$ rad/s. This leads to two periods, $T_1 = 2$ s and $T_2 = 2/3$ s. Because $T_2 = 3T_1$, the period T_2 will contain one cycle of $\sin(\omega_1 t)$ and three periods of $0.8\cos(\omega_2 t)$. So, $T_2 = 2$ s is the fundamental period of this complex signal. In general, the fundamental period of a complex signal will be the least common denominator of the contributory periods.

An interesting situation arises when two waves of equal amplitude and nearly equal frequencies are added. The resulting wave exhibits a relatively slow beat with a frequency of one-half the difference in the two nearly equal frequencies, called the **beat frequency**. In general, the sum of two sine waves of frequencies, f and $f + \Delta f$, beats with a frequency of $\Delta f/2$. This beating is displayed in the bottom plot of Figure 11.4, which shows the signal $y(t) = \sin(\pi t) + \sin(1.15\pi t)$. For this signal, $\Delta f = 0.15$, so the beat frequency is 0.075 Hz. This implies that the beat repeats itself every 13.33 s, which can be seen in the plot. This phenomenon is called **heterodyning** and is exploited in tuning musical instruments and in laser Doppler velocimeters.

11.5   COMPLEX NUMBERS AND WAVES

Complex numbers can be used to simplify waveform notation. Waves, such as electromagnetic waves that are all around us, also can be expressed using complex notation.

11.2 | MATLAB SIDEBAR

In some instances, the amount of time that a signal resides within some amplitude window needs to be determined. The MATLAB M-file epswin.m can be used for this purpose and to determine the times at which the signal's minimum and maximum amplitude occur. Figure 11.5 was generated by epswin.m applied to the case of a pressure transducer's response to an oscillating flow field. The M-file is constructed to receive a user-specified data file that consists of two columns, time and amplitude. The amplitude's window is established by a user-specified center amplitude and an amplitude percentage. epswin.m plots the signal and indicates the percentage of the time that the signal resides within the window. The number of instances that the amplitude is within the window is determined using an algorithm based on an array whose values are negative when the amplitude is within the window. The times at which the signal reaches its minimum and maximum amplitudes also are determined and indicated.

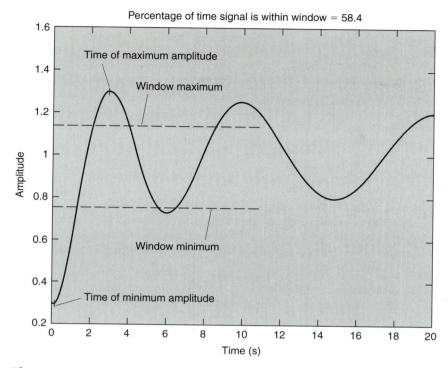

Figure 11.5   Output plot of M-file epswin.m used to examine a signal's amplitude behavior.

The complex exponential function is defined as

$$\exp(z) = e^z = e^{(x+iy)} = e^x e^{iy} \equiv e^x (\cos y + i \sin y), \qquad \textbf{[11.13]}$$

where $z = x + iy$, with the complex number $i \equiv \sqrt{-1}$ and x and y as real numbers. The **complex conjugate** of z, denoted by z^*, is $z^* = x - iy$. The **modulus** or absolute value of z is given by $|z| = \sqrt{zz^*} = \sqrt{(x+iy)(x-iy)} = \sqrt{x^2 + y^2}$, which is a real number. Using Equation 11.13, the Euler formula results,

$$e^{i\theta} = \cos\theta + i\sin\theta, \qquad \textbf{[11.14]}$$

which also leads to

$$e^{-i\theta} = \cos\theta - i\sin\theta. \qquad \textbf{[11.15]}$$

The complex expressions for the sine and cosine functions can be found from Equations 11.14 and 11.15,

$$\cos\theta = \frac{1}{2}\left(e^{i\theta} + e^{-i\theta}\right) \qquad \textbf{[11.16]}$$

and

$$\sin\theta = \frac{1}{2i}\left(e^{i\theta} - e^{-i\theta}\right). \qquad [11.17]$$

A wave can be represented by sine and cosine functions. Such representations are advantageous because (1) these functions are periodic, like many waves in nature; (2) linear math operations on them, such as integration and differentiation, yield waveforms of the same frequency but different amplitude and phase; and (3) they form complex waveforms that can be expressed in terms of Fourier series.

A wave can be represented by the general expression

$$y(t) = A_r \cos\frac{2\pi}{\lambda}(x - ct) + iA_i \sin\frac{2\pi}{\lambda}(x - ct), \qquad [11.18]$$

in which A_r is the real amplitude, A_i is the imaginary amplitude, x is the distance, λ is the wavelength, and c is the wave speed. This expression can be written in another form, as

$$y(t) = A_r \cos(\kappa x - \omega t) + iA_i \sin(\kappa x - \omega t), \qquad [11.19]$$

in which κ is the (angular) wave number and ω is the circular frequency. The **wave number** denotes the number of waves in 2π units of length, where $\kappa = 2\pi/\lambda$. The wave speed is related to the wave number by $c = \omega/\kappa$. The cosine term represents the real part of the wave and the sine term the imaginary part. Further, the phase lag is defined as

$$\alpha = \tan^{-1}\frac{A_i}{A_r}. \qquad [11.20]$$

Equations 11.19 and 11.20 imply that

$$y(t) = \sqrt{A_r^2 + A_i^2}\cos(\kappa x - \omega t - \alpha), \qquad [11.21]$$

in which the complex part of the wave manifests itself as a phase lag.

Example 11.2

STATEMENT

Determine the phase lag of the wave given by $z(t) = 20e^{i(4x-3t)}$.

SOLUTION

The given wave equation, when expanded using Euler's formula, reveals that both the real and imaginary amplitudes equal 20. Thus, according to Equation 11.20, $\alpha = \tan^{-1}(20/20) = \pi/4$ rad.

11.6   SIGNAL STATISTICAL PARAMETERS

Signals can be either continuous in time or discrete. Discrete signals usually arise from the digitization of a continuous signal, to be discussed in Chapter 12, and from sample-to-sample experiments, which were considered in Chapter 9. A large number of statistical parameters can be determined from either continuous or discrete signal information. The parameters most frequently of interest are the signal's mean, variance, standard deviation, and rms. For continuous signals, these parameters are computed from integrals of the signal over time. For discrete signals, these parameters are determined from summations over the number of samples. The expressions for these properties are presented in Table 11.1. Note that as $T \rightarrow \infty$ or $N \rightarrow \infty$, the statistical parameter values approach the true values of the underlying process.

The choice of a time period that is used to determine the statistical parameters of a signal, called the **signal sample period**, depends on the type of waveform. When the waveform is periodic, either simple or complex, the signal sample period should be the fundamental period. When the waveform is almost periodic or nondeterministic, no single signal sample period will produce exact results. For this situation, it is best to keep increasing the signal's sample period until the statistical parameter values of interest become constant to within acceptable limits.

Determining an appropriate sample period is not always straightforward. The values of the mean, variance, skewness, and kurtosis of two data samples are shown in Figure 11.6. The first sample, indicated by solid curves, was drawn randomly from a normal population having a mean value of 3.0 and a standard deviation of 0.5. The second sample, indicated by dotted curves, was the same as the first but with an additional amplitude decrease in time equal to 0.001/s. The mean of the first sample reaches its final value at approximately 100 s. The mean of the second sample exhibits a decrease in time, which is linear after approximately 100 s. The variance, skewness, and kurtosis values of both samples vary with respect to sample time and between samples during most of the entire sample time. The variances of the two samples agree up to approximately 300 s. Then, they deviate from one another because of the second sample's mean value decrease in time. This example illustrates the

Table 11.1 Statistical parameters for continuous and discrete signals.

Quantity	Continuous	Discrete
Mean	$\bar{x} = \frac{1}{T} \int_0^T x(t)\, dt$	$\bar{x} = \frac{1}{N} \sum_{i=1}^{N} x_i$
Variance	$S_x^2 = \frac{1}{T} \int_0^T [x(t) - \bar{x}]^2\, dt$	$S_x^2 = \frac{1}{N-1} \sum_{i=1}^{N} (x_i - \bar{x})^2$
Standard deviation	$S_x = \sqrt{\frac{1}{T} \int_0^T [x(t) - \bar{x}]^2\, dt}$	$S_x = \sqrt{\frac{1}{N-1} \sum_{i=1}^{N} (x_i - \bar{x})^2}$
rms	$x_{\mathrm{rms}} = \sqrt{\frac{1}{T} \int_0^T x(t)^2\, dt}$	$x_{\mathrm{rms}} = \sqrt{\frac{1}{N} \sum_{i=1}^{N} x_i^2}$

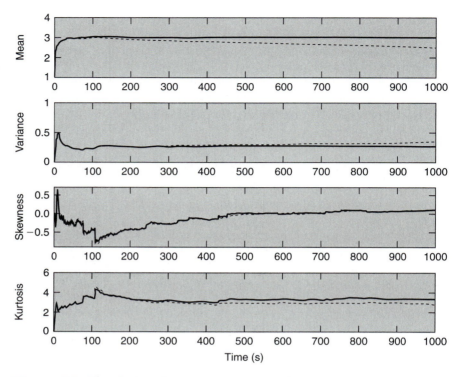

Figure 11.6   Statistical properties versus sample time.

complexity in determining an appropriate sample time, especially if the signal being sampled has a gradual change in time over the sample period in addition to short-time fluctuations.

Sometimes it is important to examine the fluctuating component of a signal. The average value of a signal (its DC component) can be removed (subtracted) from the original signal to reveal more clearly the signal's fluctuating behavior (its AC component). This is shown in Figure 11.7, in which the left plot is the complete signal (DC plus AC components); the middle plot is the DC component and the AC

11.3 │ **MATLAB SIDEBAR**

The MATLAB M-file `propintime.m` was used to generate Figure 11.6. It is constructed to read a user-specified data file and plot the values of the data's mean, variance, skewness, and kurtosis for various sample periods. This M-file can be used to determine the minimum sample time required to achieve statistical property values within acceptable limits.

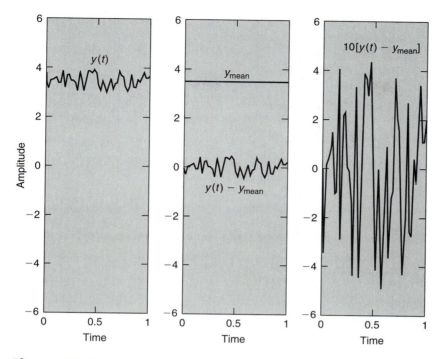

Figure 11.7 Subtraction of the mean value from a signal.

component, each shown separately; and the right plot is the AC component amplified 10 times.

The concepts of the mean, variance, and standard deviation were presented in Chapter 7. The **root mean square (rms)** is another important statistical parameter. It is defined as the positive square root of the mean of the squares. Its continuous and discrete representations are presented in Table 11.1. The rms characterizes the dynamic portion (AC component) of the signal and the mean characterizes its static

---

11.4 | **MATLAB SIDEBAR**

The M-file acdc.m was used to generate Figure 11.7. A period of time from $t = 0$ to $t = 2$ in increments of 0.02 is established first. Then $y(t)$ is computed for that period. The mean value is determined next using the MATLAB mean command. This is the DC component. Then the AC component is determined by subtracting the mean from the signal. Finally, the AC component is amplified by a factor of 10. The following syntax is used as part of the M-file:

```
t=[0:0.02:1];
y=3+rand(size(t));
dc=ones(size(t))*mean(y);
ac=y-dc;
z=10*ac;
```

portion (DC component). The magnitudes of these components for a typical signal are shown in Figure 11.8. When no fluctuation is present in the signal, $x(t)$ is constant and equal to its mean value, $\bar{x}$. So, $x_{rms} \geq \bar{x}$ always. x^2_{rms} is the temporal average of the square of the amplitude of x.

The following two applications of the rms concept show its utility:

1. The total energy dissipated over a period of time by a resistor in a circuit is

$$E_T = \int_{t_1}^{t_2} P(t)\,dt = R\int_{t_1}^{t_2} [I(t)]^2\,dt = R(t_2 - t_1)I^2_{rms}, \qquad \textbf{[11.22]}$$

where

$$I^2_{rms} = \frac{1}{t_2 - t_1}\int_{t_1}^{t_2} [I(t)]^2\,dt. \qquad \textbf{[11.23]}$$

2. The temporal-averaged kinetic energy per unit volume in a fluid at a point in a flow is

$$\bar{E} = \frac{\rho}{2(t_2 - t_1)}\int_{t_1}^{t_2} [U(t)]^2\,dt = \frac{1}{2}\rho U^2_{rms}, \qquad \textbf{[11.24]}$$

where

$$U^2_{rms} = \frac{1}{t_2 - t_1}\int_{t_1}^{t_2} [U(t)]^2\,dt. \qquad \textbf{[11.25]}$$

Sometimes, the term rms refers to the rms of the *fluctuating* component of the signal and *not* to the rms of the signal itself. For example, the fluctuating component of a fluid velocity, $u(t)$, can be written as the difference between a total velocity, $U(t)$, and a mean velocity, $\bar{U}(t)$, as

$$u(t) = U(t) - \bar{U}(t). \qquad \textbf{[11.26]}$$

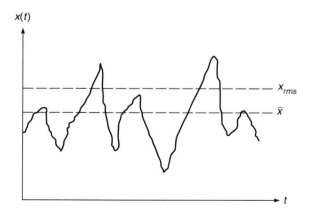

Figure 11.8   A signal showing its mean and rms values.

11.5 | MATLAB SIDEBAR

The rms can be computed using the MATLAB `norm` command. This expression `norm(x)` equals `sum[abs (x).²]¹/²`. The "." is present in this expression because of the multiplication of the vector x with itself. So,

the rms for the discrete signal x can be written in MATLAB as

```
rms=norm(x)/sqrt(length(x));
```

So, the rms of the fluctuating component is

$$u_{rms} = \left\{ \frac{1}{t_2 - t_1} \int_{t_1}^{t_2} \left[U(t) - \overline{U}(t) \right]^2 dt \right\}^{1/2},$$ **[11.27]**

where

$$\overline{U}(t) = \frac{1}{t_2 - t_1} \int_{t_1}^{t_2} U(t)\, dt.$$ **[11.28]**

By comparing Equations 11.25 and 11.27, it is evident that $U_{rms} \neq u_{rms}$.

STATEMENT

Example 11.3

Determine the rms of the ramp function $y(t) = A(t/T)$ in which A is the amplitude and T is the period.

SOLUTION

Because $y(t)$ is a deterministic periodic function, the rms needs to be computed for only one period, from $t = 0$ to $t = T$. Application of the rms equation from Table 11.1 for $y(t)$, which is a continuous signal, yields

$$y_{rms} = \left[\frac{A^2}{T^2(t_2 - t_1)} \int_{t_1}^{t_2} t^2\, dt \right]^{1/2} = \frac{A^2(t_2^3 - t_1^3)}{3T^2(t_2 - t_1)}.$$

For $t_1 = 0$ and $t_2 = T$, the rms becomes

$$y_{rms} = \frac{A}{\sqrt{3}}.$$

What is $\overline{y}(t)$? (Answer: $A/2$.) What is the rms of a sine wave of amplitude A? (Answer: $A/\sqrt{2}$.)

11.7 FOURIER SERIES OF A PERIODIC SIGNAL

Before considering the Fourier series, the definition of orthogonality must be examined. The **inner product** (dot product), $(x,\ y)$, of two real-valued functions $x(t)$ and

$y(t)$ over the interval $a \leq t \leq b$ is defined as

$$(x, y) = \int_a^b x(t)y(t)\, dt. \tag{11.29}$$

If $(x, y) = 0$ over that interval, then the functions x and y are **orthogonal** in the interval. If each *distinct* pair of functions in a set of functions is orthogonal, then the set of functions is **mutually orthogonal**.

For example, the set of functions $\sin(2\pi mt/T)$ and $\cos(2\pi mt/T), m = 1, 2, \ldots,$ form one distinct pair and are mutually orthogonal because

$$\int_{-T/2}^{T/2} \sin\left(\frac{2\pi mt}{T}\right) \cos\left(\frac{2\pi nt}{T}\right) dt = 0 \text{ for all } m,\, n. \tag{11.30}$$

Also, these functions satisfy the other orthogonality relations

$$y(t) = \int_{-T/2}^{T/2} \cos\left(\frac{2\pi mt}{T}\right) \cos\left(\frac{2\pi nt}{T}\right) dt = \begin{cases} 0 & m \neq n \\ T & m = n \end{cases} \tag{11.31}$$

and

$$y(t) = \int_{-T/2}^{T/2} \sin\left(\frac{2\pi mt}{T}\right) \sin\left(\frac{2\pi nt}{T}\right) dt = \begin{cases} 0 & m \neq n \\ T & m = n. \end{cases} \tag{11.32}$$

Knowing these facts is useful when performing certain integrals, such as those that occur when determining the Fourier coefficients.

Fourier analysis and synthesis, named after Jean-Baptiste-Joseph Fourier (1768–1830), a French mathematician, now can be examined. Fourier showed that the temperature distribution through a body could be represented by a series of harmonically related sinusoids. The mathematical theory for this, however, actually was developed by others [3]. Fourier methods enable complex signals to be approximated in terms of a series of sines and cosines. This is called the trigonometric **Fourier series**. The Fourier trigonometric series that represents a signal of period T can be expressed as

$$y(t) = \frac{A_0}{2} + \sum_{n=1}^{\infty} \left[A_n \cos\left(\frac{2\pi nt}{T}\right) + B_n \sin\left(\frac{2\pi nt}{T}\right) \right], \tag{11.33}$$

where

$$A_0 = \frac{2}{T} \int_{-T/2}^{T/2} y(t)\, dt, \tag{11.34}$$

$$A_n = \frac{2}{T} \int_{-T/2}^{T/2} y(t) \cos\left(\frac{2\pi nt}{T}\right) dt \quad n = 1, 2, \ldots, \tag{11.35}$$

and

$$B_n = \frac{2}{T} \int_{-T/2}^{T/2} y(t) \sin\left(\frac{2\pi nt}{T}\right) dt \quad n = 1, 2, \ldots. \tag{11.36}$$

The frequencies associated with the sines and cosines are integer multiples (*nth harmonics*) of the fundamental frequency. The *fundamental* or *primary* frequency, the first harmonic, is denoted by $n = 1$, the second harmonic by $n = 2$, the third harmonic by $n = 3$, and so on. A_0 is twice the average of $y(t)$ over one period. A_n and B_n are called the **Fourier coefficients** of the Fourier amplitudes. The expression for A_n can be determined by multiplying both sides of the original series expression for $y(t)$ by $\cos(2\pi nt/T)$, then integrating over one period from $t = -T/2$ to $t = T/2$. The expression for B_n is found similarly but instead by multiplying by $\sin(2\pi nt/T)$. This is called Fourier's trick.

The procedure by which the Fourier amplitudes for any specified $y(t)$ are found is called **Fourier analysis**. Fourier analysis is the analog to a prism that separates white light (a complex signal) into colors (simple periodic sine functions). **Fourier synthesis** is the reverse procedure by which $y(t)$ is constructed from a series of appropriately weighted sines and cosines. The Fourier synthesis of a signal is useful because the amplitude and frequency components of the signal can be identified.

A Fourier series representation of $y(t)$ exists if $y(t)$ satisfies the following Dirichlet conditions:

1. $y(t)$ has a finite number of discontinuities in the period T (it is piecewise differentiable).

2. $y(t)$ has a finite average value.

3. $y(t)$ has a finite number of relative maxima and minima in the period T.

If these conditions are met, then the series converges to $y(t)$ at the values of t where $y(t)$ is continuous and converges to the mean of $y(t^+)$ and $y(t^-)$ at a finite discontinuity. Fortunately, these conditions hold for most situations.

Recall that a periodic function with period T satisfies $y(t + T) = y(t)$ for all t. It follows that if $y(t)$ is an integrable periodic function with a period T, then the integral of $y(t)$ over *any* interval of length T has the same value. Hence, the limits from $-T/2$ to $T/2$ of the Fourier coefficient integrals can be replaced by, for example, from 0 to T or from $-T/4$ to $3T/4$. Changing these limits sometimes simplifies the integration procedure.

The process of arriving at the Fourier coefficients also can be simplified by examining whether the integrands are either even or odd functions. Example even and odd functions are shown in Figure 11.9. If $y(t)$ is an **even function**, where it is symmetric about the y-axis, then $g(x) = g(-x)$. Thus,

$$\int_{-T}^{T} g(x)\,dx = 2\int_{0}^{T} g(x)\,dx. \qquad \textbf{[11.37]}$$

The cosine is an even function. Likewise, if $y(t)$ is an **odd function**, where it is symmetric about the origin, then $g(x) = -g(-x)$. So,

$$\int_{-T}^{T} g(x)\,dx = 0. \qquad \textbf{[11.38]}$$

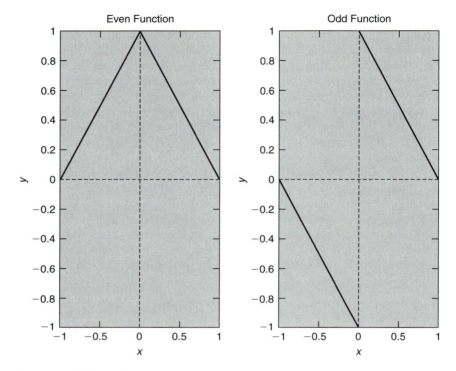

Figure 11.9   Even and odd functions.

The sine is an odd function. Other properties of even and odd functions include the following:

1. The sum, difference, product, or quotient of two even functions are even.
2. The sum or difference of two odd functions are odd.
3. The product or quotient of two odd functions are even.
4. The product or quotient of an even function and an odd function is odd.
5. The sum or difference of an even function and an odd function is neither even nor odd.
6. A general function can be decomposed into a sum of even plus odd functions.

From these properties, Equation 11.33, and the Fourier coefficient equations, it follows that, when $y(t)$ is an even periodic function, $B_n = 0$ and $y(t)$ has the Fourier series

$$y(t) = \frac{A_0}{2} + \sum_{n=1}^{\infty}\left[A_n \cos\left(\frac{2\pi nt}{T}\right)\right]. \qquad \textbf{[11.39]}$$

This is called the *Fourier cosine series*. Further, when $y(t)$ is an odd periodic function, $A_0 = A_n = 0$ and $y(t)$ has the Fourier series

$$y(t) = \sum_{n=1}^{\infty} \left[B_n \sin\left(\frac{2\pi nt}{T}\right) \right].$$ **[11.40]**

This is called the *Fourier sine series*.

STATEMENT   **Example 11.4**

Find the frequency spectrum of the step function

$$y(t) = \begin{cases} -A & -\pi \leq t < 0 \\ +A & 0 \leq t < \pi \end{cases}$$

SOLUTION

This is an odd function, therefore $A_0 = A_n = 0$.

$$B_n = \frac{2}{T} \int_{-T/2}^{T/2} y(t) \sin\left(\frac{2\pi nt}{T}\right) dt$$

$$\text{note} : \omega = \frac{2\pi}{T} \text{ where } T = 2\pi$$

$$= \frac{1}{\pi} \left[\int_{-\pi}^{0} (-A) \sin(nt)\, dt + \int_{0}^{\pi} (A) \sin(nt)\, dt \right]$$

$$= \frac{1}{\pi} \left\{ \left[\frac{A}{n} \cos(nt) \right]_{-\pi}^{0} - \left[\frac{A}{n} \cos(nt) \right]_{0}^{\pi} \right\}$$

$$= \frac{A}{n\pi} [1 - \cos(-n\pi) - \cos(n\pi) + 1]$$

$$= \frac{2A}{n\pi} [1 - \cos(n\pi)]$$

$$= \frac{4A}{n\pi} \text{ for } n \text{ odd}$$

$$= 0 \text{ for } n \text{ even.}$$

Note that B_n involves only n and constants.

$$\Rightarrow y(t) = \sum_{n=1}^{\infty} B_n \sin(nt)$$

$$= \sum_{n=1}^{\infty} \frac{2A}{n\pi} [1 - \cos(n\pi)] \sin(nt)$$

$$= \frac{4A}{\pi} \sum_{n=1,3,5,\dots}^{\infty} \left(\frac{1}{n}\right) \sin(2\pi nft)$$

$$= \frac{4A}{\pi} \left[\sin(t) + \frac{1}{3} \sin(3t) + \frac{1}{5} \sin(5t) + \cdots \right],$$

$y(t)$ involves both n and t. The frequencies of each of the sine terms are $1, 3, 5, \ldots$ in units of ω (rad/s), or $1/2\pi, 3/2\pi, 5/2\pi, \ldots$ in units of f (cycles/s = Hz). Generally, this can be written as $f_n = (2n - 1)f_1$. Likewise, the corresponding amplitudes can be expressed as $A_n = A_1/(2n - 1)$, where $A_1 = 4A/\pi$. $y(t)$ is shown shortly in Figure 11.11 for three different partial sums for $A = 5$.

The contributions of each of the harmonics to the amplitude of the square wave is illustrated in Figure 11.10. The square wave shown along the back plane is the sum of the first 500 harmonics. Only the first five harmonics are given in the figure. The decreasing amplitude and increasing frequency contributions of the next higher harmonic tend to fill in the contributions of the previously summed harmonics such that the resulting wave approaches a square wave.

The partial Fourier series sums of the step function are shown in Figure 11.11 for $N = 1$, 10, and 500. Clearly, the more terms that are included in the sum, the closer the sum approximates the actual step function. Relatively small fluctuations at the end of the step can be seen, especially in the $N = 500$ sum. This is known as the Gibbs phenomenon. The inclusion of more terms in the sum will attenuate these fluctuations but never completely eliminate them.

A plot of amplitude versus frequency can be constructed for a square wave, as presented in Figure 11.12 for the first eight harmonics. This figure illustrates in another way that the Fourier series representation of a square wave consists of multiple frequencies of decreasing amplitudes.

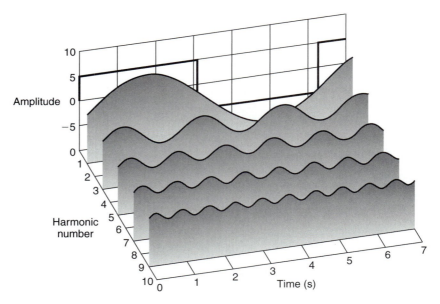

Figure 11.10   Contributions of the first five harmonics to the Fourier series of a step function.

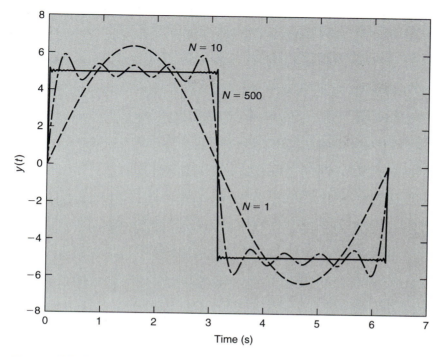

Figure 11.11 Partial Fourier series sums for the step function.

11.6 | MATLAB SIDEBAR

The M-file `fsstep.m` was used to construct Figure 11.10. This M-file uses the MATLAB `symsum` command to create the sum of the Fourier series. This is accomplished in part for the sum of 500 harmonics shown in the figure by the syntax

```
syms n t
y=(10/(n*pi))*(1-cos(n*pi))*sin(n*t);
step=symsum(y,n,1.500);
t=0:0.01:7;
ystep=eval(vectorize(step));
```

11.7 | MATLAB SIDEBAR

The M-file `sersum3.m` constructs a plot of the Fourier series sum for a user-specified $y(t)$ and three values of N. This was used to construct Figure 11.11. This M-file uses the MATLAB `symsum` command, as described in the previous MATLAB sidebar. It also was used to construct Figure 11.13, which will be discussed later.

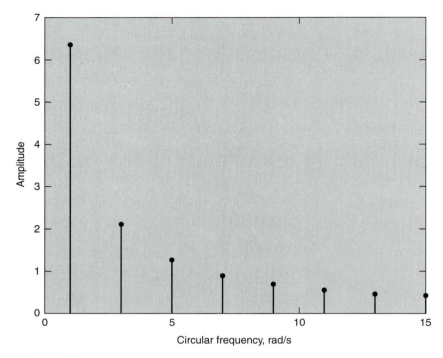

Figure 11.12   Amplitude spectrum for the first eight terms of the step function.

Example 11.5 | **STATEMENT**

Find the frequency spectrum of the ramp function:

$$y(t) = \begin{cases} 2At & 0 \le t < 1/2 \\ 0 & 1/2 \le t < 1 \end{cases}$$

with $T = 1$ s.

SOLUTION

This function is neither even nor odd.

$$A_0 = \frac{2}{T} \int_{-T/2}^{T/2} y(t)\,dt = \frac{2}{1}\left(\int_{-0.5}^{0} 0\,dt + \int_{0}^{0.5} 2At\,dt\right) = 4A\,\frac{t^2}{2}\Big|_{0}^{0.5} = \frac{A}{2}$$

$$A_n = \frac{2}{T} \int_{-T/2}^{T/2} y(t) \cos\frac{2\pi nt}{T}\,dt$$

$$= 2\int_{0}^{0.5} 2At \cos\frac{2\pi nt}{T}\,dt$$

$$= 4A\left[\frac{\cos(2\pi nt/T)}{(2\pi n/T)^2} + \frac{t\sin(2\pi nt/T)}{(2\pi n/T)}\right]\Big|_{0}^{0.5}$$

because $\int t \cos mt \, dt = (1/m^2) \cos mt + (t/m) \sin mt$ (from integration by parts).
Continuing,

$$A_n = 4A \left[\frac{\cos n\pi - \cos 0}{(2\pi n/T)^2} + \frac{0.5 \sin n\pi - 0}{(2\pi n/T)} \right]$$

$$= (4A) \left(\frac{T^2}{4\pi^2 n^2} \right) (\cos n\pi - 1) = \frac{A}{\pi^2 n^2} (\cos n\pi - 1).$$

Further,

$$B_n = \frac{2}{T} \int_{-T/2}^{T/2} y(t) \sin \frac{2\pi nt}{T} \, dt$$

$$= 2 \int_0^{0.5} 2At \sin \frac{2\pi nt}{T} \, dt$$

$$= 4A \left[\frac{\sin(2\pi nt/T)}{(2\pi n/T)^2} - \frac{t \cos(2\pi nt/T)}{(2\pi n/T)} \right] \Bigg|_0^{0.5}$$

because $\int t \sin mt \, dt = (1/m^2) \sin mt - (t/m) \cos mt$ (from integration by parts).
So,

$$B_n = 4A \left[\frac{\sin n\pi - \sin 0}{(2\pi n/T)^2} + \frac{0.5 \cos n\pi - 0}{(2\pi n/T)} \right]$$

$$= (4A) \frac{-0.5 \cos n\pi}{(2\pi n/T)} = \frac{-A}{\pi n} \cos n\pi.$$

Thus,

$$y(t) = A \left\{ \frac{1}{4} + \sum_{n=1}^{\infty} \left[\frac{(\cos n\pi - 1)}{\pi^2 n^2} \cos \frac{2\pi nt}{T} - \frac{\cos n\pi}{\pi n} \sin \frac{2\pi nt}{T} \right] \right\}$$

$$= A \left\{ \frac{1}{4} + \sum_{n=1}^{\infty} \frac{1}{n\pi} \left[\frac{(-1 + (-1)^n)}{\pi n} \cos \frac{2\pi nt}{T} - (-1)^n \sin \frac{2\pi nt}{T} \right] \right\}.$$

This $y(t)$ is shown (with $A = 1$) for three different partial sums in Figure 11.13.

11.8   **EXPONENTIAL FOURIER SERIES**

The trigonometric Fourier series can be simplified using complex number notation.
Starting with the trigonometric Fourier series

$$y(t) = \frac{A_0}{2} + \sum_{n=1}^{\infty} \left[A_n \cos \left(\frac{2\pi nt}{T} \right) + B_n \sin \left(\frac{2\pi nt}{T} \right) \right], \qquad \textbf{[11.41]}$$

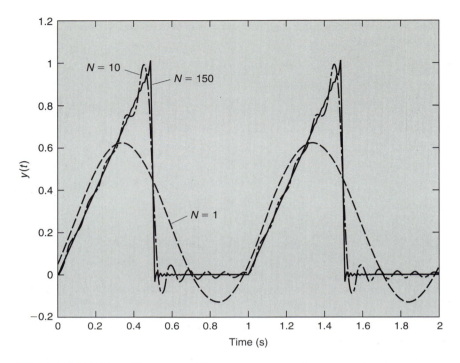

Figure 11.13   Three partial Fourier series sums for a ramp function.

and substituting Equations 11.16 and 11.17 into Equation 11.41 yields

$$y(t) = \frac{A_0}{2} + \sum_{n=1}^{\infty} \left[\frac{A_n}{2} \left(e^{in\omega_0 t} + e^{-in\omega_0 t} \right) + \frac{B_n}{2i} \left(e^{in\omega_0 t} - e^{-in\omega_0 t} \right) \right], \quad \textbf{[11.42]}$$

where $\theta = n\omega_0 t = 2\pi n t / T$. Rearranging the terms in this equation gives

$$y(t) = \frac{A_0}{2} + \sum_{n=1}^{\infty} \left[e^{in\omega_0 t} \left(\frac{A_n}{2} - \frac{i B_n}{2} \right) + e^{-in\omega_0 t} \left(\frac{A_n}{2} + \frac{i B_n}{2} \right) \right], \quad \textbf{[11.43]}$$

noting $1/i = -i$.

Using the definitions

$$C_n \equiv \frac{A_n}{2} - \frac{i B_n}{2}, \quad \textbf{[11.44]}$$

$$C_{-n} \equiv \frac{A_n}{2} + \frac{i B_n}{2}, \quad \textbf{[11.45]}$$

this equation can be simplified as follows:

$$y(t) = \sum_{n=1}^{\infty} \left(C_n e^{in\omega_0 t} + C_{-n} e^{-in\omega_0 t} \right) = \frac{A_0}{2} + \sum_{n=-\infty}^{-1} C_n e^{in\omega_0 t} + \sum_{n=1}^{\infty} C_n e^{in\omega_0 t}.$$

$$\textbf{[11.46]}$$

Combining the two summations yields

$$y(t) = \sum_{n=-\infty}^{\infty} C_n e^{in\omega_0 t}, \qquad \text{[11.47]}$$

where $C_0 = A_0/2$.

The coefficients C_n can be found by multiplying Equation 11.47 by $e^{-in\omega_0 t}$ and then integrating from 0 to T. The integral of the right-hand side equals zero except where $m = n$, which then yields the integral equal to T. Thus,

$$C_m = \frac{1}{T} \int_0^T y(t) e^{-im\omega_0 t} \, dt. \qquad \text{[11.48]}$$

What has been done here is noteworthy. An expression (Equation 11.41) involving two coefficients A_n and B_n with sums from $n = 1$ to $n = \infty$ was reduced to a simpler form having one coefficient, C_n, with a sum from $n = -\infty$ to $n = \infty$ (Equation 11.47). This illustrates the power of complex notation.

11.9 SPECTRAL REPRESENTATIONS

Additional information about the process represented by a signal can be gathered by displaying the signal's amplitude components versus their corresponding frequency components. This results in representations of the signal's amplitude, power, and power density versus frequency, which are termed the **amplitude spectrum**, **power spectrum**, and **power density spectrum**, respectively. These spectra can be determined from the Fourier series of the signal.

Consider the time average of the square of $y(t)$,

$$\left\langle [y(t)]^2 \right\rangle \equiv \frac{1}{T} \int_0^T [y(t)]^2 \, dt. \qquad \text{[11.49]}$$

The square of $y(t)$ in terms of the Fourier complex exponential sums is

$$[y(t)]^2 = \left(\sum_{m=-\infty}^{\infty} C_m e^{im\omega_0 t} \right) \left(\sum_{n=-\infty}^{\infty} C_n e^{in\omega_0 t} \right) = \sum_{m=-\infty}^{\infty} \sum_{n=-\infty}^{\infty} C_m C_n e^{i(m+n)\omega_0 t}.$$

$$\text{[11.50]}$$

If Equation 11.50 is substituted into Equation 11.49 and the integral on the right-hand side is performed, the integral will equal zero unless $m = -n$, where, in that case, it will equal T. This immediately leads to

$$\left\langle [y(t)]^2 \right\rangle = \sum_{n=-\infty}^{\infty} C_n C_{-n} = 2 \sum_{n=0}^{\infty} \left(\frac{A_n}{2} - \frac{i B_n}{2} \right) \left(\frac{A_n}{2} + \frac{i B_n}{2} \right)$$

$$= 2 \sum_{n=0}^{\infty} \left(\frac{A_n^2}{4} + \frac{B_n^2}{4} \right) = 2 \sum_{n=0}^{\infty} |C_n|^2 = \sum_{n=-\infty}^{\infty} |C_n|^2, \quad \text{[11.51]}$$

11.8 | **MATLAB SIDEBAR**

The M-file `tsfft.m` constructs a plot of a signal and its spectra using the MATLAB `fft` and `psd` commands. The M-file is written such that a two-column user-specified data file can be called. The data file contains two columns, time and amplitude. Information about the digital methods used to do this is contained in Chapter 12.

---

where $y(t)$ is assumed to be real and $\langle [y(t)]^2 \rangle$ is termed the **mean-squared amplitude** or average power. Note that this equation is an approximation to the actual mean squared amplitude whenever the number of summed terms is finite. By comparing Equation 11.50 with the definition of the rms for a discrete signal (see Table 11.1), it can be seen that the power is simply the square of the rms. Recall also that the average power over a time interval equals the product of the total energy expended over that interval and the reciprocal of the time interval.

Equations 11.49 and 11.50 can be combined to yield Parseval's relation for continuous-time periodic signals

$$\frac{1}{T} \int_0^T [y(t)]^2 \, dt = \sum_{n=-\infty}^{\infty} |C_n|^2 . \qquad [\mathbf{11.52}]$$

This states that the total average power in a periodic signal equals the sum of the average powers of all its harmonic components.

The nth amplitude equals $2\sqrt{|C_n|^2} = \sqrt{A_n^2 + B_n^2}$. The plot of $2\sqrt{|C_n|^2}$ versus the frequency is called the amplitude spectrum of the signal $y(t)$. The ordinate units are those of the amplitude. The plot of $|C_n|^2$ versus the frequency is called the power spectrum of the signal $y(t)$. The ordinate units are those of the amplitude squared. The absolute value squared of the C_n Fourier coefficient, which equals one-quarter of the amplitude squared, gives the amount of power associated with the nth harmonic. Both spectra are two-sided because they involve summations on both sides of $n = 0$. They can be made one-sided by multiplying their values by two and then summing from $n = 0$ to ∞. The power density spectrum is the derivative of the power spectrum. Its ordinate units are those of amplitude squared divided by frequency. So, the integral of the power density spectrum over a particular frequency range yields the power contained in the signal in that range. The spectra of a signal obtained by impulsively tapping a cantilever beam supported on its end, shown in Figure 11.14, displays a dominant frequency at approximately 30 Hz.

11.10   **THE CONTINUOUS FOURIER TRANSFORM**

Fourier analysis of a periodic signal can be extended to an aperiodic signal by treating the aperiodic signal as a periodic signal with an infinite period. From Equation 11.33, the Fourier trigonometric series representation of a signal with zero mean ($A_0 = 0$) is

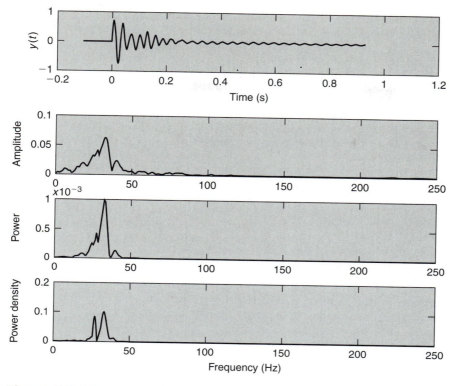

Figure 11.14   A signal and its spectra.

$$y(t) = \sum_{n=1}^{\infty} \left[A_n \cos\left(\frac{2\pi nt}{T}\right) + B_n \sin\left(\frac{2\pi nt}{T}\right) \right]$$

$$= \sum_{n=1}^{\infty} \left[\frac{2}{T} \int_{-T/2}^{T/2} y(t) \cos\left(\frac{2\pi nt}{T}\right) dt \right] \cos\left(\frac{2\pi nt}{T}\right)$$

$$+ \sum_{n=1}^{\infty} \left[\frac{2}{T} \int_{-T/2}^{T/2} y(t) \sin\left(\frac{2\pi nt}{T}\right) dt \right] \sin\left(\frac{2\pi nt}{T}\right). \quad \textbf{[11.53]}$$

Noting $\omega_n = 2\pi n/T$ and $\Delta\omega = 2\pi/T$, where $\Delta\omega$ is the spacing between adjacent harmonics, gives

$$y(t) = \sum_{n=1}^{\infty} \left[\frac{\Delta\omega}{\pi} \int_{-T/2}^{T/2} y(t) \cos(\omega_n t) \, dt \right] \cos(\omega_n t)$$

$$+ \sum_{n=1}^{\infty} \left[\frac{\Delta\omega}{\pi} \int_{-T/2}^{T/2} y(t) \sin(\omega_n t) \, dt \right] \sin(\omega_n t). \quad \textbf{[11.54]}$$

As $T \to \infty$ and $\Delta\omega \to d\omega$, $\omega_n \to \omega$ and the summations become integrals with the limits $\omega = 0$ and $\omega = \infty$,

$$y(t) = \int_0^\infty \left[\frac{d\omega}{\pi} \int_{-\infty}^{+\infty} y(t)\cos(\omega t)\, dt \right] \cos(\omega t)$$

$$+ \int_0^\infty \left[\frac{d\omega}{\pi} \int_{-\infty}^{+\infty} y(t)\sin(\omega t)\, dt \right] \sin(\omega t). \qquad \textbf{[11.55]}$$

Equation 11.55 can be simplified by defining

$$A(\omega) = \int_{-\infty}^{+\infty} y(t)\cos(\omega t)\, dt \qquad \textbf{[11.56]}$$

and

$$B(\omega) = \int_{-\infty}^{+\infty} y(t)\sin(\omega t)\, dt. \qquad \textbf{[11.57]}$$

$A(\omega)$ and $B(\omega)$ are the components of the Fourier transform of $y(t)$. Note that $A(\omega)$ and $B(\omega)$ have units of y/ω, whereas A_n and B_n in Equations 11.35 and 11.36, respectively, have units of y. $A(\omega)$ is an even function of ω and $B(\omega)$ is an odd function of ω. Substituting these definitions into Equation 11.55 yields

$$y(t) = \frac{1}{\pi} \int_0^\infty A(\omega)\cos(\omega t)\, d\omega + \frac{1}{\pi} \int_0^\infty B(\omega)\sin(\omega t)\, d\omega$$

$$= \frac{1}{2\pi} \int_{-\infty}^\infty A(\omega)\cos(\omega t)\, d\omega + \frac{1}{2\pi} \int_{-\infty}^\infty B(\omega)\sin(\omega t)\, d\omega.$$

$$\textbf{[11.58]}$$

Note that these integrals involve negative frequencies. The negative frequencies are simply a consequence of the mathematics and have no mystical significance.

The **complex Fourier coefficient** is defined as

$$Y(\omega) \equiv A(\omega) - i B(\omega). \qquad \textbf{[11.59]}$$

Substituting the definitions of the Fourier coefficients gives

$$Y(\omega) = \int_{-\infty}^{+\infty} y(t)[\cos(\omega t) - i \sin(\omega t)]\, dt = \int_{-\infty}^{+\infty} y(t)e^{-i\omega t}\, dt. \qquad \textbf{[11.60]}$$

This expression gives the Fourier transform of $y(t)$. That is, $Y(\omega)$ is the Fourier transform of $y(t)$. $|Y(\omega)|^2$ corresponds to the amount of power contained in the frequency range from ω to $\omega + d\omega$.

The inverse Fourier transform of $Y(\omega)$ is developed as follows. Note that

$$\frac{i}{2\pi} \int_{-\infty}^{+\infty} A(\omega)\sin(\omega t)\, d\omega = 0, \qquad \textbf{[11.61]}$$

because $A(\omega)$ and $\sin(\omega t)$ are orthogonal. Likewise,

$$\frac{-i}{2\pi} \int_{-\infty}^{+\infty} B(\omega) \cos(\omega t)\, d\omega = 0, \qquad \textbf{[11.62]}$$

because $B(\omega)$ and $\cos(\omega t)$ are orthogonal. These integral terms can be added to those in Equation 11.58, which results in

$$y(t) = \frac{1}{2\pi} \int_{-\infty}^{\infty} A(\omega) \cos(\omega t)\, d\omega + \frac{1}{2\pi} \int_{-\infty}^{\infty} B(\omega) \sin(\omega t)\, d\omega$$

$$+ \frac{i}{2\pi} \int_{-\infty}^{\infty} A(\omega) \sin(\omega t)\, d\omega - \frac{i}{2\pi} \int_{-\infty}^{\infty} B(\omega) \cos(\omega t)\, d\omega$$

$$= \frac{1}{2\pi} \int_{-\infty}^{\infty} [A(\omega) - i B(\omega)][\cos(\omega t) + i \sin(\omega t)]\, d\omega$$

$$= \frac{1}{2\pi} \int_{-\infty}^{\infty} Y(\omega) e^{i\omega t}\, d\omega. \qquad \textbf{[11.63]}$$

Thus, there is the **Fourier transform pair**, which consists of the **Fourier transform** of $y(t)$,

$$Y(\omega) = \int_{-\infty}^{\infty} y(t) e^{-i\omega t}\, dt, \qquad \textbf{[11.64]}$$

and the **inverse Fourier transform** of $Y(\omega)$,

$$y(t) = \frac{1}{2\pi} \int_{-\infty}^{\infty} Y(\omega) e^{i\omega t}\, d\omega. \qquad \textbf{[11.65]}$$

Equation 11.64 determines the amplitude-frequency characteristics of the signal $y(t)$ from its amplitude-time characteristics. Equation 11.65 constructs the amplitude-time characteristics of the signal from its amplitude-frequency characteristics. Taking the inverse Fourier transform of the Fourier transform always should recover the original $y(t)$.

11.11 CONTINUOUS FOURIER TRANSFORM PROPERTIES

Many useful properties can be derived from the Fourier transform pair [3]. These properties are relevant to understanding how signals in the time domain are represented in the frequency domain. Examine the Fourier transform

$$Y(\omega) = 2\pi \delta(\omega - \omega_o), \qquad \textbf{[11.66]}$$

where $\delta(x)$ denotes the delta function, which has the properties $\delta(0) = 1$ and $\delta(\neq 0) = 0$. Substitution of this expression into Equation 11.65 yields

$$y(t) = \frac{1}{2\pi} \int_{-\infty}^{\infty} 2\pi \delta(\omega - \omega_o) e^{i\omega t}\, d\omega = e^{i\omega_o t}. \qquad \textbf{[11.67]}$$

Thus, the Fourier transform of $y(t) = e^{i\omega_o t}$ is $2\pi \delta(\omega - \omega_o)$. This transform occurs in those involving sine and cosine functions, which are the foundations of the Fourier series.

As an example, consider the Fourier transform of $y(t) = \cos(\omega_o t)$ and, consequently, how that signal is represented in the frequency domain. Substituting the complex expression for the cosine function (Equation 11.17) into Equation 11.64 gives

$$Y(\omega) = \frac{1}{2} \int_{-\infty}^{\infty} \left(e^{i\omega_o t} + e^{-i\omega_o t} \right) e^{-i\omega t} \, dt$$

$$= \frac{1}{2} \int_{-\infty}^{\infty} \left(e^{i\omega_o t} e^{-i\omega t} + e^{-i\omega_o t} e^{-i\omega t} \right) dt$$

$$= \pi [\delta(\omega - \omega_o) + \delta(\omega + \omega_o)], \qquad \textbf{[11.68]}$$

using Equations 11.66 and 11.67. This implies that the signal $y(t) = \cos(\omega_o t)$ in the time domain appears as impulses of amplitude π in the frequency domain at $\omega = -\omega_o$ and $\omega = \omega_o$.

In a similar manner, the Fourier transform of $y(t) = y_1(t)e^{i\omega_o t}$ becomes

$$Y(\omega) = \int_{-\infty}^{\infty} y_1(t)e^{i\omega_o t} e^{-i\omega t} \, dt$$

$$= \int_{-\infty}^{\infty} y_1(t)e^{-i(\omega - \omega_o)t} \, dt$$

$$= Y_1(\omega - \omega_o). \qquad \textbf{[11.69]}$$

The multiplication of the signal $y_1(t)$ by $e^{i\omega_o t}$ is called **complex modulation**. Equation 11.69 implies that the Fourier transform of $y_1(t)e^{i\omega_o t}$ results in a frequency shift, from ω to $\omega - \omega_o$, in the frequency domain.

The multiplication of two functions in one domain is related to the convolution the two functions' transforms in the transformed domain. The **convolution** of the functions $y_1(t)$ and $y_2(t)$ for $t \geq 0$ is

$$y_1(t) * y_2(t) = \int_0^t y_1(\tau)y_2(t - \tau) \, d\tau, \qquad \textbf{[11.70]}$$

in which the $*$ denotes the convolution operator. This leads immediately to the multiplication and convolution properties of the Fourier transform that are presented in Table 11.2. These two properties are quite useful because one function often can be expressed as the product of two functions whose Fourier transforms or inverse transforms are known.

Table 11.2   Properties of the continuous Fourier transform.

Signal	Fourier transform	Property
$ay_1(t) + by_2(t)$	$aY_1(\omega) + bY_2(\omega)$	linearity
$y(t - t_o)$	$Y(\omega)e^{-i\omega t_o}$	time shifting
$y(t)e^{i\omega_o t}$	$Y(\omega - \omega_o)$	frequency shifting
$y(at)$	$\frac{1}{\lvert a \rvert} Y(\omega/a)$	time scaling
$\frac{dy(t)}{dt}$	$i\omega Y(\omega)$	time differentiation
$\int_{-\infty}^{t} y(\tau)\,d\tau$	$\pi Y(0)\delta(\omega) + \frac{1}{i\omega} Y(\omega)$	integration
$y_e(t)$	$\mathrm{Re}[Y(\omega)] = a(\omega)$	even signal
$y_o(t)$	$i\,\mathrm{Im}[Y(\omega)] = iB(\omega)$	odd signal
$y_1(t)y_2(t)$	$\frac{1}{2\pi} Y_1(\omega) * Y_2(\omega)$	multiplication
$y_1(t) * y_2(t)$	$Y_1(\omega) + Y_2(\omega)$	convolution

STATEMENT

Determine $3 * \sin 2\omega$.

Example 11.6

SOLUTION

Let $Y_1(\omega) = 3$ and $Y_2(\omega) = \sin 2\omega$. Thus, $Y_1(\tau) = 3$ and $Y_2(\omega - \tau) = \sin 2(\omega - \tau)$. Applying Equation 11.70 gives

$$3 * \sin 2\omega = \int_0^{\omega} 3 \sin 2(\omega - \tau)\,d\tau = \frac{3}{2}\cos 2(\omega - \tau)\big|_0^{\omega} = \frac{3}{2}(1 - \cos 2\omega).$$

REFERENCES

[1] *Merriam-Webster OnLine Collegiate Dictionary* at http://www.m-w.com/.
[2] J. S. Bendat and A. G. Piersol. 1986. *Random Data: Analysis and Measurement Procedures*, 2nd ed. New York: John Wiley and Sons.
[3] A. V. Oppenheim and A. S. Willsky. 1997. *Signals & Systems*, 2nd ed. Upper Saddle River, N.J.: Prentice-Hall.

REVIEW PROBLEMS

1. Consider the deterministic signal $y(t) = 3.8 \sin(\omega t)$, where ω is the circular frequency. Determine the rms value of the signal to three decimal places.

2. Compute the rms of the dimensionless data set in the file data10.dat. Respond to four significant figures.

3. Find the third Fourier coefficient of the function pictured in Figure 11.15, where $h = 1$. (Note: Use $n = 3$ to find the desired coefficient.) Consider the function to be periodic. Respond to three significant figures.

4. What is the value of the power spectrum at the cyclic frequency 1 Hz for the function given by the Figure 11.16?

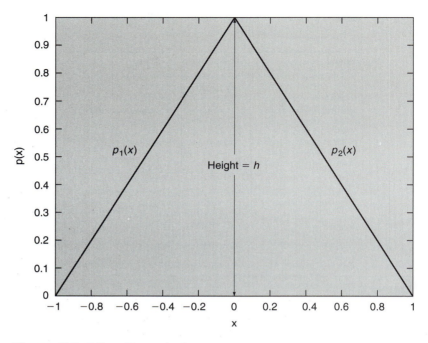

Figure 11.15   Triangular function.

Respond to three significant figures. (The desired result is achieved by representing the function on an infinite interval.)

5. Which of the following functions are periodic? [a] $x(t) = 5\sin(2\pi t)$, [b] $x(t) = \cos(2\pi t)\exp(-5t)$.

6. Which of the following are true? A stationary random process must [a] be continuous, [b] be discrete, [c] be ergodic, [d] have ensemble averaged properties that are independent of time, [e] have time-averaged properties that are equal to the ensemble-averaged properties.

7. Which of the following are true? An ergodic random process must [a] be discrete, [b] be continuous, [c] be stationary, [d] have ensemble-averaged properties that are independent of time, [e] have time-averaged properties that are equal to the ensemble-averaged properties.

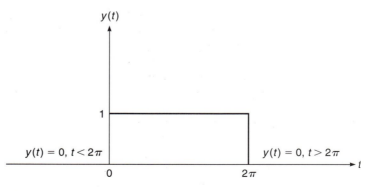

Figure 11.16   Rectangular function.

8. Which of the following are true? A single time history record can be used to find all the statistical properties of a process if the process is [a] deterministic, [b] ergodic, [c] stationary, [d] all of the above.

9. Which of the following are true? The autocorrelation function of a stationary random process [a] must decrease as $|\tau|$ increases, [b] is a function of $|\tau|$ only, [c] must approach a constant as $|\tau|$ increases, [d] must always be nonnegative.

10. Is the signal $x(t) = \cos(2\pi t)\exp(-5t)$ periodic? Answer yes or no.

11. Determine for the time period from 0 to $2T$ the rms value of a square wave of period T given by $y(t) = 0$ from 0 to $T/2$ and $y(t) = A$ from $T/2$ to T.

12. Determine the mean value of a rectified (i.e., always ≥ 0) sine wave that is given by $y(t) = |A\sin\left(\frac{2\pi t}{T}\right)|$ over the time period $0 < t < 100T$.

HOMEWORK PROBLEMS

1. Plot the autocorrelation function as its value versus $|\tau|$ for [a] $x(t) = c$, where c is a constant, [b] $x(t) = \sin(2\pi t)$, [c] $x(t) = \cos(2\pi t)$.

2. Determine the average and rms values for the function $y(t) = 30 + 2\cos(6\pi t)$ over the following time periods: [a] 0 to 0.1 s, [b] 0.4 to 0.5 s, [c] 0 to $\frac{1}{3}$ s, and [d] 0 to 20 s.

3. Consider the deterministic signal $y(t) = 7\sin(4t)$ with t in units of seconds and 7 (the signal's amplitude) in units of volts. Determine the signal's [a] cyclic frequency, [b] circular frequency, [c] period for one cycle, [d] mean value, and [e] rms value. You must put the correct units with each answer. Shown next are two integrals that you may or may not need:

$$\int \sin^2(x)\,dx = \frac{1}{2}x - \frac{1}{4}\sin(2x)$$

and

$$\int \cos^2(x)\,dx = \frac{1}{2}x + \frac{1}{4}\sin(2x).$$

4. For the continuous periodic function $y(t) = y_1(t) - y_2(t)$, where $y_1(t) = A(t/T)^{1/2}$ and $y_2(t) = B(t/T)$, determine for one period [a] the mean value of $y(t)$ and [b] the rms of $y_1(t)$.

5. Determine the Fourier series for the period T of the function described by

$$y(t) = \frac{4At}{T} + A \text{ for } -\frac{T}{2} \leq t \leq 0$$

and

$$y(t) = \frac{-4At}{T} + A \text{ for } 0 \leq t \leq \frac{T}{2}.$$

Do this by hand. Do not use any computer programs. Show all your work. Then, on one graph, plot the three resulting series for 2, 10, and 50 terms along with the original function $y(t)$.

6. Determine the Fourier series of the function

$$y(t) = t \text{ for } -5 < t < 5.$$

(This function repeats itself every 10 units, such as from 5 to 15, 15 to 25, ...). Do this by hand. Do not use any computer programs. Show all your work. Then, on one graph, plot the three resulting series for one, two, and three terms along with the original function $y(t)$.

7. Consider the signal $y(t) = 2 + 4\sin(3\pi t) + 3\cos(3\pi t)$ with t in units of seconds. Determine [a] the fundamental frequency (in Hz) contained in the signal, [b] the mean value of $y(t)$ over the time period from 0 to 2/3s, and [c] sketch the amplitude-frequency spectrum of $y(t)$.

8. For the Fourier series

$$y(t) = (20/\pi)[\sin(4\pi t/7) + 4\sin(8\pi t/7) + 3\sin(12\pi t/7) + 5\sin(16\pi t/7)],$$

determine the amplitude of the third harmonic.

9. Calculate the mean value of a rectified sine wave given by

$$y = |A\sin\frac{2\pi t}{T}|$$

during the time period $0 < t < 1000T$.

10. Determine the rms (in V) of the signal $y(t) = 7\sin(4t)$, where y is in units of V and t is in units of s. An integral that may be helpful is $\int \sin^2 ax\,dx = x/2 - (1/4a)\sin(2ax)$.

11. Determine the Fourier coefficients A_0, A_n, and B_n and the trigonometric Fourier series for the function $y(t) = At$, where the function has a period of 2 s with $y(-1) = -A$ and $y(1) = A$.

CHAPTER REVIEW

ACROSS

1. magnitude
4. type of frequency in units of Hz
5. an even trigonometric function
7. the fundamental frequency is the first (?)
10. having no period
11. mean-squared amplitude
14. procedure by which a signal is constructed of Fourier components
15. signal shape

DOWN

2. has units of radians
3. collection of time history records
4. type of frequency in units of rad/s
6. steady in time
8. the absolute value of a complex number
9. having more than one period
12. nondeterministic
13. the positive square root of the mean of the squares

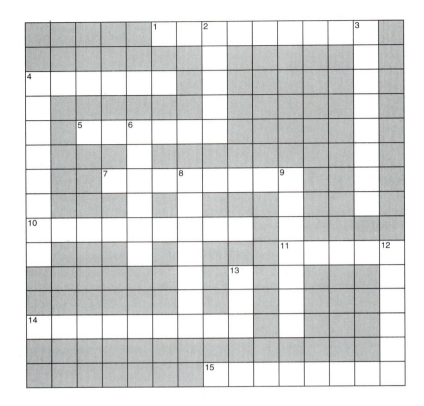

The Tide Predictor

Kelvin's tide predictor, built in 1872, utilizes digital-to-analog conversion. It was designed by William Thomson (1824–1907), later Lord Kelvin, a British physicist and mathematician. It was used to predict the patterns of harbor tides. The positions of the device's 10 dials relate to particular gear settings that correspond to the oceanographic and astronomical data of a harbor. When the handle is cranked, the device calculates the harbor's tide patterns for up to a year, taking only 4 hours to do so.

For even the most stupid of men, by some instinct of nature, by himself and without any instruction (which is a remarkable thing), is convinced that the more observations have been made, the less danger there is of wandering from one's goal.

Jacob Bernoulli. 1713.
Ars Conjectandi.

CHAPTER OUTLINE

12.1   CHAPTER OVERVIEW

Today, most data are acquired and stored digitally. This format is advantageous because of relatively rapid acquisition rates and minimal storage requirements. Digital data acquisition, however, introduces errors. Fortunately, these can be minimized with some foresight. So, how are signal acquisition and analysis done digitally? What errors are introduced? How can these errors be minimized such that the acquired information truly represents that of the process under investigation? Such questions are addressed and answered in this chapter.

12.2   LEARNING OBJECTIVES

You should be able to do the following after completing this chapter:

- Know the difference between analog, discrete, and digital signals
- Describe how a signal's amplitude, time, and frequency are related to one another
- Determine the frequency resolution, Nyquist frequency, and minimum and maximum frequencies for a given sample rate and number of discrete data
- Know the sampling theorem
- Describe aliasing
- Determine the sampling frequency to avoid aliasing, given a deterministic signal
- Use the folding diagram to determine any aliased frequencies
- Describe amplitude ambiguity
- Determine whether or not aliasing and/or amplitude ambiguity will occur for a given periodic signal
- Describe the effects of windowing and ensemble averaging on the frequency spectrum
- Know the concept of the Fourier transform
- Use provided MATLAB software to plot the spectra of discrete data
- Determine the appropriate sample period and rate, filtering, and instrument signal ranges for a typical experiment using a PC-based data acquisition system

12.3   DIGITAL SAMPLING

Consider the analog signal, $y(t)$, shown in Figure 12.1 as a solid curve. This signal is sampled digitally over a period of T seconds at a rate of one sample every δt seconds. The resulting discrete signal, $y(r\delta t)$, is comprised of the analog signal's amplitude values y_1 through y_N at the times $r\delta t$, where $r = 1, 2, \ldots, N$ for N samples. The discrete signal is represented by circles in Figure 12.1. The accurate representation

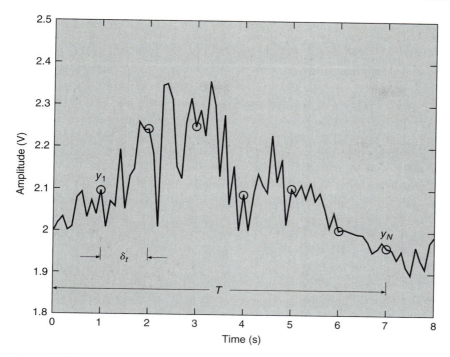

Figure 12.1 Discrete sampling of an analog signal.

of the analog signal by the discrete signal depends on a number of factors. These include, at a minimum, the frequency content of $y(t)$, the time-record length of the signal, $T = N\delta t$, and the frequency at which the signal is sampled, $f_s = 1/\delta t = N/T$.

Further assume that the signal contains frequencies ranging from 0 to W Hz, which implies that the signal's bandwidth is from 0 to W Hz. The minimum resolvable frequency, f_{min}, will be $1/T = 1/(N\delta t)$. If the sampling rate is chosen such that $f_s = 2W$, then, as will be seen shortly, the maximum resolvable frequency, f_{max}, will be $W = 1/(2\delta t)$. Thus, the number of discrete frequencies, N_f, that can be resolved

---

12.1 | MATLAB SIDEBAR

The MATLAB M-file `genplot.m` plots the analog and discrete versions of a user-specified signal. It also stores the discrete signal in a user-specified data file. Figure 12.2 was generated using this M-file for the user-specified signal $y(t) = 3\sin(2\pi t)$ for a sample period of 3 s at a sample rate of 10 samples/s. This M-file can be used to examine the effect of the sample rate and sample period on the discrete representation of an analog signal.

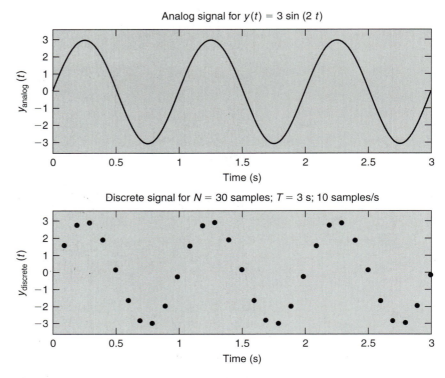

Figure 12.2   Analog and discrete representations of $y(t) = 3 \sin(2\pi t)$.

from $f_{\min}$ to $f_{\max}$ will be

$$N_f = \frac{f_{\max} - f_{\min}}{\delta f} = \frac{1/(2\delta t) - 1/(N\delta t)}{1/(N\delta t)} = \frac{N}{2} - 1. \qquad \text{[12.1]}$$

This implies that there will be $N/2$ discrete frequencies from $f_{\min}$ to *and including* $f_{\max}$. This process is illustrated in Figure 12.3.

Situations arise that introduce errors into the acquired information. For example, if the sampling frequency is too low, the discrete signal will contain false or alias amplitudes at lower frequencies, which is termed **aliasing**. Further, if the total sample period is not an integer multiple of *all* of the signal's contributory periods, **amplitude ambiguity** will result. That is, false or ambiguous amplitudes will occur at frequencies that are immediately adjacent to the actual frequency. Thus, by not using the correct sampling frequency and sampling period, incorrect amplitudes and frequencies result. This is undesirable.

How can these problems be avoided? Signal aliasing can be eliminated simply by choosing a sampling frequency, f_s, equal to *at least twice* the highest frequency, $f_{\max}$, contained in the signal. However, it is difficult to avoid amplitude ambiguity. Its effect only can be minimized. This is accomplished by reducing the magnitude of

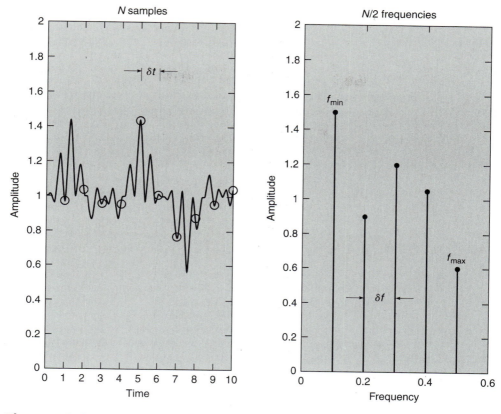

Figure 12.3 Amplitude-time-frequency mapping.

the signal at the beginning and the end of the sample period through a process called **windowing**.

Before discussing each of these effects in detail, the discrete version of the Fourier transform must be considered. This transform yields the amplitude-frequency spectrum of discrete data. The spectrum can be determined for the discrete representation of a known periodic signal, which best illustrates the effects of aliasing and amplitude ambiguity.

12.4 DISCRETE FOURIER TRANSFORM

The discrete Fourier transform is a method used to obtain the frequency content of a signal by implementing the discrete version of the Fourier transform. A more detailed discussion of the discrete and fast Fourier transforms is presented in [2].

Consider a sample of the signal $y(t)$ with a finite record length, T, which is its fundamental period. This signal is sampled N times at δt increments of time. N values of the signal are obtained, $y_n = y(r\delta t)$, where $r = 1, 2, \ldots, N$. The discrete signal becomes

$$y(r\delta t) = y(t) \cdot \tilde{\delta}(t - r\delta t). \qquad \text{[12.2]}$$

The impulse function, $\tilde{\delta}$, is defined such that $\tilde{\delta}(0) = 1$ and $\tilde{\delta}(\neq 0) = 0$.

Now recall the Fourier series representation of $y(t)$,

$$y(t) = \frac{A_0}{2} + \sum_{n=1}^{\infty} \left[A_n \cos\left(\frac{2\pi nt}{T}\right) + B_n \sin\left(\frac{2\pi nt}{T}\right) \right], \qquad \text{[12.3]}$$

with the Fourier coefficients given by

$$A_0 = \frac{2}{T} \int_0^T y(t)\, dt,$$

$$A_n = \frac{2}{T} \int_0^T y(t) \cos\left(\frac{2\pi nt}{T}\right) dt \quad n = 1, 2, \ldots, \infty, \qquad \text{[12.4]}$$

and

$$B_n = \frac{2}{T} \int_0^T y(t) \sin\left(\frac{2\pi nt}{T}\right) dt \quad n = 1, 2, \ldots, \infty. \qquad \text{[12.5]}$$

Fourier analysis of a discrete signal is accomplished by replacing the following in Equations 12.3 through 12.5:

1. The integrals over t by summations over δt;
2. Continuous time t by discrete time $r\delta t$, where $r = 1, 2, \ldots, N$;
3. $T = N\delta t$, where N is a positive even integer; and
4. n from 1 to ∞ with k from 0 to $N/2$.

In doing so, A_n becomes a_k, where

$$a_k = \frac{2}{N\delta t} \sum_{r=1}^{N} y(r\delta t) \cos\left(\frac{2\pi kr\delta t}{N\delta t}\right) \delta t$$

$$= \frac{2}{N} \sum_{r=1}^{N} y(r\delta t) \cos\left(\frac{2\pi kr}{N}\right) \quad k = 0, 1, \ldots, \frac{N}{2}. \qquad \text{[12.6]}$$

Likewise,

$$b_k = \frac{2}{N} \sum_{r=1}^{N} y(r\delta t) \sin\left(\frac{2\pi kr}{N}\right) \quad k = 1, 2, \ldots, \frac{N}{2} - 1 \qquad \text{[12.7]}$$

and

$$c_k = \frac{2}{N} \sum_{r=1}^{N} \sqrt{a_r^2 + b_r^2}. \qquad \text{[12.8]}$$

Note that k represents the discrete frequency and r the discrete sample point. Each discrete sample point can contribute to a discrete frequency. Every sample point's contribution to a particular discrete frequency is included by summing over all sample points at that frequency. This yields the corresponding discrete expression for $y(t)$,

$$y(r\delta t) = \frac{a_0}{2} + \sum_{k=1}^{(N/2)-1} \left[a_k \cos\left(\frac{2\pi rk}{N}\right) + b_k \sin\left(\frac{2\pi rk}{N}\right) \right]$$

$$+ \frac{a_{N/2}}{2} \cos(\pi r). \qquad \textbf{[12.9]}$$

The last term corresponds to $f_{max} = f_N$. The equations for a_k and b_k comprise the **discrete Fourier transform** or DFT of $y(r\delta t)$. The equation $y(r\delta t)$ is the **discrete Fourier series**.

A computer program or M-file can be written to perform the DFT, which would include the following steps:

1. Fix k.
2. Evaluate $2\pi rk/N$ for all r.
3. Compute $\cos(2\pi rk/N)$ and $\sin(2\pi rk/N)$.
4. Compute $y(r\delta t)\cos(2\pi rk/N)$ and $y(r\delta t)\sin(2\pi rk/N)$.
5. Sum these values from $r = 1$ to N to give a_k and b_k as given in Equations 12.6 and 12.7.
6. Repeat for next k.
7. After completing for all k, determine c_k using Equation 12.8.

This method involves N^2 real multiply-add operations.

Alternatively, the DFT can be written using complex notation. Using the Fourier coefficient definitions in Equation 12.5, and introducing Y_n, which was called C_n in Chapter 11, gives

$$Y_n = \frac{A_n}{2} - i\frac{B_n}{2}. \qquad \textbf{[12.10]}$$

This leads to

$$Y_n(t) = \frac{1}{T} \int_0^T y(t) \left[\cos\left(\frac{2\pi nt}{T}\right) - i \sin\left(\frac{2\pi nt}{T}\right) \right] dt$$

$$= \frac{1}{T} \int_0^T y(t) \exp[-i(2\pi nt/T)]dt. \qquad \textbf{[12.11]}$$

By making the appropriate substitutions for T, δt, and n in Equation 12.11, the discrete Fourier transform in complex form becomes

$$Y_k = \frac{1}{N\delta t} \sum_{r=1}^{N} y(r\delta t) \exp\left[-i\left(\frac{2\pi k r \delta t}{N\delta t}\right)\right] dt$$

$$= \frac{1}{N} \sum_{r=1}^{N} y(r\delta t) \exp\left[-i\left(\frac{2\pi k r}{N}\right)\right]$$

$$= \frac{1}{N} \sum_{r=1}^{N} y_r \exp\left[-i\left(\frac{2\pi k r}{N}\right)\right]. \qquad \text{[12.12]}$$

Again, k represents the discrete frequency and r the discrete sample point. This method requires N^2 complex multiplications. Note also that $2\pi rk/N$ can be replaced by $2\pi r f_k \delta t$ because $f_k = k/T = k/(N\delta t)$.

12.5   FAST FOURIER TRANSFORM

The **fast Fourier transform**, or FFT, is a specific type of DFT that is computationally faster than the original DFT. G. C. Danielson and C. Lanczos produced one such FFT algorithm in 1942. J. W. Cooley and J. W. Tukey developed the most frequently used one in the mid-1960s. Danielson and Lanczos showed that a DFT of length N can be rewritten as the sum of two DFTs, each of length $N/2$, one coming from the even-numbered points of the original N, the other from the odd-numbered points [3]. Equation 12.12 can be rearranged to conform to this format as

$$Y_k = \frac{1}{N} \sum_{r=0}^{N-1} y_r e^{-i(2\pi rk/N)}$$

$$= \frac{1}{N} \left\{ \sum_{r=0}^{(N/2)-1} y_{2r} e^{-i[2\pi(2r)k/N]} + \sum_{r=0}^{(N/2)-1} y_{2r+1} e^{-i[2\pi(2r+1)k/N]} \right\}$$

$$= \frac{1}{N} \left\{ \sum_{r=0}^{(N/2)-1} y_{2r} e^{-i[2\pi rk/(N/2)]} \right.$$

$$\left. + W^k \sum_{r=0}^{(N/2)-1} y_{2r+1} e^{-i[2\pi rk/(N/2)]} \right\}, \qquad \text{[12.13]}$$

where $W^k \equiv e^{-i(2\pi k/N)}$. Equation 12.13 can be written in a more condensed form, $Y_k = Y_k^{\text{even}} + W^k Y_k^{\text{odd}}$, where Y_k^{even} is the kth component of the DFT of length $N/2$ formed from the even-numbered y_k values and Y_k^{odd} is the kth component of the DFT of length $N/2$ formed from the odd-numbered y_k values. This approach can be

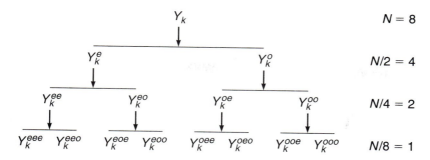

Figure 12.4   DFT sequence for $N = 8$.

applied successively until the last transforms each have only one term. At that point, the DFT of the term equals the term itself, where Y_k (for $k = 0$, $r = 0$, $N = 1$) = $(1/1)y_0 e^{-i \cdot 0} = y_0 = Y_0$.

The sequence of the computational breakdown for $N = 8$ is displayed in Figure 12.4. Symmetry is maintained when $N = 2^M$. Here Y_k^{xxx} are the DFTs of length one. They equal the values of the discrete sample points, $y(r \delta t)$. For a given N, the particular y_k values can be related to a pattern of e's and o's in the sequence. By reversing the pattern of e's and o's (with $e = 0$ and $o = 1$), the value of k in binary is obtained. This is called *bit reversal*. This process is illustrated in Figure 12.5. The speed of this FFT is $\sim O(N \log_2 N)$ vs. $O(N^2)$ for the DFT, which is approximately 40 000 times faster than the original DFT!

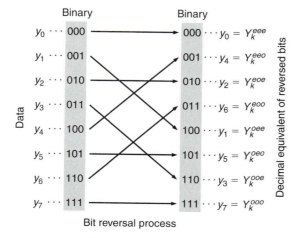

Figure 12.5   DFT sequence for $N = 8$ showing bit reversal.

12.2 | MATLAB SIDEBAR

Some data files are developed from experiments in which the values of a number of variables are recorded with respect to time. Usually such a data file is structured with columns (time and other variables) with an unspecified number of rows. Often, such as for spectral analysis, a variable is analyzed in blocks of a specified and even size, such as 256 or 512. The MATLAB M-file convert.m can be used to select a specific variable with n values and then store the variable in a user-specified data file as m sets of n/m columns. The MATLAB floor command is used by convert.m to eliminate any values beyond the desired number of sets.

Example 12.1 | **STATEMENT**

For the case of $N = 4$, determine the four DFT terms y_0, y_1, y_2, and y_3 for $r = 0, \ldots, N-1$.

SOLUTION

Direct implementation of Equation 12.13 yields

$$Y_k = \frac{1}{4} \sum_{r=0}^{3} y_r e^{-i(2\pi rk/4)}$$

$$= \frac{1}{4} \left[y_0 + y_1 e^{-i(2\pi k/4)} + y_2 e^{-i(2\pi 2k/4)} + y_3 e^{-i(2\pi 3k/4)} \right]$$

$$= \frac{1}{4} \left\{ y_0 + y_2 \cdot e^{-i(2\pi k/2)} + e^{-i(2\pi k/4)} \cdot \left[y_1 + y_3 e^{-i(2\pi k/2)} \right] \right\}$$

Thus, the DFT could be performed computationally faster in the sequence, as illustrated in Figure 12.6, by starting with the even (y_0 and y_2) and odd (y_1 and y_3) pairs.

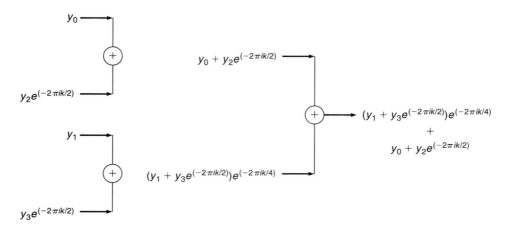

Figure 12.6   More efficient DFT sequence for $N = 4$.

12.3 | MATLAB SIDEBAR

The MATLAB command `fft(y)` is the DFT of the vector y. If y is a power of two, a fast radix-2 FFT algorithm is used. If not, a slower non-power-of-two algorithm is used. The command `fft(y,N)` is the N-point FFT padded with zeros if y has less than N points and truncated if y has more than N points. The M-file `fastyoft.m` uses the MATLAB `fft(y,N)` command to determine and plot the amplitude-frequency of a user-specified $y(t)$. It also plots the discrete version of $y(t)$ for the user-specified number of sample points (N) and sample time increment. The MATLAB commands

used to determine the amplitude (A) and frequency (f) are

```
F=fft(y.n);
A=(2/n)*sqrt(F.*conj(F));
dc=ones(size(t))*mean(y);
f=(1/(n*dt))*(0:(n/2)-1);
```

This M-file was used to construct the figures that follow in the text in which a deterministic $y(t)$ and its amplitude-frequency spectrum are plotted.

12.4 | MATLAB SIDEBAR

Using the M-file `fastyoft.m`, the discrete time series representation and amplitude-frequency spectrum of $y(t) = 5\sin(2\pi t)$ for 128 samples taken at 0.125-s increments can be plotted.

The resulting plots are shown in Figure 12.7. Note that an amplitude of 5 occurs at the frequency of 1 Hz, which is when it should occur.

12.6 ALIASING

Ambiguities arise in the digitized signal's frequency content whenever the analog signal is not sampled at a high enough rate. Shannon's sampling theorem basically states that for aliasing not to occur, the signal should be sampled at a frequency that is greater than twice the maximum frequency contained in the signal, which often is termed the maximum frequency of interest. That is, $f_s > 2f_{max}$. At the sampling frequency $f_s = 2f_{max}$, f_{max} also is known as the Nyquist frequency, f_N. So, $f_s = 2f_N$.

To illustrate analytically how aliasing occurs [1], consider the two signals $y_1(t) = \cos(2\pi f_1 t)$ and $y_2(t) = \cos(2\pi f_2 t)$, in which f_2 is chosen subject to two conditions: (1) $f_2 = 2mf_N \pm f_1$ with $m = 1, 2, \ldots$, and (2) $f_2 > f_N$. These conditions yield specific f_2 frequencies above the Nyquist frequency, all of which alias down to the frequency f_1. The resulting frequencies are displayed on the frequency map shown in Figure 12.8.

Now assume that these two periodic signals are sampled at δt time increments, r times. Then

$$y_1(t) = \cos(2\pi f_1 t) \text{ becomes } y_1(r\delta t) = \cos\left(\frac{2\pi r f_1}{f_s}\right), \qquad \textbf{[12.14]}$$

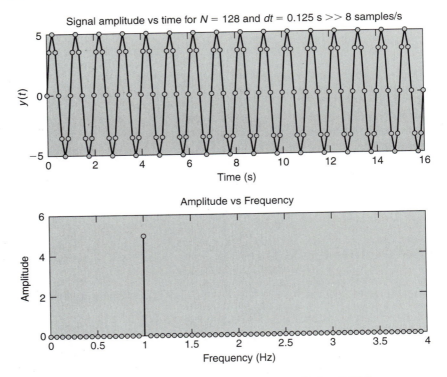

Figure 12.7 Signal and frequency spectrum with $dt = 0.125$ s.

and

$$y_2(t) = \cos(2\pi f_2 t) \text{ becomes } y_2(r\delta t) = \cos\left(\frac{2\pi r f_2}{f_s}\right).$$ **[12.15]**

Further reduction of Equation 12.15 reveals that

$$y_2(r\delta t) = \cos\left(\frac{2\pi r\,[2mf_N \pm f_1]}{f_s}\right)$$

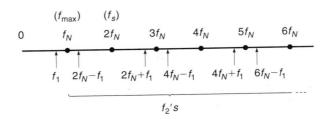

Figure 12.8 Frequency map to illustrate aliasing.

$$= \cos\left(2\pi rm \pm \frac{2\pi r f_1}{f_s}\right)$$

$$= \cos\left(2\pi r \frac{m \pm f_1}{f_s}\right)$$

$$= \cos\left(\frac{2\pi r f_1}{f_s}\right)$$

$$= y_1(r\delta t). \qquad\qquad \textbf{[12.16]}$$

Thus, the sampled signal $y_2(r\delta t)$ will be identical to the sampled signal $y_1(r\delta t)$ and the frequencies f_1 and f_2 will be indistinguishable. In other words, *all* of the signal content at the f_2 frequencies will appear at the f_1 frequency. Their amplitudes will combine in quadrature with the signal's original amplitude at frequency f_1, thereby producing a false amplitude at frequency f_1.

When aliasing occurs, the higher f_2 frequencies can be said to *fold* into the lower frequency f_1. This mapping of the f_2 frequencies into f_1 is illustrated by the folding diagram, as shown in Figure 12.9. The frequency, f_a, into which a frequency f is folded, assuming $f > f_n$, is identified as follows:

1. Determine k, where $k = f/f_N$. Note that $f_N = f_{\max} = f_s/2$.
2. Find the value k_a that k folds into by using the folding diagram. Note $0 \le k_a \le 1$.
3. Calculate f_a, where $f_a = k_a f_N$.

Example 12.2 illustrates aliasing.

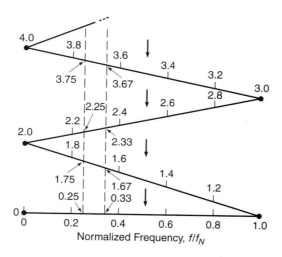

Figure 12.9 The folding diagram.

12.5 | MATLAB SIDEBAR

The MATLAB M-file `foverfn.m` determines the value of the apparent frequency, f_a, given the original frequency, f, and the sampling frequency, f_s. It is the MATLAB equivalent of the folding diagram. The frequencies determined in the previous example can be verified using `foverfn.m`.

Example 12.2 | **STATEMENT**

Assume that there is an analog signal whose highest frequency of interest is 200 Hz ($= f_N$), although there may be frequencies higher than that contained in the signal. According to the sampling theorem, the sampling frequency must be set at $f_s = 400$ Hz for the digitized signal to accurately represent any signal content at and below 200 Hz. However, the signal content above 200 Hz will be aliased. At what frequency will an arbitrary aliased frequency appear?

SOLUTION

According to the folding diagram, for example, the f_2 frequencies of 350 ($1.75 f_N$), 450 ($2.25 f_N$), 750 ($3.75 f_N$), and 850 ($4.25 f_N$) all will map into $f_1 = 50$ Hz ($0.25 f_N$). Likewise, other frequencies greater than f_N will map down to frequencies less than f_N. For example, 334 Hz will map down to 67 Hz.

Thus, for aliasing of a signal not to occur and for the digitized signal not to be contaminated by unwanted higher-frequency content, f_N first must be identified and then set such that $f_s > 2 f_N$. Second, a filter must be used to eliminate all frequency content in the signal above f_N. In an experiment, this can be accomplished readily by filtering the signal with an **anti-alias** (low-pass) **filter** prior to sampling, with the filter cutoff set at f_N.

Example 12.3 | **STATEMENT**

The signal $y(t) = \sin(2\pi 10 t)$ is sampled at 12 Hz. Will the signal be aliased and, if so, to what frequency?

SOLUTION

Here $f = 10$ Hz. For signal aliasing not to occur, the signal should be sampled at a frequency that is at least twice the maximum frequency of interest. For this case, the required minimum sampling frequency would be 20 Hz. Because the signal actually is sampled at only 12 Hz, aliasing will occur. If $f_s = 12$ Hz, then $f_N = 6$ Hz, which is one-half of the sampling frequency. Thus, $k = f/f_N = 10/6 = 1.67$. This gives $k_a = 0.33$, which implies that $f_a = 0.33 f_N = (0.33)(6) = 2$ Hz. So, the aliased signal will appear as a sine wave with a frequency of 2 Hz. This is illustrated in Figure 12.10.

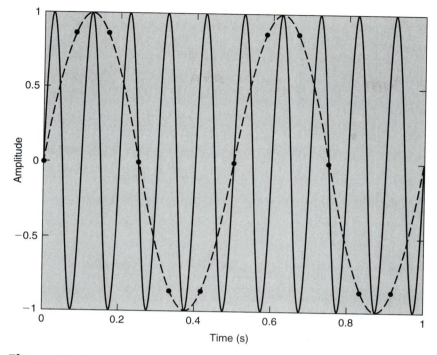

Figure 12.10 Aliasing of $y(t) = \sin(2\pi 10t)$ (solid curve is $y(t)$; dashed curve is aliased signal from sampling at 12 Hz).

12.7 AMPLITUDE AMBIGUITY

Amplitude ambiguity also arises when the sample time period, T_r, is not an integer multiple of the fundamental period of the signal. If the signal has more than one period or is aperiodic, this will complicate matters. For complex periodic signals, T_r must be equal to the least common integer multiple of all frequencies contained in the signal. For aperiodic signals, T_r theoretically must be infinite. Practically, finite records of length T_r are considered and windowing must be used to minimize the effect of amplitude ambiguity. Application of the DFT or FFT to an aperiodic signal implicitly assumes that the signal is infinite in length and formed by repeating the signal of length T_r an infinite number of times. This leads to discontinuities in the amplitude that occur at each integer multiple of T_r, as shown in Figure 12.11 at the time equal to 20 s. These discontinuities are steplike, which introduce false amplitudes that decrease around the main frequencies similar to those observed in the Fourier transform of a step function (see Chapter 11).

Thus, the amplitudes of simple or complex periodic waveforms will be accurately represented in the DFT when $f_s > 2f_{max}$ and $T_r = mT_1$, where $m = 1, 2, \ldots, T_1$

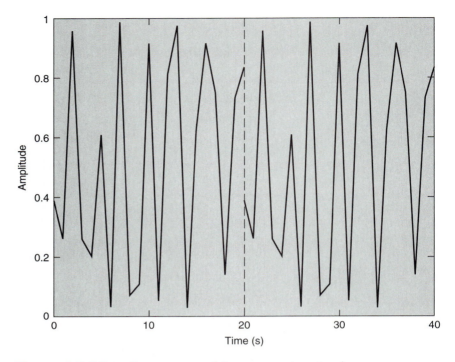

Figure 12.11 Two segments of the same random signal.

is the fundamental period $(= 1/f_1)$ and T_r the total sample period $(= N\delta t = N/f_s)$, which implies that $N = m(f_s/f_1)$. If the later condition is not met, *leakage* will occur in the DFT, appearing as amplitudes at f_1 spilling over into other adjacent frequencies. Further, for DFT computations to be fast, N must be set equal to 2^M, which yields $2^M = m(f_s/f_1)$, where m and N are positive integers. These conditions are summarized as follows:

1. Set $f_{max} = f_N \Rightarrow f_s = 2 f_{max}$, assuming that f_1 and f_{max} are known.

2. Find a suitable N by the steps:

   Choose a value for m, keeping $m \geq 10$.

   Is there an integer solution for M, where $2^M = m(f_s/f_1)$?

   If so, stop.

   If not, iterate until an integer M is found, hence $N = 2^M$ and $T_r = N\delta t$.

For aperiodic and nondeterministic waveforms, the frequency resolution $\delta f \ (= 1/N\delta t)$ is varied until leakage is minimized. Sometimes, all frequencies are not known. In that case, to minimize leakage, windowing must be used.

Example 12.4 illustrates the effect of sampling rate on the resulting amplitude-frequency spectrum in terms of either aliasing or amplitude ambiguity.

STATEMENT **Example 12.4**

Convert the analog voltage, $E(t) = 5 \sin 2\pi t$ mV, into a discrete time signal. Specifically, using sample time increments of [a] 0.125 s, [b] 0.30 s, and [c] 0.75 s, plot each series as a function of time over at least one period. Discuss apparent differences between the discrete representation of the analog signal. Also, compute the DFT for each of the three discrete signals. Discuss apparent differences. Use a data set of 128 points.

SOLUTION

$$y(t) = 5 \sin(2\pi t) \Rightarrow f = 1 \text{ Hz}$$

Aliasing will *not* occur when $f_s (= 1/dt) > 2f$ ($f = 1$ Hz). Amplitude ambiguity will *not* occur when $T = mT_1 \Rightarrow m = fNdt$ (m : integer).

For part [a] $f_s > 2f$ and $m = (1)(128)(0.125) = 16 \Rightarrow$ no aliasing or amplitude ambiguity. The result is shown in Figure 12.7, which was presented previously to illustrate the FFT.

For part [b] $f_s > 2f \Rightarrow$ no aliasing, and $m = (1)(128)(0.3) = 38.4 \Rightarrow$ amplitude ambiguity will occur. This is displayed in Figure 12.12. The amplitude, however, is less than the actual amplitude (here it is less than 4). Around that frequency the amplitude appears to leak into adjacent frequencies.

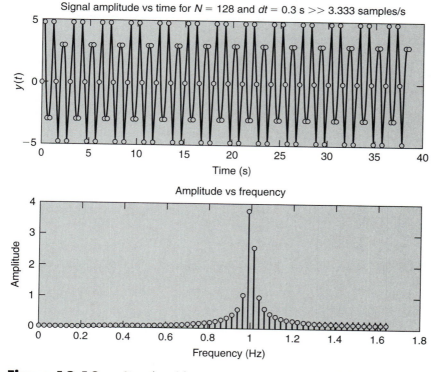

Figure 12.12   Signal and frequency spectrum with $dt = 0.3$ s.

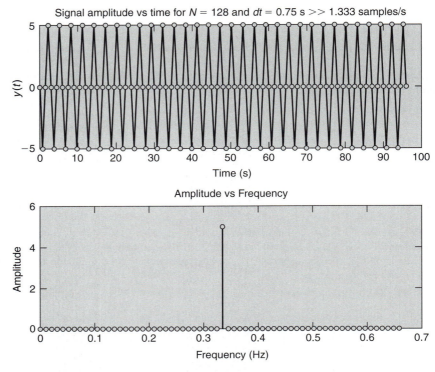

Figure 12.13   Signal and frequency spectrum with dt = 0.75 s.

For part [c] $f_s < 2f \Rightarrow$ aliasing will occur, and $m = (1)(128)(0.75) = 96 \Rightarrow$ no amplitude ambiguity will be present. This is shown in Figure 12.13. The aliased frequency can be determined using the aforementioned folding-diagram procedure. Here, $f_s = 4/3$, $f_N = 2/3$, and $f = 1$. This leads to $k = 3/2$, which implies $k_a = 1/2$ using the folding diagram. Thus, $f_a = (1/2)(2/3) = 1/3$ Hz.

Now consider an example where both aliasing and amplitude ambiguity can occur simultaneously.

Example 12.5   **STATEMENT**

Compute the DFT for the discrete time signal that results from sampling the analog signal, $T(t) = 2\sin(4\pi t)\,°C$, at sample rates of 3 and 8 Hz. Use a data set of 128 points. Discuss and compare your results.

SOLUTION

$$T(t) = 2\sin(4\pi t) \Rightarrow 2 \text{ Hz}$$

For the sample rate of 3 Hz, $f_s = 2f \Rightarrow$ aliasing will occur and $m = (1)(128)(1/3) = 42.67 \Rightarrow$ amplitude ambiguity will be present. The results are presented in Figure 12.14. The aliased frequency occurs where the amplitude is maximum, at 1 Hz. This can be determined using the aforementioned folding-diagram procedure. Here, $f_s = 3$, $f_N = 3/2$, and $f = 2$. This leads to $k = 4/3$, which implies $k_a = 2/3$ using the folding diagram. Thus, $f_a = (2/3)(3/2) = 1$. Also, note the distortion of the signal's time record that occurs because of the low sampling rate.

When the sampling rate is increased to 8 Hz, $f_s > 2f \Rightarrow$ no aliasing occurs. Also $m = (1)(128)(0.125) = 16 \Rightarrow$ no amplitude ambiguity occurs. This is shown in Figure 12.15, which is the correct spectrum.

Analysis becomes more complicated when more than one frequency is present in the signal. Next, consider an example that involves a signal containing two frequencies.

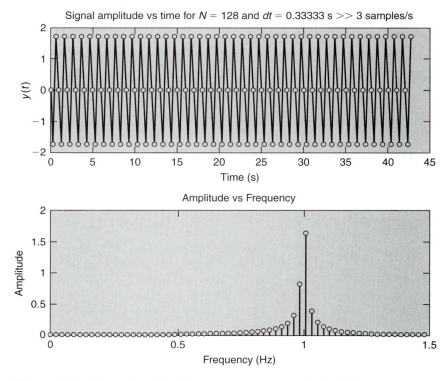

Figure 12.14   Signal and frequency spectrum with $dt = 1/3$ s.

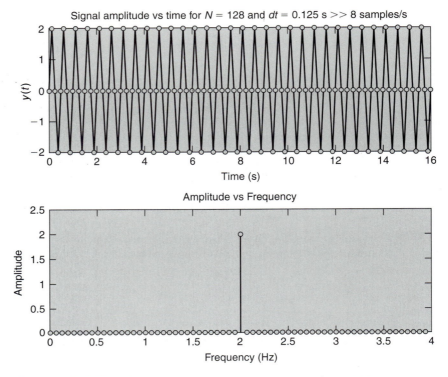

Figure 12.15   Signal and frequency spectrum with $dt = 0.125$ s.

Example 12.6 | **STATEMENT**

Consider the signal $y(t) = 3.61\sin(4\pi t + 0.59) + 5\sin(8\pi t)$. Plot $y(t)$ versus time and the resulting frequency spectrum for the following cases and discuss what is observed with respect to aliasing and amplitude ambiguity:

(i) $N = 100$, $f_s = 50$

(ii) $N = 20$, $f_s = 10$

(iii) $N = 10$, $f_s = 5$

(iv) $N = 96$, $f_s = 5$

(v) $N = 96$, $f_s = 10$

SOLUTION

$y(t) = 3.61\sin(4\pi t + 0.59) + 5\sin(8\pi t)$. So, $f_1 = 2$ Hz and $f_2 = 4$ Hz, which implies that $f_{max} = 4$ Hz. If $f_s = 5$ samples/s

$$\frac{f_s}{f_1} = \frac{5}{2} = 2.5 > 2 \Rightarrow \text{no aliasing,}$$

$$\frac{f_s}{f_2} = \frac{5}{4} = 1.25 < 2 \Rightarrow \text{aliasing will occur.}$$

To where will the 4-Hz component be aliased?

$$f_N = \frac{f_s}{2} = \frac{5}{2} = 2.5 \text{ Hz} \Rightarrow \frac{f_2}{f_N} = \frac{4}{2.5} = 1.6.$$

Using the folding diagram, $1.6 f_N$ is folded down to $0.4 f_N = (0.4)(2.5) = 1$ Hz. That is, the 4-Hz component appears as a 1-Hz component.

But what about amplitude ambiguity? $T_1 = 1/f_1 = 1/2$ s, and $T_2 = 1/f_2 = 1/4$ s. The total sample period, T, must contain integer multiples of both T_1 and T_2 so as not to have amplitude ambiguity in both components. This can be easily met by having $T = mT_1 = m/2$ s (since $T_2 = T_1/2$). In essence, the least common integer multiple of T_1 and T_2 is sought. Recalling that $T = N\delta t = N/f_s$, if $m/2 = N/f_s$ no amplitude ambiguity will be present. That is, when $N = f_s(m/2) = \frac{5}{2}m$, and m and N are integers. For example, $m = 2$ ($N = 5$), and $m = 4$ ($N = 10$) should be acceptable. However, all the frequencies of interest should be seen in the spectrum. The highest frequency of interest is $f_{max} = f_N = f_s/2$. Because there are $N/2$ discrete frequencies and assuming that $f_{min} = 0$ needs to be considered, this yields

$$f_{max} = \frac{1}{T}\left(\frac{N}{2} - 1\right) = \frac{f_s}{2} = \frac{f_s}{N}\left(\frac{N}{2} - 1\right),$$

where $T = N/f_s$. Solving for N,

$$N = \frac{2 f_s}{f_s - 4}.$$

So, when $f_s = 5$, $N = 10/(5 - 4)$. Thus, $N = 10$ is the minimum N needed to see both components.

For case (i), the discrete signal and the amplitude-frequency spectrum are correct. This is shown in Figure 12.16.

For case (ii), the spectrum remains correct and the discrete signal, although still correct, does not represent the signal well because of the lower sampling rate. This is illustrated in Figure 12.17.

For case (iii), the 4-Hz component is aliased down to 1 Hz and the 2-Hz component is correct. No amplitude ambiguity has occurred. This is displayed in Figure 12.18.

For case (iv), amplitude ambiguity has occurred for both components and only the 4-Hz component is aliased down to 1 Hz. This is shown in Figure 12.19.

For case (v), amplitude ambiguity has occurred but not aliasing. This is presented in Figure 12.20.

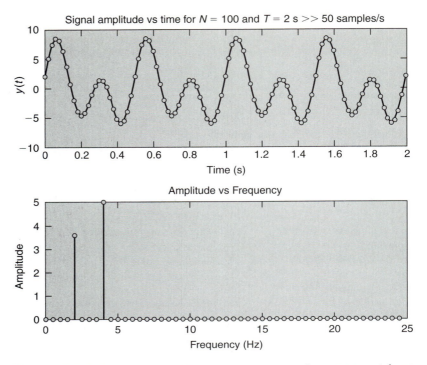

Figure 12.16   Signal and frequency spectrum with N = 100 and f = 50.

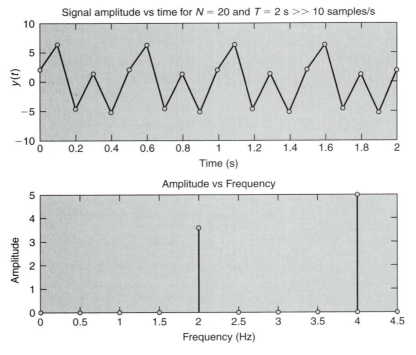

Figure 12.17   Signal and frequency spectrum with N = 20 and f = 10.

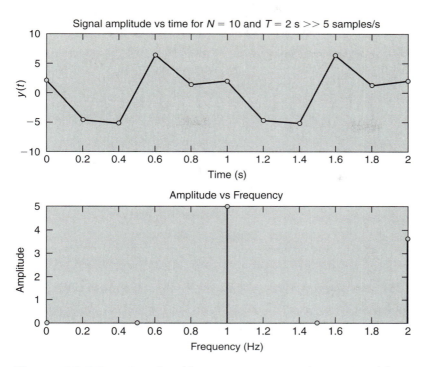

Figure 12.18 Signal and frequency spectrum with N = 10 and f = 5.

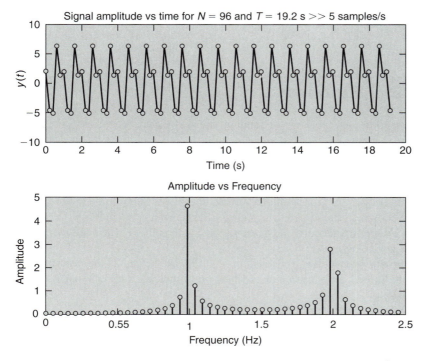

Figure 12.19 Signal and frequency spectrum with N = 96 and f = 5.

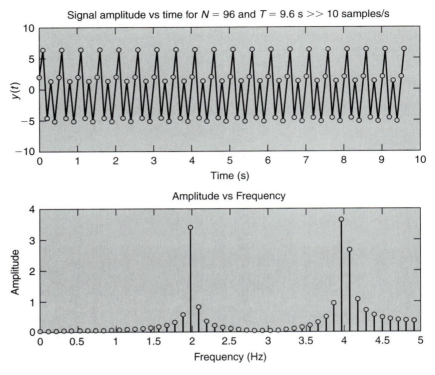

Figure 12.20   Signal and frequency spectrum with $N = 96$ and $f = 10$.

12.8   WINDOWING

Because either an aperiodic or random signal does not have a period, the Fourier transform applied to such a signal's finite record length produces leakage in its spectrum. This effect can be minimized by applying a **windowing function**. This process effectively attenuates the signal's amplitude near the discontinuities that were discussed previously in Section 12.7, thereby leading to less leakage. The windowing function actually is a function that weights the signal's amplitude in time. The effect of a windowing function on the spectrum can be seen by examining the convolution of two Fourier transforms, one of the signal and the other of the windowing function. This is considered next.

The discrete Fourier transform of $y(t)$ can be viewed as the Fourier transform of an unlimited time history record $v(t)$ multiplied by a rectangular time window $u(t)$, where

$$u(t) = \begin{cases} 1 & 0 \leq t \leq T \\ 0 & \text{otherwise.} \end{cases}$$

This is illustrated in Figure 12.21. Now,

$$Y_n(f, T) = \frac{1}{T} \int_0^T y_n(t) \exp(-i2\pi ft)\, dt,$$ [12.17]

which leads to

$$Y(f) = \int_{-\infty}^{\infty} U(\alpha) V(f - \alpha)\, d\alpha.$$ [12.18]

This is the convolution integral [3]. So, the Fourier transform of $y(t)$, $Y(f)$, is the convolution of the Fourier transforms of $u(t)$ and $v(t)$, which are denoted by $U(\alpha)$ and $V(f - \alpha)$, respectively. For the present application, $u(t)$ represents the windowing function and $y(t)$ is $v(t)$. The record length is denoted by T. Various windowing functions can be used. Each yield different amounts of leakage suppression.

The rectangular windowing function $u_{\text{rect}}(t)$ has the Fourier transform $U_{\text{rect}}(f)$, given by

$$U_{\text{rect}}(f) = T\left(\frac{\sin \pi ft}{\pi ft}\right).$$ [12.19]

The relatively large side lobes of $|U(f)/U(0)|$ produce a leakage at frequencies separated from the main lobe. This produces a distortion throughout the spectra,

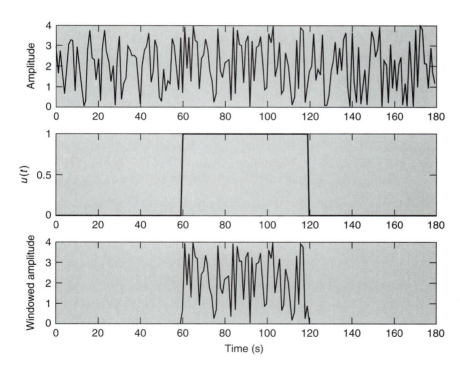

Figure 12.21 Rectangular windowing.

especially when the signal consists of a narrow band of frequencies. This is illustrated in Figure 12.22.

It is better to taper the signal to eliminate the discontinuity at the beginning and end of the data record. There are many types of tapering windows available. The cosine-squared or Hanning window is used most commonly. This is given by

$$u_{\text{hanning}}(t) = \frac{1}{2}\left(1 - \cos\frac{2\pi t}{T}\right) = 1 - \cos^2\left(\frac{\pi t}{T}\right) \qquad \textbf{[12.20]}$$

when $0 \le t \le T$. Otherwise, $u_{\text{hanning}} = 0$. Further,

$$U_{\text{hanning}}(f) = \frac{1}{2}U(f) + \frac{1}{4}U(f - f_1) + \frac{1}{4}U(f + f_1), \qquad \textbf{[12.21]}$$

where $f_1 = 1/T$ and $U(f)$ is defined as before with

$$U(f - f_1) = T\left[\frac{\sin \pi(f - f_1)T}{\pi(f - f_1)T}\right] \qquad \textbf{[12.22]}$$

and

$$U(f + f_1) = T\left[\frac{\sin \pi(f + f_1)T}{\pi(f + f_1)T}\right]. \qquad \textbf{[12.23]}$$

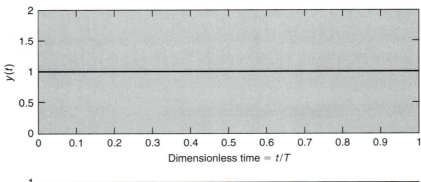

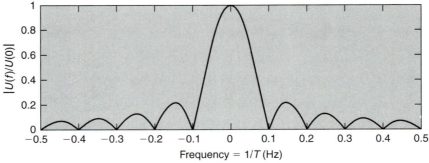

Figure 12.22   The rectangular (boxcar) window.

12.6 | MATLAB SIDEBAR

The MATLAB M-file `windowhan.m` determines and plots the Hanning window function and its spectral attenuation effect. It uses the MATLAB command `hanning(N)` for the record length of N. Other window functions can be added to this M-file. This M-file was used to generate Figures 12.22 and 12.23 using the appropriate window function.

The Hanning window is presented in Figure 12.23.

Finally, it should be noted that windowing reduces the amplitudes of the spectrum. For a given window, this loss factor can be calculated [2]. For the Hanning window, the amplitude spectrum must be scaled by the factor $\sqrt{8/3}$ to compensate for this attenuation. Thus,

$$Y_n(f_k) = \delta t \sqrt{\frac{8}{3}} \sum_{m=0}^{N-1} y_{nm}\left(1 - \cos^2\frac{\pi m}{N}\right)\exp\left(-i\frac{2\pi km}{N}\right), \qquad \textbf{[12.24]}$$

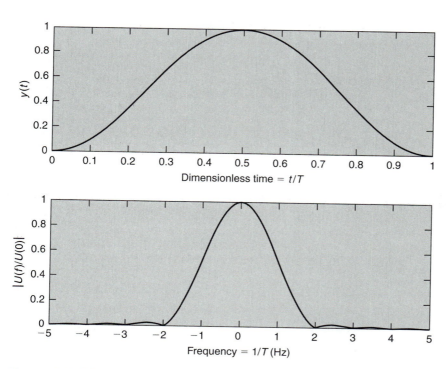

Figure 12.23   The Hanning window.

12.7 | MATLAB SIDEBAR

The M-file `fwinyoft.m` determines and plots the discrete signal and the amplitude-frequency of a user-specified $y(t)$. It also applies a windowing function, such as the MATLAB command `hanning(N)`. This M-file can be adapted to read a time record input of any signal. The command `hanning(N)` weights a discrete signal of length N by a factor of zero at its beginning and end points, a factor of 1 at its midpoint (at $N/2$) and by its cosine weighting as expressed by Equation 12.20. An example of its output is shown in Figure 12.24, in which the signal $y(t) = 3\sin(2\pi t)$ is

sampled at a frequency of 4 Hz for 128 samples. The effect of the Hanning window in this case is to reduce the actual amplitude and create some leakage. Actually, windowing functions are not needed for periodic signals. They can obscure the spectrum. The application of windowing to aperiodic or random signals would present no problem because the inherent leakage that results from finite sampling would be attenuated. The Hanning window was applied in this example simply to illustrate how windowing is implemented.

with $f_k = k/(N\delta t)$, where $k = 0, 1, 2, \ldots, N/2$, and

$$G_y(f_k) = \frac{2}{n_d N\delta t} \sum_{i=1}^{n_d} |Y_n(f_k)|^2. \qquad [12.25]$$

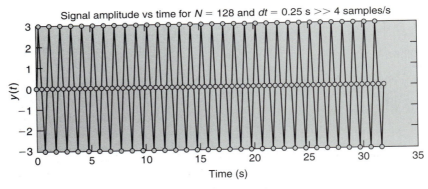

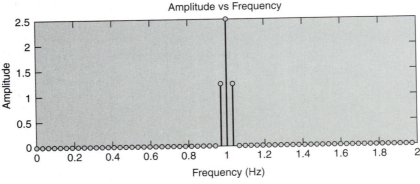

Figure 12.24   The Hanning window applied to $y(t) = 3\sin(2\pi t)$.

The recommended procedure [2] for computing a smoothed amplitude spectrum is the following:

1. Divide data into n_d blocks, each of size $N = 2^M$.

2. Taper the data values in each block $\{y_n\}$ ($n = 0, 1, 2, \ldots, N-1$) with a Hanning or other window.

3. Compute the N-point FFT for each data block yielding $Y_n(f_k)$, adjusting the scale to account for the tapering loss (for example, multiply by $\sqrt{8/3}$ for the Hanning window).

4. Compute $G_y(f_k)$ for n_d blocks.

REFERENCES

[1] T. G. Beckwith, R. D. Marangoni, and J. H. Leinhard V. 1993. *Mechanical Measurements*, 5th ed. New York: Addison-Wesley.

[2] J. S. Bendat and A. G. Piersol. *Random Data: Analysis and Measurement Procedures*, 2nd ed. New York: John Wiley and Sons.

[3] W. H. Press, S. A. Teukolsy, W. T. Vetterling, and B. P. Flannery. 1992. *Numerical Recipes*, 2nd ed. New York: Cambridge University Press.

REVIEW PROBLEMS

1. Determine the number of discrete frequencies from the minimum frequency to and including the maximum frequency that will appear in an amplitude-frequency plot of a signal sampled every 0.2 s. The signal's minimum frequency is 0.5 Hz.

2. Determine the frequency resolution of a signal sampled 256 times for 4 s.

3. Determine the alias frequency of a signal's component at 100 Hz that is sampled at 50 Hz.

4. Does windowing of a signal produce a signal with no amplitude distortion?

HOMEWORK PROBLEMS

1. A discrete Fourier transform of the signal $B(t) = \cos(30t)$ is made to obtain its power-frequency spectrum. $N = 4000$ samples are chosen. Determine [a] the period of $B(t)$ (in s), [b] the cyclic frequency of $B(t)$ (in Hz), [c] the appropriate sampling rate (in samples/s), and [d] the highest resolvable frequency, $f_{\max}$ (in Hz). Finally, [e] if $N = 4096$ samples were chosen instead, would the computations of the Fourier transform be faster or slower and why?

2. Using a computer program written by yourself or constructed from available subroutines, calculate and plot the following: one plot containing the continuous signal $y(t)$ and its discrete version versus time, and the other plot containing the amplitude spectrum of the discrete sample. Provide a complete listing of the program. Do this for each of the cases below. Support any observed aliasing or leakage of the sample by appropriate calculations. State, for each case, whether or not aliasing and/or leakage occur. The continuous signal is given by

$$y(t) = 5\sin(2\pi t + 0.8) + 2\sin(4\pi t) + 3\cos(4\pi t)$$
$$+ 7\sin(7\pi t).$$

The cases to examine are [a] $N = 100, T = 10$ s; [b] $N = 100, T = 18$ s; [c] $N = 100, T = 20$ s; [d] $N = 100, T = 15$ s; [e] $N = 50, T = 15$ s; where N represents the number of sample points and T the sample period.

3. Consider the signal $y(t) = 5 + 10\cos(30t) + 15\cos(90t)$. Determine [a] the frequencies (in Hz) contained in the signal, [b] the *minimum* sample rate (in samples/s) to avoid aliasing, and [c] the frequency resolution of the frequency spectrum if the signal is sampled at that rate for 2 s. Finally, sketch [d] the amplitude-frequency spectrum of $y(t)$ and [e] the amplitude-frequency spectrum if the signal is sampled at 20 samples/s.

4. A velocity sensor is placed in the wake behind an airfoil subjected to a periodic pitching motion. The output signal of the velocity transducer is $y(t) = 2\cos(10\pi t) + 3\cos(30\pi t) + 5\cos(60\pi t)$. Determine [a] the fundamental frequency of the signal (in Hz), [b] the maximum frequency of the signal (in Hz), [c] the range of acceptable frequencies (in Hz) that will avoid signal aliasing, and [d] the minimum sampling frequency (in Hz) that will avoid *both* signal aliasing and amplitude ambiguity if 20 samples of the signal are taken during the sample period. Finally, if the signal is sampled at 20 Hz, determine [e] the frequency content of the resulting discrete series, $y(\delta nt)$, and [f] the resulting discrete series $y(\delta nt)$.

5. The signal $y(t) = 3\cos(\omega t)$ has a period of 4 s. Determine the following for the signal: [a] its amplitude, [b] its cyclic frequency, [c] the *minimum* sampling rate to avoid aliasing, [d] its mean value over *three* periods, and [e] its rms value over *two* periods. The formula

$$\int [\cos(ax)]^2 dx = \frac{1}{a}\left[-\frac{1}{2}\cos(ax)\sin(ax) + \frac{1}{2}ax\right]$$

may or may not be useful.

CHAPTER REVIEW

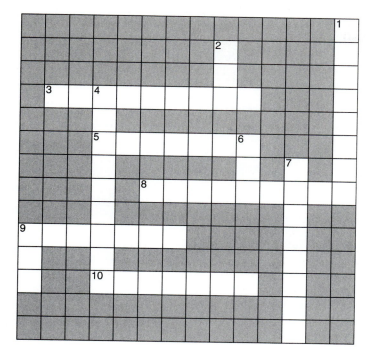

ACROSS

3. occurs whenever the sample period is not an integer multiple of the signal's frequencies
5. the sampling frequency should be twice this frequency
8. signal reduction at the beginning and end of a signal
9. type of diagram used to determine aliased frequencies
10. type of windowing filter

DOWN

1. occurrence of false amplitudes at lower frequencies
2. the discrete version of the FFT
4. a range of signal frequencies
6. ratio of number of sample points to number of discrete frequencies
7. sampled at fixed time increments
9. Danielson and Lanczos developed one type of it

Chapter 4

AC	alternating current
C	capacitance
DC	direct current
E	electric field
e_{load}	loading error
E_{Th}	Thévenin equivalent voltage
F_e	electric force
I	current
L	inductance
P	electric power
q	charge
R	resistance
T	temperature
V	voltage
X	reactance
Z	impedance
Z_{Th}	Thévenin equivalent impedance
$\Delta\Phi$	potential difference
ρ	electric resistivity
Φ	electric potential

Chapter 5

a_{cal}	calibration accuracy
C_v	specific heat at constant volume
E	total energy
e_{abs}	absolute error
$F(t)$	input forcing function
h	convective heat transfer coefficient

I	current
K	static sensitivity
k	spring constant
n	order
M	magnitude ratio
q	charge
Q	rate of heat transfer
R	resistance
y_h	homogeneous solution
y_p	particular solution
β	lag time
δ	fractional dynamic error
ϵ	absolute error
τ	time constant
ϕ	phase shift
ω_d	ringing frequency

Chapter 6

C	capacitance
E_i	input signal
E_o	output signal
f_c	cutoff frequency
G	gain
L	inductance
R	resistance
R_0	resistance at reference temperature
T	temperature
T_0	reference temperature
α, β, γ	coefficients of thermal expansion

Δt	time lag		$\{S_x\}$	pooled standard deviation
ϵ	relative permittivity		$\{S_{\bar{x}}\}$	pooled standard deviation of the means
ϵ_0	permittivity of free space			
η	material constant		$\{S_x^2\}$	pooled variance
ν	Poisson's ratio		$\{\bar{x}\}$	pooled weighted mean
ρ	resistivity		$\{S_x\}_w$	pooled weighted standard deviation
ϕ	phase lag			
			$\{S_{\bar{x}}\}_w$	pooled weighted standard deviation of the means

Chapter 7

$\cup$	union		α	level of significance
$\cap$	intersection		Γ	gamma function
$\mid$	independent		ν	degrees of freedom
C_m^n	combinations of n objects taken m at a time		σ	standard deviation
			σ^2	variance
$E[\ ]$ or $\langle\ \rangle$	expected value		χ^2	chi-squared
f_j	frequency density		$\tilde{\chi}^2$	reduced chi-squared
f_j^*	normalized frequency density			
Ku	kurtosis			

Chapter 9

P_m^n	permutations of n objects taken m at a time		B_i	source bias limit
			B_x	systematic uncertainty
$p(x)$	probability density function		C	confidence level
$P(x)$	probability distribution function		$\frac{\Delta B_x}{B_x}$	relative systematic uncertainty
Pr[ ]	probability of an event		e_i	elemental instrument error (see text for specific types)
S_x^2	sample variance			
Sk	skewness		e_I	overall instrument error
x'	mean value of population		FSO	full-scale output
μ_m	mth central moment		N	number of measurements in a sample
σ^2	variance			
			P_i	source precision index

Chapter 8

$p(x)$	probability density function		P_x	random uncertainty
$P(x)$	probability distribution function		$P_{\bar{x}}$	random uncertainty in the average value of the measurand x
$p(z_1)$	normal error function			
S_x	sample standard deviation		R	result
$\pm t_{\nu,P}$	precision interval of the true mean		t	Student's t factor
			u_0	zeroth-order instrument uncertainty
$\bar{x}$	sample mean		u_I	instrument uncertainty
x'	mean value of population		u_d	design-stage uncertainty
$\{\bar{x}\}$	pooled mean		u_N	nth order uncertainty

u_r^2	combined estimated variance	$\bar{y}_i$	estimate of the sample mean
u_R	uncertainty in a result	y_i'	true mean of y_i
u_x	uncertainty in x	α	true intercept
U_x	overall uncertainty	β	true slope
x'	population mean	Γ	gamma function
x_{true}	true value of x	ρ	population correlation coefficient
$\bar{x}$	sample mean	ρ_{xx}	population autocorrelation coefficient
ν	degrees of freedom		
ν_{B_x}	degrees of freedom for the systematic uncertainty	ρ_{xy}, ρ_{yx}	population cross-correlation coefficients
ν_{P_x}	degrees of freedom for the random uncertainty	σ_{xy}	population covariance
		σ_x	population variance of x
θ_i	absolute sensitivity coefficient	$\sigma_{y_i}^2$	population variance of y_i

Chapter 10

D	the sum of the squares of all the vertical distances between the measured and calculated values used in regression analysis
d_i	the ith vertical distance between the i measured and calculated values
e_i	the error between the ith calculated and measured values
G	a coefficient matrix used in multivariable linear analysis
R	a coefficient matrix used in multivariable linear analysis
r	sample correlation coefficient
R_{xx}	autocorrelation function
R_{xy}, R_{yx}	cross-correlation functions
S_{xy}	standard error of the fit
$\bar{S}_x^2$	variances of the x errors
SSE	sum of the squares of the error
SSR	sum of the squares of the regression
SST	sum of the squares of the total error
$t_{\nu,P}S_{yx}$	precision interval
$\bar{x}$	sample mean
$\bar{x}^2$	sample mean of the sum of the squares

Chapter 11

$A_0, A_n,$ B_n, C_n	Fourier coefficients
$\check{A}$	complex amplitude
$A(\omega)$	real component of the Fourier transform
$B(\omega)$	imaginary component of the Fourier transform
c	wave speed
C	amplitude
f	cyclic frequency
i	imaginary number $= \sqrt{-1}$
n	integer
$R_x(\tau)$	temporal-averaged autocorrelation coefficient
$R_x(t_1, t_1 + \tau)$	ensemble-averaged autocorrelation coefficient
S_x	sample standard deviation of x
S_x^2	sample variance of x
t	time
T	fundamental period
$\bar{x}$	sample mean value of x
x_{rms}	root-mean-square of x
$Y(\omega)$	complex Fourier coefficient

z	complex number	f_1	fundamental frequency
z^*	complex conjugate	f_N	Nyquist frequency
α	phase angle resulting from complex amplitude	f_s	sampling frequency
		k	folding diagram factor
κ	wave number	k_a	folding diagram aliased factor
λ	wavelength	m	positive integer
μ_x	temporal-averaged mean value	N	number of sample points
$\mu_x(t_1)$	ensemble-averaged mean value	T_1	fundamental period
ω	circular frequency	T_r	time-record length
ϕ	phase angle	$u(t)$	windowing function
		$U(f)$	Fourier transform of the windowing function

Chapter 12

$A_0, A_n,$		Y_k	discrete Fourier transform
B_n, C_n	Fourier coefficients	$Y(f)$	Fourier transform
$a_0, a_k,$		$y(r\delta t)$	discretized signal
b_k, c_k	discrete Fourier coefficients	δf	frequency increment
f	frequency	δt	time increment
f_a	aliased frequency		

GLOSSARY

A

absolute error   the absolute value of the difference between an indicated value and its true value

absolute quantization error   one-half the instrument resolution

absolute sensitivity coefficient   change in a result due to an incremental change in a particular variable

absolute uncertainty   uncertainty in a particular variable

accuracy   closeness of agreement between a measured and true value

active   requiring no external power supply to produce a voltage or current

active filter   filter composed of operational amplifiers, resistors, and capacitors

actuator   transducer whose output provides motion

aliasing   false lower frequencies created by a sampling rate less than twice the highest frequency of interest

almost-periodic   comprised of two or more sinusoids of arbitrary frequencies

alternating current (AC)   current varying cyclically in time

ammeter   meter that measures current

ampere   amount of current that must be maintained between two wires separated by one meter in free space in order to produce a force between the two wires equal to 2×10^{-7} N/m of wire length

amplitude   magnitude or size

amplitude ambiguity   false amplitudes occurring in the amplitude spectrum at the fundamental and adjacent frequencies

amplitude spectrum   plot of a signal's amplitude versus frequency

analog   continuous in time and magnitude

anode   positively charged terminal that loses electrons

analysis of variance   ANOVA; statistical analysis that tests the significance of differences among three or more sample means to ascertain if they come from the same population

anti-alias filter   filter that prevents signal aliasing

autocorrelation   correlation of a signal with itself at various time delays

autocorrelation coefficient   number between -1 and 1 that characterizes the extent of autocorrelation

autocorrelation function   function used to determine the autocorrelation

B

balanced bridge   Wheatstone bridge when the products of the cross-bridge resistances are equal

band-pass filter   filter passing a signal's amplitude over a range of frequencies, but not above and below that range

bandwidth   range of frequencies over which the output amplitude of a system remains above 70.7 % of its input amplitude

base   center element of a transistor

base unit   dimensionally independent unit

beat frequency   relatively low frequency equal to one-half the difference in two nearly equal frequencies

bias error   *see* systematic error

binomial distribution   discrete distribution describing the probability of one of two possible outcomes

C

calibration   process of comparing the response of an instrument to a standard input over some range

calibration curve   plot of a calibration's input versus output data

candela   luminous intensity in a given direction of a source that emits radiation at only the specific wavelength of 540×10^{12} hertz and that has a radiant intensity in that direction of 1/683 watts per steradian

capacitance   ratio of the charge on one of a pair of conductors to the electrical potential difference between them

cathode   negatively charged terminal that gains electrons

center frequency   frequency equal to one-half the sum of the low and high cutoff frequencies

central moments   statistical moments calculated with respect to the centroid or mean

central tendency   tendency to scatter about an average value

charge   electrical quantity representative of the excess or deficiency of electrons

class interval   interval range of values

coefficient of determination   square of the sample correlation coefficient

coherent   no numerical factors other than unity occur in all unit equations

collector   one of the end elements of a transistor

combination   number of possible ways that members of a set can be arranged irrespective of their order

combined estimated variance   *see* combined standard uncertainty

combined standard uncertainty   combination of all individual uncertainties

common-mode rejection ratio (CMRR)   the ratio of a system's differential-to-common-mode voltage gains

complement   *see* null set

complex   having more than one period

complex conjugate   complex number with a complex part that is opposite in sign to its complex number counterpart

complex Fourier coefficient   complex coefficient defined in terms of the Fourier transform components

complex modulation   multiplication of the signal $x(t)$ by $e^{i\omega_o t}$

conditional probability   probability of an event given that specified events have occurred previously

consistent   no numerical factors other than 1 occur for all unit equations in a system of units

continuous   without a break or cessation

control experiment   experiment nearly identical to the subject experiment as possible

controlled experiment   experiment in which all the variables involved in the process under investigation are identified and controlled

conventional current   current moving from anode to cathode

conventionalism process in which experiments are performed to illustrate an aspect of nature

convolution integral of two functions over an interval, where one function is delayed in time or sample number with respect to the other function

correlation analysis process of calculating the correlation coefficient

correlation coefficient number between -1 and 1 that characterizes the amount of correlation

covariance mathematical function that characterizes the relationship between two random variables

coverage factor number representing of an assured probability of occurrence that multiplies the combined standard uncertainty to determine the overall uncertainty

critically damped having a damping ratio of unity

cross-correlation correlation between two variables

cross-correlation coefficient number between -1 and 1 that characterizes the extent of cross-correlation

cross-correlation function expectation of the product of two variables' values, in which one of the variables is offset in time or sample number with respect to the other

cumulative probability distribution function *see* probability distribution function

current charge per unit time

cutoff frequency frequency at which a signal's amplitude is attenuated

D

damping ratio nondimensional parameter characterizing a second-order system

decade frequency ratio of ten-to-one

deflection method method using a balanced Wheatstone bridge to achieve an output voltage proportional to a change in resistance

degrees of freedom number of data points minus the number of constraints used for the required calculations

dependent affected by changes in an independent variable

derived unit type of unit composed of base and supplementary units

design of experiments statistical method that determines the sensitivity of an outcome to levels of factors

detailed uncertainty analysis uncertainty analysis, which identifies the systematic and random errors contributing to each measurand's overall uncertainty and then propagates them into the final result

deterministic signal signal that is predictable in time or space, such as a sine wave or a ramp function

digital having discrete values at discrete times or locations

dimension measure of spatial extent

direct current (DC) current constant in time

discrete having values at distinct times or locations

discrete Fourier series discrete representation of a Fourier series

discrete Fourier transform (DFT) discrete representation of the Fourier transform

domain sensor category that is chemical, electrical, magnetic, mechanical, radiant, or thermal

dynamic varying in time

dynamic calibration calibration using a time-dependent input

dynamic error error related to the amplitude difference between a system's input and output

E

electric field electric force acting on a positive charge divided by the magnitude of the charge

electric force force acting between two charged bodies separated at a distance

electric potential   potential energy per unit charge

electric noise   anything that obscures a signal

electric power   electrical energy transferred per unit time

electric resistivity   material property related to its resistance

electromotive force   the work per unit charge done by an electrical device to raise its electric potential

emitter   electrode in a transistor where electrons originate

empiricism   practice of relying on observing and experiment

engineering gage factor   the ratio of small, finite changes in relative resistance to relative length

ensemble   collection of time history records

ergodic   ensemble-averaged values equal the corresponding average values computed over time from an arbitrary, single time history in the ensemble

even function   function symmetric about the ordinate

event   outcome

exhaustive   space spanned by a set and its complement

expectation   *see* expected value

expected value   probabilistic average value

experiment   act in which one physically intervenes with the process under investigation and notes the results

explorational   conducted to explore an idea or possible theory

extraneous   variable that cannot be controlled

F

factorial design   statistical method that identifies factors having the greatest influence on a result

fallibilism   process in which experiments are performed to test the validity of a conjecture

fast Fourier transform (FFT)   method that recursively divides the sample points in one-half down to two-point samples before it performs the Fourier transform

finite   bounded or limited in magnitude or in spatial or temporal extent

first central moment   mean

first-order replication level   level that considers the additional random error resulting from small uncontrolled factors

Fourier analysis   procedure that identifies the Fourier amplitudes of a signal

Fourier coefficients   coefficients in a Fourier series

Fourier series   series represented by sines and cosines of different periods and amplitudes that are added together to form an infinite series

Fourier synthesis   procedure that constructs a signal representation from a series of appropriately weighted sines and cosines

Fourier transform   mathematical transformation of a signal that gives the signal's amplitude versus frequency

Fourier transform pairs   pair of equations consisting of the Fourier transform and the inverse Fourier transform

fourth central moment   *see* kurtosis

fossilized   treated as a fixed systematic error

fractional uncertainty   uncertainty in a result divided by the value of the result

frequency   measure of a signal's temporal variation

frequency density distribution   plot of the number of occurrences of a certain value divided by the product of the total number of occurrences and the class interval width versus the value of the occurrence

frequency distribution   plot of the number of occurrences of a certain value divided by the total number of occurrences versus the value of the occurrence

frequency response   response of a system as a function of input waveform frequency

fundamental dimensions   length, mass, time, temperature, electric current, amount of substance, and luminous intensity

fundamental period   smallest value of time for which a signal repeats itself

G

Gaussian distribution   normal distribution

general uncertainty analysis   simplified approach to uncertainty analysis that considers each measurand's overall uncertainty and its propagation into the final result

H

heterodyning   when two waves of equal amplitude and nearly equal frequencies are added, resulting in a slow-beat frequency equal to one-half the difference in the nearly equal frequencies

high cutoff frequency   the higher frequency of a bandwidth at which cutoff occurs

high-pass filter   filter that passes a signal's amplitude above but not below a specific frequency

histogram   literally a *picture of cells*; plot of the number of occurrences of a certain value versus the value of the occurrence

hypothesis testing   statistical method to assess the plausibility of an assumption

hysteresis   difference in the indicated value obtained when approaching a particular input value in increasing versus decreasing directions

I

impedance   electrical resistance of a circuit containing linear passive components (resistors, capacitors, and inductors)

impedance matching   when the output impedance of a device connected to another device equals the input impedance of that device

impulse   rapid change of a variable in time

independent   not dependent on another variable (in experiments); not dependent on another event (in probability)

inductance   ratio of magnetic flux to current

inductivism   process arriving at the laws and theories of nature based on facts gained from experiments

infer   estimate statistically

inner product   integral of the product of two variables over a specified interval

instrument error   sum of an instrument's elemental errors identified through calibration

interference noise   electrical noise caused by another signal

intersection   set of all members common to both sets

intrinsic noise   electrical noise caused by inherent fluctuations in a material's state

intrinsically linear variables   variables in a higher-order equation that can be transformed to yield a linear equation

invasive   located within the environment under investigation

inverse Fourier transform   inverse of the Fourier transform that gives the signal's amplitude versus time

K

kelvin   1/273.16 of the thermodynamic temperature of the triple point of pure water

kilogram   unit of mass defined as the mass of a cylinder of platinum-iridium alloy kept by the International Bureau of Weights and Measures in Sèvres, France

Kirchhoff's current (or first) law   current into a circuit junction equals current out of the junction

Kirchhoff's voltage (or second) law   sum of potential differences around a circuit loop equals zero

kurtosis   fourth central moment normalized by the square of the variance

L

least significant digit   rightmost nonzero digit

level of significance   one minus the χ^2 probability

linear device   device in which the output amplitude is linearly proportional to its input amplitude

loading error   one minus the ratio of measured output impedance to true output impedance

local gage factor   ratio of differential changes in the relative resistance to the relative length

lognormal distribution   continuous distribution of the logarithm of a normally distributed variable

loop   closed path in a circuit going from one node back to itself without passing through any intermediate node more than once

low cutoff frequency   the lower frequency of a bandwidth at which cutoff occurs

low-pass filter   filter that passes a signal's amplitude below but not above a specific frequency

M

magnitude   extent of dimension; size

magnitude ratio   ratio of a dynamic system's output amplitude to its input amplitude

mean   average; first central moment

measure   *see* magnitude

mean-squared amplitude   expectation of the square of the amplitude

measurand   measured variable

measurement error   true, unknown difference between measured value and true value

measurement uncertainty   estimate of the error in a measurement

meter   length that light travels in a vacuum during a time interval equal to 1/299 792 458 seconds

mode   most frequently occurring value

modulus   absolute value of a complex number

mole   amount of substance of a system that contains as many elementary units as there are the number of atoms ($6.022\ 142 \times 10^{23}$) in 0.012 kilograms of carbon-12

most significant digit   leftmost nonzero digit

mutually exclusive   two sets not sharing any common members

mutually orthogonal   set in which each distinct pair of functions is orthogonal

N

node   point in a circuit where two or more elements meet

nondeterministic   random

noninvasive   located outside of the environment under investigation

nonrecursive   when the input to a device is the signal of interest and the device's output that is *not* fed back to the input

nonstationary   not stationary (*see* stationary)

normal distribution   continuous distribution caused by a very large number of small, uncontrollable factors that influence the outcome

normal equations   equations resulting from the method of least squares

normalized z-variable   a nondimensional variable indicating the number of standard deviations that a specific value deviates from the mean value

Norton equivalent circuit theorem   any two-terminal network of linear impedances and current sources can be replaced by an equivalent circuit consisting of an ideal current source in parallel with an impedance

notch filter filter that passes a signal's amplitude over a range of frequencies above and below a specified range

nth-order replication level level at which more than one random error beyond that in the first-order replication level is considered

null method use of a Wheatstone bridge to determine an unknown resistance by having two other resistances fixed and varying the fourth until the bridge is balanced

null set set of all occurrences in which a desired event is not the outcome

numerical equation equation containing only the measures of physical quantities

O

octave frequency ratio of two-to-one

odd function function symmetric about the origin

ohmmeter meter that measures resistance

order degree in a continuum of size or quantity

orthogonal property of two functions whose inner product is equal to zero over an interval

outcome result of a test

overall uncertainty measure of the uncertainty in a variable; the product of the coverage factor and the combined standard uncertainty

overdamped having a damping ratio greater than unity

P

parameter variable or function of variables that is fixed (in experiments); a measure of a population's characteristics (in probability)

passive requiring an external power supply to produce a voltage or current

pedagogical class of experiment designed to teach the novice or to demonstrate something that is already known

periodic repeating itself in time

permutations number of ways that a set can be arranged respective of its members' order

phase lag lag of an output signal with respect to an input signal

Poisson distribution continuous distribution describing rarely occurring events

pooled formed into one set from a set of replicated experiments each involving multiple measurements

population collection of all possible values of a random variable

population correlation coefficient ratio of the population covariance to the square root of the product of the population standard deviations

population covariance expectation of the product of two variables, each evaluated with respect to their respective mean

potential difference difference between two electric potentials

power density spectrum plot of a signal's power density versus frequency

power spectrum plot of a signal's power versus frequency

precision variation of a variable's values obtained by repeated measurements

precision error *see* random error

precision interval interval characterized by the product of a coverage factor and a random uncertainty

precision interval of the true mean interval that contains the estimate of the true mean, as characterized by the product of Student's t factor and the sample standard deviation

precision interval of the true variance interval that contains the estimate of the true variance, as characterized by degrees of freedom, chi-square, and sample variance values

probability   number of specific occurrences over the total number of occurrences

probability density function (pdf)   function when integrated over a range of values yields the probability of occurrence of those values

probability distribution function (PDF)   integral of the probability density function; also known as the cumulative probability distribution function

R

ramp method   method to perform electronically analog-to-digital conversion by increasing a voltage and comparing it to the analog input signal's voltage

random   having no particular order

random error   error related to the scatter in the data obtained under fixed conditions; also known as the precision error

random uncertainty   estimate of the random error

random variable   variable whose value has no deterministic relation to any of its other values

range   lower to upper limits of an instrument or test

rationalism   theory that reason in itself is a source of knowledge superior to and independent of sensory perceptions

reactance   influence of a coil of wire on an alternating current passing through it that impedes the current

readability   closeness with which the scale of the instrument is read

rectifier   electrical component that changes alternating current into direct current

recursive   when the input to a device is the signal of interest *and* the device's output connected back to its input

reduced chi-squared variable χ^2 variable normalized by the number of degrees of freedom

regression analysis   process identifying the regression coefficients in the method of least squares

regression coefficients   coefficients found in the method of least squares

relative accuracy   accuracy divided by the true value

relative error   absolute error divided by the absolute value of the true value

relative systematic uncertainty   ratio of the reliability of the systematic uncertainty to the systematic uncertainty

reliability   estimate of the accuracy of a systematic uncertainty

repeatability   ability to achieve the same value upon repeated measurement

repetition   repeated measurements made during the same test under the same operating conditions

replacement   return of members to their set after selection, thereby allowing for their reoccurrence

replicates   experiments identical to the original

replication   duplication of an experiment under similar operating conditions

resistance   defined by Ohm's law as the ratio of voltage to current

resistivity   material property defined as the ratio of electric field at a point to current density at that point; reciprocal of conductivity

resolution   smallest physically indicated division that an instrument displays or is marked

result   variable that is a function of one or more measurands

ringing frequency   frequency at which a second-order system rings or continually oscillates

rise time   time required for a first-order system to respond to 90 % of a step change

root mean square (rms) positive square root of the mean of the squares

round off truncate a number to its desired length

s

sample subset of the population

sample correlation coefficient ratio of the sample covariance to the square root of the product of the sample standard deviations

sample mean mean of a sample

sample variance variance of a sample

sample-to-sample measurand values are recorded for multiple samples

scattergram discrete representation of an analog signal

scientific method method of investigation involving observation and theory to test scientific hypotheses

scientific notation system in which numbers are expressed as products consisting of a number between 1 and 10 multiplied by the appropriate power of 10

second duration of 9 192 631 770 cycles of the radiation associated with the transition between two hyperfine levels of the ground state of cesium-133

second central moment variance

sensor device that senses a physical stimulus and converts it into an impulse

sequential systematically increased

set group of all occurrences in which a desired event is the outcome

settling time time beyond which a second-order system's response remains within $\pm$ 10 % of its steady-state value

signal measurement system's representation of the temporal variation of a measurand

signal conditioning preparing the signal in its final form to be processed optimally and then recorded

signal processing operating on a signal to obtain desired results

signal sample period time period used to determine the statistical properties of a signal

significant figures number of digits required to express a result

simple having one period

simple *RC* filter filter comprised of a resistor and a capacitor

skewness third central moment normalized by the cube of the standard deviation

source groups groups that help to categorize sources of error, which are typically grouped as calibration, data acquisition, and data reduction

standard known value usually used as a basis of calibration

standard deviation square root of the variance, which characterizes the width of the probability distribution

standard deviation of the means (SDOM) standard deviation of the mean values obtained from groups of repeated measurements

standard error of the fit error characterizing the differences between data and its curve fit

standardized normal variate nondimensional variable indicating the number of standard deviations that a variable deviates from its mean value

static steady in time

static calibration calibration performed when the system is static

static sensitivity slope of a static calibration curve at a particular input value

stationary see **strongly stationary** and **weakly stationary**

statistics branch of applied mathematics concerned with the collection and interpretation of quantitative data and the use of probability theory to estimate population parameters

steady-state response   part of a second-order system's response that is periodic

strongly stationary   having all ensemble moments invariant with respect to the record's time

Student's *t* distribution   continuous distribution representing a small sample of a normally distributed variable

successive approximation method   method to perform electronically analog-to-digital conversions by subtracting the analog input signal from a digital-to-analog converter's output signal

supplementary unit   nondimensional unit that does not represent a fundamental dimension

system of units   system in which physical quantities can be expressed and related to one another through physical laws

systematic error   error related to the difference between a measured and true value; sometimes called the bias error

systematic uncertainty   estimate of the systematic error

T

third central moment   *see* skewness

Thévenin's equivalent circuit theorem   any two-terminal network of linear impedances and voltage sources can be replaced by an equivalent circuit consisting of an ideal voltage source in series with an impedance

Thévenin's equivalent impedance   ratio of Thévenin equivalent voltage to short-circuit current

Thévenin's equivalent voltage   open-circuit voltage

time constant   characteristic system-response time to step-input forcing

time history record   plot of a signal's amplitude versus time for a given period of time

time lag   delay in time between a signal's input and output through a device

timewise   experiment in which measurand values are recorded sequentially in time

transducer   device that changes an impulse into a desired quantity

transient response   part of a second-order system's response that decays in time

transient solutions   homogeneous solutions to a differential equation that decay to zero in time

true mean value   mean value of the population

true value   error-free value of a variable

true variance   variance of the population

U

uncertainty   estimate of error in a variable

underdamped   having a damping ratio less than unity

union   set of all members of two sets that are in only one, only in the other, or in both

unit   precisely specified quantity in terms of which the magnitudes of other quantities of the same kind can be stated

unit equation   equation in which only units are used or defined

V

validational   experiment conducted to validate a specific hypothesis

variables   physical quantities involved in the process that can undergo change and thereby affect the process

variability   variation in value with time

variance   statistical measure of the spread of values with respect to their mean

variational   experiment quantifying the mathematical relationships between experimental variables

voltmeter   meter that measures voltage

W

waveform actual shape of a signal

wave number number of waves in 2π units of length

weakly stationary having the ensemble mean and autocorrelation invariant with respect to the record's time

Weibull distribution continuous distribution describing the time to failure of a physical system

Wheatstone bridge electrical circuit consisting of four resistors in a specific configuration and a voltage source

windowing mathematical method that reduces the magnitude of a signal record at its beginning and end

windowing function a mathematical function that windows a signal record

Z

zeroth central moment integral of the probability density function; equals unity if the probability density function is normalized

zeroth-order replication level level at which only measurement system errors are present

C

CONVERSIONS

Table C.1   Mechanical and thermal dimensions, units, and conversion factors.

Dimension [fundamental]	SI Units	TE Units	Conversion Factor
Length $[L]$	m	ft	1 ft = 0.3048 m
Mass $[M]$	kg	slug	1 slug = 14.5939 kg
Time $[T]$	s	s	*same*
Temperature $[\theta]$	K	°R	1 °R = K/1.8
Area $[L^2]$	m^2	ft^2	1 ft^2 = 0.09290 m^2
Volume $[L^3]$	m^3	ft^3	1 ft^3 = 0.02832 m^3
Velocity, speed $[LT^{-1}]$	m/s	ft/s	1 ft/s = 0.3048 m/s
Acceleration $[LT^{-2}]$	m/s^2	ft/s^2	1 ft/s^2 = 0.3048 m/s^2
Momentum $[MLT^{-1}]$	kg · m/s	slug · ft/s	1 slug · ft/s = 4.448 kg · m/s
Plane angle [dimensionless]	rad	rad	*same*
Solid angle [dimensionless]	sr	sr	*same*
Cyclic frequency $[T^{-1}]$	Hz = cycles/s	Hz	*same*
Wavenumber $[L^{-1}]$	1/m	1/ft	ft^{-1} = 3.2808 m^{-1}
Force $[MLT^{-2}]$	N = kg · m/s^2	lbf	1 lbf = 4.448 N
Pressure, stress $[ML^{-1}T^{-2}]$	Pa = N/m^2 = kg/(m · s^2)	lbf/ft^2	1 lbf/ft^2 = 47.88 Pa
Energy, work, heat, force moment $[ML^2\,T^{-2}]$	J = N · m = kg · m^2/s^2	lbf · ft	1 lbf · ft = 1.3558 J
Power, radiant flux $[ML^2\,T^{-3}]$	W = J/s = kg · m^2/s^3	lbf · ft/s	1 lbf · ft/s = 1.3558 W
Mass density $[ML^{-3}]$	kg/m^3	slug/ft^3	1 slug/ft^3 = 515.4 kg/m^3
Energy density $[L^2\,T^{-2}]$	J/kg = m^2/s^2	ft^2/s^2	1 ft^2/s^2 = 0.09290 m^2/s^2
Viscosity (absolute) $[ML^{-1}\,T^{-1}]$	Pa/s = kg/(m · s)	slug/(ft · s)	1 slug/(ft · s) = 47.88 kg/(m · s)
Viscosity (kinematic) $[L^2\,T^{-1}]$	m^2/s	ft^2/s	1 ft^2/s = 0.09290 m^2/s
Thermal conductivity $[MLT^{-2}\theta^{-1}]$	W/(m · K) = kg · m/(s^2 · K)	slug · ft/(s^2 · °R)	1 slug · ft/(s^2 · °R) = 2.4712 W/(m · K)
Specific* heat, specific* entropy $[L^2\,T^{-2}\theta^{-1}]$	J/(kg · K) = m^2/(s^2 · K)	ft^2/(s^2 · °R)	1 ft^2/(s^2 · °R) = 0.1672 m^2/(s^2 · K)
Specific* enthalpy $[L^2\,T^{-2}]$	m^2/s^2	ft^2/s^2	1 ft^2/s^2 = 0.09290 m^2/s^2
Radiant intensity $[ML^2\,T^{-3}]$	W/sr = kg · m^2/(s^3 · sr)	lbf · ft/(s · sr)	1 lbf · ft/(s · sr) = 1.3558 W/sr

Adapted from F. M. White. 1999. *Fluid Mechanics*, 4th ed. New York: McGraw-Hill.
* The term "specific" means per unit mass.

Table C.2   Rotational dimensions, units, and conversion factors.

Dimension [fundamental]	SI Units	TE Units	Conversion Factor
Angular displacement [dimensionless]	rad	rad	*same*
Angular velocity $[T^{-1}]$	rad/s	rad/s	*same*
Angular momentum $[ML^2T^{-1}]$	$kg \cdot m^2 \cdot rad/s$	$slug \cdot ft^2 \cdot rad/s$	$1\ slug \cdot ft^2 \cdot rad/s = 1.3558\ kg \cdot m^2 \cdot rad/s$
Angular acceleration $[T^{-2}]$	rad/s^2	rad/s^2	*same*
Torque $[ML^2\ T^{-2}]$	$kg \cdot m^2/s^2$	$slug \cdot ft^2/s^2$	$1\ slug \cdot ft^2/s^2 = 1.3558\ kg \cdot m^2/s^2$
Rotational inertia $[ML^2]$	$kg \cdot m^2$	$slug \cdot ft^2$	$1\ slug \cdot ft^2 = 1.3558\ kg \cdot m^2$
Rotational energy $[ML^2\ T^{-2}]$	$kg \cdot m^2 \cdot rad^2/s^2$	$slug \cdot ft^2 \cdot rad^2/s^2$	$1\ slug \cdot ft^2 \cdot rad^2/s^2 = 1.3558\ kg \cdot m^2 \cdot rad^2/s^2$

| Adapted from D. Halliday and R. Resnick. 1966. *Physics*. Combined edition. Parts I and II. New York: John Wiley and Sons.

Table C.3   Photometry dimensions, units, and conversion factors.

Dimension [fundamental]	SI Units	TE Units	Conversion Factor
Luminous intensity $[\mathcal{K}]$	$cd = lm/sr$	candlepower/sr	1 candlepower/sr = 12.566 cd
Luminous flux $[\mathcal{K}]$	$lm = cd \cdot sr$	candlepower	1 candlepower = 12.566 lm
Illuminance $[\mathcal{K}L^{-2}]$	$lx = lm/m^2$	footcandle = lm/ft^2	1 footcandle = 10.764 lx
Luminance $[\mathcal{K}L^{-2}]$	cd/m^2	footlambert = cd/ft^2	1 footlambert = $10.764\ cd/m^2$
Irradiance $[MT^{-3}]$	W/m^2	$lbf/(ft \cdot s)$	$1\ lbf/(ft \cdot s) = 14.5939\ W/m^2$
Radiance $[MT^{-3}]$	$W/(m^2 \cdot sr)$	$lbf/(ft \cdot s \cdot sr)$	$1\ lbf/(ft \cdot s \cdot sr) = 14.5939\ W/(s^3 \cdot sr)$

| Information from S. Strauss. 1995. *The Sizesaurus*. New York: Kodansha International; and C. F. Bohren. 1991. *What Light Through Yonder Window Breaks?* New York: John Wiley and Sons.

Table C.4   Acoustic dimensions, units, and conversion factors.

Dimension [fundamental]	SI Units	TE Units	Conversion Factor
Pressure $[ML^{-1}\ T^{-2}]$	$Pa = kg/(m \cdot s^2)$	lbf/ft^2	$1\ lbf/ft^2 = 47.88\ Pa$
Power $[ML^2\ T^{-3}]$	$W = kg \cdot m^2/s^3$	hp	1 hp = 745.7 W
Intensity $[MT^{-3}]$	$W/m^2 = kg/s^3$	$lbf \cdot ft/ft^2$	$1\ lbf \cdot ft/ft^2 = 14.5939\ W/m^2$
Intensity level* [dimensionless]	dB	dB	*same*
Pressure level# [dimensionless]	dB	dB	*same*

| Information from 1976. *About Sound*. U.S. Environmental Protection Agency, Office of Noise Abatement and Control, Washington, DC.
* Intensity level (dB) = 10 log (sound intensity/10^{-12} W/m^2).
# Pressure level (dB) = 20 log (sound pressure/20 μPa).

Table C.5   SI chemical dimensions and units.

Dimension [fundamental]	SI Units
Molar mass $[MM^{-1}]$	kg/mol
Molecular weight $[MM^{-1}]$	kg/kg-mol
Moles $[M]$	mol
Molarity $[ML^{-3}]$	mol(solute)/m^3(solution)
Molar concentration $[ML^{-3}]$	mol/m^3
Molar concentration rate, molar reaction rate $[ML^{-3}T^{-1}]$	mol/(m$^3 \cdot$ s)
Molar flow rate $[MT^{-1}]$	mol/s
Volumetric flow rate $[L^3T^{-1}]$	m^3/s
Mass flow rate $[MT^{-1}]$	kg/s

| Information from L. Pauling. 1970. *General Chemistry,* 3rd ed. San Francisco: W.H. Freeman.

Table C.6   SI electrical and magnetic dimensions and units.

Dimension [fundamental]	Name	Symbol	Base Unit s
Electric capacitance $[M^{-1}L^{-2}T^4A^2]$	farad	F	s$^4 \cdot$A^2/(kg $\cdot$ m^2)
Electric charge $[TA]$	coulomb	C	A $\cdot$ s
Electric conductance $[M^{-1}L^{-2}T^3A^2]$	siemens	S	s$^3 \cdot$A^2/(kg $\cdot$ m^2)
Electric field strength $[MLT^{-3}A^{-1}]$	volt/m	V	kg $\cdot$ m/(s^{3}A)
Electric inductance $[ML^2T^{-2}A^{-2}]$	henry	H	kg $\cdot$ m^2/(s$^2 \cdot$ A^2)
Electric potential $[ML^2T^{-3}A^{-1}]$	volt	V	kg $\cdot$ m^2/(s^{3}A)
Electric resistance $[ML^2T^{-3}A^{-2}]$	ohm	Ω	kg $\cdot$ m^2/(s^{3}A^2)
Electromotive force $[ML^2T^{-3}A^{-1}]$	volt	V	kg $\cdot$ m^2/(s^{3}A)
Magnetic field strength $[L^{-2}A]$	ampere/m^2	A/m^2	A/m^2
Magnetic flux $[ML^2T^{-2}A^{-1}]$	weber	Wb	kg $\cdot$ m^2/(s$^2 \cdot$ A)
Magnetic flux density $[MT^{-2}A^{-1}]$	tesla	T	kg $\cdot$ /(s$^2 \cdot$ A)

LEARNING OBJECTIVE NOMENCLATURE

The following lists the action verbs and their specific meanings related to the learning objectives as stated in this text.

In the main text:

1. Know: memorize and give an example
2. Convert: calculate the equivalent quantity in another system of units
3. Identify: choose from and specify
4. Round off : self-explanatory according to its defined method
5. Express: write out in numerical form
6. Describe: explain in words, being as specific as possible
7. Perform: go through the procedure as specified in the notes
8. Plot: self-explanatory, but according to the format specified in the notes
9. Determine: calculate using equations and information given or provided in tables to determine a specified result
10. Calculate: use equations to determine a specified result
11. Differentiate: compare and contrast in words, being as specific as possible
12. Apply: use standard approach in a new situation
13. Estimate: approximate through the use of simple calculations

In the laboratory exercises:

1. Operate: use according to directions to achieve desired output
2. Calibrate: operate to achieve desired calibration data

PHYSICAL PRINCIPLES

The following material is taken with permission from pages 451–455 of D. G. Alciatore and M. B. Histand. 2003. *Introduction to Mechatronics and Measurement Systems*, 2nd ed. ISBN 0-07-119557-2. New York: McGraw-Hill.

APPENDIX OBJECTIVES

After you read, discuss, study, and apply ideas in this appendix, you will be able to:

1. Identify possible relationships between various physical quantities

2. Identify approaches for measuring nearly all physical quantities

Sensor and transducer design always involves the application of some law or principle of physics or chemistry that relates the variable of interest to some measurable quantity. The following list summarizes many of the physical laws and principles that have potential application in sensor and transducer design. Some examples of applications are also provided. This list is extremely useful to a transducer designer who is searching for a method to measure a physical quantity. Practically every transducer applies one or more of these principles in its operation. The parameters related by the respective principles are highlighted.

- *Ampere's law:* A **current**-carrying conductor in a **magnetic field** experiences a **force**.

    Based on this law, a galvanometer measures current by measuring the deflection of a pivoted coil in a permanent magnetic field.

- *Archimedes' principle:* The buoyant **force** exerted on a submerged or floating object is equal to the weight of the fluid displaced. The **volume** displaced depends on the fluid **density**.

    A ball submersion hydrometer uses this effect to measure the density of a fluid (e.g., automotive coolant).

- *Bernoulli's equation:* Conservation of energy in a fluid predicts a relationship between **pressure** and **velocity** of the fluid.

    A pitot tube uses this effect to measure air speed of an aircraft.

- *Biot-Savart law:*   A conductor carrying a **current** is surrounded by a **magnetic field**.

  A magnetic pickup sensor uses this effect as a nonintrusive method of measuring current in a conductor.

- *Biot's law:*   The rate of **heat conduction** through a medium is directly proportional to the **temperature** difference across the medium.

  This principle is basic to time constants associated with temperature transducers.

- *Blagdeno law:*   The freezing **temperature** of a liquid drops and the boiling temperature rises with **concentration** of impurities in the liquid.

- *Boyle's law:*   An ideal gas maintains a constant **pressure-volume** product with constant **temperature**.

- *Bragg's law:*   The intensity of an X-ray beam diffracted by a **crystal lattice** is related to the crystal plane separation and the **wavelength** of the beam.

  An X-ray diffraction system uses this effect to measure the crystal lattice geometry of a crystalline specimen.

- *Brewster's law:*   The **index of refraction** of a material is related to the angle of **polarized light** reflection or transmission.

  A Brewster's window on a laser tube is used to extract some of the power in the form of a laser beam. Lasers are used extensively in measurement systems.

- *Butterfly effect:*   Chaotic nonlinear systems exhibit a sensitive dependence on initial conditions.

- *Centrifugal force:*   A body moving along a curved path experiences an apparent outward **force** in line with the radius of curvature.

- *Charles' law:*   An ideal gas maintains a constant **pressure-temperature** product with constant **volume**.

- *Christiansen effect:*   Powders suspended in a liquid (i.e., a colloidal solution) result in altered fluid **refraction** properties.

- *Corbino effect:*   **Current** flow is induced in a conducting disk rotating in a **magnetic field**.

- *Coriolis effect:*   A body moving relative to a rotating frame of reference (e.g., the earth) experiences a **force** relative to the frame.

  A coriolis flow meter uses this effect to measure mass flow rate in a u-tube in rotational vibration.

- *Coulomb's law:*   **Electric charges** exert a **force** between each other.

- *Curie-Weiss law:*   There is a transition **temperature** at which ferromagnetic materials exhibit **paramagnetic** behavior.

- *d'Alembert's principle:*   **Acceleration** of a **mass** is equivalent to an equal and opposite applied **force**.

- *Debye frequency effect:*   The **conductance** of an electrolyte increases (i.e., the **resistance** decreases) with **frequency**.

- *Doppler effect:*   The **frequency** received from a wave source (e.g., sound or light) depends on the **speed** of the source.

    A laser doppler velocimeter (LDV) uses the frequency shift of laser light reflected off of particles suspended in a fluid to measure fluid velocity.

- *Edison effect:*   When metal is heated in a vacuum, it emits charged particles (i.e., **thermionic emission**) at a rate dependent on **temperature**.

    A vacuum tube amplifier is based on this effect, where electrons are emitted and controlled to produce amplification of current.

- *Faraday's law of electrolysis:*   The rate of **ion deposition** or depletion is proportional to the electrolytic **current**.

- *Faraday's law of induction:*   A coil resists a change in **magnetic field** linkage with an **electromotive force**.

    The induced voltages in the secondary coils of a linear variable differential transformer (LVDT) are a result of this effect.

- *Gauss effect:*   The **resistance** of a conductor increases when **magnetized**.

- *Gladstone-Dale law:*   The **index of refraction** of a substance is dependent on **density**.

- *Gyroscopic effect:*   A body rotating about one axis resists rotation about other axes.

    A navigation gyroscope uses this effect to track the orientation of a body with the aid of a gimbal-mounted flywheel that maintains constant orientation in space.

- *Hall effect:*   A **voltage** is generated perpendicular to **current** flow in a **magnetic field**.

    A Hall effect proximity sensor detects when a magnetic field changes due to the presence of a metallic object.

- *Hertz effect:*   **Ultraviolet light** affects the discharge of a spark across a gap.

- *Johnsen-Rahbek effect:*   **Friction** at interfaces between a conductor, semiconductor, or insulator increases with **voltage** across the interfaces.

- *Joule's law:*   **Heat** is produced by **current** flowing through a **resistor**.

    The design of a hot-wire anemometer is based on this principle.

- *Kerr effect:*   Applying a **voltage** across a substance can cause **optical polarization**.

    Liquid crystal displays (LCDs) function as a result of this principle.

- *Kohlrausch's law:* An **electrolytic** substance has a limiting conductance (minimum **resistance**).

- *Lambert's cosine law:* The reflected **luminance** of a surface varies with the cosine of the **angle of incidence**.

- *Lenz's law:* A **current**-carrying conductor moving in a **magnetic field** experiences a **force**.

- *Lorentz's law:* There is a **force** on a charged particle moving in an **electric** and **magnetic field**.

- *Magnus effect:* When fluid flows over a rotating body, the body experiences a **force** in a direction perpendicular to the flow.

- *Meissner effect:* A **superconducting** material within a **magnetic field** blocks this field and experiences no internal field.

- *Moore's law:* The density of transistors that can be manufactured on an integrated circuit doubles every 18 months.

- *Murphy's law:* Whatever can go wrong will go wrong and at the wrong time and in the wrong place.

    Your experiments in the laboratory will often demonstrate this law.

- *Nernst effect:* **Heat flow** across **magnetic field** lines produces a **voltage**.

- *Newton's law:* **Acceleration** of an object is proportional to force acting on the object.

- *Ohm's law:* **Current** through a **resistor** is proportional to the **voltage** drop across the resistor.

- *Parkinson's law:* Human work expands to fill the time allotted for it.

- *Peltier effect:* When **current** flows through the junction between two metals, **heat** is absorbed or liberated at the junction.

    Thermocouple measurements can be adversely affected by this principle.

- *Photoconductive effect:* When **light** strikes certain semiconductor materials, the **resistance** of the material decreases.

    A photodiode, which is used extensively in photodetector pairs, functions based on this effect.

- *Photoelectric effect:* When **light** strikes a metal cathode, electrons are emitted and attracted to an anode, resulting in **current** flow.

    The operation of a photomultiplier tube is based on this effect.

- *Photovoltaic effect:* When **light** strikes a semiconductor in contact with a metal base, a **voltage** is produced.

    The operation of a solar cell is based on this effect.

- *Piezoelectric effect:*   **Charge** is displaced across a crystal when it is strained.

   A piezoelectric accelerometer measures charge polarization across a piezoelectric crystal subject to deformations due to the inertia of a mass.

   A piezoelectric microphone's ability to convert sound pressure waves to a voltage signal is a result of this principle.

- *Piezoresistive effect:*   **Resistance** is proportional to an applied **stress**.

   This effect is partially responsible for the response of a strain gage.

- *Pinch effect:*   The cross section of a liquid conductor reduces with **current**.

- *Poisson effect:*   A material deforms in a direction perpendicular to an applied **stress**.

   This effect is partially responsible for the response of a strain gage.

- *Pyroelectric effect:*   A crystal becomes **polarized** when its **temperature** changes.

- *Raleigh criteria:*   Relates the **acceleration** of a fluid to bubble formation.

- *Raoult's effect:*   **Resistance** of a conductor changes when its length is changed.

   This effect is partially responsible for the response of a strain gage.

- *Seebeck effect:*   Dissimilar metals in contact result in a **voltage** difference across the junction that depends on **temperature**.

   This is the primary effect that explains the function of a thermocouple.

- *Shape memory effect:*   A deformed metal, when heated, returns to its original shape.

- *Snell's law:*   Reflected and refracted rays of **light** at an optical interface are related to the angle of incidence.

- *Stark effect:*   The **spectral lines** of an electromagnetic source split when the source is in a strong **electric field**.

- *Stefan-Boltzmann law:*   The **heat** radiated from a black body is proportional to the fourth power of its **temperature**.

   The design of a pyrometer is based on this principle.

- *Stokes' law:*   The **wavelength** of light emitted from a fluorescent material is always longer than that of the absorbed photons.

- *Tribo-electric effect:*   Relative motion and **friction** between two dissimilar metals produces a **voltage** between the interface.

- *Wiedemann-Franz law:*   The ratio of **thermal** to **electrical conductivity** of a material is proportional to its absolute **temperature**.

- *Wien effect:*   The **conductance** of an electrolyte increases (i.e., the **resistance** decreases) with applied **voltage**.

- *Wien's displacement law:*   As the **temperature** of an incandescent material increases, the spectrum of emitted **light** shifts toward blue.

REVIEW PUZZLE SOLUTIONS

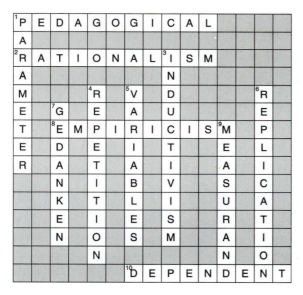

Chapter 1   Experiments puzzle solution.

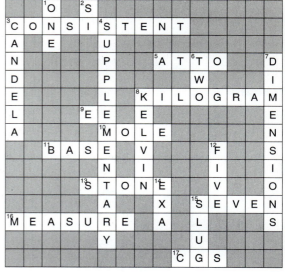

Chapter 2   Units and significant figures puzzle solution.

Chapter 3   Technical communications puzzle solution.

Chapter 5   Calibration and response puzzle solution.

Chapter 4   Basic electronics puzzle solution.

Chapter 6   Measurement systems puzzle solution.

Chapter 7   Probability puzzle solution.

Chapter 9   Uncertainty puzzle solution.

Chapter 8   Statistics puzzle solution.

Chapter 10   Regression and correlation puzzle solution.

Chapter 11   Signals characteristics puzzle solution.

Chapter 12   Digital signal analysis puzzle solution.

PROBLEM SOLUTIONS

The following lists the answers to all Review Problems and some selected Homework Problems.

G.1   REVIEW PROBLEMS

Chapter 1

[1] independent; [2] validational; [3] a beaker of ice water; [4] independent = changed by the experimenter; dependent = affected by a change made by the experimenter; extraneous = not controlled during the experiment; parameter = fixed throughout the experiment; measurand = measured during the experiment

Chapter 2

[1] 1 slug; [2] 1; [3] the diagonal of a unit cell of iron; [4] Pa · m^2; [5] 7; [6] 7.3 E3 N; [7] 2.76 E2 N; [8] 73.5 kg; [9] second, ampere, mole, candela; [10] 291.3 N; [11] 0.211 ft lbf; [12] 12.04 E23; [13] 13.7 slugs; [14] 8; [15] 12 000 N; [16] 10; [17] 4578; [18] 8; [19] 7

Chapter 3

[1] e. summary, findings, references; [2] c. past

Chapter 4

[1] 4.0 μF; [2] d. 1 C · V; [3] 13 V; [4] 1000 Ω; [5] 3.47 V; [6] a and b; [7] current = Ampere; charge = Coulomb; electric field work = Joule; electric potential = Volta; resistance = Ohm; power = Watt; inductance = Henry; capacitance = Farad; [8] 10.3 A; [9] 30.9 Ω; [10] 0.003 V; [11] 200; [12] 4.95

V; [13] a. deflection method; [14] d. reciprocals of parallel resistances add; [15] d. compression on the underside gage causes an increase in resistance; [16] 0.038; [17] 11 Ω; [18] 47 Ω

Chapter 5

[1] the larger-diameter thermocouple has the larger time constant; [2] 0.11; [3] 25 %; [4] 9.5; [5] zero; [6] 0.06; [7] no; [8] c. [1 − (1/e)]B; [9] b. 1.0 s; [10] b. −146; [11] d. 0.1B; [12] a. −0.15 s

Chapter 6

[1] radiator fluid temperature = physical variable; thermistor = sensor; Wheatstone bridge = transducer; car computer = matches signal processor; [2] A/D converter; [3] 1.1897; [4] 0.47 Ω; [5] 49.4; [6] 0.471; [7] 0.06 s; [8] 34.2 μF; [9] 875 Ω

Chapter 7

[1] August birthday = 0.0849; Feb. 29, 1979, birthday = 0; month-with-30-days birthday = 0.329; 31st-of-the-month birthday = 0.0192; [2] 1/12; 1/4; 1/20; 7/10; [3] 0.0625; [4] 12 bins; [5] 1.0; [6] 0; [7] 2.4; [8] 0; [9] 0.4082; [10] approximately 2 %; [11] 1/3; [12] 1/volts; [13] the kurtosis is the fourth central moment

Chapter 8

[1] 0.9544; [2] 0.8644; [3] 0.1621; [4] 0.0160;
[5] 0.8742; [6] 5.3290; [7] 0.149; [8] 0.047;
[9] 99.94 %; [10] 99 %; [11] 1.00; [12] 10.82;
[13] 0.51; [14] 28.6 %; [15] 0.23 %

Chapter 9

[1] −0.50; [2] 0.60; [3] Legendre; [4] 0.49; [5] 0.98;
[6] any of the above; [7] 9.1 V

Chapter 10

[1] readability; [2] 0.026; [3] player 3; [4] metric
ruler is 0.29375 mm more accurate; [5] 0.00459
$kg \cdot m^2$; [6] 0.000131; [7] 0.024 in.; [8] temperature
fluctuations in the laboratory = random error; physically pushing the pendulum through its swing =
experimental mistake; the numerical bounds of the
scatter in the distance traveled by the racquetball =

uncertainty; releasing the pendulum from an initial
position so that the shaft has a small initial angle =
systematic error; [9] 0.6 %; [10] 1 %; [11] 0.013;
[12] random error; [13] 6; [14] 12.83 s; [15] 0.002;
[16] 1; [17] 0.20 V; [18] d; [19] 6 %; [20] 0.5 %;
[21] 0.1 %

Chapter 11

[1] 2.687; [2] 4.914; [3] 0.0450; [4] 0.0611;
[5] [a] periodic with period of 1 s; [b] aperiodic
because of the exponential term; [6] d; [7] c, d, and
e; [8] a and b; [9] b and c; [10] no because of the
exponential term; [11] 0.707 A; [12] 2 A/π

Chapter 12

[1] 5; [2] 0.25 Hz; [3] 0 Hz; [4] No. All common
windowing functions reduce the signal's amplitude

| **G.2** **HOMEWORK PROBLEMS**

Chapter 2

[6] [a] 244 N; [b] 4.50 slug; [c] 65.7 kg; [d] 65.7 kg;
[10] [a] 1.19 kg/m^3; [b] 138 ft/s

Chapter 4

[1] [a] 100 Ω; [b] 0 V; [c] 0.53 V; [5] 49 500 Ω

Chapter 5

[1] [a] 0.80 ms; [b] −45°; [5] 2 s; [b] $E(t) =$
$49.23exp[−t/2] + 100 + 6.15 \sin 4t + 0.77 \cos 4t$;
[12] −172.4°

Chapter 6

[1] [a] 5970 Ω; [b] yes because the input impedance
of an op amp is far greater than 5970 Ω; [c] 48.6;
[5] [a] 0.0995; [b] 1.59 ms; [c] 15.9 μF; [8] [a] 15
Hz; [b] 10.6 ms; [c] 6.64 V; [d] −9.35 ms

Chapter 7

[3] 2 %; [5] [a] 61.9 %; [b] 15; [c] 2

Chapter 8

[1] [a] 15.87 %; [b] 0.135 %; [c] 3; [5] [a] 22.55 %;
[b] from 43.77 to 44.63; [c] from 3.03 to 5.55;
[10] 28.6 %

Chapter 9

[3] intercept = 0.2 and slope = 0.498; [5] 1.5

Chapter 10

[1] 1.96 %; [5] 2.94 psi; [9] [a] 0.225 in. H_2O;
[b] 22.5 %; [c] no; [15] 1.11 E−4 in.

Chapter 11

[2] [a] 31.02; [b] 29.00; [c] 30.03; [d] 30.03;
[7] [a] 1.5 Hz; [b] 2; [c] 2 at 0 Hz and 5 at 1.5 Hz;
[11] [a] 3; [b] 0.25 Hz; [c] 0.5 Hz; [d] 0; [e] 2.121

Chapter 12

[1] [a] 0.21 s; [b] 4.78 Hz; [c] 1.91 E4 Hz; [d] 9550
Hz [e] faster; FFT algorithm could be used instead
of a DFT algorithm

LABORATORY EXERCISES

This appendix presents 12 laboratory exercises that were designed to supplement the material in this text. Each section describes a particular laboratory exercise. An accompanying *Laboratory Exercises Manual* presents the actual student handout for each exercise. A companion *Laboratory Exercises Solutions Manual* provides the answers to all of the questions posed in the laboratory exercise handout and the data acquired for each exercise. In this manner, a virtual laboratory exercise can be performed by providing students with data for analysis and reporting. The laboratory exercises can be performed as written. All have been tested many times by students over the past several years and have been refined. One intent in offering these descriptions is to provide a base for instructors to extrapolate from and generate new exercises.

Typically 6 to 10 exercises are conducted during a one-semester, three-credit-hour undergraduate measurements course. The purpose of these exercises is to introduce the student to the process of conducting experiments and analyzing their results. Some exercises are oriented toward learning about instrumentation and measurement system hardware; others toward examining an actual physical process. The overall objective is to provide students with a variety of measurement and data analysis experiences such that they are fully prepared for subsequent laboratory courses that focus on investigating physical processes, such as those in fluid mechanics, aerodynamics, or heat transfer laboratory courses.

Some of the exercises were designed to be performed in series, although each exercise stands alone. In particular, Exercises 2 through 6 progressively introduce the student to the foundational concepts and use of strain gages for both static and dynamic force measurements. Exercises 1, 7, and 10 involve the comparison of measurements with theory within the context of uncertainty. Exercises 8 through 11 introduce the student to various instrumentation and measurement systems. Finally, the Exercise 12 focuses on postexperiment data analysis using files of provided data.

Table H.1 lists the instrumentation used for each exercise.

Table H.1   Laboratory exercise instrumentation.

Instrumentation	Laboratory Exercise Number											
	1	2	3	4	5	6	7	8	9	10	11	12
Multimeter		√		√	√							
Dial indicator		√										
Wheatstone bridge			√		√	√						
Cantilever beam			√			√						
Oscilloscope			√			√	√	√	√	√		
Strain gage			√		√	√						
Manometer				√								
Barometer				√								
Dynamometer				√								
DC power supply				√								
Calibration weights				√	√							
Stroboscope				√								
Function generator								√	√	√		
Data acquisition system		√				√		√				
Thermocouples										√		
RLC circuit										√		
Helium-neon laser											√	
Diode/detector pair											√	
Optics											√	

H.1   EXERCISE 1—MEASUREMENT, MODELING, AND UNCERTAINTY

H.1.1   INTRODUCTION AND OBJECTIVES

This laboratory exercise demonstrates the roles that modeling and empirical uncertainties play in determining the outcome of a simple experiment. The experiment involves launching a ball from a pendulum apparatus and measuring the vertical and horizontal distances that the ball travels. The experiment is repeated at different pendulum head release angles, with a specified number of times for each angle. The average results for each angle are compared with the theoretical predictions within the context of the uncertainties involved in the model and in the experiment.

As part of this exercise, a model must be developed that predicts the horizontal distance, x, $\pm$ its uncertainty (estimated at 95 % confidence) where the ball will land based on the release angle of the pendulum head with respect to the vertical top position, θ_{rel}. The values of some of the model's variables, such as the coefficient of restitution of the ball, in turn, rely on other empirical information that may need to be gathered by performing subsidiary experiments. A subsidiary experiment is any experiment other than the actual one that needs to be performed to obtain input information for the model.

A schematic of the pendulum apparatus is shown in Figure H.1. There is a large pendulum that consists of an Al 2024 shaft (46.60 cm long ± 0.05 cm; 0.95 cm diameter ± 0.01 cm; mass = 89.8 g ± 0.5 g) that extends into a rectangular yellow-brass strike head (length = 6.36 cm ± 0.01 cm; width = 3.18 cm ± 0.01 cm; height = 3.18 cm ± 0.01 cm; mass = 528.8 g ± 0.5 g). The pendulum is swung about a top pivot point, which contains an angle indicator (resolution = 1°). The ball having mass m_2 is located on a tee and placed such that contact with the strike head is made at the bottom of the swing. The distance between the center of the strike head at the top of the swing and the bottom of the swing, h_1, is 90.0 cm ± 0.2 cm. The center of mass of the system consisting of the rod and the strike head lies at 42.00 cm ± 0.01 cm from the center bearing. Figure H.2 shows schematically the pendulum at the top and bottom of its swing with its nomenclature.

H.1.2   INSTRUMENTATION

The following equipment will be used:

- Pendulum apparatus to launch the ball
- A standard racquet ball with a mass of 40.60 g ± 0.05 g and a diameter of 5.61 cm ± 0.05 cm or a standard golf ball with a mass of 45.30 g ± 0.05 g and a diameter of 4.27 cm ± 0.05 cm
- English/metric tape measure

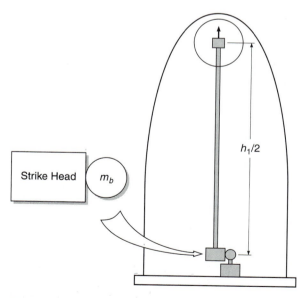

Figure H.1   The pendulum apparatus.

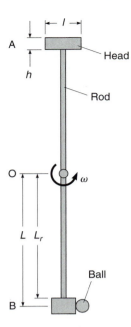

Figure H.2   Pendulum nomenclature.

H.1.3   MEASUREMENTS

The objective is to obtain data relating the horizontal distance that a golf ball travels to the pendulum head release angle, which then can be compared to theoretical predictions. The experiment should be repeated at each release angle five times to obtain an average horizontal distance. A minimum of four release angles must be examined. More data will give a better comparison between experiment and theory.

The pendulum-launching apparatus should be placed in a stable position on a tabletop such that the pendulum can travel freely in a complete circle and that there are no obstructions for approximately 20 ft in front of it for the ball to travel. Start by performing a trial experiment to identify where the ball will land on the lab floor. Have one person launch the ball and the other note where it lands on the floor. Repeat the trial several times to identify the approximate impact point. Tape a piece of carbon paper on the floor to mark on the paper where the ball lands each time. Now perform the experiment five times, always noting the impact point. When complete, repeat for another pendulum head release angle. Measure the vertical distance and all horizontal distances using the provided tape measure. Make sure that all the necessary subsidiary experiments and measurements have been performed.

H.1.4 WHAT TO REPORT

Present the results of this effort in a technical memo. The memo at a *minimum* must include the following: (1) a statement that summarizes the agreement or disagreement between the measured and predicted distances and plausible, scientifically based reasons for any disagreement; (2) a table of the predicted and the measured (average) horizontal distances from the launch point (in cm) to the center of the landing impact point and their associated distance uncertainties. The uncertainty estimates must be supported by detailed calculations (present these in an attached appendix) using standard uncertainty analysis at 95 % confidence (a plot of the same would also be helpful); (3) a brief description of the model developed, stating all parameters and assumptions (any detailed calculations can be put in an appendix); and (4) a table that presents the symbols and lists the values in both SI and Technical English units of all of the model inputs and outputs.

H.2 EXERCISE 2—RESISTANCE AND STRAIN

H.2.1 INTRODUCTION AND OBJECTIVES

The primary purpose of this exercise is to determine the relationship between the relative change in resistance of a fine wire and the relative change in its length. This concept is the fundamental principle by which strain gages operate. A strain gage (see Figure 6.2 in Chapter 6) basically consists of a metallic pattern bonded to an insulating backing. This can be attached to a surface to provide a method to measure strains induced by loading. The important concept is that when a wire is stretched (strained), its resistance changes.

These objectives will be accomplished by stretching a wire and measuring its resistance at various lengths. In the process of performing this experiment, a digital multimeter will be used to measure the resistance. The experimental results will be examined by plotting the relative change in resistance versus the relative change in length. From this information, a gage factor can be determined for this wire. The uncertainties in the measurements also will be estimated and related to the results. Local and engineering gage factors are explained in Chapter 6. The engineering gage factor will be determined here.

H.2.2 INSTRUMENTATION

The following instruments will be used:

- Hewlett Packard 3468A Multimeter (resolution: 1 $\mu\Omega$ in the ohm range)
- Starrett dial indicator (resolution: 0.0005 in.)
- A metal meterstick (resolution: 0.5 mm)
- A wire stretcher for wires approximately 1 m in length

H.2.3 MEASUREMENTS

First and foremost—a safety note. It is imperative that safety glasses be worn when stretching the wire. It can break and snap back.

A length of wire is to be mounted between the two clamps of the tension device and loaded using the screw mechanism. First, cut a piece of wire approximately 4 ft long from the spool. Figure H.3 shows a schematic of the clamping mechanism. With the top of the clamp removed, loop the wire once about the end post before directing it to the other clamp. Then replace the top of the clamp and tighten the set screw to prevent slipping. Do likewise with the other clamp. Wire slippage while applying tension will result in an indication of displacement without an expected increase in resistance. When the wire is mounted correctly, it should be straight, not sagging, and the tension end with the brass thumb wheel should have about 0.5 in. of travel available.

The four-wire resistance measurement will be used (see Chapter 4). Connect the multimeter to the wire as follows:

1. Using one pair of banana-alligator test leads, connect the HI and LO pair of terminals of the multimeter under INPUT to the wire near the clamps, one at each end.

2. Using a second pair of test leads, connect the HI and LO terminals under Ω-Sense to the wire just inside of the two leads of step 1.

3. Place the multimeter into the four-wire resistance mode by pressing the "4 WIRE" button (4 Ω should appear on the LCD display). Also depress with the AUTO/MAN button to put the meter in the manual mode (M RNG should appear on the display). This will yield a resistance reading resolution of 10 mΩ. The KOHM range should be displayed. If not, depress the up arrow button to obtain its indication. Finally, depress the blue button, then the INT TRIG button to set the multimeter in the auto zero mode. If the auto zero mode is NOT set, then AZ OFF will appear on the display (no indication means it's set correctly). The indicated reading should be approximately 150 Ω to 200 Ω. If a negative resistance is indicated, the two inner wires can be switched to make it positive. Allow a couple of minutes for the meter to warm up before starting to take data.

Slowly add tension to the wire by turning the brass thumb wheel until the resistance starts to increase (watch the least and second-least significant digits for some consistent increase). This will be the zero point. Record the reading on the dial gage

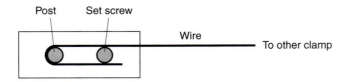

Figure H.3 Clamping of wire in mechanism.

and the resistance. (The dial indicator scale goes from 0 to 50, corresponding to 0.000 in. to 0.050 in. of travel). Measure the initial length of wire between the measuring points, the leads of step 2, using the metal meterstick.

At increments of approximately 0.01 in. (increments of 10 on the dial indicator), record the elongation (inches) and the resistance (ohms) until the wire has been stretched about 0.20 in. The resistance changes approximately 0.1 Ω for every 0.01 in. of stretch. Return to a couple of data points and repeat those measurements to see if they have changed at all. Now, try repeating the experiment all the way out to failure of the wire. This should take on the order of 0.50 in. of travel. Try taking around 20 data points, with larger intervals at the beginning, becoming smaller as the wire approaches failure.

When finished, turn off the multimeter, bring the dial indicator back to the zero starting point, disconnect the test leads from the wire and the multimeter, and remove the wire.

H.2.4 WHAT TO REPORT

Plot the relative resistance change versus the relative length change, for both the first case and the case when the wire was stretched to failure. Estimate the uncertainties of $\Delta R/R$ and of $\Delta L/L$ following the procedures detailed in Chapter 9. Calculate the engineering gage factors (see Chapter 6) for both cases. Are the values the same? Explain this in the context of the measurement uncertainties. Try approximating some local gage factors by calculating slopes over a few data points in the data sets, especially at lower strains. What can be said about the relation between these local gage factors and the extent to which the wire was strained at that point? Plot the local gage factor versus the strain to illustrate this. Compare the local with the engineering gage factors from the two cases, always being aware of the uncertainties involved.

Perform a least-squares linear regression analysis of the relative resistance change versus strain (see Chapter 10). Determine the correlation coefficient and the percent confidence associated with that correlation coefficient. How does the slope of the best-fit line compare with the gage factors that were calculated earlier?

All important experimental results and answers to the posed questions must be presented as a technical memo. Answers to the posed questions should be contained in the explanation of the results and *not* listed item-for-item.

H.3 EXERCISE 3—STRAIN-GAGE-INSTRUMENTED BEAM: CALIBRATION AND USE

H.3.1 INTRODUCTION AND OBJECTIVES

This laboratory exercise involves the static calibration of a system consisting of four strain gages mounted on a cantilever beam. Once calibrated, this system can be used in either a static or a dynamic mode to determine the weight of an object, the mass flow rate of a material, and the frequency of a vibrating beam. In this exercise, we will see how uncertainties enter into the calibration process and how they subsequently enter into the uncertainty when determining such quantities weight, mass flow rate, and frequency.

The concept that the change of a wire's resistance with strain can be utilized in a practical measurement system will be examined. It is possible to take small changes in resistance and, using an electrical circuit, transform the signal into a change in voltage. By making a strain gage one of the resistors in a Wheatstone bridge circuit, a voltage is measured that is proportional to strain. Specifically, in this lab exercise, four strain gages bonded to a cantilever beam will be used. Each of the four gages will serve as one resistor in a leg of a Wheatstone bridge.

A static calibration will be performed to obtain a mathematical relationship between bridge output voltage and force. Once this expression is known, the measurement system can be used for many static and dynamic applications.

H.3.2 INSTRUMENTATION

The following equipment will be used:

- Wheatstone bridge and operational amplifier measurement system, as shown in Figure H.4
- Cantilever beam load cell
- Tektronix TDS 210 two-channel digital real-time oscilloscope
- Four Micro-Measurement 120 Ω CEA-13-125UW-120 strain gages (bonded to the beam)
- Four unknown materials: circular pipe, rectangular pipe, cylinder, and hexagonal cylinder
- Plastic bottle with fine-grained sand
- 1000 mL plastic beaker

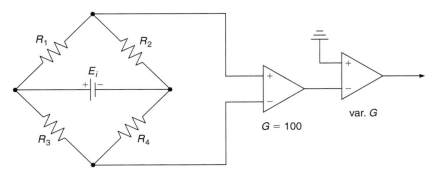

Figure H.4   Wheatstone bridge and op amp configuration.

H.3.3   MEASUREMENTS

Part 1: Static Calibration

1. Record the lab setup number. The setup number is located on the top of the cantilever beam.

2. Make sure that the power in ON (the warm-up period for stable readings is about 1 h).

3. Connect the two strain gages on the tension side of the beam to the Wheatstone bridge as follows: connect one between Bridge Excitation and + Bridge Out, and then other between − Bridge Out and Ground. The tension side gages have yellow end connectors.

4. Connect the two strain gages on the compression side of the beam to the Wheatstone bridge as follows: connect one between Bridge Excitation and − Bridge Out, and then other between + Bridge Out and Ground. The compression side gages have blue end connectors.

5. Set the Bridge Excitation voltage to 4.50 V ± 0.01 V by adjusting the Bridge Excitation dial and observing the output on the panel meter (be sure that the selector switch is set to Bridge Excite).

6. Set the Amplifier Gain dial position to 4.

7. With no weights attached to the beam end, adjust the Balance dial to obtain an Amplifier Out signal of 0.00 V ± 0.02 V (be sure that the selector switch is set to Amplifier Out). Wait for about 30 s to 60 s after the dial is adjusted to be sure that the voltage stays where it was set. Reset the dial if necessary.

8. Add 450 g of weights to the hanger at the end of the beam. Note that the mass of the hanger is 50 g, so the total mass is 500 g. Make sure not to apply any total mass greater than 1 kg to avoid damaging the strain gages.

9. Adjust the Amplifier Gain dial to achieve 4.00 V ± 0.02 V. Wait for about 30 s to 60 s after adjusting the dial to be sure that the voltage stays where it was set. Reset the dial if necessary.

10. Take the hanger and weights off of the end of the beam. Check the Amplifier Output. It should be 0.00 V ± 0.02 V. If not, repeat steps 6 through 8.

11. Now begin a full calibration. First record the actual zero weight reading. Then, start with just the hanger attached. Record the Amplifier Out reading after waiting about 30 s to 60 s. Make sure that the hanger is not moving (small movements will cause variations in the readings). Now proceed to add weights progressively up to 450 g, recording the mass and the Amplifier Out reading each time. Next, progressively take off the weights and record the values, ending up with no weight.

12. Then repeat a few of the measurements to determine how repeatable the measurements are.

13. Remove the weights and hanger and record the voltage. It should be at 0 V (the initial bridge-balanced no-weight condition). If it is not within an acceptable

range about 0 V ± 0.02 V, the bridge may have to be rebalanced and the measurements taken over again.

Part 2: Unknown Object Measurements

1. Once all of the calibration data has been taken, select one of the objects of unknown weight. Using the can at the end of the beam to hold the object (make sure the object is close to the center of the can), determine and record the Amplifier Output voltage. Repeat for each object.

2. Obtain a plastic beaker. Fill the beaker with 600 mL of water. Place the object in the water and measure the displacement of the water. Repeat for *every one* of the objects.

3. Measure and record the dimensions of each of the unknown objects with calipers. This, in conjunction with the previous water displacement measurements, will give two separate measurements of the volume for each of the four objects.

Part 3: Measurement of Mass Flow Rate and Oscilloscope Setup

1. Obtain a squeeze bottle full of sand and a plastic bag.

2. Turn on the oscilloscope and connect CH. 1 to the Amplifier Output from the bridge using a cable. Now press the Autoset button. Next, press the CH. 1 button under the vertical section twice. A menu should appear in the display window. Under the menu heading "Coupling," choose "DC." This is done by pressing menu keys next to the display to the immediate right of the display.

3. Using the Volts/Div knob for CH. 1, change the voltage per division to 200 mV/div. The volts per division is displayed in the lower left-hand corner of the display.

4. Now change the seconds per division to 5 s/div using the SEC/DIV knob under the Horizontal section of the oscilloscope. Again, the seconds per division is displayed at the bottom of the screen.

5. Move the vertical position of CH. 1 down to one division above the bottom of the screen using the vertical position knob for CH. 1.

6. Place the plastic bag in the can.

7. Hold but do not invert the squeeze bottle filled with sand over the can at the end of the beam. Wait until the trace on the oscilloscope reaches the end of the first time division. Remove the end cover of the spout from the squeeze bottle, invert the bottle, and let the sand flow freely (do *not* squeeze the bottle) into the bag lining the can.

8. When the signal has reached the end of the oscilloscope's display area or the sand begins to run out, press the Run/Stop button located at the top right of the oscilloscope. The data should be frozen on the screen. Press the Cursor button. Set the "type" to time using the top menu button. Two vertical lines should appear on the screen. Using the vertical position knobs for CH. 1 and CH. 2 to move the

lines, measure the horizontal displacement (time difference) of the data. Measure only the linear region. The time difference is displayed on the screen under the heading "Delta Record."

9. On the cursor menu, select "Voltage" under the menu heading "Type." Using the vertical knobs for CH. 1 and CH. 2 to move the horizontal lines, measure and record the horizontal displacement (voltage difference) of the data.

10. Carefully remove the bag from the beam, ensuring that sand does not spill out of the bag. Empty all sand back into the squeeze bottle.

11. Using a dry plastic beaker, measure 100 mL of sand.

12. Place the plastic bag in the can.

13. Place the 100 mL of sand in the plastic bag lining the can and record the Amplifier Out Voltage.

14. Carefully remove the bag from the beam, ensuring that sand does not spill out of the plastic bag. Empty all sand back into the squeeze bottle.

Part 4: Dynamical Measurement

1. Press the Run/Stop button if necessary to view the signal again. Move the vertical position of the signal to the center of the screen using the vertical position knob for CH. 1.

2. Now change the SEC/DIV under the Horizontal section of the oscilloscope to 50 ms/div. Gently tap the end of the beam. Wait until the entire signal is visible within the oscilloscope display, then press the Run/Stop button on the oscilloscope to freeze the signal.

3. Press the Cursor button and select "Time" under the "Type" menu. Using the vertical knobs, measure and record the time for several periods of the signal. The frequency of vibration (in Hz) is the inverse of *one* period (in seconds) of the signal.

H.3.4 WHAT TO REPORT

The following must be included in a technical memo (use SI units and 95 % confidence throughout):

1. Plot up the Amplifier Out (V) versus Weight (N), where Amplifier Out is on the ordinate and Weight is on the abscissa. Use different symbols for the "up" calibration sequence (when weights were *added*), and for "down" sequence (when weights were *removed*). Note that the masses (in g) must be converted into force units (N). Note in the technical memo any differences between the "up" and "down" calibration sequences.

2. Plot a linear least-squares regression analysis of the data (see Chapter 10).

3. Based on the information provided by the least-squares regression analysis, determine the uncertainty in the voltage that is related to the standard error of the fit (the "y estimate").

4. Find the uncertainty in determining weight from a voltage measurement (the "x-from-y estimate"). To determine this uncertainty use a voltage near the midpoint of the calibration voltages and project it back through the calibration curve to the x-axis to determine the expected x-value. Also project back through the appropriate confidence limits to the x-axis to find the minimum and maximum possible x-values. Find the differences between these values and the expected x-value. Use the greater of the two as the uncertainty.

5. Determine the volume of each unknown material using both the water displacement and caliper measurements.

6. Use both water-displacement and caliper-determined volumes and the weights determined using the calibration, determine the density of each material. For three of the objects there will be two density values (one from each measurement method).

7. In a table, report the volume determined by each method and the density for each object. Compare the density values with those found from the literature or on the Web. Be sure to cite any sources.

8. Determine the uncertainty in the volume and density values. This should be done for both types of volume measurement. Which method has the least uncertainty?

9. Assume that there is no access to calipers and a low-cost experiment must be designed to determine the volume and density of the unknown objects. Design a new water-displacement-based experiment that costs less than $100.00 and has a volume uncertainty less than 0.061 in.^{3}. Be specific and cite sources.

10. Determine the mass flow rate of the sand using the calibration information and the recorded time difference.

11. Determine the uncertainty in the mass flow rate.

12. Present the weight, volume, density, and mass flow rate uncertainties all in one table.

13. Determine the density of the dry sand and calculate the uncertainty of the dry sand density.

14. Assume an hourglass is to be made based on the collected data. How much sand would be needed to measure an hour?

15. Determine the natural frequency of the beam using the data collected from lab (see Supplemental Information). Calculate a theoretical natural frequency for the beam. Compare the experimental and theoretical values and give reasons for any possible differences.

All of the important experimental results and answers to the posed questions must be presented as a technical memo. Answers should be contained in the explanation of the results and *not* listed item-for-item.

H.3.5 **Supplemental Information**

All solid objects vibrate to some extent when they are hit. Determining the frequency of vibration is very important in design. Vibration can cause wear and reliability problems and can induce unwanted noise. In some instances, vibration is desired and even necessary. A vibrating conveyor belt is a good example of this. The frequency at which an object vibrates depends on both its shape and material properties. An object's unforced (natural) frequency can be determined experimentally by giving the object a short impulse and measuring its frequency of vibration, as is done in this lab. For simple configurations, it is also possible to determine the natural frequency using theoretical approximations. For the cantilever beam configuration used in this lab, the frequency of vibration can be determined by

$$\omega_r = \frac{A_r}{l^2}\left(\frac{EI}{\rho A}\right)^{1/2}, \qquad\qquad \textbf{[H.1]}$$

where ω_r is the natural frequency, E is the modulus of elasticity, I is the moment of inertia, A is the cross-sectional area, A_r is the mode coefficient, l is length, and ρ is the density. The cantilever beams used in this lab are made of 6061-T65 aluminum with a modulus of elasticity of approximately 10^7 psi and a density of 0.00305 slugs/in.^{3}. The length, l, is the distance for the fixed point to the free end of the beam. For the lab set-up, l is

$$l = L - D, \qquad\qquad \textbf{[H.2]}$$

where L and D are the distances shown in Figure H.5.

Recall that the moment of inertia for a rectangular cross section is

$$I = \frac{bh^3}{12}, \qquad\qquad \textbf{[H.3]}$$

where b is the width of the beam and h is its height. Figure H.6 shows a diagram of relevant dimensions. Every beam in this lab has slightly different dimensions; this is true with any engineered object. Table H.2 contains the measurements for each experimental set-up. When the theoretical values are calculated for this lab, use the appropriate dimensions for the particular station. The accuracy of each measurement is also given.

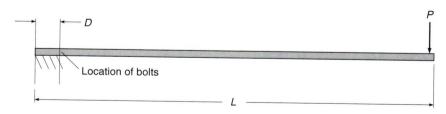

Figure H.5   Cantilever beam diagram.

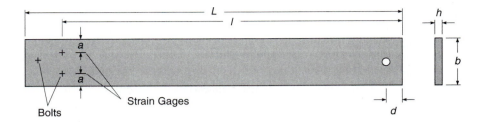

Figure H.6   Cantilever beam with four strain gages.

Table H.2   Beam dimensions (all units are in inches).

Beam	L (±0.002)	l (±0.020)	d (±0.0015)	b (±0.001)	h (±0.0005)
1	11.913	9.930	0.491	1.504	0.1275
2	11.933	9.910	0.492	1.504	0.1275
3	11.919	9.920	0.495	1.503	0.1275
4	11.924	9.920	0.492	1.504	0.1275
5	11.943	9.950	0.488	1.503	0.1275
6	11.944	9.940	0.491	1.503	0.1275

Table H.3   Mode coefficients for the first three vibration modes.

Mode 1	A_1	1.875
Mode 2	A_2	4.694
Mode 3	A_3	7.854

Notice that the dimensions for a and d are not listed in the table. These dimensions are not critical to the calculations and were not measured.

The mode coefficient, A_r, depends on the mode of vibration, r. The natural frequency is the first mode ($r = 1$). As the mode of vibration increases so does the frequency of vibration and the mode coefficient. Table H.3 gives the mode coefficients for the first three vibration modes.

H.4   EXERCISE 4—PROPELLER DYNAMOMETER: STATIC THRUST, TORQUE, AND RPM MEASUREMENT

H.4.1   INTRODUCTION AND OBJECTIVES

In this lab exercise, two strain gages, a Wheatstone bridge, and amplifiers will be used to determine the thrust and torque generated by a radio-controlled aircraft propeller under *static* operating conditions. More specifically, the power into the motor, the thrust and torque output of the system, and the RPM of the propeller will be determined.

The data taken from a thrust stand such as this can be used to gather performance data on different propellers. This in turn can help engineers make more informed design choices when selecting a propeller for a given airframe and propulsion system. If this experiment were taken a step further, the measurement system could be easily modified and be placed in a wind tunnel in order to gather *dynamic* propeller data. Also, such a measurement system could be used to examine the performance of a fan in a heating, ventilating, and air-conditioning system. More information regarding propellers and how they work is included in the Supplemental Information section.

H.4.2   INSTRUMENTATION

The following equipment will be used:

- Zinger 11-7 propeller ($d = 11$ in.)
- Two Wheatstone bridge and operational amplifier instrument systems
- Propeller dynamometer
- Motor power supply
- Voltmeter/ammeter readout box
- Calibration weights and hanger
- Stroboscope

H.4.3   MEASUREMENTS

Before starting on any of the measurements, a word about safety; this is the first exercise that has the potential of being extremely dangerous. The propeller turning at several thousand RPM will not hesitate to remove fingers if one is careless enough to place them in the plane of the blade. Therefore, the rule of operation for the dynamometer is: *never remove the safety cage without first disconnecting the power supply to the motor.* Do this by disconnecting the red and black banana plug power leads from the motor power supply.

1. Connect and adjust the bridges and amplifiers.
2. Calibration:

   - First check to make sure that the motor power supply is disconnected. Once this is done, remove the safety cage from the propeller. Attach the hanging wire to the screw at the center and immediately in front of the propeller, and hang it over the pulley at the front of the test stand.

- Zero the bridges for both the thrust and torque readouts by adjusting the "BRIDGE BALANCE ADJUST" knobs on the appropriate panels.
- Perform both thrust and torque calibrations, one at a time, by adding weights to the respective hanger and recording the corresponding voltage output. These data will be used later to convert the voltage measurements taken during the actual running of the propeller to thrust and torque readings. For the thrust calibration, add weights in 50-g increments up to around 1.0 kg. For torque calibration, increase by about 5 g to 10 g up to around 100 g. The little blue basket should be used as the initial torque calibration weight. Its mass is 7 g. The length of the torque calibration moment arm is 8.50 in. ±0.05 in.
- After completing the calibration, remove the wire hanger from the front of the propeller shaft and replace the safety cage.

3.   Thrust, Torque, and RPM measurements:

- Reconnect the motor power supply. Turn on the supply. If the red overload light is flashing, depress the button on the voltmeter/ammeter box. Turn the voltage increase dial up just enough to get the propeller spinning. Now, depress the voltmeter/ammeter button again. You should hear the propeller increase its RPM.
- Going in 1-V increments from 1 V to 12 V, record the following data: voltage into the motor, current into the motor, thrust voltage, torque voltage, and propeller RPM. Repeat the measurements at several motor voltages after the first set is completed. The voltages and currents in are read from the voltmeter/ammeter box panel meters. Thrust and torque voltages are the output voltages from the appropriate bridge/amplifier system. The RPM is measured using the strobe. It is best to turn the room lights out during these measurements so that the standing image of the propeller can be seen better.

  The stroboscope has three scales: LOW (100 RPM–700 RPM), MEDIUM (600 RPM–4200 RPM), and HIGH (3600 RPM–25 000 RPM). During the course of the measurements, all three scales will be used. One can shift from one to another by depressing the appropriate button on the back of the stroboscope. The propeller is marked near its tip with distinct black lines. Basically, the strobe light will be adjusted until a stationary image of the propeller is seen, with the marks on the ends of the propeller appearing identical to what they would be if the propeller were not moving. This is a little tricky because standing images with the correct marks (two horizontal lines on one side; two vertical lines on the other) occur at even integer *fractions* of the correct RPM for a two-blade propeller as well as at the correct RPM. However, a standing image obtained at a strobe frequency of twice the correct propeller RPM will not show the correct marks. So, once a correct image is obtained, keep doubling the strobe frequency until the correct RPM is identified. For the present system, at a motor voltage of

1 V, standing images should be seen at approximately 170 RPM, 340 RPM, and 680 RPM, with correct images at 170 RPM and 340 RPM. So, the correct propeller RPM is approximately 340 RPM. The correct RPM at 2 V should be approximately 1000 RPM.

- Disconnect the motor power supply once all measurements are finished.

H.4.4   WHAT TO REPORT

Submit the information in the form of a technical memo. Be sure to include at the very least the following (with some discussion of each): (1) plots of the thrust and torque calibrations (T and Q versus the voltage outputs of the measurement system); (2) plots of T, Q, and $P_{\text{prop,in}}$ versus N_{prop}; (3) a plot of η_m versus N_{prop}; and (4) plots of C_T, C_Q, and C_P versus N_{prop}. Decide which plots are the important ones to put in the body of the memo. Put the other plots in an appendix. Remember to construct all plots according to the format presented in the text. Be very careful with the units when calculating the values of all these parameters. Include a sample calculation of each parameter in an appendix to demonstrate proper unit conversion.

H.4.5   SUPPLEMENTAL INFORMATION

An understanding of how a propeller produces thrust rests in a knowledge of how an airfoil generates lift, L, and drag, D. An airfoil in motion generates lift and drag. If we consider the propeller as a rotating airfoil, we can understand how it generates a forward thrust, T, and a torque, Q, where Q results from a force, F_Q, acting perpendicular to the forward direction.

Examine this in more detail. Refer to Figure H.7, which shows a propeller consisting of two blade elements of pitch angle, β, each located at distance R from the

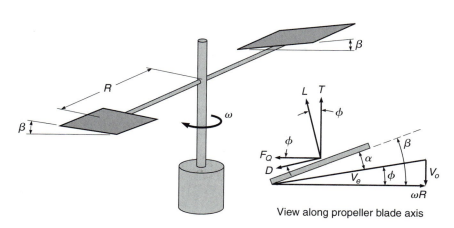

Figure H.7   Velocities and forces acting on a propeller blade element.

axis of rotation. The velocity V_o is that of the air through which the propeller advances. Because the blade element also is rotating with an angular velocity ω, it will have a rotational velocity of ωR. The velocities V_o and ωr are vectors that combine to yield the relative velocity V_e. This is the velocity of the air relative to the rotating blade element. Its approach angle is ϕ. This implies that the actual angle of attack α equals $\beta - \phi$. Therefore

$$\phi = \tan^{-1}\left(\frac{V_o}{\omega R}\right) \qquad \text{[H.4]}$$

and

$$V_e = \sqrt{V_o^2 + (\omega R)^2}. \qquad \text{[H.5]}$$

Now examine the view along the propeller blade axis. From trigonometry

$$T = 2(L\cos\phi - D\sin\phi) \qquad \text{[H.6]}$$

and

$$Q = 2RF_Q = 2R(L\sin\phi + D\cos\phi). \qquad \text{[H.7]}$$

Further, the power required to turn the propeller is

$$P_{\text{req}} = Q\omega. \qquad \text{[H.8]}$$

When $V_o = 0$, $\phi = 0$, giving $\alpha = \beta$. This leads to

$$T_{\text{static}} = 2L_{\text{static}}, \qquad \text{[H.9]}$$

$$Q_{\text{static}} = 2RD_{\text{static}}, \qquad \text{[H.10]}$$

and

$$P_{\text{req,static}} = Q_o\omega = 2\omega RD_{\text{static}}. \qquad \text{[H.11]}$$

Thus, for both static ($V_o = 0$) and dynamic ($V_o \neq 0$) conditions, a rotating propeller generates thrust and torque from its lift and drag. Further, the power required to turn the propeller is related to its torque and rotational velocity and, hence, to its lift and drag.

Now examine the power required to turn the propeller in the experimental setup. Some expressions that relate the measured variables to those that characterize the performance of some of its components need to be developed. Refer to Figure H.8.

The power into the motor is simply the product of its input current, i, and voltage, V, both of which are measured. That is,

$$P_{\text{motor,in}} = i \cdot V. \qquad \text{[H.12]}$$

Some of this power will be lost inside the motor and eventually dissipated as heat. This is quantified by the motor efficiency, η_m, where

$$\eta_m = \frac{P_{\text{motor,out}}}{P_{\text{motor,in}}}. \qquad \text{[H.13]}$$

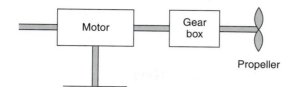

Figure H.8   Schematic of the motor-gear box-propeller system.

Now, the power out of the motor equals the power into the gear box, $P_{\text{gearbox,in}}$. In a similar manner, a gear box efficiency, η_g, can be defined where

$$\eta_g = \frac{P_{\text{gearbox,out}}}{P_{\text{gearbox,in}}}. \qquad \textbf{[H.14]}$$

Assume that $\eta_g = 0.95$; η_m will be determined through experiment.

The purpose of the gear box is to reduce the RPM of the motor to one that allows the propeller to operate within its most efficient range. The gear box reduces the motor RPM, N_{motor} to the propeller RPM, N_{prop}, by a factor known as the gear ratio, GR. This is given by

$$N_{\text{motor}} = \text{GR} \cdot N_{\text{prop}}, \qquad \textbf{[H.15]}$$

where for these experiments GR= 2.21.

From the preceding equations it can be determined that

$$P_{\text{motor,out}} = \frac{Q_{\text{prop,in}} \cdot N_{\text{prop}}}{\eta_g}. \qquad \textbf{[H.16]}$$

This yields

$$\eta_m = \frac{Q_{\text{prop,in}} \cdot N_{\text{prop}}}{\eta_g \cdot i \cdot V}. \qquad \textbf{[H.17]}$$

Every term on the right-hand side is known or determined from measurements. Hence, the motor efficiency can be determined using the data and Equation H.17.

Finally, the data can be used to determine the values of three coefficients that are commonly employed to characterize propeller performance. These are the thrust coefficient, C_T; the torque coefficient, C_Q; and the power coefficient, C_P. These are defined by the following equations:

$$C_T = \frac{T}{\rho \cdot n^2 \cdot d^4}, \qquad \textbf{[H.18]}$$

$$C_Q = \frac{Q_{\text{in}}}{\rho \cdot n^2 \cdot d^5}, \qquad \textbf{[H.19]}$$

and

$$C_P = \frac{P_{\text{in}}}{\rho \cdot n^3 \cdot d^4}, \qquad \textbf{[H.20]}$$

where T, Q_{in}, and P_{in} are for the propeller; d is its diameter; ρ is the density of air; and n is the propeller's revolutions per second.

H.5   EXERCISE 5—SOLID ROCKET MOTOR: TRANSIENT THRUST MEASUREMENT

H.5.1   INTRODUCTION AND OBJECTIVES

In this exercise, a load cell consisting of four strain gages mounted on a cantilever beam will be utilized to make dynamic thrust measurements of a solid rocket motor. These measurements in turn will be used in Exercise H.6 to make altitude predictions of a model rocket during ascent.

The first step in this process will be to calibrate the measurement system. Then, a digital oscilloscope will be used in conjunction with a Wheatstone bridge setup to record the thrust-time history for the model rocket engine.

H.5.2   INSTRUMENTATION

The following instrumentation will be used:

- Wheatstone bridge and operational amplifier measurement system
- Cantilever load cell
- Calibration weights and hanger
- Fluke PM3380-A Combiscope (Analog and Digital Oscilloscope)
- Estes A-83 solid propellant rocket motor
- Estes launch controller and igniters

H.5.3   MEASUREMENTS

Unlike any of the previous lab exercises, these thrust measurements will be made in teams. Before the start of the lab, all groups will be given their rocket motor, and will take turns in acquiring their thrust data. For safety, it is important that no one be near the rocket motors when they are ready to be lit.

1. Before firing, weight the motor and record its initial mass.

2. The first step will be a calibration of the load cell. Calibrate by placing the calibration masses in the cup at the end of the beam and recording the output voltage of the system. Calibrate up to a mass of 1 kg in order to cover the full range of the rocket motor thrust output.

3. Once the system is calibrated, make sure that the rocket motor sleeve is in the can at the end of the beam. Now configure the scope to make the appropriate measurements. The appropriate settings are 0.1 V DC, with a time base of 100 ms. As with the dynamic beam response, set the trigger settings to trigger off of the beam response by adjusting the trigger level to around one division above ground. Give a delay of about one division as well. Depress the "SINGLE" button

to arm the scope. Tap on the beam to check that the system is connected properly and triggering as planned. Then arm the scope again.

4. Now we are ready to set up the firing. Insert the motor into the sleeve at the end of the beam. Insert an igniter into the engine and cap it off with an engine plug. Make sure the safety key is removed from the launch controller, and then attach the launch leads to the igniter wires.

5. Once everything is connected, arm the scope and clear out of the firing area. Insert the safety key into the launch controller. The lightbulb on the front of the controller should be lit. Fire the motor by depressing the button on the controller.

6. Check the scope to see if a reasonable thrust trace was obtained. Save the signal into one of the scope's memory locations. Download this information onto the laboratory computer to save the data for subsequent analysis.

7. Finally, weigh the empty motor casing.

H.5.4   WHAT TO REPORT

Report the results in a technical memo.

H.6   EXERCISE 6—ROCKET LAUNCH: ALTITUDE PREDICTION AND MEASUREMENT

H.6.1   INTRODUCTION AND OBJECTIVES

Using the thrust curve for the solid rocket motors in Exercise H.5, predictions can be made of the maximum altitude that a model rocket can reach. This can be accomplished by deriving the appropriate equations of motion, developing appropriate models for all of the force terms, and solving this differential equation for the altitude. An actual model rocket also can be launched and its maximum altitude measured.

H.6.2   INSTRUMENTATION

The following equipment will be used:

*   Estes model rocket
*   Estes launch pad
*   Estes altimeter
*   Video camera

H.6.3   MEASUREMENTS

Several launches of similar rockets will be conducted. For each launch, the altitude will be determined using the altimeter, which uses a triangulation method to determine the height.

H.6.4 EQUATIONS OF MOTION

Simplify the analysis by assuming that the rocket travels only along a straight line in the vertical direction. Drawing an appropriate free-body diagram for the rocket and applying Newton's second law yields

$$\sum F = T - mg - D = m\frac{d^2y}{dt^2}. \qquad \text{[H.21]}$$

In Equation H.21, T is thrust, m is the mass, g is gravity, D is the drag, and y is the vertical displacement. If the drag is written in terms of a drag coefficient, C_D,

$$T - mg - \frac{1}{2}\rho S C_D \left(\frac{dy}{dt}\right)^2 = m\frac{d^2y}{dt^2}. \qquad \text{[H.22]}$$

The quantity S is the appropriate reference area, which in this case is the body tube cross-sectional area. The drag coefficient can be obtained from the handouts on the drag of model rockets. Note that during the burn phase of the rocket motor, the thrust varies with time, as determined in Exercise H.5. After the burn phase, the thrust becomes zero. The rocket, however, still continues to travel vertically upward until it reaches its maximum altitude where its velocity equals zero. Hence, there are two phases in the rocket's ascent, the burn phase and the coast phase. Both phases can be described by the same equation of motion by specifying the value of the thrust to become zero at the end of the burn phase.

Equation H.22 is a nonlinear, second-order differential equation. This equation is complicated by the fact that both the thrust and mass are quantities that are changing with time. This equation cannot be solved analytically unless some simplifying assumptions are made. Alternatively, through the use of numerical techniques and using MATLAB, this equation can be integrated directly to give us displacement as a function of time.

H.6.5 SOLUTION APPROACHES

Analytical Solution   An analytical solution to the preceding equation of motion can be obtained if the simplifying assumptions that the mass and thrust are constant in time during the burn phase of the rocket motor are made. These constant values can be calculated from the data that was acquired in the previous solid rocket motor firing lab exercise. The average thrust value should be obtained using the MATLAB `trapz` function and the thrust data file. The following equations result from integrating the equation of motion by parts.

The altitude, h_b, at the end of the burn phase will be

$$h_b = \frac{\beta_o}{g} \log_e \left[\cosh\left(g t_b \sqrt{a_o/\beta_o}\right) \right], \qquad \text{[H.23]}$$

where a_o denotes the drag-free acceleration in g's as given by

$$a_o = \frac{T}{W} - 1, \qquad \text{[H.24]}$$

with T being the thrust and W the weight. Also β_o, known as the density ballistic coefficient, is defined as

$$\beta_o = \frac{W}{0.5\rho C_D S}.$$ [H.25]

Further, the velocity, V_b, at the end of the burn phase is

$$V_b = \sqrt{a_o \cdot \beta_o}\,\tanh\!\left(gt_b\sqrt{\frac{a_o}{\beta_o}}\right).$$ [H.26]

The altitude gained up to the maximum altitude during the coast phase, h_c, will be

$$h_c = \frac{\beta_o}{2g}\ln\!\left(1 + \frac{V_b^2}{\beta_o}\right).$$ [H.27]

Thus, the maximum altitude, $h_{\max}$, is

$$h_{\max} = h_b + h_c.$$ [H.28]

Numerical Solution The second-order differential equation can be written as a system of two first-order equations. If y_2 is defined as the vertical displacement and y_1 as the vertical velocity, the preceding second-order equation reduces to

$$\dot{y}_1 = \frac{1}{m}\left(T - mg - \frac{1}{2}\rho S C_D y_1^2\right),$$ [H.29]

and

$$\dot{y}_2 = y_1.$$ [H.30]

These equations can be integrated using a numerical-integration algorithm.

The problem of the changing mass and thrust values with time still has to be addressed. The mass can be approximated by assuming that the motor burns at a constant rate. Thus, the mass will decrease linearly from the initial mass of the rocket to the final mass after the motor is spent. If m_o is the initial rocket mass and m_f the final mass, then

$$m(t) = m_o - \frac{m_o - m_f}{t_b}t,$$ [H.31]

where t_b is the burn time of the rocket motor. This expression is valid only up until the motor stops firing. After the thrust stops, the mass is constant and equal to m_f. This expression easily is incorporated into any solution algorithm. The thrust presents a unique problem in that the thrust data is at discrete values of time. There are several ways to approach this problem. The easiest method would be to obtain some average thrust value and assume that the thrust assumes this constant average value for the duration of the burn time. More involved methods might be to curve fit a polynomial curve to the thrust data to arrive at a continuous representation of the thrust over the time period of interest. Finally, an interpolation algorithm could be written to find an approximation for the thrust value at any time during the calculation.

There are MATLAB commands that will help demonstrate how to make these calculations. The command ode23 is a numerical integration algorithm that can be used to solve systems of differential equations. Additional information, as well as additional examples, can be found in the MATLAB manual.

To proceed, first create an M-file called `launch.m`. In that M-file, type the command `[t,y] = ode23('alt',t0,tf,y0)`. Before that command, specify the values for t0 (the initial time, here set equal to zero), tf (the final computation time, on the order of several seconds based on the type of rocket motor used), and y0 (the initial altitude, here set equal to zero). The command calls another M-file (name it `alt.m`) that will numerically integrate the equations that are set up in `alt.m` and then pass the results back to `launch.m` for subsequent plotting. Altitude can be plotted versus time in `launch.m` by the command plot(t,y(:,2)). The maximum altitude is given by the command max(y(:,2)).

Remember to create `alt.m`. The essential lines in `alt.m` are the first line: function ydot = alt(y,t) and the last line: ydot = [(1/m)*(T-m*g-(0.5*ρ*C$_D$*S)* y(1).^{2};y(1)]. In between the first and last lines, T, m, g, ρ, C_D, and S must be defined. Here, use average values for T and m or compute them.

Finally, to obtain the solution plot, simply type "launch."

H.6.6 WHAT TO REPORT

The results of the rocket motor firing and rocket launch laboratory exercises are to be submitted together in the form of a technical report by each team. The report should highlight the altitude predictions and how the data that relates to these calculations. A comparison between the predicted and actual measured altitudes must be made. Rational, scientifically based explanations, supported by additional calculations, must be presented to explain any differences between the predicted and measured altitudes.

H.7 EXERCISE 7—CYLINDER IN CROSS-FLOW: PRESSURE AND VELOCITY MEASUREMENT

H.7.1 INTRODUCTION AND OBJECTIVES

The main objective of this lab is to become familiar with the techniques and equipment for making pressure measurements on a circular cylinder. The cylinder is placed in a cross-flow in a subsonic in-draft wind tunnel. Velocity measurements also are made. In addition, concepts of uncertainty are addressed both in the taking of the measurements and the propagation of these uncertainties to obtain estimates for the lift and drag coefficients and the drag of the cylinder.

H.7.2 INSTRUMENTATION

- Dwyer Model 246, 0 in. H_2O to 6 in. H_2O inclined manometer (resolution: 0.02 in. H_2O)
- Microswitch Model 163PC01D36, -5 in. H_2O to $+5$ in. H_2O differential pressure transducer

- Tenma Model 72-4025 digital multimeter (resolution: 0.01 V on 20 V full scale; 0.001 V on 2 V full scale)
- Princo barometer (resolutions: 0.01 in. Hg and 1 °C)
- Wind tunnel RPM indicator (resolution: 20 RPM)
- Cylinder rotating position indicator (resolution: 1° angle)

The test section for this exercise contains a pitot-static tube and a cylinder fitted with pressure taps. The pitot-static probe is located in the front of the test section and will be used to determine the free-stream centerline velocity of the wind tunnel. The cylinder is 1.675 in. ±0.005 in. diameter and 16.750 in. ±0.005 in. length, and has several pressure taps located in a line along its span. For this experiment, the tap in the middle of the cylinder will be used to minimize any possible wind tunnel wall effects. The cylinder and, more importantly, the pressure tap can be rotated through 360° using the position indicator on the side of the test section.

In this exercise, both an inclined manometer and differential pressure transducers will be used for measuring pressure. Each pressure transducer is connected to a voltmeter to measure its output. The transducers used have a *linear* 1.01-V to 6.05-V DC range (corresponding to a range of −5.0 in. H_2O to +5.0 in. H_2O). Thus, the transducer output will be 3.52 V when the pressure difference is 0 in. H_2O. Any negative differential pressure will be less than 3.52 V. It is important to remember that both the inclined manometer and the transducers measure the difference in the pressure between the two lines connected to them. Be sure that all of the pressure lines are connected in the appropriate manner.

H.7.3 MEASUREMENTS

First check to see that all of the lines are connected properly. One of the pressure transducers will be connected hydraulically in parallel with the inclined manometer to measure the pressure difference from the pitot-static probe. The other transducer will measure the pressure difference between the pressure tap on the cylinder and an adjacent static port. The static port for the cylinder can be located on the side wall of the test section, just above the cylinder.

Adjust the voltmeters to the appropriate scales. With the tunnel off, what should the voltmeters display? If the output is not as expected, be sure to make a note of it so that the bias can be accounted for later when reducing the data. Now, check the level at the top of the inclined manometer and adjust the manometer until it is level. If necessary, zero the manometer by loosening and sliding the scale until the bottom of the meniscus is set at zero.

Record the room temperature and pressure using the Princo barometer. Record the room temperature in °C and the pressure in in. Hg. Also record the % correction factor. This factor corrects for the thermal expansion of the metal scale that is used to determine the pressure. For example, the correction factor is 0.38505 % at 22 °C. So, the actual pressure equals the recorded pressure times $(1 - 0.0038\,505)$. The actual pressure and temperature values will be used later to compute the density of the air in the lab assuming ideal gas behavior. Subsequently, the density value is needed to compute velocities and the Reynolds number.

First, calibrate the wind tunnel RPM indicator with respect to the wind tunnel velocity. Do this by setting the tunnel fan at various RPM and record the pressure difference measured using both the inclined manometer and the pressure transducer connected to the pitot-static tube. Start the wind tunnel fan by following the directions on the control panel stand. Make sure the circuit breaker is turned to on and push the start button. Set the RPM indicator to 100 RPM and wait a minute for the tunnel to come to steady state. Record the pressure difference indicated on the inclined manometer and the voltage from the voltmeter connected to the output of the pressure transducer that is connected to the pitot-static tube. Proceed through all the RPM settings in increments of 100 RPM up to and including 900 RPM. Repeat several RPM measurements to assure reproducibility. While taking data, convert a recorded pressure transducer voltage to in. H_2O. Are the inclined manometer and pressure transducer readings in agreement?

Now perform pressure measurements on the cylinder. Check that the cylinder's pressure tap orientation is at $0°$, as indicated on the rotating position indicator. Now set the tunnel RPM to that RPM specified. Record the dynamic pressure from the pitot-static tube from both the voltmeter and the inclined manometer. Then, in increments of $10°$, rotate the cylinder and record the output from the pressure transducer connected to the cylinder. Also record the voltage from the pressure transducer connected to the pitot-static tube. This reading is a good indication of how constant the wind tunnel velocity is during the measurements. When the entire $360°$ range has been covered, go back and make a couple of spot checks at various angles to check repeatability.

When finished collecting the data, turn the dial indicator on the wind tunnel fan control panel back to zero and stop the tunnel. Again, follow the instructions on the control panel. Finally, make sure that all equipment is turned off.

H.7.4 WHAT TO REPORT

Submit the results in a technical memo. Be sure to include (as a minimum) the following information:

- Calculations of the air density, operating tunnel velocity (both in SI units), and the Reynolds number.

- Calculation of the temporal precision error ($= S_x / \sqrt{N}$) of the pitot-static tube pressure transducer voltage taken at the various θ during the cylinder measurements. Compare this value to the mean value by determining the percentage of the precision error with respect to the mean value. Ideally, this should be zero if the wind tunnel velocity remained constant during the measurements period.

- Two plots of the wind tunnel RPM calibration, one with the velocity (in ft/s) calculated from the measured in. H_2O from the inclined manometer and the other with that calculated from the pressure transducer voltage. Plot each manometer measurement or voltage along the ordinate versus the tunnel RPM along the abscissa. Perform the necessary regression analysis and display the proper error bars. In performing the regression analysis, keep in mind the actual relationship between differential pressure and velocity.

- Plot of the pressure coefficient, C_p, on the y-axis as a function of azimuthal angle, θ, on the x-axis (include on this plot the analytical, inviscid solution for comparison (see the Supplemental Information section).

- Calculations of the lift and drag coefficient of the cylinder, C_L and C_D, and the drag force on the cylinder, D, in units of N. Does this calculated drag force appear reasonable?

- Uncertainty estimates presented in the form of tables supported by example calculations. Two tables are required, one for the measurand uncertainties and the other for the result uncertainties. Supporting calculations of all of uncertainty estimates should be contained in an appendix.

Calculations of the lift and drag coefficients will require some numerical integrations of the C_p data. The appropriate equations are included in Section H.7.5. A spreadsheet can be used to perform a simple trapezoidal rule integration. What should be the C_L value of the cylinder? Use the C_L calculation to check the calculations.

Some possible areas of discussion for this lab might include: What does the analytical solution predict for a drag coefficient? Does the experimental value confirm this? What might some possible reasons be for this? Try to think of the assumptions made in the analytical solution. Are these assumptions valid?

H.7.5   SUPPLEMENTAL INFORMATION

Velocity Calculation   If incompressible, inviscid, irrotational flow is assumed, then the complete form of the momentum equation reduces to Bernoulli's equation,

$$P - P_\infty = \frac{1}{2}\rho u_\infty^2, \qquad \textbf{[H.32]}$$

where $P - P_\infty$ is the pressure difference measured by the pitot-static tube, ρ is the density, and u_∞ is the free stream velocity. Solving for u_∞ gives

$$u_\infty = \sqrt{\frac{2\Delta P}{\rho}}. \qquad \textbf{[H.33]}$$

This relation can be used to calculate the wind tunnel velocity for a given RPM setting. This relation is also used to estimate the uncertainty in the calculated velocity based on the experimental uncertainties in both the density and pressure difference. Combining uncertainties, the appropriate formula for this velocity calculation would be

$$\partial u_\infty = \sqrt{\left(\frac{\partial u_\infty}{\partial \Delta P}u_{\Delta P}\right)^2 + \left(\frac{\partial u_\infty}{\partial \rho}u_\rho\right)^2}. \qquad \textbf{[H.34]}$$

Reynolds Number   The Reynolds number is defined as

$$\text{Re} = \frac{\rho u_\infty D}{\mu}. \tag{H.35}$$

This is based on the cylinder diameter, D, the free stream velocity, u_∞, the density, ρ, and the absolute (dynamic) viscosity, μ. For air, the absolute viscosity is given by the Equation 2.4 in Chapter 2,

$$\mu = \frac{b \cdot T^{3/2}}{T + S}, \tag{H.36}$$

where μ is in units of $N \cdot s/m^2$, T in K, $S = 110.4$ K, and $b = 1.458 \times 10^{-6}$ kg/$(m \cdot s \cdot K^{1/2})$.

Pressure Coefficient   The pressure coefficient is defined as

$$C_p(\theta) = \frac{P_\theta - P_\infty}{\frac{1}{2}\rho u_\infty^2}. \tag{H.37}$$

The pressure difference in this equation is the ΔP measured for each individual rotation angle, θ. Thus, for every angle, a C_p value can be calculated. At $\theta = 0$, a stagnation point exists, for which $C_p = 1$. If C_p is calculated based on the dynamic pressure measured by the pitot-static tube upstream of the cylinder, at $\theta = 0$, the C_p value may be slightly less than one. This mainly is due to a small pressure decrease through the tunnel between the pitot-static tube and the cylinder. In calculating the C_p values from the data, it is often easiest to simply assume $C_p = 1$ at $\theta = 0$, and calculate the rest of the C_p values based on the $\theta = 0$ pressure measurement. That is, the corrected C_p value is given by

$$C_{p,\text{ corrected}}(\theta) = \frac{C_p(\theta)}{C_p(\theta = 0)} = \frac{V_{\text{trans}}(\theta)}{V_{\text{trans}}(\theta = 0)}, \tag{H.38}$$

where V_{trans} denotes the voltage of the pressure transducer after being corrected for its offset voltage at zero velocity. Equation H.38 is valid because the differential pressure is related linearly to the transducer voltage after the offset correction.

The derivation of the analytic solution for C_p for this situation can be found in any standard aerodynamics text. The final result is

$$C_p(\theta) = 1 - 4\sin^2\theta. \tag{H.39}$$

Lift and Drag Coefficients   The formulas for the lift and drag coefficients of a circular cylinder in cross-flow are derived in many aerodynamic texts. The results are presented here:

$$C_D = -\frac{1}{2}\int_0^{2\pi} C_p(\theta)\cos(\theta)\, d\theta, \tag{H.40}$$

$$C_L = -\frac{1}{2} \int_0^{2\pi} C_p(\theta) \sin(\theta)\, d\theta. \qquad \textbf{[H.41]}$$

These can be calculated for the data using a numerical integration algorithm such as the `trapz` function in MATLAB. The command `z=trapz(x,y)` computes the integral of y with respect to x using trapezoidal integration, where x and y are vectors of the same length. Once C_D is known, the actual drag force on the cylinder can be found using C_D, the dynamic pressure, and the frontal area of the cylinder.

H.8   EXERCISE 8—DIGITAL OSCILLOSCOPE AND FUNCTION GENERATOR

H.8.1   OBJECTIVES

The objective of this laboratory exercise is to introduce the capabilities of a function generator and digital oscilloscope, and their use in basic measurements.

H.8.2   INSTRUMENTATION

The following equipment will be used:

- Hewlett Packard HP 33120A Function Generator
- Fluke PM3380-A CombiScope (Analog and Digital Oscilloscope)

H.8.3   MEASUREMENTS

In this laboratory exercise the capabilities of a function generator (FG) and a digital oscilloscope (DO) will be demonstrated. The DO will be used to observe and analyze various signals produced by the FG. The triggering capabilities of the DO also will be studied.

The FG is an electronic instrument that generates waveforms of preset shape, amplitude, and frequency. Often it is used in a laboratory setting to provide a known input to data acquisition devices such as the DO or a computer. This helps the investigator debug and calibrate measurement systems. The DO is perhaps the most used piece of electronic equipment in a laboratory. It is the experimenter's electronic eye. It has the basic capability to acquire, store, display, and analyze signals, and download them to other devices. The typical DO has at two amplifiers (with variable gains), a sample and hold circuit, and an A/D converter. The digitized signal is stored in its random access memory (RAM), and the output is sent to the video display. Most digital scopes use a single CCD (charge-coupled device) array per channel to sample the signal and hold that value until the A/D has time to convert the signal. Once data have been acquired by the DO, the data can be overwritten to display a new signal or data can be saved and analyzed using onboard software programs. Also, data can be downloaded to another device such as a plotter or computer for further analysis.

The front panel of the DO is divided into seven functional areas, as shown in Figure H.9. First, examine these areas, referring to their functions listed next.

Area 1: Basic screen and power controls with self-explanatory labels.

Area 2: Screen text control buttons and menu buttons.

Area 3: Basic controls for input channels 1 and 2. There are controls for amplitude scaling (volts per division), positioning the signal, establishing the signal coupling (AC, DC, or GND), turning the channel ON or OFF, scaling the signal, AUTO RANGE, and determining whether the signal will be increasing or decreasing for triggering. Each area also has two additional keys that have special application. The first is the VERT MENU key (not used at this time) and an AVERAGE key, which averages the signals on both channels simultaneously. The last key is the INV key, which is applicable to channel 2 only (it inverts the signal).

Area 4: The time and trigger control section for the main time base (to be presented later). Again, there is an AUTO RANGE control plus controls for time scaling (per division), trace position, magnification, and several trigger controls that will be discussed later.

Area 5: The cursor control section. The TRACK control knob has a dual purpose. If it is used for measurements of a voltage versus time trace, it sets the reference x-cursor, while the delta control knob (the one with the Δ above it) positions the measurement x-cursor. A reading of the value of the reference x-cursor or the difference between the x-cursors (in volts, time, or both) is provided in text on the screen if chosen. The TRACK control also is used for selection in the menu items and the requirement for their use is indicated by a small "T" inside of a circle.

Area 6: The delayed time base control area has a special application that won't be covered in this exercise.

Area 7: External trigger section that enables one to use a signal of choice as the control for initiating acquisition of data other than the signals of either channels 1 or 2 or the line signal.

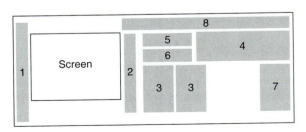

Figure H.9   Schematic of the front panel of the PM3380-A CombiScope.

Area 8: This is the extended function area. The simplest function is the AUTOSET button. This automatically finds the signal and adjusts the settings to produce a properly proportioned signal on the screen. The other buttons provide the user with a host of powerful built-in functions of math, measurement, and presentation.

Last, are the hard-wired inputs along the bottom that are clearly labeled.

Part 1: Viewing a Periodic Signal on the DO Connect the OUTPUT of the FG to channel 1 of the DO. Set the FG to deliver a square wave having 150 Hz and VPP (peak-to-peak) amplitude of 4 V, as observed on the DSO (which is 2 VPP on the FG). (*Note*: The VPP amplitude set on the FG appears as twice that amplitude on the DSO. This is because of an impedance mismatch that is not dealt with at this point—see Chapter 4 for an explanation). Make sure that the DO is in digital mode by pressing the yellow ANALOG button, which toggles between the two modes and indicates briefly the mode on the screen. The trace should show one or at most two complete cycles of the signal while maximizing the vertical display. Do *not* change the FG setting from what was set initially. Use the DO's vertical gain control, specified in V/div, and horizontal time control, specified in s/div. Center the trace vertically using the position control for that channel. Place the start of the trace on the left edge of the screen grid by adjusting the X POS knob. Record the following using visual observation, *not* using the cursors.

1. Vertical scaling per division (V peak-to-peak)
2. Time scaling per division (s)
3. Frequency (Hz)

Did the display show an actual square wave, top and bottom parallel with the horizontal grid with little connection between the two (very faint compared to the horizontal lines)? If not, correct it by changing the signal coupling on the DO. Immediately to the right of the vertical scaling value displayed on the screen is an = sign for DC, an ~ sign for AC, and a ⊥ sign for ground. Observe the signal first with DC coupling and then with AC coupling. Sketch each of the two traces.

Now repeat the measurements using the cursors. To do so, simply press the CURSORS key. Select the second of the =, ∥, #, or "auto" choices. Then select and follow the READOUT menu, selecting reading of V1 (voltage measured using first cursor) and 1/ΔT (frequency). Use the cursors to determine the following:

4. Minimum voltage (V)
5. Maximum voltage (V)
6. Maximum − minimum voltage (V)
7. Frequency (Hz)

Now examine more closely the function of the types of couplings, AC, DC, and GND. First set the coupling on the DO to GND. This grounds the input to the channel, resulting in a horizontal line trace at 0 V. Next set the coupling to DC. Note the shape and position of the trace. Now add a DC offset of +0.5 V on the FG to the generated signal. Adjust the scaling or position of the signal on the DO as necessary to keep it in view. What has happened to the trace?

Now, change the coupling to AC. What change in the display occurred? Sketch the DC and AC coupled traces.

Often the signal being observed has a midpoint, zero-voltage level that is not as obvious. Then, it is necessary to establish a zero reference line. On the DO, this reference is arbitrary. To establish a zero reference, simply ground the input signal. Establish a zero reference on the center horizontal graticule from the bottom by rotating the vertical position knob and then unground the signal.

Now go back and observe the original signal with DC coupling. Is it centered on the reference line? Now set the FG offset back to 0 V. Is it centered on the reference line? Try this also with AC coupling with and without offset. How are these different than with DC coupling?

An easier approach is to use the AUTOSET button which, based on some preset criteria, will select the appropriate time and amplitude scaling and coupling. Another function is the magnification function, MAGNIFY. Before using this, change the time base to get approximately 20 cycles on the screen without changing the FG settings. Now press the right directional arrow MAGNIFY button. Notice that a horizontal bar temporarily indicates what portion of the trace in memory is currently being displayed. Now keep pressing the right arrow MAGNIFY button until only one to two cycles is seen on the screen. The amount of time amplification is indicated temporarily on the screen by an * followed by a number indicating the amount. Note that the signal no longer looks like a perfect square wave. This is because the signal stored in the RAM consists of a fixed number of digital points. Magnifying it produces a signal constructed by drawing lines in between the points. Finally, using the left directional arrow MAGNIFY button, bring the time amplification back to 1.

Part 2: Using the Trigger on the DO   In most situations, the user wants to see what the continuous signal looks like. Therefore, the DO is normally operated in a continuous sweep mode. However, there is something happening in the background that is not readily apparent to the user. In most instances, if the user has selected a time scale that allows viewing of several cycles of a standing wave, it appears that the signal is starting at the same point in its cycle. This is because the DO, in its default mode, is being triggered by the signal itself. The trigger on a DO determines when the trace will begin. On most DOs, there are two conditions that must be met for a trigger to occur: a specified voltage relative to ground and a direction of change. In addition, there are several choices for the source of the trigger, be it the signal into channel 1 or 2, an external signal, or the (power) line signal (in this country, a 60-Hz signal). In all cases, the trigger initiates the trace, which then continues until it is completed. Then, based on the trigger options selected, the trace will start over on the next trigger received, stop all together, or execute some other user-selected option.

To explore the trigger, start with something already familiar. Establish a square wave as done before with the same characteristics (150 Hz, 4 V peak-to-peak on the DO, no DC offset). Check the settings (press the STATUS button).

The first thing is to observe the effect of direction of change, or slope, on the trace output. In the control box of the channel selected for input, press the TRIG button and observe the change. The slope direction is indicated on the far right, bottom corner of the screen. Keep the STATUS text on screen during the process to observe the settings.

Next, in the time and trigger control section, press the TRIGGER button. Scan through the menu to see its options. Note that when the trigger source (the second item down) is changed, an important thing happens. If the trigger source and the input signal do not have a common frequency, the trace is no longer stable. For each trigger source, observe the stability of the trace. Now connect the SYNC of the FG to the EXT TRIG INPUT of the DO. Go back and put the trigger source on "ext.trig". The square wave should be stable now on the screen. This is because the FG sends out a sharply rising pulse at the instant the square wave begins, which is an excellent source for an external trigger.

Now, set the trigger for edge, ch1, level-pp off, noise off, and AC coupling. Then press the TB MODE button (time base) and toggle through the selectable trigger options for the top item of the menu. Note that when on "single", the trace freezes. As the name indicates, it will select a trigger only *once* in this mode. To obtain a new trace, press the SINGLE button in the time and trigger section. As this is done, note that a red light to the right of the button briefly comes on. When it is lit, a trigger is armed, indicating that a trigger signal has not been received. After the trigger occurs, the light goes off.

As stated earlier, the DO is normally used to observe periodic signals and the trigger not as much. But, with the trigger option, the DO can be a valuable data acquisition and analysis tool. Consider the following situation. Assume that a signal from a single-event test, such as the firing of a rocket motor, is expected to have an amplitude of 1.5 V to 5 V and a duration of 1.5 ms to 6 ms. To see the complete event from start to finish, view the trace starting 1 ms before the event occurs and ending 1 ms after it ends.

To do this, set up a known signal having an initially rapid amplitude change, a square wave. Change the settings on the FG to obtain a peak-to-peak amplitude of 2.5 VPP on the FG display with a DC offset of 1.0 V. Set the FG frequency to 250 Hz to obtain a period of one square wave equal to 4 ms. Be sure that the TB MODE is in "auto". Verify these requirements using the cursors and record the information below.

8. Maximum voltage (V)

9. Minimum voltage (V)

10. Period of one cycle (ms)

Note that if the trace is moving in an apparent random fashion, the action can be stopped by pressing the RUN/STOP button in the time and trigger section. If any changes are made to the generated signal, allow the DO to run to allow the changes to be displayed. Another method is to use the MEASURE function with MEAS 1 set to measure "pkpk" and MEAS 2 set to "freq". Both must be turned on. Once everything is set, press the MEASURE button again to turn the menu display off. Choose a time scale to obtain at least one period of the signal plus 1 ms before and after. Record this.

11. Time scale chosen (ms/div)

Remember, the DO provides ranges in a 1, 2, 5 format. Now set the pretrigger (the time-record length prior to receiving the trigger). To do this, turn the TRIGGER POSITION knob until the appropriate reading (this will be indicated by a "-dv" on the screen). The position will be shown by a small triangle only for "-dv"'s. All data gathered to the left of this symbol occurs prior to the trigger; all the data to the right occurs after the trigger. The time to the left of the symbol should be 1 ms. Now the trigger level must be set. First, make sure that the conditions of the trigger used above are set, namely edge, ch1, level-pp off, noise off, and AC coupling. Set "auto" in the TB MODE. Determine the lowest trigger level that will produce an active, stable trace (constantly being refreshed and remaining steady) by turning the knob labeled TRIGGER LEVEL. The value will be indicated by "Level=" on the screen. Record the value below. Now find the highest trigger level and record its value.

12. Lowest trigger level (V)

13. Highest trigger level (V)

14. Difference between trigger levels (V)

Now repeat the process using a DC coupling on the trigger. Record the values.

15. Lowest trigger level (V)

16. Highest trigger level (V)

17. Difference between trigger levels (V)

Now repeat the process again for DC coupling but with no DC offset on the FG. Record the values below. The AC coupling case would be the same as before.

18. Lowest trigger level (V)

19. Highest trigger level (V)

20. Difference between trigger levels (V)

How do the *ranges* of the trigger levels compare for the three cases? How do the absolute trigger levels compare for the three cases?

Next, with a properly triggered signal, press the SINGLE button in the time and trigger section of the front panel. If all was done correctly, a stationary trace of the square wave signal is seen. If not, go back and repeat before proceeding to the next step.

Now return the DO to "auto" under the TB MODE. Set the FG to display an amplitude setting of 2 VPP, a frequency of 1 Hz, and no DC offset. The low frequency will allow one to see the actual trigger occur. Adjust the amplitude setting to 1 V DC and the time base setting to 100 ms. Center the trace on the screen by establishing a zero reference. View the signal. It should be repeating over and over in time on the screen. Now set "single" under TB MODE and "ch1", "level-pp off" and "dc" under TRIGGER. A "T-" should be on the screen, indicating the amplitude level of the trigger. Move it up and down by rotating the TRIGGER LEVEL knob, then finally position it about one division above the top level of the square wave. Press the SINGLE button in the time and trigger area. Only a horizontal line should be present on the screen, indicating that the signal has not triggered. The red "ARM'D" light should be on. Now gradually rotate the knob slowly to bring down the trigger level while watching the "Level=" value on the screen. Observe and record below the value indicated when the signal triggers. If desired, the triggering process can be repeated by pressing the SINGLE button to arm the trigger and then moving the knob in smaller increments to get a better estimate of the trigger level. Now move the indicator below the bottom of the signal and find the minimum value (by moving the indicator up) where the signal will trigger. Record the value below. Thus, determine the range of the trigger level that will properly trigger the DO for this square wave.

21. Lowest trigger level (V)

22. Highest trigger level (V)

23. Difference between trigger levels (V)

H.8.4   What to Report

Turn in this document with all questions answered. No technical memo is required for this exercise.

H.9   Exercise 9—Digital Data Acquisition

H.9.1   Objectives

The objective of this laboratory exercise is to investigate several aspects of digital data acquisition using both a digital oscilloscope and a computer data acquisition system. Specifically, several ways to acquire, store, and analyze various waveforms will be explored and some of the limitations of digital data acquisition will be examined.

H.9.2   Instrumentation

The following equipment will be used:

- Hewlett Packard HP 33120A Function Generator
- Fluke PM3380-A CombiScope (Analog and Digital Oscilloscope)
- Personal computer with a United Electronics Incorporated (UEI) 12-bit, 16-channel A/D board for −5-V to +5-V input, and associated software

H.9.3 MEASUREMENTS

This laboratory exercise is divided into three sections. The first considers the use of a computer data acquisition system to digitally sample a simple periodic wave of known frequency. The second and third involve the use of the digital oscilloscope to capture, store, and subsequently analyze various waveforms.

H.9.4 SAMPLING A PERIODIC WAVEFORM

For this section, the FG will be used to generate a simple sine wave. The output of the FG will be sent to both the DO and the computer data acquisition system (DAS). The FG and DO will be used to ascertain the amplitude and frequency of the wave. The waveform acquired by the DAS will be viewed using the graphical display of a software package. Then, the sample rate of the DAS will be varied and the waveform frequency recorded. This will enable investigation of the relation between the sampling frequency and the frequency of the input wave (and thus to observe the effects of signal aliasing). To begin, check to see that the BNC cable from the FG to the DAS is *disconnected* (input signal amplitudes greater than 7 V will destroy the input circuitry of the A/D board). Also check to see that the FG output is *connected* to channel 2 of the DO. Using the FG, generate a 500-Hz sine wave with 100 mV peak-to-peak with zero DC offset. Press the light green AUTOSET button on the DO to view the waveform. Acquire a single trace by pressing the SINGLE button in the time base settings area. This freezes a single trace on the display. Using the cursors (press the CURSORS button), record the observed signal frequency and peak-to-peak amplitude. Connect the DAS BNC cable to the T-connector at the FG output. From this point on in this section the FG or DO settings remain the same. Only the sampling frequency of the DAS is varied and its output recorded. Set the sampling rate on the configuration page of the software package and then measure Δt from the software package's graph. The second and third columns will be recorded during the lab. The fourth and fifth columns should be filled in *before* coming to lab. $f_{\Delta t}$ is computed directly from Δt, f_N is one-half of f_{sample} and f_{calc} from f_1 and f_{sample} (see Chapter 12 on how to use the folding diagram to calculate the f_{calc} values). f_{calc} is the aliased frequency expected to occur at f_{sample}.

Now take data using the DAS. Open the data acquisition software by double-clicking on the UEI icon, then on the UEI Status for Windows icon. Under File select Load configuration and then lab8.cfg. Then under Analog select "Configure." This brings up the data acquisition parameter display. Set the sampling rate by typing it in or selecting it if available. The samples per channel should be set and remain at 128. The duration is simply the samples per channel divided by the sampling rate when only one channel is used, as in this case. After the desired sampling rate is set, press enter, then the F7 key to start the acquisition process. When completed, a graph will come up on the screen displaying the acquired signal. Remember that this is the digital representation of the signal, so it will not always look exactly like the input signal. Observe the graph. Many periods of the wave are acquired. Now examine only a few of the periods (between 2 and 5) by using the expansion icon shown immediately under the word "Graph". Select a region of interest by dragging the mouse and then clicking it. The selected region should now occupy the entire graph. Next, position the two cursors by single-clicking the mouse at each of two points on the graph. Try to position the cursors at the same position on the wave (its top) over several periods such that a more accurate value of the Δt for *one* period is obtained and recorded. Record the Δt for $f_{sample} = 20\,000$ Hz and its peak-to-peak amplitude. How does this value compare to the DO peak-to-peak amplitude recorded earlier? Now go back to the parameter display and change the sampling rate. Repeat this process until all of the listed sampling rates have been investigated and have filled in the raw data values in the table. When finished, exit the UEI software. Also, *disconnect* the DAS BNC cable from T-connector. When all of the columns in Table H.4 are completed after lab, state how f_{calc} and $f_{\Delta t}$ compare for each f_{sample}.

H.9.5 EXAMINING FREQUENCY SPECTRA

Now set the FG to deliver a 1-kHz sine wave with a 100-mV peak-to-peak amplitude with zero DC offset. Press the AUTOSET button on the DO and then adjust the settings on the DO to display around 10 cycles on the screen. Acquire a single trace by pressing the SINGLE button in the time base settings area. This provides a frozen signal to perform the fast Fourier transform (FFT).

Enter the MATH menu by pressing the MATH button along the top of the scope. Then under the MATH 1 feature select "fft" and "ch2," then press ENTER. Then select "on" (the FFT of the sine wave should appear on the screen) and "no" for DISPLAY source (this will remove the sine wave trace from the display). The FFT of the input sine wave should be displayed on the screen. Press the CURSORS button and then use the cursors to measure the frequency and amplitude (in dB) of the largest peak in the frequency spectrum. Note that the scope displays the amplitudes by referencing all of the values to the largest peak present. In essence, the amplitude at each peak corresponds to the Fourier coefficient at that frequency (see Chapter 11). If the amplitude of the largest peak is defined as A_1 and that of a subsequent ith peak

Table H.4   DAS sampling data.

f_{sample} (Hz)	Δt (ms)	$f_{\Delta t}$ (Hz)	f_N (Hz)	f_{calc} (Hz)
20 000				
1500				
1200				
1000				
800				
600				
400				
200				
120				

as A_i, then the value reported by the scope (in dB) is found from the definition of the decibel:

$$dB = 20 \log \left(\frac{A_i}{A_1} \right). \qquad \text{[H.42]}$$

Thus, it can be seen that the largest peak will be reported by the scope as having a value of 0 dB because, for this case, $A_i = A_1$. Further, if the actual value of A_1 is known, the values of each of the A_i's can be computed using Equation H.42. Also remember that the DO displays digital information, so there may be two adjacent frequencies having maximum amplitudes. In this case, the actual frequency at maximum amplitude lies in between the two frequencies. Record the single frequency value at maximum amplitude (or the average value if there are two local maxima) in Table H.5. Is this frequency expected?

Press the AUTOSET button to start over and then repeat the above procedure for a square wave (the second-from-the-left button on the FG) for the same frequency, peak-to-peak amplitude and zero DC offset. Record both the frequencies and the amplitudes of the first five major peaks in the spectra in Table H.5. After lab, compute the amplitudes (in dB) of the first five major peaks in the spectra of a 1-kHz square wave. The amplitudes (in dB) can be found in the same manner as done in the class notes for a step function. Compare these calculated amplitudes with those obtained above. If there are any differences, explain what could be the cause(s).

H.9.6   SAMPLING AN APERIODIC WAVEFORM

Next, the digital oscilloscope will be used to capture a transient waveform. The event to record is the oscillatory response of the cantilever load cell to an impact loading. This will be accomplished by dropping a golf ball into the can at the end of the cantilever beam.

Table H.5   Amplitude-frequency data for sine and square waveforms.

Waveform	Frequency (kHz)	Amplitude (dB)
Sine wave	—	—
Peak 1		
Square wave	—	—
Peak 1		
2		
3		
4		
5		

To start off, make sure that the load cell is connected to the bridge circuit correctly. The panel meter wire with its end connector should be connected to the output of the second op amp on the bridge circuit box. Balance the bridge to zero by adjusting the ZERO ADJUST knob on the panel. Test to make sure the load cell is connected properly by lightly depressing the beam and ensuring the panel meter is responding (there should be approximately a 0.1-V to 0.2-V indication).

Check that the BNC output of the bridge circuit is connected to channel 1 of the DO. Set the scope to the following settings: DC coupling, 0.2 V, 100 ms. This will ensure capture of the full signal. Now, go into the TRIGGER menu and set "edge," "ch1," "level-pp" to off, and "dc." If "ch2" appears instead of "ch1," simply press the TRIG 1 button in the Ch1 area on the panel. Then "ch1" should appear in the TRIGGER menu. The trigger level should now be marked on the scope with a "T." Set it at about one division above the centerline of the display using the TRIG-GER LEVEL knob at the far right of the DO panel. Now set a delay for the trigger such that a part of the signal prior to the trigger event will be displayed. Do this by turning the TRIGGER POSITION knob counterclockwise. A small Δ should appear on the screen. Set it at approximately -1.00 dv. Finally, press the TB MODE button and select "single." Then press the TB MODE button again to exit that menu. Depress the SINGLE button on the scope such that the red arming light comes on and the scope is waiting for the event to trigger. The trigger level knob may have to be adjusted slightly higher such that the scope does not trigger off of electronic noise.

Once the scope settings are correct and the trigger level is set properly, arm the scope again (if needed) by pressing SINGLE. Now, take the golf ball and drop it into the can from a height just above the top of the can. Did the scope trigger and was the desired signal captured? If not, adjust the scope settings until a good oscillatory response from the beam is obtained.

The response of the beam should be an oscillation damped in time. When hit, the beam vibrates at its natural frequency, which can be measured using the strain gage and Wheatstone bridge configuration. As done in the previous section, use the math function to calculate the FFT of the acquired trace. What is the dominant frequency in this signal?

Set the MATH PLUS menu to "off" and then press the MATH button to exit that menu. The stored trace of the signal should be the only item remaining on the screen.

Now download the data to the laboratory computer to save the information in a text file. Then, using the text file, plot the data to reproduce the trace as seen on the scope screen. Also determine and plot the amplitude-frequency spectrum of the signal.

H.9.7 WHAT TO REPORT

Report the results as a technical memo, being sure to include the plots requested, answers to all questions posed, and the calculations of the square wave amplitudes and frequencies for comparison with the measured values.

H.10   EXERCISE 10—DYNAMIC RESPONSE OF MEASUREMENT SYSTEMS

H.10.1   INTRODUCTION AND OBJECTIVES

The main objective of this laboratory exercise is to investigate the dynamic response characteristics of first-order and second-order measurement systems. First, the dynamic responses of two different-size thermocouples (first-order systems) to step input changes in temperature will be studied. Then, the second-order system dynamic response characteristics of an *RLC* circuit to a sinusoidal input will be investigated. All data will be acquired, stored, and analyzed using a digital oscilloscope.

H.10.2   INSTRUMENTATION

A schematic of the setup for part 1 is shown in Figure H.10. The instrumentation consists of a thermocouple (TC), an ice bath (IB), a thermocouple reference junction and amplifier (TRJA), and a digital oscilloscope (DO).

- Digital oscilloscope
- Two type-K (chromel-alumel) thermocouples of different size
- Analog Devices AD595AQ type-K thermocouple reference junction, linearizer, and amplifier chip in a box
- Ice bath (beaker filled with crushed ice and water)

A schematic of the setup for part 2 is shown in Figure H.11. The instrumentation consists of a function generator (FG), a *RLC* circuit box (RLC), and a digital oscilloscope (DO).

- Function generator
- Digital oscilloscope
- *RLC* circuit box

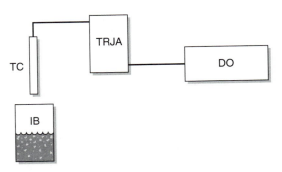

Figure H.10   First-order system response experimental setup.

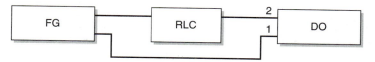

Figure H.11   Second-order system response experimental setup.

H.10.3   MEASUREMENTS

First-Order System Response   In this part of the lab exercise, the DO will be used to determine the time constants of two thermocouples. A thermocouple is a passive temperature sensor. It consists of two dissimilar wires connected together at two junctions, namely, the hot junction and the cold junction. When one junction is hotter than the other, an emf (electromotive force: a voltage difference) is developed between the two junctions. This voltage difference is proportional to the temperature difference between the two junctions and is generally in the millivolt range. The TRJA simulates a cold junction, amplifies the thermocouple output, and linearizes it with the temperature at the hot junction at 10 mV/°C.

In fact, the thermocouple behaves as a first-order system. The equation that describes the energy exchange between the thermocouple's tip and the environment is

$$m C_v \frac{dT}{dt} = h A_s [T_\infty - T(t)], \qquad \textbf{[H.43]}$$

where m is the mass of the tip, C_v is the specific heat at constant volume of the tip, A_s is the surface area of the tip, h is the heat transfer coefficient, $T(t)$ is the temperature of the tip with respect to time, and T_∞ is the temperature of the liquid at "infinity." The solution to this first-order, linear differential equation is

$$T(t) = T_\infty + (T_o - T_\infty) \exp\left(\frac{-t}{\tau}\right), \qquad \textbf{[H.44]}$$

where τ is the time constant of the thermocouple, equal to $m C_v / h A_s$. This equation can be rearranged to yield

$$\ln \frac{T(t) - T_\infty}{T_o - T_\infty} = \frac{-t}{\tau.} \qquad \textbf{[H.45]}$$

A plot of the ln term versus time, t, will yield a line with the decreasing slope equal to $1/\tau$. The time constant is the characteristic measure of the thermocouple's rate of response. The time constant τ can be determined readily by exposing the thermocouple to a step input in temperature. In this section, two thermocouples (A and B), each having different-sized tips, will be exposed to an "instantaneous" (step) decrease in temperature (from room temperature to 0 °C).

To start, turn on the DO and the TRJA. Connect thermocouple A to the TRJA and the TRJA's output to channel 1 on the DO. Note that the TRJA gives a linear output of 10 mV/°C referenced from 0 mV at 0 °C. Thus, its output voltage will decrease

~200 mV for a ~20 °C decrease in temperature. Press auto set on the DO. A steady line with some noise that reads the current temperature in the lab should be observed. Adjust the voltage scale on channel 1 to be 50 mV/div and adjust the time scale to be 250 ms for the thin thermocouple and 2.5 s for the thick one. This ensures that the signal can be captured with good details. Adjust the ch1 vertical position knob such that the signal baseline is displayed just approximately one division *below* the top of the DO display.

Now set up the DO to trigger correctly in response to a step-input forcing: Press the trigger menu button. Make sure to read edge, slope = falling (the DO will trigger when the slope falls), source = ch1 (the trigger function is looking for signal from channel 1), mode = single (the trigger is waiting for a single event to occur), coupling = DC.

Move the horizontal position button to point on one division from the left side of the screen. This ensures that the original level of the signal on one division is seen on the left and the triggered signal on the remaining part of the screen. Move the trigger level knob to point at about 0.4 divisions below the signal level. This ensures that the DO will not trigger until the signal slope falls to this level (decreasing the 0.4 to a smaller value has the risk that the DO can be triggered with noise). Make sure that the thermocouple is away from the ice bath to avoid triggering the DO.

Immerse the thermocouple into the ice bath. After ~30 s, the screen should display the triggered signal starting from the room temperature to the ice temperature. Press the cursor button, adjust type = voltage (the cursors will be horizontal to measure voltage), source = ch1, and use the voltage cursors and the screen grid lines to take 10 readings of voltage versus time. Record these values in Table H.6. The time constant is the time needed by the thermocouple to reach 63.2 % of the final voltage, which can be read directly from a plot of voltage versus time.

Another method to obtain the time constant uses a least-squares regression fit of the data. Because the thermocouple can be represented by a first-order system, the voltage changes with time is governed by

Table H.6   Thermocouple response data.

No.	V_{tcA} (mV)	Time$_{tcA}$ (ms)	V_{tcB} (mV)	Time$_{tcB}$ (ms)
1				
2				
3				
4				
5				
6				
7				
8				
9				
10				

$$V(t) = V_\infty + (V_o - V_\infty) \exp\left(\frac{-t}{\tau}\right). \qquad \textbf{[H.46]}$$

By taking the logarithm of both sides of Equation H.46, the variables can be transformed such that a least-squares linear regression analysis can be performed. From this information the time constant can be determined.

Note any obvious physical differences between thermocouples A and B.

Finally, turn *off* the TRJA's power and disconnect the TRJA BNC from ch1 of the DO when finished with this part of the exercise.

Second-Order System Response   In this part a FG and a DO will be used to determine the response characteristics (the magnitude ratio and the phase lag as functions of the input frequency) of an electrical *RLC* circuit. This circuit consists of a resistor (R), an inductor (L), and a capacitor (C) and has the response characteristics of a second-order system. The circuit will be characterized by providing an input sinusoidal wave of known amplitude and frequency from the FG to the *RLC* circuit and measuring the circuit's output amplitude and time delay using the DO, as depicted schematically in Figure H.11.

The electrical diagram of the *RLC* circuit is shown in Figure H.12. The input to the circuit is between the resistor and ground and the output is measured across the capacitor connected to ground. The resistor is the parallel combination of a 1-Ω to 10-kΩ variable resistor and a fixed 1-kΩ resistor, yielding an effective variable resistance between approximately 1 Ω and 910 Ω using a knob. The capacitance is fixed at 0.68 µF and the inductance at 5 mH. The passive resistance of the inductor is 9.0 Ω, so the lowest effective resistance that the circuit can have is approximately 10 Ω (9.0 Ω + 1 Ω).

The voltage differences, V, across each component in an AC circuit are $V = RI$ for the resistor, $V = L(dI/dt)$ for the inductor and $V = Q/C$ for the capacitor, where $I = dQ/dt$. In this circuit, all three components are in series. Thus, application of Kirchhoff's voltage law for the circuit gives

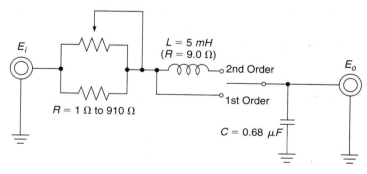

Figure H.12   *RLC* circuit diagram.

$$L\left(\frac{d^2 Q}{dt^2}\right) + R\left(\frac{dQ}{dt}\right) + \frac{Q}{C} = E_i \sin(\omega t). \qquad \textbf{[H.47]}$$

This second-order, linear differential equation can be solved for Q to yield the steady-state output voltage amplitude

$$E_o = \frac{Q}{C} = \frac{E_i}{C\sqrt{\left(1/C - L\omega^2\right)^2 + (R\omega)^2}}. \qquad \textbf{[H.48]}$$

From Equation H.48 and the solution equation for Q, the magnitude ratio is

$$M(\omega) \equiv \frac{E_o}{E_i} = \frac{1}{\sqrt{\left[1 - (\omega/\omega_n)^2\right]^2 + \left[2(R/R_c)(\omega/\omega_n)\right]^2}} \qquad \textbf{[H.49]}$$

and the phase lag is

$$\phi(\omega) \equiv \tan^{-1}\left[\frac{2(R/R_c)(\omega/\omega_n)}{1 - (\omega/\omega_n)^2}\right]. \qquad \textbf{[H.50]}$$

This equation yields positive values of $\phi(\omega)$. By convention, because $\phi(\omega)$ is a phase *lag*, it is plotted as having negative values. Further, for $\omega > \omega_n$, the phase shift must be referenced correctly. Thus, the conventional plot of $\phi(\omega)$ (in °) versus ω would actually be $-\phi(\omega)$ for $\omega \le \omega_n$ and $-180° - \phi(\omega)$ for $\omega > \omega_n$. Also note that in Equations H.49 and H.50 the resonant frequency is given by $\omega_n = \sqrt{1/LC}$ and the critical resistance by $R_c = 2\sqrt{L/C}$.

To start, make sure that the output cable from the FG is attached to the input of the RLC box and in parallel to channel 1 on the DO. The output of the RLC box should be connected to channel 2 of the DO. In that way, both the input and output signals of the RLC box can be viewed on the DO. Make sure that the toggle switch is set to "2nd order." Now turn the R knob on the RLC box fully counterclockwise (to MAX). This sets the resistance in the circuit to its highest value, corresponding to a high damping ratio. Then set the FG and the DO to their initial prescribed settings. These are, for the FG, sine wave with 100 Hz frequency, 4 V peak-to-peak (V_{pp}) amplitude, and no DC offset; for the DO, chs 1 and 2, both AC with divisional settings of 2 V and 2 ms (change these settings if needed or press auto set to let the DO select the best settings).

Data will be analyzed in the final form of $M(\omega)$ and $\phi(\omega)$, each versus the normalized frequency ratio, ω/ω_n. These values will be determined from the raw data. This includes the input and output amplitudes, E_i and E_o, and the phase lag time, Δt, which is the time between the peak of E_i and the corresponding peak of E_o. The phase lag in degrees equals $-(360°)(\Delta t/T_i)$, where T_i is the inverse of the input frequency in Hz and the minus sign indicates a lag in time.

Once a satisfactory set of signals has been captured on the DO display, use the cursors to record the data. Enter all the raw data in the first four columns in Table H.7. The last two columns can be filled in after the lab. When done with an input frequency, set the next one on the FG and repeat the process.

Table H.7 RLC-high-resistance-response data.

Freq. (Hz)	E_i (V)	E_o (V)	Δt (s)	ω (rad/s)	$M(\omega)$	$\phi(\omega)$ (°)
10						
100						
500						
1000						
1200						
1600						
2000						
2200						
2500						
2650						
2800						
3100						
3600						
4000						
5000						
7000						
10 000						

Finally, when done with all the frequencies, rotate the R knob on the RLC box clockwise such that the mark on the knob points to the top of the "I" in "MIN." This sets the resistance in the circuit to another value, corresponding to a different damping ratio. The damping ratio, ζ, equals R/R_c for this circuit. Then repeat the whole procedure again for all frequencies, recording the raw data in Table H.8. When done using the voltmeter, measure the total resistance of the *RLC* circuit (between the center pins of the IN and the OUT connectors). Subtract 9 Ω (the resistance of the inductor) from this value and record it. This is the value of R for this case, which will be needed later in the calculations.

H.10.4 WHAT TO REPORT

Turn in the results in a technical memo along with answers to the questions posed. Also attach any pertinent plots and M-file listings.

1. Using the data in Table H.6, plot the data for each thermocouple and determine the time constant directly. Then transform the variables and perform a linear least-squares regression analysis. From that determine the each time constant. Compare the time constants obtained from the two methods. Which method is more accurate and why?

2. Compare the time constant of thermocouple A to that of thermocouple B. How and why are the time constants different? Provide a plausible physical explanation for their difference.

Table H.8   *RLC*-low-resistance-response data.

Freq. (Hz)	E_i (V)	E_o (V)	Δt (s)	ω (rad/s)	$M(\omega)$	$\phi(\omega)(°)$
100						
500						
1000						
1200						
1600						
2000						
2200						
2500						
2800						
3100						
3600						
4000						
5000						
7000						
10 000						

3. Complete the columns for ω, $M(\omega)$, and $\phi(\omega)$ in both Tables H.7 and H.8.

4. Using these results and a program or M-file, construct two plots, one of $M(\omega)$ and other of $\phi(\omega)$ versus the normalized frequency ratio, where each of these two plots contains both of the high R and low R cases. These plots must contain the data along with the theoretical curves given by Equations H.49 and H.50, substituting the appropriate values for R, L, and C for each case. Plot the data for each of the two cases using a different set of symbols for each case.

5. Does the data support the conclusion the *RLC* circuit behaves as a second-order system in both cases? Finally, compare the values of the damping ratio found for each case with each value of R/R_c. Do this by comparing the data with the corresponding values determined using various values of R/R_c in Equations H.49 and H.50. How well do the experimental and theoretical values of ζ compare?

H.11   EXERCISE 11—OPTICS OF LENSES, LASERS, AND DETECTORS

H.11.1   INTRODUCTION AND OBJECTIVES

The main objective of this exercise is to become familiar with concepts in optical design and to apply basic design techniques to several model problems. The exercises will involve incoherent- and coherent-light sources, lenses, and optical detectors.

H.11.2   INSTRUMENTATION

The following equipment will be used:

- Small flashlight, to be used as a white-light source

- Ruler and protractor

- Lens, double-convex, $\phi = 65$ mm, unknown f

- Metrologic ML-211, diode-based laser, of unknown λ, with mount

- Holographic, diffraction grating, 750 lines/mm, with mount

- Diode/detector pair, with amplifier circuit and battery

- Digital, minitachometer

- Digital oscilloscope

- Spinning wheel, of unknown rotation rate, N

- Dispersing prism, with unknown index, n

The lab exercises will use an optical rail and a full optical bench as convenient platforms to mount various components. Several of the exercises will depend on the careful alignment of the optical elements.

This investigation will be broken into several sections. In the first exercise, a lens will be used to image a source onto a screen. The source will be the bulb of a flashlight. With sufficient magnification, fine details of the filament may be seen. This exercise will demonstrate the mounting and positioning of lenses, based on the thin-lens equation.

In the second exercise, the use of a laser will be explored by passing the laser radiation through a diffraction grating. Two of the most important aspects of laser light are its coherence and its monochromaticity. Both of these features will be utilized in the measurement of laser wavelength, based on the grating equation.

In the third exercise, a light-emitting diode (LED)-photodiode pair will be provided. This circuit, along with an oscilloscope, will allow for the measurement of the rotation rate of a spinning wheel. The measurement can be compared to that of the minitachometer, also provided.

In the fourth exercise, the properties of a prism will be investigated. The laser will be used to make a measurement that will allow for the calculation of the index of refraction of the glass.

H.11.3   LASER SAFETY

The laser to be used is a relatively safe, low-power, laser pointer. As such, special eye protection is not needed. However, direct, prolonged exposure of the eye to this laser can still cause damage, so common sense and caution should be exercised. The following steps can help provide for a safe lab experience.

- The laser should never be aimed directly into a person's eyes.

Table H.9   Imaging of flashlight.

Reading	l (cm)	l' (cm)	Obj. Height (cm)	Image Height (cm)	Mag.	f (cm)
1						
2						
3						

- The laser beam should be blocked off so that it cannot extend beyond the limits of the individual laboratory section. Dull, nonreflective barriers such as dark-colored paper or stacks of books can be used for this.
- Reflective objects in the area should be covered with cloth or blocked, to prevent secondary reflections in the lab.

H.11.4 MEASUREMENTS

The goal of the first exercise is to become proficient in the use of simple lenses. To this end, the flashlight will be imaged onto a viewing screen. In order to properly image the filament, the distances, or conjugates, between the source and lens, and between the lens and screen, must satisfy the thin-lens equation, which is provided in the Supplemental Information section.

Using the optical rail provided, experiment with the positioning of the lens versus the source and viewing screen. For each choice of conjugates, a magnification of the image can be calculated. Try a variety of conjugate-distance combinations, and adjust the optical system to be sure the image is in focus at the screen for each combination. Record the results in Table H.9, and calculate the magnification of each attempt. You should find that the calculation for the focal length of the lens is the same in each case; the focal length is a constant property of the lens, not of the system.

In the second exercise, measure the wavelength, λ, of a laser by utilizing a diffraction grating. By using the grating equation, which is provided in the Supplemental Information section, and given the spatial frequency of the grating to be 750 lines/mm, the wavelength can be calculated. Carefully measure the position of the first fringe (i.e., for $m = 1$ or $m = -1$), and average several readings. Record the results in Table H.10.

Table H.10   Measurement of laser wavelength.

Reading	x (cm)	z (cm)	d (mm/line)	λ (nm)
1			1/750	
2			1/750	
3			1/750	

Table H.11   Measurement of wheel rotation rate.

Reading	Chopping freq.	rot/s	N (RPM)	Minitach. (RPM)
1				
2				
3				

In the third exercise, calculate the rotation rate of the spinning wheel. The rotation rate, N, of a wheel can be measured by configuring a light source on one side of the wheel, and a detector on the other side, and preparing the wheel so that it periodically obstructs the beam of light. Such a sensor is representative of many simple sensors that are used in industrial areas to anchor process-control loops. In order to calculate N, capture the chopped signal on an oscilloscope. Estimate the chopping frequency, and use the geometry of the wheel to infer the rotations per minute. Compare the result to that of the minitachometer. Capture the oscilloscope plot, and import it to the final report. Record the results in Table H.11.

In the fourth exercise, investigate the properties of a dispersing prism. As detailed in the Supplemental Information section, a prism can be used to separate, or disperse, the various frequencies of the signal. Using the equation given, and knowing the wavelength of the laser light from the second exercise, estimate what must be the refractive index of the prism. Record the results in Table H.12.

H.11.5   WHAT TO REPORT

Submit the results in a technical memo. Be sure to include (as a minimum) the following information:

- Calculation of the focal length, f, of the double-convex lens
- Calculation of laser wavelength, λ
- Calculation of wheel-rotation rate, N
- Digitized plot of the oscilloscope trace of the rotation-rate sensor

Table H.12   Measurement of prism index of refraction.

Reading	$i_1(°)$	a (°)	d (°)	Index, n
1				
2				
3				

Table H.13   Measureand uncertainties.

Measureand	Units	Uncertainty
Conjugates l and l'	cm	
Object and image height, h, h'	cm	
Diffraction-angle parameters, x, z	cm	
Wheel chopping frequency	Hz	
Index-of-refraction angles, i_1, a, d	rad	

- Calculation of refractive index of the prism, n
- Uncertainty estimates presented in the form of Tables H.13, H.14, and H.15, supported by example calculations

Be sure to include any interesting observations from any of the four sections of the lab.

H.11.6   SUPPLEMENTAL INFORMATION

Imaging an Incoherent Source The thin-lens equation is given by Smith (p. 20, W. J. Smith, *Modern Optical Engineering*, McGraw-Hill, New York, 1966) as

$$\frac{-1}{l} + \frac{1}{l'} = \frac{1}{f} \qquad \textbf{[H.51]}$$

according to Figure H.13, where l is the distance from the lens to the object (typically a negative number, as in the above diagram), l' is the distance from the lens to the image, and f is the focal length of the lens. Note that in this case, for simplicity, the object is the source of radiation, the lamp. Thus, if the system is arranged with an unknown lens such that the image is in sharp focus, the conjugates, l and l', can be measured, and the focal length of the lens can be calculated.

In addition, the magnification of the system can be calculated from the conjugates as

$$m = \frac{l'}{l} = \frac{h'}{h}, \qquad \textbf{[H.52]}$$

Table H.14   Results.

Result	Units	Calculation
Focal length of lens, f	cm	
Laser wavelength, λ	nm	
Rotation rate of wheel, N	RPM	
Prism index of refraction, n	dimensionless	

Table H.15   Result uncertainties.

Result	Units	Uncertainty
Focal length of lens, u_f	cm	
Laser wavelength, u_λ	nm	
Rotation rate of wheel, u_N	RPM	
Prism index of refraction, u_n	dimensionless	

where h is the object height and h' is the image height. The image height will typically be negative, by convention, where a negative magnification indicates an inverted image. Note that for larger magnifications, the irradiance of the image will decrease.

As an example, if the distance from the lens to the object is -20 mm, and the distance from the lens to the image is 100 mm, then the focal length will be 16.7 mm, and the magnification will be -5. A negative magnification indicates that the image is inverted, compared to the object.

Characterizing a Coherent Source   Laser light in general is both monochromatic and spatially coherent, and these are the reasons the laser is such an important tool in many fields. The spatial coherence of the laser means that the beam is perfectly collimated, as if it had originated infinitely far away. It is this property that allows for strong interference fringes when the beam is crossed with itself, and this is the basis of interferometry. The monochromaticity of the laser is the purity of its wavelength. Because the laser can deliver significant power at such a narrow bandwidth, it is useful in fields ranging from spectroscopy to fiber-optic communications.

The wavelength of the laser diode used in this lab can be calculated by measuring the positions of the fringes in the diffraction pattern beyond a grating. The grating equation is given by Metrologic (p. 22, "Laser-Pointer Education Kit," Metrologic Instruments, Bellmawr, NJ, 1996) as

$$m\lambda = d \sin\theta, \qquad\qquad \textbf{[H.53]}$$

according to the Figure H.14, where m is the order of the fringe, λ is the wavelength of radiation, d is the grating spacing, and θ is the angle of deviation off-axis of the diffraction fringe.

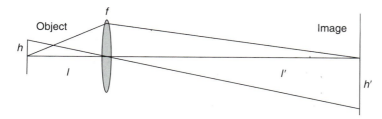

Figure H.13   Imaging with a thin lens.

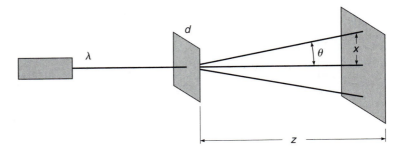

Figure H.14   Laser-beam diffraction.

Now, for the simplified case of $m = 1$, and given that for small angles, $\sin\theta$ can be approximated by θ, the wavelength of the laser can be calculated using the relation

$$\lambda = d\frac{x}{z},$$ **[H.54]**

where x is the deflection of the fringe off of the optical axis, and z is the distance from the grating to the screen.

Optical Measurement of Wheel-Rotation Rate   In this exercise, a source-detector pair will be used for the measurement of the rotation rate of a spinning wheel. Some simple source and detector circuits are provided by Mims (F. M. Mims, *Engineer's Mini-Notebook, Optoelectronic Circuits*, Printed by Forrest Mims, 1986). Often LEDs are used as the source of a simple sensor. An LED is typically DC powered, and, when biased, it radiates light at a relatively narrow bandwidth. This type of source can be controlled very precisely, compared to an incandescent lamp. A simple detector will often be a silicon photodiode, followed by an operational amplifier. Such a detector is based on the photoelectric effect, where a small current is generated by the photodiode on exposure to light, and the current is amplified and converted to voltage by the op amp.

In this exercise, an LED source will be coupled with a phototransistor for signal detection. The phototransistor output need not be amplified, but rather just fed directly into an oscilloscope for analysis. The circuits are shown in Figure H.15.

Connect the spinning wheel (a muffin fan) to the wall outlet. Couple the detector circuit to channel 1 of the oscilloscope, and switch the detector circuit "on." When not in use, the detector should be switched off, to conserve the battery. Align the detector circuit, minitachometer, and wheel such that both sensors can view the chopping at the same time. The minitachometer will need an illuminating source (like a flashlight) on the other side of the fan. Acquire the signal on the oscilloscope; the following notes may be of help:

- Trigger the oscilloscope off of the detector signal itself; use the buttons "trig" and "chan 1."

- Use AC coupling to optimize the signal on the scope.

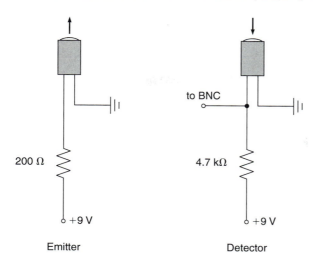

to BNC

200 Ω

4.7 kΩ

+9 V

+9 V

Emitter

Detector

Figure H.15   Emitter and detector circuits.

- Set the amplitude on the scope to 0.1 V/div.
- If at this point a nice square wave is not seen, check the circuit's battery with a voltmeter; a replacement may be needed.
- Use the cursors to measure the chopping frequency; use the buttons, "cursors" "on" and then the "track" and "arrow" buttons.

By measuring the frequency of the chopped signal, estimate the rotation rate of the wheel. Save the scope plot, and also record the measurement of the minitachometer. How close are the two measurements? If time allows, try a couple different measurement positions along the wheel, and see if the result is repeatable.

Measurement of Refractive Index of Prism   Because the index of refraction of glass is dependent on the wavelength of the radiation, prisms have been used as the basis of simple monochrometers. The dispersion of light by a prism is given by Smith (p. 72, W. J. Smith, *Modern Optical Engineering*, McGraw-Hill, New York, 1966) as

$$d = i_1 - a + \arcsin\left\{[n(\lambda)^2 - \sin^2(i_1)]^{1/2}\sin(a) - \cos(a)\sin(i_1)\right\}, \qquad \textbf{[H.55]}$$

according to Figure H.16.

In Figure H.16, d is the final angle of exit of the beam, i_1 is the angle of incidence, a is the characteristic angle of the prism, and $n(\lambda)$ is the wavelength-dependent index of refraction of the glass. Direct the laser through the prism, and carefully measure the various geometric quantities. Then, knowing the wavelength of radiation, calculate the index of refraction of the prism. Try a couple different orientations of the prism, varying the angle of incidence slightly, to see if the result is repeatable.

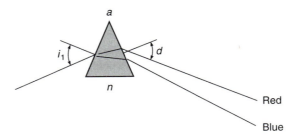

Figure H.16   Refraction by a prism.

H.12   EXERCISE 12—STATISTICAL ANALYSIS OF DATA USING MATLAB

H.12.1   INTRODUCTION AND OBJECTIVES

The overall objectives of this exercise are to solidify an understanding of several statistical and probabilistic methods and to provide the opportunity to learn how to use MATLAB to retrieve, analyze, and display actual experimental data on a computer system.

There are three different data files that will be used in this exercise. The first file, shed.dat, consists of two columns of data acquired during an experiment in a subsonic wind tunnel. The first column is the time (in s), t, at which a value of the velocity (the second column, in m/s), $U(t)$, was acquired using a hot-wire sensor located immediately behind a cylinder positioned in the wind tunnel. The second file, gball.dat, is a single column of the golf ball weights (in N). The third file, voltage.dat, is a single column of voltage readings from an amplifier/Wheatstone bridge system taken at fixed time increments.

In this exercise, MATLAB must be used to analyze this data. Several m-files must be written. Be sure to provide statements at the beginning of each M-file describing its particular function. The following explains specifically what must be done. Turn in the results in a technical memo.

Some helpful information on using MATLAB for probability and statistical calculations is presented throughout the text. Other information can be obtained using MATLAB's "help" command. Some of the commands needed are *not* contained in the student version of MATLAB.

H.12.2   TEMPORAL REALIZATION OF TIME SERIES DATA

Using shed.dat, determine the following: (1) the period (in s) of the signal's *third* cycle, (2) the average period (in s) of one cycle, (3) the signal's average cyclic frequency (in Hz), (4) the mean velocity (in m/s), (5) the standard deviation of the velocity (in m/s), (6) the minimum velocity (in m/s), and (7) the maximum velocity (in m/s). Write an M-file to read in the data from `shed.dat` and then plot the velocity on the ordinate versus time on the abscissa. Include in that M-file the commands to automatically identify on the plot the mean velocity, the mean velocity +1 standard deviation and the mean velocity −1 standard deviation. *Hints*: These MATLAB commands coded into an M-file may be useful: [a] eval(['load info.dat']) loads the data from `info.dat` into the MATLAB workspace; [b] col1 = eval(['info(:,1)']) assigns the first column of `info.dat` the name col1; [c] text(xpos,ypos,'zz') places "zz," such as a line or an arrow, at the coordinates xpos,ypos on a plot. For this part, hand in the results, figure, the written M-file, and proof of the calculations of the seven above quantities (this could be, for example, the printout of the MATLAB session).

H.12.3 DISTRIBUTION COMPARISONS

Using `gball.dat`, write an M-file to read in the data and then plot the histogram and frequency distribution side by side on one page (*Hint*: use MATLAB's "subplot" command). Follow the rules for determining the number of bins as described in Chapter 7. Assume that u_w, the uncertainty in the measurement of the weight, w, equals 0.01 N. Check to make sure that the bin width is greater than this value. Next, by either writing another M-file or adding on to the previous one, determine and then plot the histogram of the data and the expected values as determined for a normal distribution (use the sample mean and standard deviation for the mean and standard deviation of the Normal pdf). *Hints*: The following MATLAB commands coded into an M-file may be useful: [a] hist(x,k) plots the histogram of x with k bins; [b] [a,b] = hist(x,k) produces the column matrices a and b, where a contains the counts in each bin and b contains the center coordinates for each bin; [c] bar(b,a/N) will plot the frequency distribution, where N is the total number of x values; [d] q = c:dq:e will produce values of q ranging from c to e in increments of dq; and [e] plot(q,sin(q)) will plot q on the abscissa versus sin(q) on the ordinate as a continuous and smooth curve provided that the increment dq is much smaller than the range of q. For this part, hand in the figures and the written M-files.

H.12.4 FINITE VERSUS INFINITE SAMPLES

Using `voltage.dat`, write an M-file to plot the running mean of the data from 1 up to all 1000 points (running mean on the ordinate; the number of points on the abscissa). The running mean of N points is simply the mean of those N points. As N gets larger, the running mean should approach a constant value equal to the true mean of the underlying population from which the N points were drawn. On the plot, indicate the running mean of 1000 points and its value using the previously described "text" command. Next, determine the number of measurements, $N*$, required for the running mean to stay *continually* within 1 % of the running mean of 1000 points. Then, compute the sample mean, the sample standard deviation and the standard deviation of the means for all 1000 points. Write a statement for the estimate of the true mean value of the parent population from which this data was drawn at the 95% confidence level based on these values. Put the values of N^*, the sample mean, the sample standard deviation, the standard deviation of the means, and the true mean estimate statement in a table. For this part, hand in the figure, the written M-file, and the table.

H.12.5 χ^2-ANALYSIS

Continue with the analysis to construct the plot of the histogram of the gball.dat weights versus the values expected for a normal distribution. Perform a χ^2-analysis to determine the % confidence that the golf ball weights are Normally distributed. *Hint*: The following MATLAB command coded into an M-file may be useful: alpha = 100-100*chi2cdf(chisq,nu). Report the % confidence value and hand in the M-file written to calculate this.

H.12.6 WHAT TO REPORT

Turn in a technical memo with the answers to the posed questions, plots, and listings of all written M-files.

APPENDIX

DERIVATIONS

STEP-INPUT FORCING

The second-order differential equation for step-input forcing is

$$\left(\frac{1}{\omega_n}\right)^2 y'' + \left(\frac{2\zeta}{\omega_n}\right)y' + y = K \cdot F(t).$$

SOLUTION

Characteristic equation:

$$\left(\frac{1}{\omega_n}\right)^2 r^2 + \left(\frac{2\zeta}{\omega_n}\right)r + 1 = 0.$$

The quadratic formula gives

$$r_{1,2} = \frac{-2\zeta/\omega_n \pm \sqrt{4\zeta^2/\omega_n^2 - 4/\omega_n^2}}{2/\omega_n^2},$$

which can be simplified to

$$-\zeta\omega_n \pm \omega_n\sqrt{\zeta^2 - 1}.$$

There are three possible solutions to this differential equation:

1. If $\zeta^2 - 1 > 0$, there are only real roots:

$$r_1 = -\zeta\omega_n + \omega_n\sqrt{\zeta^2 - 1}; \quad r_2 = -\zeta\omega_n - \omega_n\sqrt{\zeta^2 - 1}.$$

The homogeneous solution is of the form

$$\begin{aligned}
y_h &= c_1 e^{r_1 t} + c_2 e^{r_2 t} \\
&= c_1 e^{\left(-\zeta\omega_n + \omega_n\sqrt{\zeta^2 - 1}\right)t} + c_2 e^{\left(-\zeta\omega_n - \omega_n\sqrt{\zeta^2 - 1}\right)t} \\
&= e^{-\zeta\omega_n t}\left(c_1 e^{\omega_n\sqrt{\zeta^2 - 1}\, t} + c_2 e^{-\omega_n\sqrt{\zeta^2 - 1}\, t}\right).
\end{aligned}$$

523

For the step input $K \cdot F(t) = KA$, so assume a particular solution of the form

$$y_p = KA.$$

The general solution becomes

$$y = y_h + y_p = KA + e^{-\zeta \omega_n t}\left(c_1 e^{\omega_n \sqrt{\zeta^2 - 1}\,t} + c_2 e^{-\omega_n \sqrt{\zeta^2 - 1}\,t}\right).$$

Now apply the initial conditions $y'(0) = y(0) = 0$:

$$y(0) = 0 \Rightarrow 0 = KA + c_1 + c_2.$$

The first derivative of the general solution is

$$y' = \left(-\zeta \omega_n + \omega_n \sqrt{\zeta^2 - 1}\right)c_1 e^{\omega_n \sqrt{\zeta^2 - 1}\,t} + \left(-\zeta \omega_n - \omega_n \sqrt{\zeta^2 - 1}\right)c_2 e^{-\omega_n \sqrt{\zeta^2 - 1}\,t},$$

$$y'(0) = 0 \Rightarrow \left(-\zeta \omega_n + \omega_n \sqrt{\zeta^2 - 1}\right)c_1 = -\left(-\zeta \omega_n - \omega_n \sqrt{\zeta^2 - 1}\right)c_2,$$

or

$$k_1 c_1 = -k_2 c_2 \Rightarrow c_1 = \frac{-k_2}{k_1}c_2$$

$$\Rightarrow c_1 = \frac{\zeta \omega_n + \omega_n \sqrt{\zeta^2 - 1}}{-\zeta \omega_n + \omega_n \sqrt{\zeta^2 - 1}}c_2 = k_3 c_2.$$

So,

$$c_2 = -KA - c_1 = -KA - k_3 c_2,$$

or

$$c_2 = -\frac{KA}{1 + k_3},$$

and

$$c_1 = -KA \cdot \frac{k_3}{(1 + k_3)}.$$

Solving for c_2:

$$c_2 = \frac{-KA}{1 + k_3} = -KA \left(\frac{-\zeta \omega_n + \omega_n \sqrt{\zeta^2 - 1} + \zeta \omega_n + \omega_n \sqrt{\zeta^2 - 1}}{-\zeta \omega_n + \omega_n \sqrt{\zeta^2 - 1}}\right)^{-1}$$

$$= -KA \left[\frac{2\omega_n \sqrt{\zeta^2 - 1}}{\omega_n (-\zeta + \sqrt{\zeta^2 - 1})}\right]^{-1}$$

$$= -KA \left(\frac{2\sqrt{\zeta^2 - 1}}{-\zeta + \sqrt{\zeta^2 - 1}}\right)^{-1}$$

$$= -KA \left(\frac{-\zeta + \sqrt{\zeta^2 - 1}}{2\sqrt{\zeta^2 - 1}}\right).$$

And then solving for c_1:

$$c_1 = -KA\left(\frac{-\zeta + \sqrt{\zeta^2 - 1}}{2\sqrt{\zeta^2 - 1}}\right)\left(\frac{\zeta + \sqrt{\zeta^2 - 1}}{-\zeta + \sqrt{\zeta^2 - 1}}\right)$$

$$= -KA\left(\frac{\zeta + \sqrt{\zeta^2 - 1}}{2\sqrt{\zeta^2 - 1}}\right).$$

Substituting,

$$y = KA - KA \cdot e^{-\zeta \omega_n t} \cdot \left[\left(\frac{\zeta + \sqrt{\zeta^2 - 1}}{2\sqrt{\zeta^2 - 1}}\right)e^{\omega_n \sqrt{\zeta^2 - 1}t}\right.$$

$$\left. + \left(\frac{-\zeta + \sqrt{\zeta^2 - 1}}{2\sqrt{\zeta^2 - 1}}\right)e^{-\omega_n \sqrt{\zeta^2 - 1}t}\right]$$

$$= KA - KA \cdot \frac{e^{-\zeta \omega_n t}}{2\sqrt{\zeta^2 - 1}} \cdot \left[\left(\zeta + \sqrt{\zeta^2 - 1}\right)e^{\omega_n t \sqrt{\zeta^2 - 1}}\right.$$

$$\left. + \left(-\zeta + \sqrt{\zeta^2 - 1}\right)e^{-\omega_n t \sqrt{\zeta^2 - 1}}\right]$$

$$= KA - KA \cdot \frac{e^{-\zeta \omega_n t}}{2\sqrt{\zeta^2 - 1}} \cdot \sqrt{\zeta^2 - 1} \cdot \left[\left(1 + \frac{\zeta}{\sqrt{\zeta^2 - 1}}\right)e^{\omega_n t \sqrt{\zeta^2 - 1}}\right.$$

$$\left. + \left(1 - \frac{\zeta}{\sqrt{\zeta^2 - 1}}\right)e^{-\omega_n t \sqrt{\zeta^2 - 1}}\right]$$

$$= KA - KAe^{-\zeta \omega_n t} \cdot \left[\frac{1 + (\zeta/\sqrt{\zeta^2 - 1})}{2}e^{\omega_n t \sqrt{\zeta^2 - 1}}\right.$$

$$\left. + \frac{1 - (\zeta/\sqrt{\zeta^2 - 1})}{2}e^{-\omega_n t \sqrt{\zeta^2 - 1}}\right].$$

Now, $\sinh x = (e^x - e^{-x})/2$ and $\cosh x = (e^x + e^{-x})/2$, so the general solution becomes

$$y = \cosh(\omega_n t \sqrt{\zeta^2 - 1}) + \frac{\zeta}{\sqrt{\zeta^2 - 1}}\sinh(\omega_n t \sqrt{\zeta^2 - 1}).$$

2. If $\zeta^2 - 1 = 0$, there are two real and equal roots: $r_{1,2} = -\zeta \omega_n$. Thus, the homogeneous solution is of the form

$$y_h = c_1 e^{r_1 t} + c_2 t e^{r_1 t}$$

$$= c_1 e^{-\omega_n t} + c_2 t e^{-\omega_n t}.$$

Assume a particular solution of the form

$$y_p = KA.$$

Thus, the general solution is

$$y = KA + c_1 e^{-\omega_n t} + c_2 t e^{-\omega_n t},$$

and its first derivative is

$$y' = -\omega_n c_1 e^{-\omega_n t} - \omega_n c_2 t e^{-\omega_n t} + c_2 e^{-\omega_n t}.$$

Applying the initial conditions, c_1 and c_2 can be found:

$$y(0) = 0 \Rightarrow 0 = KA + c_1 \Rightarrow c_1 = -KA,$$
$$y'(0) = 0 \Rightarrow 0 = -\omega_n c_1 + c_2$$
$$\Rightarrow c_2 = \omega_n c_1 = -\omega_n KA.$$

So, the general solution becomes

$$y = KA - KA e^{-\omega_n t}(1 + \omega_n t).$$

3. If $\zeta^2 - 1 < 0$, there are only complex roots:

$$r_{1,2} = \lambda \pm i\mu = -\zeta \omega_n \pm i \omega_n \sqrt{1 - \zeta^2}.$$

A homogeneous solution for complex roots has the form

$$y_h = e^{\lambda t}(c_1 \cos \mu t + c_2 \sin \mu t)$$
$$= e^{-\zeta \omega_n t}\left(c_1 \cos \omega_n t \sqrt{1 - \zeta^2} + c_2 \sin \omega_n t \sqrt{1 - \zeta^2}\right).$$

Again assume a particular solution

$$y_p = KA,$$

which gives the general solution

$$y = KA + e^{-\zeta \omega_n t}\left(c_1 \cos \omega_n t \sqrt{1 - \zeta^2} + c_2 \sin \omega_n t \sqrt{1 - \zeta^2}\right).$$

Solving for the first derivative and applying initial conditions,

$$y' = -\zeta \omega_n e^{-\zeta \omega_n t}\left(c_1 \cos \omega_n t \sqrt{1 - \zeta^2} + c_2 \sin \omega_n t \sqrt{1 - \zeta^2}\right)$$
$$+ e^{-\zeta \omega_n t}\left(-\omega_n \sqrt{1 - \zeta^2} c_1 \sin \omega_n t \sqrt{1 - \zeta^2} + \omega_n \sqrt{1 - \zeta^2} c_2 \cos \omega_n t \sqrt{1 - \zeta^2}\right),$$

$$y'(0) = 0 = -\zeta \omega_n c_1 + \omega_n \sqrt{1 - \zeta^2} c_2$$

$$\Rightarrow c_2 = \frac{c_1 \zeta}{\sqrt{1 - \zeta^2}}$$

$$y(0) = 0 = KA + c_1 \Rightarrow c_1 = -KA \quad \text{and} \quad c_2 = -\frac{KA\zeta}{\sqrt{1 - \zeta^2}}.$$

So,

$$y = KA - KA \cdot e^{-\zeta \omega_n t} \left(\cos \omega_n t \sqrt{1 - \zeta^2} + \frac{\zeta}{\sqrt{1 - \zeta^2}} \sin \omega_n t \sqrt{1 - \zeta^2} \right).$$

Now, $C \cos \omega t + B \sin \omega t = \sqrt{C^2 + B^2} \sin(\omega t + \phi^*)$ and $\phi^* = \tan^{-1}(C/B)$. Here $C = 1$ and $B = \zeta/\sqrt{1 - \zeta^2}$. Thus,

$$\left(\cos \omega_n t \sqrt{1 - \zeta^2} + \frac{\zeta}{\sqrt{1 - \zeta^2}} \sin \omega_n t \sqrt{1 - \zeta^2} \right)$$

$$= \sqrt{1 + \frac{\zeta^2}{1 - \zeta^2}} \sin \left(\omega_n t \sqrt{1 - \zeta^2} + \phi \right)$$

$$= \frac{1}{\sqrt{1 - \zeta^2}} \sin \left(\omega_n t \sqrt{1 - \zeta^2} + \phi \right),$$

where $\phi = \tan^{-1} \left(\sqrt{1 - \zeta^2}/\zeta \right) = \sin^{-1} \left(\sqrt{1 - \zeta^2} \right)$. Finally,

$$y = KA - KA \cdot e^{-\zeta \omega_n t} \left[\frac{1}{\sqrt{1 - \zeta^2}} \sin \left(\omega_n t \sqrt{1 - \zeta^2} + \phi \right) \right].$$

SINUSOIDAL-INPUT FORCING

The second-order differential equation for sinusoidal-input forcing is

$$\left(\frac{1}{\omega_n} \right)^2 y'' + \left(\frac{2\zeta}{\omega_n} \right) y' + y = KA \sin \omega t.$$

SOLUTION

Let

$$y_p = c_1 KA \sin \omega t + c_2 KA \cos \omega t,$$

which gives

$$y_p' = \omega c_1 KA \cos \omega t - \omega c_2 KA \sin \omega t$$

and

$$y_p'' = -\omega^2 c_1 KA \sin \omega t - \omega^2 c_2 KA \cos \omega t.$$

Substitution yields

$$-\left(\frac{\omega}{\omega_n} \right)^2 KA[c_1 \sin \omega t + c_2 \cos \omega t] + 2\zeta \left(\frac{\omega}{\omega_n} \right) KA[c_1 \cos \omega t - c_2 \sin \omega t]$$

$$+ KA[c_1 \sin \omega t + c_2 \cos \omega t] = KA \sin \omega t.$$

Grouping like sine terms:

$$-\left(\frac{\omega}{\omega_n}\right)^2 c_1 - 2\zeta\left(\frac{\omega}{\omega_n}\right)c_2 + c_1 = 1,$$

and like cosine terms:

$$-\left(\frac{\omega}{\omega_n}\right)^2 c_2 + 2\zeta\left(\frac{\omega}{\omega_n}\right)c_1 + c_2 = 0.$$

Now solving for c_1 and c_2:

$$\left[\left(\frac{\omega}{\omega_n}\right)^2 - 1\right]c_2 = 2\zeta\left(\frac{\omega}{\omega_n}\right)c_1$$

$$\Rightarrow c_1 = \frac{[(\omega/\omega_n)^2 - 1]}{2\zeta(\omega/\omega_n)}c_2.$$

Block substitution gives

$$\left\{\frac{[1 - (\omega/\omega_n)^2][(\omega/\omega_n)^2 - 1]}{2\zeta(\omega/\omega_n)}\right\}c_2 = 1 + 2\zeta\left(\frac{\omega}{\omega_n}\right)c_2,$$

$$\Rightarrow \left\{\frac{[1 - (\omega/\omega_n)^2][(\omega/\omega_n)^2 - 1]}{2\zeta(\omega/\omega_n)}\right\}c_2 - 2\zeta\left(\frac{\omega}{\omega_n}\right)c_2 = 1,$$

$$\Rightarrow c_2 = 1 \Big/ \left\{\frac{[1 - (\omega/\omega_n)^2][(\omega/\omega_n)^2 - 1]}{2\zeta(\omega/\omega_n)}\right\} - 2\zeta\left(\frac{\omega}{\omega_n}\right).$$

Now expand the denominator

$$\left\{\frac{[1 - (\omega/\omega_n)^2][(\omega/\omega_n)^2 - 1]}{2\zeta(\omega/\omega_n)}\right\} - 2\zeta\left(\frac{\omega}{\omega_n}\right)$$

$$= \frac{[1 - (\omega/\omega_n)^2]^2}{-2\zeta(\omega/\omega_n)} - 2\zeta\left(\frac{\omega}{\omega_n}\right)$$

$$= -\frac{[1 - (\omega/\omega_n)^2]^2 + [2\zeta(\omega/\omega_n)]^2}{2\zeta(\frac{\omega}{\omega_n})}$$

$$\Rightarrow c_2 = -\frac{2\zeta(\omega/\omega_n)}{[1 - (\omega/\omega_n)^2]^2 + [2\zeta(\omega/\omega_n)]^2}.$$

Also,

$$c_1 = -\frac{[(\omega/\omega_n)^2 - 1]}{[1 - (\omega/\omega_n)^2]^2 + [2\zeta(\omega/\omega_n)]^2}.$$

The particular solution is

$$y_p = \frac{KA}{[1 - (\omega/\omega_n)^2]^2 + [2\zeta(\omega/\omega_n)]^2}$$

$$\cdot \left\{ -\left[1 - \left(\frac{\omega}{\omega_n}\right)^2\right] \sin \omega t - \left[2\zeta\left(\frac{\omega}{\omega_n}\right)\right] \cos \omega t \right\}.$$

This can be simplified. Now, $C \cos \omega t + B \sin \omega t = \sqrt{C^2 + B^2} \sin(\omega t + \phi^*)$.

$$\phi^* = \tan^{-1}\left(\frac{C}{B}\right)$$

Here,

$$C = -2\zeta\left(\frac{\omega}{\omega_n}\right) \quad \text{and} \quad B = -\left[1 - \left(\frac{\omega}{\omega_n}\right)^2\right].$$

So,

$$\sqrt{C^2 + B^2} = \left\{ \left(2\zeta\frac{\omega}{\omega_n}\right)^2 + \left[1 - \left(\frac{\omega}{\omega_n}\right)^2\right]^2 \right\}^{1/2}$$

and

$$\phi^* = \tan^{-1}\left[\frac{2\zeta(\omega/\omega_n)}{1 - (\omega/\omega_n)^2}\right].$$

Thus,

$$y_p = \frac{KA \sin(\omega t + \phi)}{\{[1 - (\omega/\omega_n)^2]^2 + [2\zeta\omega/\omega_n]^2\}^{1/2}},$$

where

$$\phi = \tan^{-1}\left[\frac{2\zeta(\omega/\omega_n)}{1 - (\omega/\omega_n)^2}\right].$$

Hence,

$$y = y_h + y_p,$$

where y_h depends on $\sqrt{\zeta^2 - 1}$ as before.

Now all y_h solutions involve terms that are multiplied by e^{rt}, where r is always negative (examine all three possible solution cases to see this). Thus, at $t \to \infty$ all y_h's $\to 0$. That is, the transient part decreases to zero. So, for steady state (when t becomes large),

$$y \to y_p$$

$$\Rightarrow M(\omega) = \frac{A_0}{A_i} = \frac{KA/\{[1 - (\omega/\omega_n)^2]^2 + [2\zeta(\omega/\omega_n)]^2\}^{1/2}}{KA}$$

$$= \frac{1}{\{[1 - (\omega/\omega_n)^2]^2 + [2\zeta(\omega/\omega_n)]^2\}^{1/2}}$$

$$= \frac{1}{\{[1 - (\omega/\omega_n)^2]^2 + [2\zeta(\omega/\omega_n)]^2\}^{1/2}}.$$

INDEX